MANAGING LOCAL AREA NETWORKS

Thomas Case
Larry Smith

Georgia Southern University
Statesboro, Georgia

McGraw-Hill
New York St. Louis San Francisco Auckland Bogotá Caracas
Lisbon London Madrid Mexico Milan Montreal New Delhi
Paris San Juan Singapore Sydney Tokyo Toronto

McGraw-Hill
San Francisco, CA 94133

Managing Local Area Networks

1 2 3 4 5 6 7 8 9 0 FRG FRG 9 0 9 8 7 6 5

ISBN 0-07-059225-X

Sponsoring editor: Frank Ruggirello
Editorial assistant: Debra Yohannan
Production supervisor: Richard DeVitto
Project manager: Cecelia Morales
Copyeditor: Ryan Stuart
Interior designer: Cecelia Morales
Cover designer: Janet Bollow
Composition: Arizona Publication Service
Printer and binder: Quebecor/Fairfield

Library of Congress Card Catalog No. 94-72774

International Edition

Copyright © 1995
Exclusive rights by McGraw-Hill Inc. for manufacture and export.
This book cannot be re-exported from the country to which it is consigned by McGraw-Hill. The International Edition is not available in North America.

When ordering this title, use ISBN 0-07-113567-7

# Contents

Part Two　　LAN Fundamentals

CHAPTER 3

LAN Alternatives　　　　　　　　　76

CHAPTER 6

Middleware   **213**

CHAPTER 7

LAN Application Software   **243**

Part Three	Network Planning and Installation

CHAPTER 8

CHAPTER 9

Part Four Network Management

CHAPTER 12

CHAPTER 13

Part Five   **Internetworking**

Part Six   Networking Careers

 **Preface**

Local area networks (LANs) are an amazing phenomenon in both large and small businesses. Because they enable users to share computing resources and provide a platform for supporting collaborative and workgroup computing, more LANs are being installed on a daily basis than any other type of computer network. LANs have become the fundamental building blocks in enterprise-wide network infrastructures and are playing an increasingly important role in the emergence of client/server networks.

The growing importance of LANs in today's computing environment mandates that students in management information systems, computer information systems, and computer science programs have a fundamental understanding of LANs and LAN applications. Such knowledge is essential to be competitive in the current, and future, computing workplace.

PURPOSE

The primary focus of this text is on LANs and the management issues associated with evaluating, installing, and administering them. While most of the discussion addresses LANs, interconnection technologies, wide area networks (WANs), and the other components of enterprise-wide networks are not overlooked. The emergence of client/server computing and client/server networks is discussed and is especially apparent in the chapters devoted to middleware, downsizing, and internetworking.

Comprehensive Coverage of Current Technologies

The book is designed to provide students with a comprehensive overview of the most popular LAN hardware and software being used today. It also identifies emerging network technologies that are likely to impact LANs and their role in enterprise-wide networks in the years ahead. The coverage is also up-to-date as reflected in the following sample of topics:

Topic	Primary Coverage
ATM and cell relay systems	Chapter 16
Backbone networks	Chapter 15
Capacity planning	Chapter 13
Client/server computing	Throughout text

Topic	Primary Coverage
Diagnostic tools	Chapter 13
Downsizing to LANs	Chapter 11
EDI	Chapter 16
Groupware	Chapter 6
Internetworking technologies	Chapter 15
LAN installation	Chapter 10
LAN switches	Chapter 15
LAN-to-WAN connections	Chapter 16
Managing printing operations	Chapter 13
Media access control	Chapter 5
Middleware	Chapter 9
Mobile computing	Chapter 15
Multimedia applications	Chapter 12
Networking careers	Chapter 17
Network management systems	Chapters 13 and 15
Network planning	Chapter 8
Network security	Chapter 14
Novell NetWare	Throughout text
Open systems	Chapter 5
Packet distribution networks	Chapter 16
Peer-to-peer LANs	Chapter 6
Selecting LAN components	Chapter 9
Superhubs	Chapter 15
Telecommuting	Chapters 15 and 16
Wireless LANs	Chapter 2 and throughout text

Focus on LAN Management

As should be evident from the foregoing sample of topics, LAN management is also a central focus of this book. This is clearly apparent in the book's coverage of network planning, selecting LAN components, network installation, administering network operations, and managing network security. The network management theme is also reflected in the chapter devoted to networking careers. Hence, students who are eager to develop network management insights that are tempered by the realities of technology should be particularly pleased with this book.

ORGANIZATION

The text is divided into six major parts.

◆ *Part One: Introduction to Data Communication Networks*, provides an overview of computer networks and network applications in today's organizations.

◆ *Part Two: LAN Fundamentals*, provides an overview of LAN hardware, software, media, network topologies, and protocols.

◆ *Part Three: Network Planning and Installation*, focuses on network planning, selecting network components, and installing networks.

◆ *Part Four: Network Management*, explores the activities associated with administering LANs on a daily basis. It focuses on managing LAN operations, monitoring network performance, troubleshooting network problems, and managing network security.

◆ *Part Five: Internetworking*, addresses how LANs can be interconnected with other networks including other LANs and WANs.

◆ *Part Six: Networking Careers*, addresses both traditional and nontraditional career tracks in networking. It also provides valuable guidelines for keeping up and advancing in this dynamic field.

PEDAGOGICAL FEATURES

A number of learning aids are used to assist students in mastering the concepts that are presented. These features include chapter objectives, emphasized key terms, end-of-chapter review questions, discussion questions, problems and exercises, and bibliographic references. In addition, a continuing case study is used to highlight some of the most important concepts addressed in each chapter.

Chapter Objectives

Each chapter begins with a list of objectives that provides a preview of the major concepts in the chapter and outlines what the student should understand by the end of the chapter.

Key Terms

Key terms are boldface and italic when they first occur in the chapter; italics are used to emphasize related concepts of secondary importance. The collective set of key terms for the chapter is listed after the chapter summary.

Summary

The end-of-chapter material begins with a summary of the most important concepts addressed by the chapter.

Review Questions

Review questions have been included to assist students in recalling the chapter's important concepts.

Discussion Questions

A set of discussion questions is included to help students synthesize and critically evaluate chapter concepts. These questions are more complex than the Review Questions and integrate concepts learned in earlier chapters.

Problems and Exercises

A variety of problems and exercises are found at the end of each chapter to encourage students to apply their understanding of the concepts and to identify information sources that reinforce the Chapter's concepts. Many of these suggest methods for structuring course content and can be used for individual or group projects.

References

A list of references is provided identifying sources of information that could be used as outside reading and for further study of the concepts presented in the chapter.

Running Case

The Southeastern State University case study, found at the end of each chapter, provides further emphasis on chapter concepts. The set of questions included with each segment of the case can be used for in-class discussions or written assignments.

Glossary

An extensive glossary covering both network technologies and network management concepts is included as a comprehensive and easy-to-use reference guide.

SUPPLEMENTS

Instructor's Manual

The Instructor's Manual includes recommendations for structuring the course to fit both semester and quarter system schedules, a lecture outline for each chapter, teaching tips, suggestions for supplemental activities, and answers to end-of-chapter review questions, discussion questions, selected problems and exercises, and case questions.

Test Bank

An author-developed test bank in the Instructor's Manual consists of at least 100 questions (multiple-choice, true-false, fill-in, and matching) for each chapter.

Overhead Masters

Masters for overhead transparencies are also available in the Instructor's Manual.

RELATED MATERIALS

To provide a hands-on component to the LAN course, instructors are encouraged to couple the use of this text with *LAN Basics with Hands-On NetWare 3.11/3.12*, by Patricia Harris. Use of both books will ensure that students get both an overview of LANs and experience with the most widely used LAN network operating system.

ACKNOWLEDGMENTS

A hearty word of thanks goes to the following reviewers for their excellent input and candid suggestions: Belinda T. Jones, Kentucky Technical College; Stephen Jordan, Cooke County Community College; Dr. Guillermo A. Francia, III, Kansas Wesleyan University; Daris Howard, Ricks College; Kathy Cupp, Oklahoma City Community College; Nancy Hoyt, Weber State University; Craig Bodkin, Trident Technical college; Seth Hock, Columbus State Community College; Michael Harris, DelMar College; Bay Arinze, Drexel University; and Pat Fenton, West Valley College.

A special word of thanks also goes to Edna Dixon, Instructor of Management at Georgia Southern University, for her efforts in compiling bibliographical references and assistance in preparation of the manuscript.

Thanks is also due to several people at McGraw-Hill who directly contributed to keeping this project on course. These include consulting editor Peter Keen for his suggestions for structuring the topical coverage of the book, editor Frank Ruggirello and editorial assistant Debra Yohannan for coordinating all aspects of development of the manuscript, production Editor Rich DeVitto who, along with his staff, ensured that the manuscript got to press in a timely fashion.

We are especially thankful to Cecelia Morales of Arizona Publication Service, who orchestrated the final layout of the discussion and art in the text, and to copyeditor Ryan Stuart for adding style and grace to our prose.

Of course, we are most thankful to our families for all their patience, support, and encouragement.

Thomas Case
Larry Smith

PART ONE

Introduction to Data Communications

Local Area Networks: An Introduction

CHAPTER OBJECTIVES

After completing this chapter, you will be able to:

- Describe the main components of a data communications system.

- Describe the differences between host-based computer networks and distributed computer networks.

- Describe the differences between wide area networks (WANs) and local area networks (LANs).

- Briefly discuss how users can have remote access to WANs and LANs.

- Identify the components of client/server networks.

- Identify the types of servers found in client/server networks

- Briefly discuss the different varieties of client/server computing.

- Describe the major characteristics of wireless LANs.

- Describe the major types of end-user, document management, workgroup computing, and database management applications found in today's networks.

- Describe the impact of groupware, videoconferencing, and multimedia on computer networks.

- Describe the differences among strategic, tactical, and operational network management.

- Discuss the reasons that downsizing, distributed processing, fault tolerance, open systems, and internetworking are important issues in network environments.

◆ WHAT IS DATA COMMUNICATION?

Communication systems, abundant in businesses and other organizations, are, for example, telephone systems, facsimile (fax) systems, and computer networks. Communication systems are composed of four essential elements: sender, receiver, medium, and message (see Figure 1-1).

For communication to occur, a sender transmits a message over a medium to a receiver. For effective communication, the message must be in a form which the receiver can accept and understand. A feedback mechanism is usually included in a communication system to check that the receiver has gotten the sender's message.

Data communication systems are special types of communication systems allowing the transmission of messages among computers. Like other communication systems, data communication systems are made up of senders, receivers, media, and messages.

SENDERS

In a data communication system, the *sender* may be a microcomputer, video display terminal, or a specialized input device such as a bar code scanner or sensor. Any type of information technology with sufficient intelligence to compose and transmit a message can be a sender.

RECEIVERS

The *receiver* in a data communication system may be a computer or terminal, an output device (such as a printer), manufacturing equipment (such as an industrial robot), climate control devices, or any of a variety of other devices possessing sufficient intelligence to decode and react to the sender's message.

Most receivers in data communication networks can transmit some sort of response back to the sender, indicating receipt of the message. This

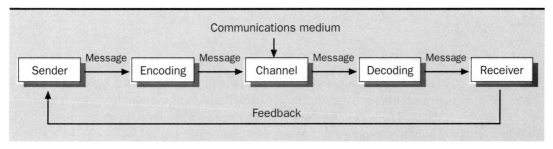

FIGURE 1-1   A communication system

feedback is often a key determinant in whether the sender continues to transmit messages to the receiver.

MEDIA

In data communication systems, messages can be transmitted from senders to receivers over a variety of *media*. Among the most common media are everyday telephone wires (called *twisted pair wire*), coaxial cable such as is found in cable television systems, fiber optic cables, and various types of "wireless" media including radio waves, microwave signals, cellular communication signals, and infrared light.

MESSAGES

Messages also take many forms in data communication systems. Many consist of text or numeric data, but the transmission of document images, graphic images, photographs, and video is also becoming more common. Complete files, electronic mail, and network status messages sent to all users by network managers are just a few of the other types of messages transmitted over data communications media.

MESSAGE ENCODING AND DECODING

To ensure effective communication within data communication networks, the receiver must understand the message that the sender has transmitted. For two humans to communicate, they must generally use the same language (or employ a translator). The same is true for sending and receiving devices in data communication systems.

Data codes are the "languages" used in data communications. Many different data codes exist, but the two most common are the *American Standard Code for Information Interchange (ASCII)* and the *Extended Binary Coded Decimal Interchange Code (EBCDIC)*. If the sender transmits in one data code (say, ASCII), but the receiver understands only another data code (say, EBCDIC), a translation is needed. Such translation can be done using hardware, software, or some combination of hardware and software.

FEEDBACK

Data communication systems typically include some type of *feedback* mechanism to inform the sender whether the receiver has gotten the message. The receiving device usually transmits a very brief response to the sender noting receipt of the message or that some type of communication error has occurred. Such feedback mechanisms are often called *error detection mechanisms*. When receivers detect errors, they usually transmit a short message to the sender indicating that all or part of the message was not received. The sender typically responds to the receiver's feedback by retransmitting the entire message or only the parts for which the receiver detected errors.

◆ COMPUTER NETWORKS

This book is about computer networks and focuses primarily on the computer networks now rapidly gaining popularity: local area networks (LANs). A *computer network* can be described as two or more computers able to communicate with one another through a communication medium. The devices in a computer network must possess processors or processing capabilities, and the devices must be able to transmit messages to one another. Such devices are also commonly classified as *information technology (IT)* because they have both computing and communication capabilities.

This book provides fairly comprehensive coverage of a wide variety of hardware and software found in LANs. The increasing use of LANs now requires computer professionals to be familiar with these technologies and how they function.

This book also covers the management of LANs. It deals with how they are planned, implemented, operated day-to-day, maintained, controlled, and upgraded. As many computer network specialists have learned, knowledge and understanding of networking technologies is often not enough; one must also be aware of the management challenges associated with LANs and computer networks.

◆ TYPES OF COMPUTER NETWORKS

Of the computer networks found in organizations today, some are centered around large computers such as mainframes and minicomputers; others consist solely of microcomputers, printers, and other forms of end-user devices. This section covers some of the major types of computer networks found in organizations. Learning about the various types of networks should help the reader understand why the term *computer network* can mean different things to different organizations.

HOST-BASED NETWORKS

For some organizations, the *computer network* is a *host-based network* comprised of terminals and equipment attached to a large, central computer such as a mainframe or minicomputer. This was the most common type of computer network in the 1960s and 1970s (before the introduction of microcomputers in organizations). In these types of systems, there are computer terminals and other peripheral equipment (such as printers) housed in different parts of the organization's facility. While these terminals and peripherals may be quite far apart from one another, all are directly connected to a host computer, typically a centrally located mainframe or minicomputer. This type of network is illustrated in Figure 1-2.

In host-based networks, the host computer does all or most of the processing work; the terminals permit users to access the host's applications. In fact, many users can use the host's applications simultaneously; thus these networks are sometimes called *time sharing networks*.

The terminals in host-based systems vary in sophistication from *dumb terminals*, which have no (or extremely limited) processing capabilities of their own, to *intelligent terminals*, which include their own central processing units (CPUs) and are able to do some processing without the host's help. Between dumb and intelligent terminals are *smart terminals*, which have limited processing capabilities, but not as many as those found in intelligent terminals. Since the early 1980s, many organizations have found that using microcomputers in place of intelligent terminals can be a cost-effective option in host-based systems.

DISTRIBUTED NETWORKS

Distributed networks consist of two or more computers connected by a data communication medium. One way to create a distributed computer network is to connect two (or more) host-based systems to one another.

The network illustrated in Figure 1-3 is an example of a distributed computer network in which several host-based networks (those for Divisions A, B, C, and D) are connected to one another through a mainframe located at corporate headquarters. This arrangement allows messages to be exchanged among divisions as well as between headquarters and one, some, or all of the divisions.

FIGURE 1-2
A centralized
host-based
network

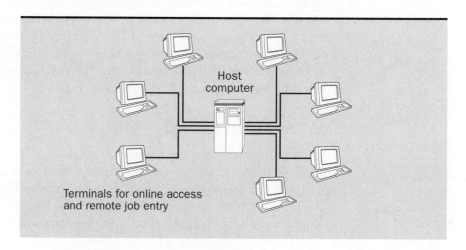

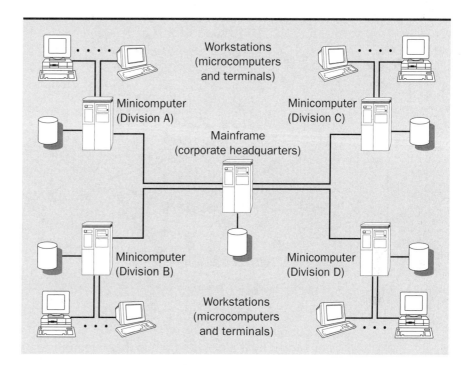

FIGURE 1-3
A distributed
network

In distributed networks, the hosts are often called ***nodes***. Messages can be transmitted from a terminal to its host, from a terminal to another node, from a terminal attached to one node to a terminal attached to another node, and from node to node.

One of the major trends of the 1990s is away from host-based systems toward distributed networks. Although mainframes and minicomputers are commonly found in distributed networks, such large computers need not be included. With the increase in the power and processing capabilities of microcomputers, they are more commonly used as nodes in distributed networks.

Distributed networks vary considerably in complexity and scope. As a result they can be categorized in a number of different ways. However, one of the most commonly used ways to classify distributed networks focuses on how large a geographic area the network covers.

WIDE AREA NETWORKS

Distributed networks that can cover broad geographic areas are typically called ***wide area networks (WANs)***. WANs may cover a significant region of a state, an entire state, several states, the entire U.S., and even the entire world. Figure 1-4 shows a WAN that connects sites in several states.

Generally, a distributed network can be considered a WAN if an individual located at one site served by the network would normally have to

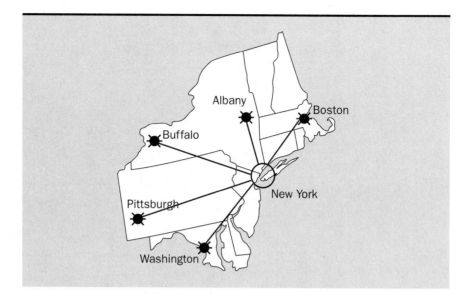

FIGURE 1-4

A wide area
network (WAN)

use long-distance services to call people at one or more of the other locations served by the network. While this rule of thumb might not apply in
very large local calling areas, in most other instances it is a useful guideline for distinguishing WANs from more geographically limited networks.

Because WANs can cover very broad areas, the communication media
connecting the computers to one another are often slower than those
found in more geographically restricted networks. In addition, the nodes
interconnected to form WANs are often larger computers such as minicomputers and mainframes.

Media speeds for WANs are expected to steadily increase during the
1990s as long-distance telephone service providers offer new and faster
services. In addition, the percentage of WAN nodes that are mainframes or
minicomputers is expected to decline during the 1990s as the power and
capabilities of microcomputers continue to increase. We will be exploring
these trends and a number of other issues related to WANs in Chapter 16.

LOCAL AREA NETWORKS

A *local area network (LAN)* covers a much smaller geographic area than a
WAN. A LAN may be restricted to a single room (such as a LAN found in a
university computer lab), to a set of rooms or offices found in the same
building, or to the set of buildings that make up a single "campus" of a
school or business.

While some believe that the term *LAN* should be restricted to networks
confined to a fixed distance (for example, a radius of five miles or ten kilometers), no hard-and-fast rule exists. In fact, the geographic area covered
by a LAN is more likely to be determined by the technology used to create

it than by anything else. For example, Apple Computer, Inc. recommends that its AppleTalk networks (used to connect Macintosh computers) should be limited to a radius of 300 meters. For Ethernet networks, the recommended working radius is 2,500 meters (which is less than 2 miles) or less. Fiber optic networks, especially those using the Fiber Distributed Data Interface (FDDI), can be used to cover distances up to 200 kilometers. The difference between 300 meters and 200 kilometers is substantial, yet all of these could be considered local area networks.

Most computers used in LANs are microcomputers. Also, the media used to connect the computers to one another are usually faster than those found in WANs. Most LANs have data transmission speeds of more than one million bits per second (bps). However, LAN data transmission speeds typically range from several hundred thousand bps (in AppleTalk LANs) to well over 100 million bps (in fiber optic LANs). In fact, some LANs have speeds over a billion bps. Similar to WANs, the average data transmission speed for LANs is expected to increase.

METROPOLITAN AREA NETWORKS

Another geographically defined network is the ***metropolitan area network (MAN)***. A MAN, as its name implies, is a distributed network spanning a metropolitan area (a city and its major suburbs). In terms of geographic scope, a MAN is considered to be somewhere between a LAN and a WAN. Figure 1-5 shows an example of a hypothetical bank's MAN.

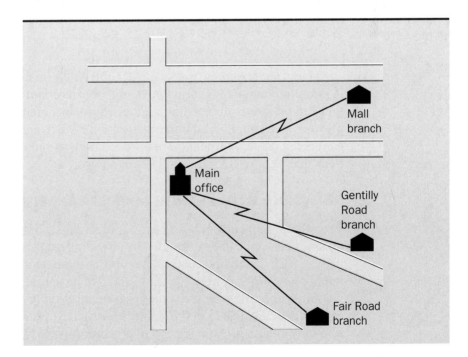

FIGURE 1-5

A metropolitan area network (MAN)

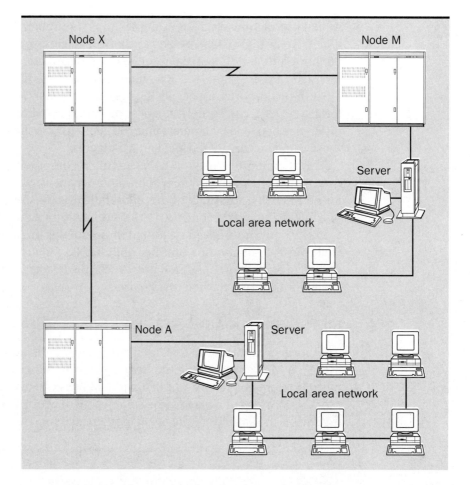

FIGURE 1-6

A WAN intercon-
necting LANs

Many organizations use both LANs and WANs. In fact, there is a growing trend to merge LANs and WANs so that all computers in an organization can be interconnected no matter where they are located. Figure 1-6
provides an example of how an organization's WAN and LANs can be interconnected. Such internetworking is becoming so important in organizations that Chapter 15 is devoted to this subject.

NETWORKS WITH REMOTE ACCESS CAPABILITIES

The geographic scope of both host-based and distributed networks (both
LANs and WANs) can be expanded by providing users *remote access* capabilities. Such capabilities make it possible to connect to the network from
remote locations not normally considered to be within the network's geographic scope. For example, a manager might be able to connect a home
computer to the host in the company's host-based system; such a connection may enable the manager to telecommute—that is, to use company

computers while working at home. A salesperson out in the field might be able to use a notebook computer to connect to the office's LAN to find out about new price changes or to get instructions from the boss.

In most instances, the computers that users at remote locations connect to have some type of "dial-out" or "direct dial" capability. The media used to carry the message are usually communication circuits provided by telephone companies (both local and/or long distance), or wireless connections such as those available in cellular telephone and data networks. The computers used at remote locations typically use communication software (such as ProComm or Crosstalk) and a modem.

The modem is a device that converts computer-transmitted communications signals to a form that can be sent over communications circuits. Another modem attached to the receiving equipment converts the message back to computer-readable form.

When remotely connecting to a host-based system, the communications software usually utilizes terminal emulation, which causes the host to recognize the user's computer as a terminal similar to the ones connected to it at the home office. Essentially, the user is then able to interact with the host just as he or she would when working at one of the host's real terminals. Because the host in a host-based system is often a mainframe (or minicomputer), the connection between it and a computer at a remote location is often called a *micro-to-mainframe link*. This type of remote access is illustrated in Figure 1-7.

Communication software is also commonly used to establish a dial-in connection to a LAN. As we will see in Chapter 15, there are several ways in which remote connections to a LAN can be accomplished. These include access through communication servers and modems (called LAN

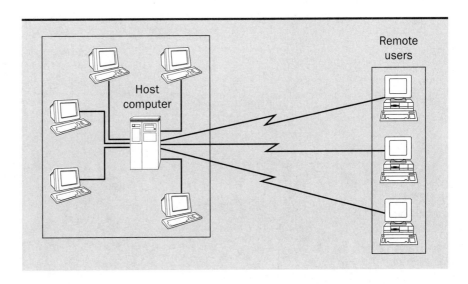

FIGURE 1-7

A host-based network with remote access capabilities

modems) connected to the LAN; another access method involves connecting to the LAN by establishing a dial-in connection to a LAN workstation. A generic example of a remote connection to a LAN is shown in Figure 1-8.

Remote users typically do not have all the processing privileges and capabilities as users at the LAN site itself. However, this is expected to change, and the range of processing options available to remote users is expected to become virtually identical to those of users at the LAN location.

CLIENT/SERVER NETWORKS

Client/server computing is one of the hottest buzzwords of the 1990s. The emergence of LANs and LAN/WAN interconnections has created a computing environment in which two or more computers can cooperate to solve a problem. In this text, client/server computing is considered a processing environment in which the hardware, software, and data resources of two or more computers are combined to solve a problem or to process a single application. A *client/server network* enables client/server computing to occur; one way to think of a client/server network is as the network becoming one computer used to solve a particular problem or to process an application.

A client/server network may be a LAN, interconnected LANs, a WAN, interconnected WANs, or interconnected LANs and WANs. The computers included in client/server networks can range from microcomputers to mainframes (or even to supercomputers). The software and data resources to be shared among the computers in a client/server network are placed at computers called *servers*. Servers manage these resources on behalf of the other computers (called *clients*) that share them.

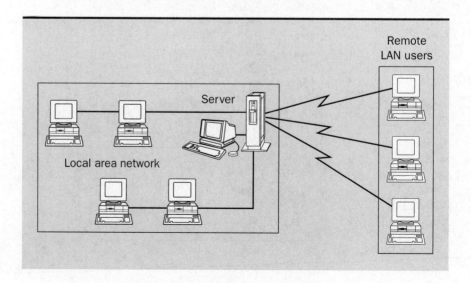

FIGURE 1-8
A local area network with remote access capabilities

Servers vary. They include: file servers, used to store application software such as word processing, spreadsheets, and so on; print servers, used to manage printing operations such as spooling and print queues for shared printers; database servers, which allow users to access the data in shared databases; and communication (or network) servers, used to manage communications between two or more networks. Several are illustrated in Figure 1-9. There may also be image scanner servers, e-mail servers, video servers, voice mail servers, credit card servers, expert system servers, and so on, depending on the functions that the network performs. Note that in practice, a computer in a client/server network can be a server for one type of application and a client for other types of applications.

The level of interaction varies between clients and servers when solving a problem or processing an application. Five types of client/server computing follow.

1. *Host-driven terminal emulation* occurs when the client connects to the server in the same manner that a terminal connects to a host. The server does virtually all processing.

2. *Host-driven front-ending* occurs when the client intercepts the messages transmitted by the server (host) and converts them to a more user-friendly interface than the standard interface provided by the server. A friendly interface is also provided by the client for users to send messages to the server. All important processing is done at the

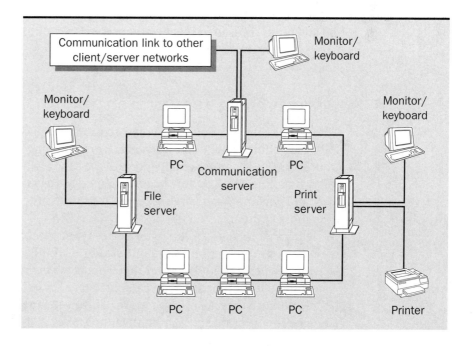

FIGURE 1-9

A client/server LAN

server and the creation of user-friendly interfaces is the only real work done by the client.

3. *Host-driven client/server* computing occurs when the client runs an application permitting it to carry out specific actions for the server (host). For example, the server may cause the client to display a menu, window, or dialog box available on the server. Typically, one or more of the displayed options can be carried out by the client's computer under the direction of the server (host). Users at the client computer can also send information back to the server.

4. *Client-driven client/server* computing occurs when the server carries out part of a processing task for the client upon request. The client essentially requests the server to perform a processing operation; the server then carries out the operation and sends the results to the client. The client uses these results to complete the processing task it is working on.

5. *Peer-to-peer client/server* computing occurs when the client and server cooperate in performing a processing operation. The roles of clients and servers may switch several times during the processing operation; both clients and servers may request services from one another.

   In peer-to-peer arrangements, host-driven and client-driven client/server computing structures are combined. This is the most cooperative of the various types of client/server computing and also the most sophisticated.

Information systems professionals must understand client/server computing and how it is used in organizations. Because of this, client/server networks and client/server computing will be a running theme throughout this textbook.

WIRELESS LOCAL AREA NETWORKS

Wireless communications are already common and come in a variety of forms including cellular telephones, pagers, building-to-building microwave links, and the very small aperture (VSAT) satellite dishes found on many buildings. With people becoming increasingly mobile both in their work and personal lives, wireless communications are expected to become even more popular.

Expect an increasing variety of devices that combine computing and wireless communications capabilities. For example, Apple Computer's Newton is a pen-based computer that can fax over wireless channels. The Newton and its competitors are categorized as *personal digital assistants* (*PDA*s) because they are easily transportable by the user and can carry out a range of computing and communications tasks.

Between now and the turn of the century, *personal communication services (PCS)* are expected to continue to spring up around publicly-available wireless WANs. Such services will make it possible to transmit telephone conversations, computer-based information, e-mail, voice mail, and other personal services.

Among the offshoots of wireless communications are **wireless LANs** (see Figure 1-10). These typically use microwave signals, radio waves, or infrared light to transmit messages between computing devices. Wireless data communications are often a good option when mobility is required. It can also make remote mobile computing a possibility, especially now that cellular data networks can be accessed from just about anywhere in the U.S.

While wireless connections are currently more expensive than most types of physical cabling, often their use may be cost effective. For example, in some existing buildings, wireless media are often less expensive than the costs (and difficulty) of renovating to install connecting wiring. Wireless LANs may also be desirable when a building's available wiring ducts are already full, for temporary operating locations, in hazardous environments, and for disaster recovery.

There are two basic types of wireless LANs in use today.

1. *Central controller-based* wireless LANs use a dedicated server (controller) that manages message transmissions among the workstations in the LANs. The controller handles all communications among devices in the LAN.

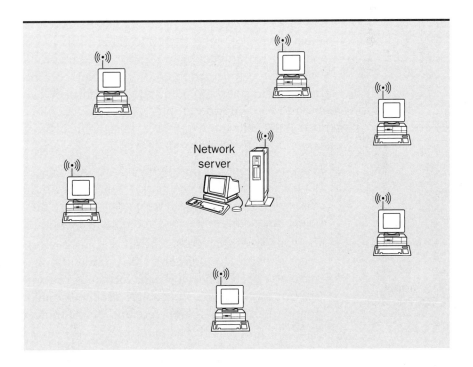

FIGURE 1-10
A wireless LAN

2. *Peer-level systems* allow each computing device to communicate with every other computing device in the LAN. These are cheaper to implement than are central controller-based systems because master control devices (and software) are not needed. However, these slow down under heavy use. They are also less secure than are controller-based systems.

As we have also noted, wireless technologies are still relatively new but are expected to become more commonplace. Wireless technologies, like most computing and communications technologies, are expected to continue to drop in price for the remainder of this decade.

◆ EXAMPLES OF NETWORK APPLICATIONS

LANs are installed primarily to allow end users to share computing resources, both hardware and software. For example, LANs can permit users to share expensive peripheral equipment such as high resolution document scanners or color laser printers. LANs also enable users to share sophisticated (and expensive) application software. The application software found in LANs is so varied (and becoming more varied every day) that it would be impossible to describe in a single book. However, this section covers some of the most common types of LAN applications. These include end-user applications, document management applications, workgroup computing applications, database applications, and a few others.

END-USER APPLICATIONS

The major types of end-user applications found in LANs generally correspond to the major types of application software available for stand-alone microcomputers (microcomputers not networked). Such software includes word processing, desktop publishing, spreadsheet, database/file management, and presentation graphics software. Each of these is briefly described in the following list. Table 1-1 lists some of the best-selling software packages of each of the four major types.

◆ *Word processing* software found in LANs is usually as versatile as the major packages found on the market today. Workgroup versions of many of these packages are being developed by vendors; these allow two or more users to concurrently access and update the same document.

◆ *Desktop publishing (DTP)* software permits users to combine text, typefaces and sizes, and graphics (including photographs) on the same page. It is used to develop page formats for newsletters, newspapers, brochures, and books. Both single-user and group versions of these are found in LANs. The group versions enable multiple users to edit the same document at the same time.

Application Type	Product	Vendor
Word processing	WordPerfect	WordPerfect Corporation
	Word	Microsoft Corporation
	Ami Pro	Lotus Development Corporation
	DeScribe	DeScribe, Inc.
	MacWrite	Claris Corporation
Desktop publishing	PageMaker	Aldus Corporation
	Quark Express	Quark, Inc.
	GEM Desktop Publisher	Digital Research, Inc.
	The Office Publisher	Laser Friendly, Inc.
Spreadsheet	Lotus 1-2-3	Lotus Development Corporation
	Excel	Microsoft Corporation
	Quattro Pro	Borland International
	Pro Plan	Software Publishing Corporation
	SuperCalc	Computer Associates International
Database management	dBase	Borland International
	Paradox	Borland International
	FoxPro	Microsoft/Fox Software
	Dataease	Software Solutions
	R:Base	Microrim, Inc.

TABLE 1-1
Examples of some of the leading software packages

◆ *Spreadsheet* software, like word processing software, is quite common in LANs. Spreadsheets, used for applications that require the manipulation of numbers, can help monitor organizational performance and forecast future performance levels. They can also be used to prepare budgets and financial statements, to perform cost analyses, to perform statistical analyses, and to conduct "what-if" analyses. Workgroup versions make it possible for users to exchange spreadsheets and simultaneously access and work on the same spreadsheet.

◆ *Database/file management* software makes it possible for users to create files and databases, update and manipulate the data stored in them, and develop useful reports. Some versions of these found in LANs make it possible for multiple users to share and be working with the same file or database at the same time.

◆ *Presentation graphics* software is used to create graphic displays such as pie charts, bar charts, and three-dimensional charts. Spreadsheet packages typically have some graphics capabilities, and many database packages have some as well. The range of graphics-oriented functionality possessed by these packages is superior to that of spread-sheets and end-user database software. Because the hardware platforms (and software) needed to support sophisticated presentation graphics applications can be expensive, it is often desirable to share this technology in LAN environments.

DOCUMENT MANAGEMENT APPLICATIONS

The manner in which documents are received, stored, and reproduced has been changing dramatically over the last decade. Multiple-part forms, paper shuffling, and file cabinets are rapidly being replaced by document scanners, image databases stored on CD-ROM, and intelligent copier systems. Users can now electronically transmit complete documents to one another. It is also becoming more common for multiple users to access and view the same document image at the same time. An example of a document management system is illustrated in Figure 1-11.

LANs are becoming increasingly common in office settings. Office workers are often able to exchange e-mail between networked computers and print out the documents created on shared printers. Some office workers' computers are directly connected to intelligent copier systems capable of receiving documents from users and printing out multiple copies of the documents. Intelligent copiers can be connected to other intelligent copiers to allow document exchanges and the reproduction of the same document in multiple locations. The combination of computing and communications

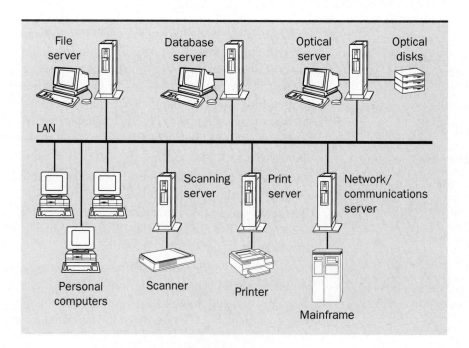

FIGURE 1-11
A LAN-based document management system

capabilities in office settings, such as the ones described, is often called *office information systems.*

Many of the LANs found in office settings include the end-user applications previously described. Many of the leading software vendors have developed "office suites" and special versions of their products which function on the network platforms that are most common in office settings.

Besides microcomputers and printers, office LANs may include document scanners, fax transmission capabilities, and electronic data interchange (EDI) capabilities. EDI makes it possible for one organization to exchange specially formatted business documents with other organizations over telephone lines.

WORKGROUP COMPUTING APPLICATIONS

Organizations use computer networks mainly to enhance computer-to-computer communications. LANs have made it possible for users to have computer-to-computer communications in addition to the traditional types of communications. LANs enable users to exchange e-mail, engage in computer conferencing, participate in electronic meetings, and arrange group meetings through group calendaring systems. Each of these is briefly described below.

◆ *Electronic mail.* **Electronic mail (e-mail)** systems send messages or documents from one user workstation to another. E-mail is one of the most common applications found in office information systems. Those systems are controlled by software capable of accepting and delivering electronic correspondence. E-mail systems can be implemented on LANs and WANs, as well as on interconnected LANs and WANs.

   Most e-mail systems use electronic mailboxes . Each user is assigned an electronic mailbox in which other users can place messages or documents. E-mail systems typically provide users with word processing capabilities for composing and editing messages; most systems also allow users to send the same message to both single or multiple mailboxes. Also, most e-mail systems allow users to store, print, and delete the messages in their mailboxes. Figure 1-12 provides an example of an electronic mailbox system.

◆ *Computer conferencing.* **Computer conferencing** systems allow users to compose messages and have these messages displayed on the screens of other users. This makes it possible for two or more users to have interactive dialogue or "conversations" in real time. Similar to e-mail systems, once messages are created, they can be stored by the computer conferencing system until the intended recipient(s) log in. These systems are less common in LANs than they are in interconnected LANs and WANs.

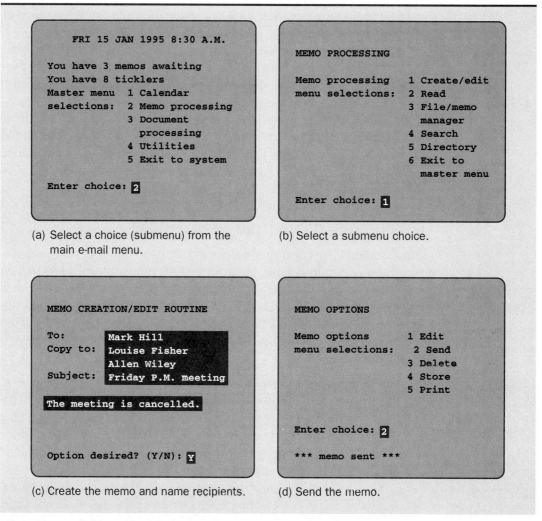

(a) Select a choice (submenu) from the main e-mail menu.

(b) Select a submenu choice.

(c) Create the memo and name recipients.

(d) Send the memo.

FIGURE 1-12 Using an e-mail system

◆ *Electronic meeting systems.* **Electronic meeting systems (EMSs)** typically combine decision support and electronic communications capabilities; they are generically called *group support systems (GSSs)*. These are designed to facilitate decision making in group settings.

EMSs allow users to communicate with one another through their workstations. Many EMSs allow users to share messages and ideas anonymously; an idea, comment, or criticism of one user's work can be displayed for everyone else in the group to see without indicating its source. Many EMSs and GSSs incorporate effective meeting management processes (such as majority voting systems and parliamentary procedure) and decision making approaches such as the Nominal Group Technique, the Delphi approach, and brainstorming.

◆ *Group calendaring systems.* Group calendaring systems facilitate the scheduling of meetings and appointments for workgroup members. Each user has his or her own calendar stored in the system. Both the user and other members of the workgroup can update each calendar with new information such as group meetings, appointments with customers, and so on.

E-mail, computer conferencing, electronic meeting systems, and group calendaring systems are just a few of the workgroup computing applications possible through the creation of LANs. The variety and sophistication of workgroup computing applications are expected to increase dramatically. These applications are being designed to help members of a workgroup coordinate their work activities more effectively and enhance teamwork and group productivity.

DATABASE APPLICATIONS

As LANs become more common in organizations, organizational data is being stored in more places. In the past, large databases were almost always stored in central locations on host-based systems. However, with increased emphasis on LANs and client/server computing, many of these large databases are being moved to LANs and distributed to multiple locations. Database servers that provide users with access to the data in a database are becoming widely used. Some database systems even have SQL (Structured Query Language) servers that translate SQL commands into database processing functions to be carried out by the servers. SQL is the most commonly used database processing language for modern database systems.

LANs and interconnected LANs and WANs make it possible to distribute organizational data to locations where it is needed the most. This can be done in many ways, but two of the most common methods are through partitioning and duplicating the database. With partitioning, the database is divided into parts, and the parts are stored where they are most needed. With duplicating, a copy of the entire database is stored at each location. Both partitioning and duplicating are illustrated in Figure 1-13.

Distributed data presents some special challenges to those in charge of keeping the database up to date. However, distributed databases are seen as being the wave of the future and are expected to become increasingly common as client/server computing matures.

Another inportant trend related to databases is the rising availability of and access to external databases and online information systems. More than 4,000 onlinc databases and information systems are accessible through dial-up connections or through interconnected LANs and WANs. Many of these services have access charges but are essential for some types of businesses (for example, stockbrokerages).

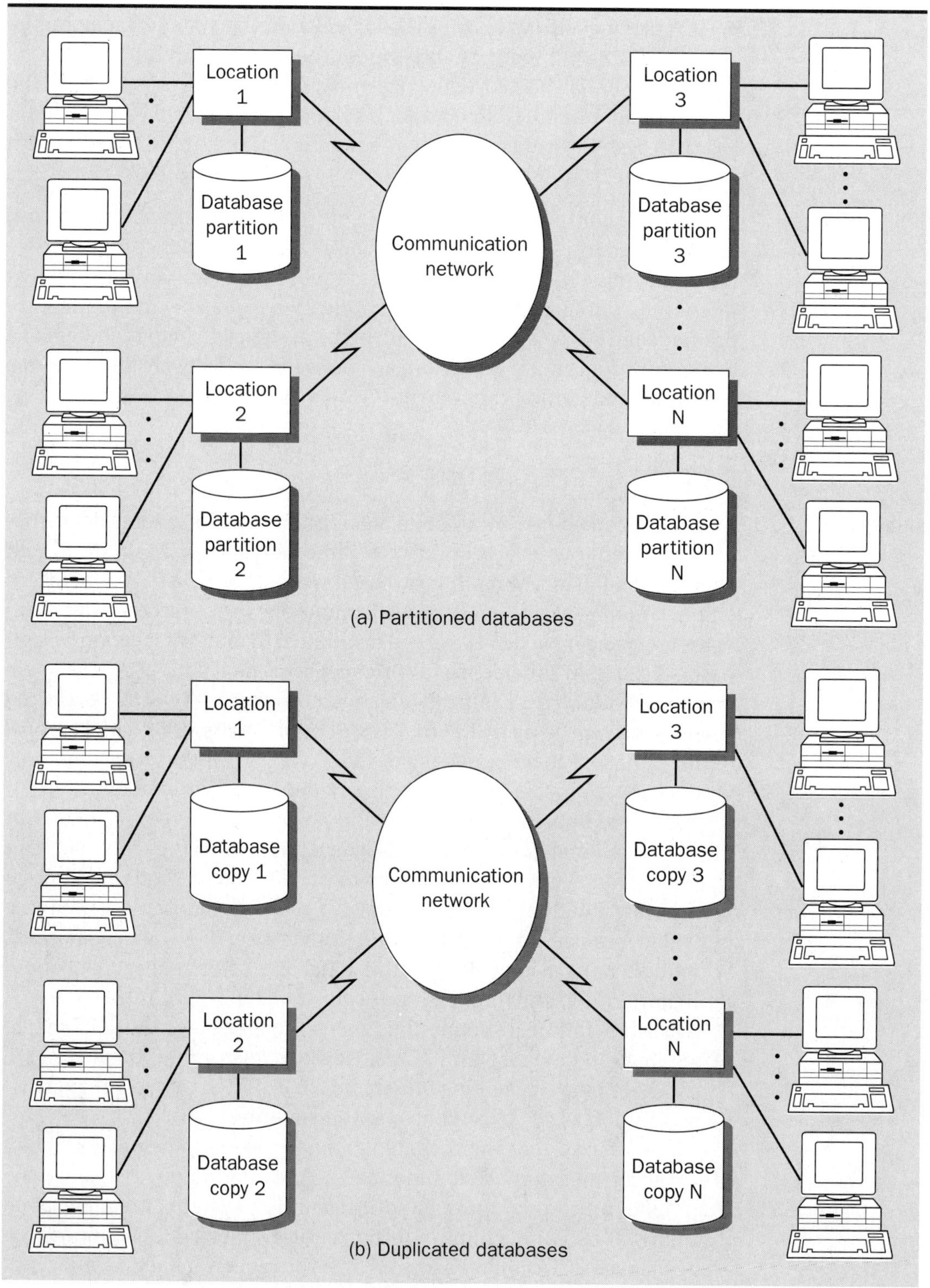

(a) Partitioned databases

(b) Duplicated databases

FIGURE 1-13   Partitioned and duplicated databases

◆ EMERGING TECHNOLOGIES AND APPLICATIONS

Computer networks have been evolving since the 1960s and will continue to evolve during the 1990s. Thus far, this chapter has provided a very brief overview of some of the changes that have taken place in computer networks. It has also discussed some of the major types of computer networks being used today and some of the ways in which they are being used. This section discusses three important new uses of networks: groupware, videoconferencing, and multimedia applications. Each of these is expected to have a major impact on LANs.

GROUPWARE

Groupware consists of software packages designed to support collaborative efforts. In other words, groupware is designed to help members of a workgroup coordinate work on common tasks. These software packages provide integrated support for a number of previously described applications including word processing, e-mail, computer conferencing, and access to external databases and information services. Groupware packages may also include group calendaring features, fax services, voice mail services, project management services, and electronic bulletin board services.

Groupware products are available from an increasing number of vendors. One of the first widely used groupware products is Lotus Notes. The use of groupware is expected to steadily expand; groupware products are also expected to become both more user-friendly and sophisticated.

VIDEOCONFERENCING

Videoconferencing refers to electronic meetings that involve people at different locations who can both see and hear each other. Videoconferencing combines voice and video images to allow users at two or more locations to interact with one another in real time. Currently, most videoconference participants gather in rather expensive, specially equipped rooms that can handle the simultaneous transmission and reception of video and audio signals. However, advances in video signal processing and video compression technologies are expected to make desktop-to-desktop videoconferencing commonplace by the year 2000.

MULTIMEDIA APPLICATIONS

The growing popularity of multimedia systems (systems that combine data, text, graphics, video, and sound) is expected to present some real challenges for LANs and LAN managers. Educational and reference tools such as multimedia encyclopedias are already available on CD-ROM, and there is pressure to make such resources available on LANs, especially in educational and research settings.

Ideally, users could access and exchange multimedia data between workstations. At the present time, such exchanges are only possible on the fastest and most sophisticated LANs. However, as is true for videoconferencing, advances in video signal processing and video compression technologies should bring multimedia applications within reach of many LAN users by the turn of the century.

◆ WHAT IS NETWORK MANAGEMENT?

As noted, this text's emphasis on network management makes it different from other textbooks about LANs. Network management activities are planned for the long term; management involves network planning, implementation, handling day-to-day operations, maintenance, and expansion.

Network management, especially LAN management, will be a running theme throughout this book. This theme is particularly evident in Parts Three and Four of the book (Chapters 8–14).

WHY IS NETWORK MANAGEMENT IMPORTANT?

Network management has become more important over the years. Similar to most other information systems specialists, network managers are primarily concerned with delivering accurate, relevant, and timely information to decision makers. Network managers are also concerned with delivering the information consistently, reliably, and cost-effectively.

Network managers must consider many factors when designing, operating, and maintaining a network. However, response times, downtime, and error rates are some of the most important factors that network managers take into account. These are especially important because they are the factors that users are most likely to focus on when judging a network's quality. If network users receive the information they need to make decisions and if the information is relatively free of errors, they are likely to be reasonably satisfied with the network. However, if the network is down (nonoperational) too frequently, if it takes a long time to receive requested information, or if messages are garbled when transmitted over the network, users are not likely to perceive the network as useful. Of course, users are not likely to *use* networks that they do not trust or value.

As computer networks become more common, the role of network manager has become more important. A LAN manager often faces daily challenges to keep the network up and running (to keep it from "crashing") and to make sure that software and hardware resources are shared by network users effectively. Some of the other challenges facing LAN managers come from adding new hardware and software to the LAN, and from connecting the LAN to other LANs and/or WANs.

SPECIALIZATIONS IN NETWORK MANAGEMENT

The two main specializations in network management covered in this textbook are LAN management and WAN management. As the titles imply, LAN managers are charged with managing one or more LANs while WAN managers are responsible for managing an organization's WAN(s). In many organizations (but not all), LAN managers report to WAN managers who, in turn, typically report to the organization's highest ranking information systems manager.

While WAN management is extremely important, this book will primarily focus on LAN managers and LAN management. As the number of LANs (as well as interconnected LANs) continues to increase, not surprisingly, so does the number of LAN managers. Indeed, many organizations have found that they just can't install a LAN and expect it to manage itself. The creation of LANs usually necessitates hiring a LAN manager (or LAN administrator). Hence, if there are many LANs in an organization, there are likely to be many LAN managers.

Network operating system vendors have created certification programs for LAN managers. For example, to become a Certified NetWare Engineer (CNE) or Certified NetWare Administrator (CNA), the Novell Corporation requires individuals to pass certification examinations. CNAs are most focused on system administration functions such as backing up servers, adding and deleting users, and maintaining LAN security. The CNE is the more advanced certification that covers system administration, system management, networking technologies, network support, and network service. Both types of certifications are valued by today's organizations. However, as you might expect, CNEs typically command higher salaries than CNAs. While it is not essential for a LAN manager to have one or both of these certifications, it is often desirable, especially as nearly 60 percent of the LANs used in organizations use Novell's NetWare as their network operating system.

Certified or not, network managers play a key role in the development, operation, and maintenance of the computer networks in organizations. Three major types of functions carried out by network managers are strategic management, tactical management, and operational management functions. These are briefly described in the next three sections.

STRATEGIC NETWORK MANAGEMENT

Strategic network management considers the role of networks and networking over the long term. Good strategic management ensures that the organization has the computer networks needed to help it achieve its long-term goals and objectives. This involves planning and building the network infrastructure that will help the organization achieve its strategic goals. Strategic network management plans typically specify the networks (including interorganizational networks) to be implemented or interconnected, the new technologies to be included in the networks, how data will be distributed among the networks, and how network security is to be enhanced.

Time frames for most strategic network plans being developed now typically range from 3 to 5 years. However, some extend to 10 years or more. Usually, the highes-ranking network managers in the organization are most directly involved in strategic network management. Chapters 8, 11, and 15 will cover further activities associated with strategic network management.

TACTICAL NETWORK MANAGEMENT

Tactical network management translates strategic network plans into more detailed, "doable" actions. Essentially, tactical network management is responsible for realizing strategic network plans; for example, tactical network managers develop implementation plans for the new networks specified in the strategic network plans. Managers also draw up sets of plans and timetables for interconnecting networks, for adding new technologies, for actually distributing data, and for making the networks more secure.

Tactical network plans typically have a time frame of 1 to 3 years. LAN and WAN managers at all levels may be involved in tactical network management. Tactical network management will be discussed more fully in Chapters 7, 8, 9, and 14.

OPERATIONAL NETWORK MANAGEMENT

Operational network management deals wtih day-to-day activities of LANs and WANs. For LAN managers, this typically involves troubleshooting (finding and correcting problems), monitoring LAN performance, adding and deleting users, adding new workstations or devices (printers, scanners, and so on) to the LAN, installing new software, backing up servers, and maintaining LAN security. The activities involved with operational network management will be discussed more fully in Chapters 11–14.

IMPORTANT NETWORK MANAGEMENT ISSUES FOR THE 1990s

A number of issues are creating special challenges for network managers. These include downsizing, distributed processing, fault tolerance, open systems, internetworking, building international networks, and how the general trend toward computer networks is redistributing power and influence within the information systems areas of organizations.

Downsizing

To general managers, downsizing typically means reducing the size of the workforce and reducing the number of managers (especially middle managers) within organizations. Within information systems departments, downsizing also means moving applications traditionally run on mainframe computers to other computing platforms, especially LANs.

Increases in microcomputers' computing power, as well as advances in client/server computing, have made downsizing an increasingly attractive option for many organizations because of substantial savings.

Distributed Processing

As noted earlier in the chapter, distributed networks and distributed processing means having computer resources and information processing activities at more than one location. It often means moving particular data processing operations close to the users who most need them. This often improves response time, reduces long-distance data communication costs, and improves data management activities; users can also share expensive computer resources more effectively. In most organizations, the development of LANs and the interconnection of LANs and WANs makes distributed processing more feasible.

Fault Tolerance

As organizations become increasingly dependent on computer networks to satisfy core information processing needs, the need for the networks to be reliable and available to users 24 hours a day has increased. Network managers often have backups in place for nodes, servers, and other key system elements so that network availability can be maximized.

A network is said to be fault tolerant when it can resist crashing or when it can recover from a crash quickly. In LAN environments, fault tolerance is often achieved by duplexed servers and disk mirroring. For example, if the file server crashes, network control functions are automatically switched to a backup file server. This allows the network to remain operational even though a key server has failed. We will be discussing fault tolerance and issues related to it in Chapters 13 and 14.

Open Systems

The term *open systems* is widely used among network managers. While there is no universal agreement on exactly what the term means, open systems are generally networks of computing devices and software designed to facilitate communications among devices. In other words, open systems' hardware and software are constructed to make it easy to connect to (interface with) other hardware and software. Often an internationally accepted set of standards is used to facilitate the creation of open systems; one of the best known is the Open Systems Interconnection (OSI) reference model. The OSI model and other widely accepted standards will be discussed in more detail in Chapter 5.

Internetworking

Internetworking makes it possible for the devices in one network to communicate with the devices in other networks. Connecting one LAN to another, connecting a LAN to a WAN, or connecting one WAN to another WAN are examples of internetworking. Many organizations have created "backbone" networks, such as that depicted in Figure 1-14, to interconnect their LANs and WANs. Internetworking is the focus of Chapter 15.

Building International Networks

As organizations expand operations beyond the borders of their native countries, they are faced with the very challenging task of building international computer networks to support their expanded operations. American companies expanding to Europe, the Pacific Rim, and other parts of the world have encountered differing standards, government policies, and available technologies that make the task of creating an international network very difficult. These and other challenges will be discussed more fully in Chapter 16.

Redistribution of Power in IS Departments

Another important issue facing network managers is their increasing importance in IS (information systems). In the past, when centralized host-based processing (and networks) was the norm, the managers of data processing and computer operations had considerable clout, mainly because these managers were responsible for the organization's core information processing activities. As computer networks have grown in importance, so has network managers' power, often at the expense of data processing and computer operations managers. Network managers have become more numerous and, because of downsizing and client/server computing, they are now often responsible for core information processing activities. Along with this responsibility has come increased power within the IS areas.

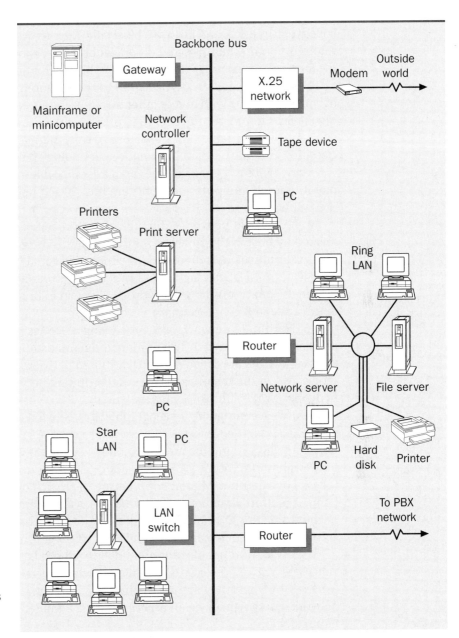

FIGURE 1-14
A backbone
network inter-
connecting LANs
and other
resources

◆ SUMMARY

Computer networks, like other communications networks, are made up of
four fundamental components: senders, receivers, messages, and media. In
data communications, senders transmit messages to receivers over communications media. The messages sent are encoded in data representation

codes such as ASCII and EBCDIC. Most computer networks also include error detection mechanisms and other feedback mechanisms to ensure that the messages transmitted by senders have been accurately received.

A computer network can be defined as two or more computers capable of communicating with one another through a communications medium. The devices found in computer networks often have both computing and communications capabilities; because of this, many of these devices are classified as being types of information technology (IT).

Many types of computer networks exist in today's organizations. Two major types of computer networks are host-based networks and distributed networks.

In host-based networks, the bulk of the processing activities are performed by host mainframes or minicomputers. Users gain access to the host via terminals. Terminals lacking processing capabilities are dumb terminals; terminals with significant processing capabilities are intelligent terminals; smart terminals have processing capabilities between those of dumb and intelligent terminals.

In distributed networks, the processing activities needed by an organization are performed by two or more interconnected hosts. In these networks, the hosts are called *nodes*. Distributed networks can vary widely in scope. Two of the major categories of distributed networks are wide area networks (WANs) and local area networks (LANs).

WANs cover a significant geographic area. Small WANs may cover a region of a state; the largest WANs are international in scope. LANs cover much smaller geographic areas than do WANs. Some LANs are restricted to a single room; other LANs connect computing devices within a few miles of one another. Microcomputers are the main computing devices in LANs. A metropolitan area network (MAN) is a moderate-sized computer network that could span a city and its major suburbs; a MAN is usually considered to be between a LAN and WAN in geographic scope.

A growing number of computer networks (both LANs and WANs) provide users with remote access capabilities; users can gain access to the networks even when they are not in the geographic region covered by the network. Users typically gain access by using a modem and communications software.

Client/server computing, one of the hottest trends of the 1990s, is a processing environment in which the hardware, software, and data resources of two or more computers are combined to solve a problem or to process a single application. A client/server network enables client/server computing to occur; the computers in client/server networks can range from microcomputers to mainframes (or even to supercomputers). The software and data resources shared among the computers in a client/server network are placed at particular computers called *servers*. Servers manage resources on behalf of the other computers (clients) sharing them.

The many kinds of servers found in client/server networks include file servers, print servers, database servers, and communication servers. The variety of client/server computing options used includes host-driven terminal emulation, host-driven front-ending, host-driven client/server computing, client-driven client/server computing, and peer-to-peer client/server computing.

Wireless networks are also becoming more popular both for WANs and LANs. Wireless WANs often involve the use of satellites or microwave communications. Wireless cellular data networks are also available to provide users with remote access to WANs and LANs. Wireless LANs typically use microwave signals, radio waves, or infrared light to transmit messages between devices. The two most common types of wireless LANs are controller-based and peer-level systems.

LANs are often implemented so that end users can share computing resources. LANs make it possible for users to share expensive peripheral equipment such as scanners or color printers. LANs also enable users to share sophisticated application software. The variety of application software available for LANs is large and growing. Among the major categories of application software for LANs are end-user application software (including word processing, desktop publishing, spreadsheet, database/file management, and presentation graphics software), document management applications, workgroup applications (including e-mail, computer conferencing, electronic meeting systems, and group calendaring systems), and database management systems (including distributed database systems).

Important organizational databases are now often stored and managed in LANs. Database and SQL (Structured Query Language) servers are often found in client/server LANs. Two of the major approaches to storing data in LANs are partitioning and replicating. With partitioning, parts of the database are stored at the locations where they are most likely to be used. With replication, a copy of the database is stored at each major node in the network.

Some of the emerging technologies expected to have a major impact on LANs are groupware, videoconferencing, and multimedia applications. Groupware software packages are designed to support the work activities of work team members. Videoconferences are electronic meetings of people at different locations who can see, hear, and interact with one another in real time. Multimedia systems that combine data, text, graphics, video, and sound are expected to present some real challenges for LANs and LAN managers.

Network management consists of the activities needed to manage a computer network. It involves network planning, implementation, day-to-day operations, maintenance, and expansion. The growing number of networks has made network management one of the top priorities in many organizations. The role of network managers has become more important and powerful in many organizations.

Both WAN and LAN managers may be found in organizations. Two important certifications for LAN managers are the Certified NetWare Administrator (CNA) and the Certified NetWare Engineer (CNE).

Three major functions carried out by network managers are strategic management, tactical management, and operational management functions. Strategic network management concerns the long-term management of networks and networking. Long-range planning is one of the chief activities associated with strategic network management; this is likely to be carried out by the organization's highest-ranking network managers. Tactical network management makes strategic network plans become realities. Operational network management deals with day-to-day activities. For LAN managers, this typically involves troubleshooting, monitoring LAN performance, adding and deleting users, adding new workstations or devices, installing new software, backing up servers, and maintaining LAN security.

Among the major issues facing network managers are downsizing, distributed processing, fault tolerance, open systems, internetworking, building international networks, and the redistribution of power and influence within the organization's information systems areas. Downsizing moves applications traditionally run on mainframe computers to other computing platforms, especially LANs. Distributed processing often improves response time, reduces long-distance data communication costs, and improves data management activities. A network is fault tolerant when it can resist crashing or recover from a crash quickly. Open systems are networks of computing devices and software designed to facilitate communications among devices. Internetworking makes it possible for the devices in one network to communicate with devices in another. Differing standards, government policies, and available technologies make the task of creating an international network very difficult. As computer networks have grown in importance, so has the power of network managers.

✴ KEY TERMS

American Standard Code for
   Information Interchange (ASCII)
Client/server computing
Client/server network
Computer conferencing
Computer network
Distributed network
Dumb terminal

Electronic mail (e-mail)
Electronic meeting system (EMS)
Extended Binary Coded Decimal
   Interchange Code (EBCDIC)
Feedback
Groupware
Host-based network
Information technology (IT)

Intelligent terminal
Local area network (LAN)
Media
Message
Metropolitan area network (MAN)
Node

Receiver
Remote access
Sender
Smart terminal
Wide area network (WAN)
Wireless LAN

✳ REVIEW QUESTIONS

1. Describe the components of a communication system.
2. What functions are carried out by senders, receivers, and media in data communication networks?
3. Identify the different types of messages found in data communication networks.
4. How are messages typically encoded in data communication networks? How does message feedback typically occur?
5. How do host-based computer networks differ from distributed computer networks?
6. What are the differences among dumb, smart, and intelligent terminals?
7. What are the differences between wide area networks (WANs) and local area networks (LANs)? What is a MAN?
8. What is meant by *remote access*? How can users access WANs and LANs from remote locations?
9. What is client/server computing?
10. What is a client/server network? What is a server? What is a client?
11. What different types of servers may be found in client/server networks?
12. Briefly describe the different types of client/server computing identified in this text.
13. What are the characteristics of wireless LANs? How do controller-based wireless LANs differ from peer-level wireless LANs?
14. Describe the major types of end-user applications found in LANs.
15. Describe the major types of workgroup computing applications found in LANs.
16. Describe the major types of document management and database management applications found in today's LANs. How may data be distributed in networks?
17. Describe the impacts that groupware, videoconferencing, and multimedia are having on LANs.
18. What is network management? Why is it important?
19. What is the difference between a CNE and a CNA?
20. How does strategic network management differ from tactical network management?

21. What activities are associated with operational network management?
22. What is downsizing? Why is it important?
23. Why is distributed processing an important issue to today's network managers?
24. What is fault tolerance in network environments?
25. What is meant by the term *open systems*?
26. What does *internetworking* mean? Why is this an important issue for today's network managers?

✳ DISCUSSION QUESTIONS

1. Why is feedback important in a communications network? Discuss why it is beneficial to combine feedback and error detection functions in data communication networks.
2. How has the emergence of networking changed the importance of network managers in information systems departments?

✳ PROBLEMS AND EXERCISES

1. Contact the network or information systems manager for a local company. Ask him or her to describe downsizing, fault tolerance, open systems, and internetworking. Compare the responses that you get to those obtained by the other members of your class.
2. List as many examples of senders and receivers found in data communication networks as you can think of. Identify devices that can be both senders and receivers.
3. Ask a local car dealer to come to your class to describe the WANs that they use to conduct business.
4. Ask a network manager to speak to your class and to describe what his or her day-to-day work entails.
5. Identify local companies using videoconferencing or multimedia. Identify how these technologies are being used.

✳ REFERENCES

Baum, D. "Developing Serious Apps with Notes." *Datamation* (April 15, 1994): p. 28.
Caldwell, B. "Leading the Charge." *Informationweek* (February 7, 1994): p. 38.
Dortch, M. "Does Workgroup Computing Work?" *Atlanta Computer Currents* (September 1994): pp. 34, 41.
Fitzgerald, J. *Business Data Communications: Basic Concepts, Security, and Design.* 4th ed. New York: John Wiley & Sons, 1993.

Flanagan, P. "Future Industry Directions: The 10 Hottest Technologies in Telecom." *Telecommunications* (May 1994): pp. 31, 70.

——. "Videoconferencing: A Status Report." *Telecommunications* (April 1993): p. 41.

Francis, B. "MIS Meets Multimedia on the Network." *Datamation* (July 15, 1993): p. 32.

Frazier, D. and K. Herbst. "Get Ready to Profit from the InfoBahn." *Datamation* (May 15, 1994): p. 50.

Frenzel, C. W. *Management of Information Technology*. Boston, MA: Boyd & Fraser, 1992.

Goldman, J. E. *Applied Data Communications: A Business-Oriented Approach*. New York: John Wiley & Sons, 1995.

Illingworth, M. M. "Virtual Managers." *Informationweek* (June 13, 1994): p. 42.

Jones, T. E. "Technologies That Could Change Your Company." *Chief Information Officer Journal* (July/August 1993): p. 56.

Kay, E. "A Platform to Build On." *Informationweek* (February 21, 1994): p. 56.

Keen, P. G. W. and J. M. Cummins. *Networks in Action: Business Choices and Telecommunications Decisions*. Belmont, CA: Wadsworth, 1994.

Korzeniowski, P. "Calendars Get the Message." *Informationweek* (August 29, 1994): p. 57.

McCusker, T. and P. Strauss. "Managing the Document Management Explosion." *Datamation* (July 1, 1993): p. 41.

McMullen, M. "Personal Space." *LAN Magazine* (April 1994): p. 47.

Pastore, R. "Paths of Least Resistance." *CIO* (April 1, 1993): p. 37.

Pepper, J. C. "Here Comes Multimedia." *Informationweek* (February 21, 1994): p. 26.

Parker, C. and T. Case. *Management Information Systems: Strategy and Action*. 2nd ed. New York: Mitchell/McGraw-Hill, 1993.

Pickering, W. "Outfitting the Ultimate Road Warrior." *Datamation* (March 1, 1994): p. 37.

Ricciuti, M. "The Best in Client/Server Computing." *Datamation* (March 1, 1994): p. 26.

Semich, J. W. "Can You Orchestrate Client/Server Computing?" *Datamation* (August 15, 1994): p. 36.

——. "Information Replaces Inventory at the Virtual Corp." *Datamation* (June 15, 1994): p. 37.

Schatt, S. *Data Communications for Business*. Englewood Cliffs, NJ: Prentice Hall, 1994.

Schnaidt, P. "Cellular Hero." *LAN Magazine* (April 1994): p. 38.

Soat, J. "Betting on the Server." *Informationweek* (November 15, 1993): p. 68.

Stahl, S. "Electronic Mail: First-Class Delivery." *Informationweek* (August 29, 1994): p. 48.

Stamper, D. A. *Business Data Communications*. 3rd ed. Redwood City, CA: Benjamin/Cummings, 1991.

——. *Local Area Networks*. Redwood City, CA: Benjamin/Cummings, 1994.

Wilder, C. "Video Jumps Into the Mainstream." *Informationweek* (June 13, 1994): p. 75.

Rains, A. L. and M. J. Palmer. *Local Area Networking with NOVELL Software*. 2nd ed. Danvers, MA: Boyd & Fraser, 1994.

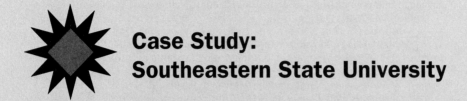

Case Study:
Southeastern State University

Note to readers: This is the introduction to a case study about a fictional university. Parts of the case study are at the end of each chapter, along with questions, problems, and exercises related to topics covered in the chapter. There are no absolutely correct answers for the questions, problems, and exercises because there are many alternatives in the design, management, maintenance, and expansion of LANs and other computer networks. For case questions that seem ambiguous, you should make (and state) any assumptions.

SOUTHEASTERN STATE UNIVERSITY: AN INTRODUCTION

Southeastern State University (SSU) is a regional university with approximately 12,000 students. It offers degrees in 95 different undergraduate majors; it also has 12 graduate degree programs including 2 doctoral programs. The academic units at SSU are organized into five colleges: Arts and Sciences, Business, Education, Engineering, and Health Sciences. The deans of the five colleges report directly to SSU's vice president for Academic Affairs.

Another large subunit at SSU is headed by the vice president for Business and Finance. Among the activities reporting to this vice president are auxiliary services (food services, stores and shops, parking and transportation, and so on), maintenance (janitorial services, painting, renovation services, and so on), facilities (facilities planning and scheduling, architectural services, landscaping), central stores (purchasing, warehousing), and the controller's office, which handles the collection of student tuition and fees, the distribution of financial aid, and so on.

The vice president for Student Services heads another major division at SSU. The SSU staff responsible for admissions, operations within the registrar's office, resident hall management, counseling, career services, health services, student programs, recreation and intramurals, and so on, report to this vice president.

The Computer Services division is responsible for supporting the computing needs of users in each of three major divisions: academic affairs, business and finance, and student services. The director of Computer

Services, the highest-ranking information systems manager at SSU, reports to each of the three vice presidents.

The current director of Computer Services has held this title since 1977. As you might expect, computing at SSU has changed dramatically while the director has been in charge.

Until 1983, most computing was performed at the main computer center. The computer center itself contained a Control Data Corporation mini-computer—one of the Cyber family. Users from the different divisions were required to drop off their programs (on cards) at a window in the computer center; they would also go to this window to pick up the printed output of their programs. Computer operations staff were responsible for placing the cards in the card readers and for distributing all printed output. Of course, they were also responsible for ensuring that the minicomputer remained operational. There was also a room at the computer center containing keypunch machines for users to create their own cards for data and programs. In addition, a data entry service for faculty researchers would transfer research data on standard coding sheets to cards.

Before 1983, other computing activities were performed at a stand-alone LAN, at computers at the state's "flagship" university, and from remote locations over telephone lines. The stand-alone LAN was the main student computer lab in the early 1980s. It consisted of 24 Apple computers (Apple IIs, IIes, and IIcs acquired as part of a federal grant), each of which was connected to a central machine that provided shared printer support and a few other low-level communications functions. This was the Corvus system. Some of the computers in the Corvus network also had direct connections to the Cyber minicomputer in the computer center. These connections allow users to create data and program files online without having to use keypunches and cards; the output for these programs could be printed out either on the printers connected to the Corvus system or on those in the computer center.

The Cyber minicomputer at the computer center could be connected to another CDC computer at the state's main university through long-distance telephone lines. The flagship university's CDC machine was, in turn, connected to an IBM mainframe via a "tie-line." This arrangement made it possible to transfer data and program files from the Cyber minicomputer at SSU to the Cyber mainframe at the flagship university, and over the tie-line to the IBM mainframe for processing. Program output could then follow a reverse route back to the SSU Cyber, where it could be printed out. In order to minimize long-distance telephone line costs, the computer center at SSU would "batch" (collect) program and data files on tape and then transfer the entire batch at one time; this occurred three times a day. After SSU had transmitted all of its files, the output from the files transmitted in the previous batch would be transmitted from the Cyber at the flagship university back to SSU.

The Cyber at SSU was also equipped with a limited number of dial-in ports that allowed users remote access. In the early 1980s, remote access was usually done through teleprinters and acoustic couplers. A teleprinter is essentially a terminal device consisting of a keyboard and a printer; it did not have a monitor and had no processing capabilities of its own. In many ways, teleprinters resemble typewriters—typewriters with extended keyboards and continuous paper feed. Acoustic couplers were among the earliest types of modems. For users at SSU, the acoustic coupler was connected (or built into) the teleprinter and was designed to accept the handset of a telephone. Typically, the user would dial the number of a dial-in port, listen for the Cyber's carrier signal, insert the telephone handset in the acoustic coupler, logon to the computer, and then interactively work on building program or data files or receiving output. All user commands were printed on the teleprinter's continuous-feed paper, as were all of the Cyber's responses. When the user completed processing activities, she would logoff the computer and remove the handset from the acoustic coupler; the telephone could then be used as normal.

If all dial-in ports were in use when the user called, he would get a busy signal; if the computer was down, the call was not answered. The remote interactive sessions through teleprinters were possible because the Cyber recognized the teleprinter as another attached terminal. Because telephone connections were used, transmission speeds were typically limited to a maximum of 2400 bps.

Since 1983, computing has significantly changed at SSU. IBM and IBM-compatible microcomputers appeared in increasing numbers in all divisions of the university, most initially as stand-alone systems. In the mid-1980s, computer labs became more common; these were initially rooms containing stand-alone microcomputers. However, by the late 1980s, a variety of LANs had been implemented in diverse academic units; a large lab that all students could use was also created. Almost all of these LANs were classified as stand-alone networks designed to allow users to share software and printers. However, many also enabled users to connect to the Cyber and thus to the computers at the flagship university.

In 1986, administrative computing functions (such as payroll) were moved from the Cyber to a cluster of Texas Instruments minicomputers. In 1987, SSU gained access to all other public colleges and universities in the state through the creation of a statewide university computing network. The state network connected to what is now the Internet; this gave users at SSU access to information and computing resources in other states and countries.

In 1990, the university decided to implement a new student information system that would enable the school to use a central relational database to coordinate admissions, registration, financial aid, student services, and student records. This necessitated the purchase of a cluster DEC (Digital Equipment Corporation) VAX minicomputers, which has become the main source of academic computing at SSU. Almost all the important student and faculty applications have been moved to the VAX system; many administrative applications have also been moved to one of the minicomputers in the VAX cluster. The main function of the Cyber was shifted to external data communications support.

Soon there were several LANs to support activities in business and finance subunits; in addition, LANs became common in clerical and administrative areas. As these networks developed within the major subunits at SSU, the need for network managers and technical support specialists became painfully apparent. Because most of the equipment and software found in LANs at SSU have been funded by the subunits in which the LANs are found, these subunits have also been responsible for hiring their own LAN managers. Today, the vast majority of the LAN managers at SSU are not employed by Computer Services, although technically, Computer Services is still responsible for supporting computing activities in all university subunits. In practice, Computer Services develops and maintains applications available to users on the VAX, applications for the student information system, and for external data communications. Computer Services does not have the budget or staff to support all of the LANs created, but will help develop, implement, and maintain connections between the networks and the university's minicomputers (as well as with external connections).

The university has continued to upgrade previously installed LANs and create specialized LANs (for example, a CD-ROM LAN in the library, a fiber optic LAN in engineering for computer-assisted design (CAD), an electronic document management system in the admissions area, and an executive information system linking key administrators across campus). Remote access to the VAX and to some of the LANs is possible. For example, faculty can access the library's "card catalog" or the CD-ROM LAN from their offices or homes; students can also access these resources from remote locations. Among students and faculty, e-mail, access to external information services and electronic bulletin boards, and "cruising" the Internet have become immensely popular. As you might expect, Windows applications have become commonplace and everyone wants multimedia capabilities.

In response to these changes, SSU's administration desires to make it possible for both on-campus and remote users to be able to access the computing resources found on the VAX and within any SSU LAN. SSU's administrators are also interested in integrating SSU's voice and data communication systems; they are not currently integrated. The administration at SSU is also equipping several classrooms with the computing and communications technology needed to deliver live, interactive television courses to remote locations. Such "distance learning" systems should enable students at other sites to earn a degree from SSU.

CASE STUDY QUESTIONS AND PROBLEMS

1. What was the primary type of computer network in place at SSU prior to 1983? What kinds of terminals do you think were most common at SSU prior to 1983? Why?
2. To what extent were distributed, wide area, and local area networks found at SSU in the early 1980s?
3. Which of the five types of client/server computing best characterizes computing at SSU in the early 1980s? Why?
4. Which types of network applications are found at SSU? Which are not mentioned?
5. Which of the emerging applications are found at SSU?
6. What types of network managers work at SSU?
7. What strategic network elements seem to be in place at SSU? What elements of tactical and operational management seem to be present?
8. Which of the important networking issues of the 1990s is SSU most focused on?

Data Communication Fundamentals

2

CHAPTER OBJECTIVES

After completing this chapter, you will be able to:

◆ Describe the different ways that data is moved electronically from one place to another.

◆ Explain the difference between analog and digital data transmission.

◆ Describe the physical characteristics of a voice grade telephone line.

◆ Discuss the characteristics of synchronous and asynchronous transmission and how these are related to the efficiency of the data transmission.

◆ Describe the nature of serial and parallel transmission and how these are applied.

◆ Compare the various physical media used to carry data transmissions.

◆ Discuss the characteristics of wireless transmission and applications.

◆ Identify the hardware components of a communications system and briefly describe the function of each.

◆ Discuss the categories of data communication software available in contemporary networks.

◆ Identify and briefly describe the communications protocols used in modern data transmission.

◆ Describe how smaller networks can be interconnected with larger networks.

◆ Compare the various network topologies used in local area networks and wide area networks.

◆ Discuss the network architectures and the role they play in various network configurations.

Today, the term data communications has such a broad scope that entire textbooks can be written on small portions of the discipline. The focus of this discussion is on information transmission between computers in a business environment. This chapter is designed to give you enough background in telecommunications that you will understand the concepts encountered in later chapters.

The world is full of examples of the electronic transmission of data and information. The context of this discussion is limited to information formatting by electrical encoding. Early examples of data communications include the telegraph key and Morse code. The familiar "dit-dah" of the boy scout or amateur radio operator was a means of data communications. Later, a machine called the teletype permitted the operator to press keys associated with letters of the alphabet. The teletype sent on/off signals that were decoded at the receiving station into meaningful characters. Today, computers send and receive signals as streams of 0s and 1s called *bits* (short for *binary digits*) at increasingly higher speeds.

A modern-day example of data communication applications is the automated time clock at a local factory that records data from the magnetic card of a worker entering and leaving the facility. This information is compiled and used to prepare of the worker's payroll check. Another example is the architect who uses a modern graphics program to produce the preliminary layout of a new building on his computer system. The layout can then be transmitted to the computer of the project manager over a telephone line. The project manager loads the file in the computer to inspect the architect's proposed configuration of the building.

More and more, businesses no longer print physical checks or drafts for the payment of accounts. A specially formatted data transmission is constructed by the computer to be sent to a clearinghouse where payment is made to the proper account and a corresponding amount is deducted from the payor's account. All of this is accomplished without any physical transfer of currency. This technique is called *electronic funds transfer* (or *EFT*). A similar approach called *electronic data interchange (EDI)* enables businesses to exchange business documents such as purchase orders and shipping invoices; EDI enables such exchanges directly without the assistance of a clearinghouse.

◆ TYPES OF DATA COMMUNICATIONS

Information is carried from one place to another in a variety of ways. The diversity of methods is partially a product of historical events and partially a product of the ever-increasing research and development efforts of companies and governmental agencies to produce faster and more efficient methods of transferring data and information.

In data communication history a continuous parade of techniques have first become feasible and then obsolete. Many became obsolete because managers found a more cost-effective way to electronically transmit data and information between computers. This evaluation of transmission and communications alternatives is an important part of network management. The material in this text should help you understand why some transmission methods are better than others for certain applications. As a network manager, you will be called upon to sift through alternatives to select which is best for your organization.

In this chapter, the types of data communications are categorized by the physical characteristics of the transmission as well as the media that carry the transmission. You will need to know these fundamental categories when interviewing vendors or consultants who may assist you in constructing your organization's system. As these technologies change, you will need to continually update your knowledge in order to make appropriate choices.

In the U.S., telephone and telegraph companies have provided many data communication services. The following sections are some physical descriptions of the signal types and data transfer techniques that are generally used.

◆ EARLY DATA COMMUNICATION SYSTEMS

Early data communication efforts were concentrated around the railroads and their need for telegraphy. Aerial metallic lines were supported by wooden poles that followed the railroads. Gradually the railroad companies consolidated into fewer entities and created the telegraph. One of the early data communication products was the telegram, which allowed a customer to write a short message on a slip of paper. The message was transmitted using Morse code by an operator using a mechanical telegraph key that emitted short and long signals on a communications line. Morse code enabled a second operator at the receiving end to translate the message into a form readable by the recipient. For the first time, a written message could be sent over long distances without being carried physically by horseback, wagon, or ship.

The types of signals of the world's telephone systems were not originally designed to carry data. Their primary mission was to transmit the spoken word, or **audio** transmissions.

The components of a telephone are designed to convert audio signals into electrical signals that can be transmitted over telephone lines; the telephone is also designed to convert transmitted electrical signals back to audio signals. There are two fundamental types of data transmission over telephone lines. Most of the world's current telephone system is **analog** in nature. Analog transmission occurs by the changing *amplitude, frequency,* or *phase* of the signal to represent different values. Analog transmission is characterized by a continuously varying signal that may take a value anywhere in the spectrum allowed by the line's transmission capability.

The term *line* as it is used here means the media used to carry the signal from sender to receiver. For example, the telephone line that normally comes to the telephone in your home or office is *twisted pair*. It usually has at least four copper wires covered with plastic for insulation, although only two may be needed for the communications circuit.

Most twisted pair wire is made of copper, a conductor of electricity, meaning that a flow of electrons can be easily created when copper is used as a medium. Insulation is a material that resists the flow of electrons; it also protects signals on the wire from external interference. Copper is used in many applications because of its conductive properties.

In an environment where the transmission signal varies continuously, many *frequencies* may be used. Frequency is the term used to describe differences in the wavelength of an analog signal. Figure 2-1 illustrates the difference between two analog signals with different frequencies (wavelengths). Like different amplitudes, different frequencies can be used to represent different data. For example, one frequency could be used to represent one (1) bit while a second frequency could be used to represent a zero (0) bit.

One of the important limitations that relates to analog transmissions is the limitation on the **bandwidth** of the media. Bandwidth relates directly to the volume of data the line can carry per time unit; that is, the greater the bandwidth, the greater the transmission capacity. Technically speaking, bandwidth is measured as the range of frequencies that can be transmitted on a circuit. For example, the bandwidth of a voice grade circuit used to make a typical phone call is 3,000 **Hertz** (Hz, 1 Hz = 1 cycle/second) because the frequencies that can be transmitted range from 300 Hz to 3,300 Hz.

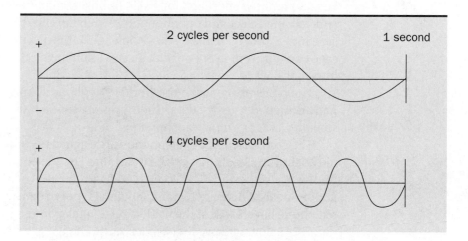

FIGURE 2-1
Frequency of
analog signals

The telephone company keeps your conversation distinct and private from other conversations by separating the 3,000 Hz bands used to transmit signals by *guardbands* that are usually 1,000 Hz wide. Crosstalk (hearing another telephone conversation during your call) occurs when the guardbands fail to keep one conversation separate from another. For efficiency and cost effectiveness, the telephone companies combine several transmissions on the same long-distance line using a process called ***multiplexing***. Thousands of conversations can be multiplexed on one long-distance line, and when analog transmission is used, each is transmitted as a continuous signal within its 3,000 Hz band.

In contrast to analog transmission, ***digital*** signaling is described as a flow of staccato pulses similar to the motion of turning a switch off and on (see Figure 2-2). Digital transmission is distinguished from analog transmission in that each switching signal produces a defined binary digit—either on or off. Most computers send information internally from one location to another in this digital, stop/start manner.

To send digital signals from one computer to another, you may have to use the telephone lines. Because telephone lines in many locations carry analog signals, a translation from digital signaling to analog signaling (and vice-versa) must take place on both ends of the circuit. This conversion requires modulator-demodulators to convert the digital signals to analog first, and then to convert the signal back to digital at the destination end of the circuit. The name of this hardware has been shortened to ***modem***. An example of a modem is presented in Figure 2-3. Recently the word modem has become a household word with the rise in usage of the personal computer as a data terminal and the proliferation of laptop computers used by persons traveling on business.

There are several reasons why digital transmission is more desirable than analog. In fact, if the telephone companies could rewrite history, they would probably have constructed a digital network in the first place.

Digital transmission is more accurate than is analog transmission. An array of factors can cause a signal to be lost or modified as it travels along the

FIGURE 2-2
The difference between analog and digital signals

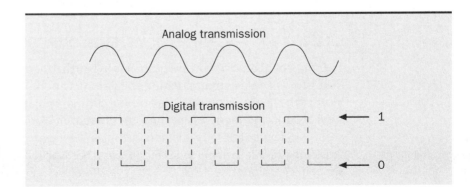

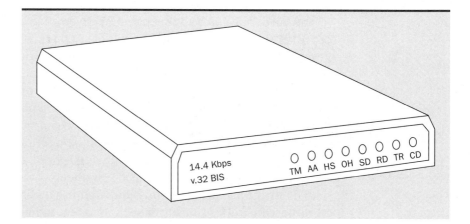

FIGURE 2-3
A modem

medium. Lightning and other weather extremes can cause noisy circuits and transmission errors. Electronic interference from lights and electronic devices can cause errors. Signals can become too weak to be recognized by receivers. In analog transmissions, the signals have to be constantly renewed in order to retain their strength. Digital signals also have to be strengthened, but the method is much different. In analog transmission, when the signal is amplified, signal distortions or irregularities are also amplified. Digital signaling has the capability of total signal recreation without the amplification of distortion. This process is accomplished by a repeater. This recreation of digital signals is far superior to analog amplification.

A related benefit is the unique ability of digital transmission to be secured by sophisticated mathematical *encryption* (scrambling) techniques. Analog transmissions have limited ability to be secured by encryption. Digital transmissions, however, are much easier to encrypt and more methods are available to make the transmission inaccessible and secure.

Currently, technology allows digital signals to be transmitted much faster per time unit than analog signals. Digital transmission speeds are increasing and very high speeds are important for applications such as multimedia. Also, by establishing a direct connection between computers, conversion from digital to analog may be unnecessary.

SIMPLEX VS. HALF-DUPLEX VS. FULL DUPLEX

The services offered by common carriers in North America evolved through the history of data communications beginning with the telegraph services. Most of the common carriers (telephone companies) offer similar types of services since, in many cases, they are interconnected with each other in order to offer various long-distance services. There are some variations internationally, but in many areas of the world the North American standards are being emulated.

Functionally there are three data transmission modes: *simplex, half-duplex*, and *full duplex* circuitry. The differences between these are illustrated in Figure 2-4. Simplex lines in North America transmit in one direction only. These circuits were used in years past for *telemetry* applications. Telemetry is used to monitor pipelines and to open and close valves in remote areas. The volume of data transmitted along these lines only required one-way traffic because the instruments used in telemetry did not issue responding signals. Although simplex circuits may still be listed in the tariffed offerings of some common carriers, in practice there are very few simplex circuits in service. An everyday example of simplex transmission is radio stations. They transmit; you receive their signals.

Half-duplex circuits were once priced at a lower rate than the full duplex circuits since the half-duplex circuit could only carry data in one direction at any point in time. The economics of modern telephony require that almost all telephone company facilities be capable of full duplex transmission. Therefore most of the offerings listed in the tariff books are based on a full duplex voice grade circuit. A full duplex circuit can carry data in both directions simultaneously. An everyday example of half-duplex is a walkie-talkie that can transmit and receive, but not at the same time. An everyday example of full duplex is the telephone; both sender and receiver may be talking at the same time.

Another reason for the movement to full duplex circuitry was the drastic change in pricing for leased data lines beginning in the early seventies. The breakup of AT&T caused dramatic changes to occur in the pricing methods of leased data lines that emphasized more than ever before the costs of the equipment at the terminations of the circuit. Previously the

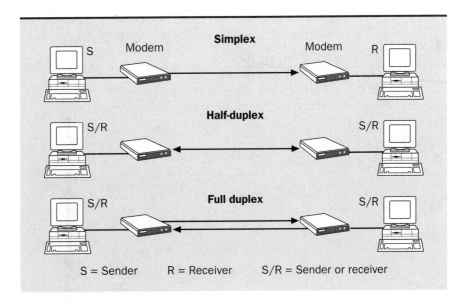

FIGURE 2-4
Simplex, half-duplex, and full duplex circuitry

pricing was based on mileage almost exclusively, with distinct differences in cost related to the categories of simplex, half-duplex, and full duplex capabilities.

ASYNCHRONOUS VS. SYNCHRONOUS TRANSMISSION

Asynchronous data transmission has been used since the days of the telegraph and Morse code. The teletype was another driving force behind the widespread use of asynchronous transmission early in this century and the emergence of terminals has further increased its popularity. The teletype was an electro-mechanical typewriter with a communications interface for attaching to a communications line. Most microcomputer-to-mainframe links used today, especially those from remote locations, utilize asynchronous communication. In asynchronous communication, to send a character (for example, a letter of the alphabet, a digit, or a special symbol), a group of seven or eight bits representing the character is transmitted. A start bit is transmitted before the bits representing the character and stop bits are attached to signify the end of the character transmission. The start and stop bits serve as separators for the character and enable the receiving machine to know which bits are associated with the character. Figure 2-5 illustrates this character framing. The disadvantage of asynchronous transmission is reduced transmission efficiency from overhead caused by adding start and stop bits to each character.

In the middle of this century, it became necessary to develop an efficient means for transferring large files between computers. *Synchronous* transmission was developed whereby a clocking mechanism on one end of the circuit could be synchronized with the clock on the other end of the line. This is accomplished through a set of synchronization characters sent before a block of data characters. Through such synchronization, the stop

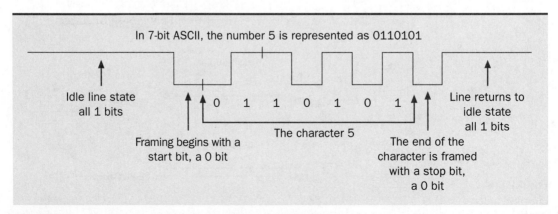

FIGURE 2-5 Character framing in asynchronous transmission

and start bits surrounding each character in asynchronous systems became unnecessary. Thus synchronous transmission became a more efficient means of communicating large volumes of data in a single direction.

Synchronous transmission utilizes a *block* or *frame* of data as a means of organizing the data, as opposed to the character-by-character orientation of asynchronous transmission. Synchronous frames of blocks may contain hundreds of individual characters. Error checking is accomplished by attaching one or more error checking characters to the end of each frame (see Figure 2-6).

In the past, speed was a determining factor between the usage of synchronous and asynchronous transmission. However, advances in telephone switching technologies have meant that asynchronous transmission can now be accomplished at much higher speeds. While teletypes communicate at speeds of 100 and 300 words per minute (with a word equaling five characters and a space), other asynchronous applications between computers are now clocked at speeds from 9600 to 56,000 bps. In fact, ATM (asynchronous transfer mode) transmissions are emerging that have speeds in excess of 100 million bps; ATM is expected to enable more multimedia applications to be used in networks.

In general, smaller personal computers that generate interactive (or two-way) traffic with mainframes or minicomputers use asynchronous transmission. However, communications between large computers involving large files or high volumes of data are most likely to use the synchronous method.

When selecting between asynchronous and synchronous transmissions, network managers often use mathematical calculations. For example, if you have a 40 kilobyte file (40,000 bytes or 320,000 bits assuming 8 bits per byte), and your transmission rate is 9600 baud or bps, how long will the transmission take? 320,000 divided by 9600 is 33.334 seconds. If the transmission method is asynchronous and you have a stop bit and a start bit for each byte, the number of bits would increase to 400,000, an increase of 25 percent. If the transmission speed were increased to 14,400 bps, the transmission time with start and stop bits would be 27.77 seconds (400,000/14,400).

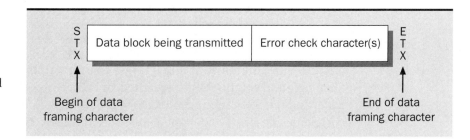

FIGURE 2-6
Framing control characters in synchronous transmission

For synchronous communications, the overhead is typically less than that for asynchronous. For example, in a synchronous frame containing 1,000 bytes that are preceded by two synchronous characters (8 bits each) and followed by two error check characters (8 bits per each), the overhead is 4 characters x 40 frames x 8 bits per character, or only 1,280 bits, which is only a fraction of that for asynchronous. At any of the speeds mentioned in the example, synchronous transmission is more efficient.

There are two important assumptions in these simple calculations. First, you are assuming that there is no connection or operation time involved in the transmission time. That assumption is not valid, but at this point you have no basis to calculate that time and therefore it is ignored. Secondly, this example assumes that one baud is always equivalent to one bit per second. This is not always true, but for a beginning point the example serves the purpose.

SERIAL VS. PARALLEL

Another physical characteristic of data transmission is the transmission mode, which refers to whether the transmission is *serial transmission* or *parallel transmission*. The majority of data communications transmissions from one location to another are serial. The term *serial* means that a stream of bits is transmitted over the media one bit at a time. Examples of this transmission include most asynchronous applications, particularly those involving modems.

The other mode of transmission is called *parallel transmission*. This mode refers to the transfer of data where all the bits of a character are transferred together. Parallel mode is most often used internally within a computer; it is also frequently used between a microcomputer and an attached printer. Both serial and parallel are illustrated in Figure 2-7.

Serial transmission is well adapted to long-distance data communications, as well as through long cables from one end of an office to the other. Parallel transmissions are limited to short distances, usually not more than 10 to 15 feet.

WIDEBAND DATA TRANSMISSION

Most telephone companies offer wideband services primarily used for data transmission. In the past, these services have offered large capacity analog circuits with a bandwidth much larger than the traditional voice grade circuits found on the public switched voice network. Speeds for analog wideband services range from 19,200 to 256,000 bps. Sometimes telephone companies describe wideband services in terms of a bundle of 12 voice grade circuits called a master group. You may also hear of a service with the

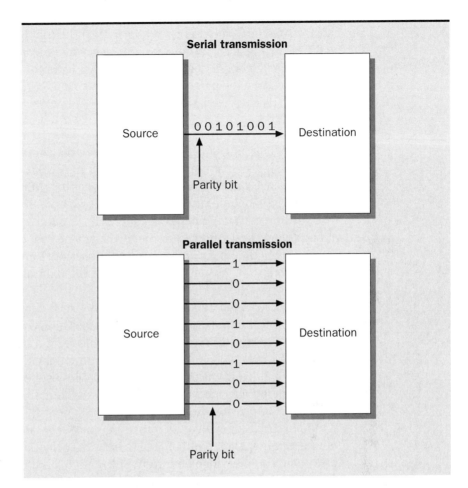

FIGURE 2-7

Parallel vs. serial transmission

equivalent of 60 voice grade channels called a supergroup. Typically these circuits are offered as both point to point services or as multipoint services. Modems must be used to take advantage of these analog wideband services.

Now telephone companies offer *dedicated* digital wideband services. The term *dedicated* means that the communication line is not shared with any other organization and that this exclusive use is available 24 hours a day. Digital services offer the advantage of not requiring modems on each end of the media for data transmission.

These wideband digital offerings provide the same benefits generally allowed by digital transmission. Digital transmission is more reliable and accurate. Most telephone companies design the service to be 99.5 percent error free at 56,000 bps. These services are also more secure than dial-up telephone lines because the public has no access to these dedicated lines. Digital wideband facilities are generally available in the heavy traffic routes, which include the industrial and business district areas of major cities, as well as in other cities with a large industrial base or government facilities. Switched digital services are also available at 38,400 bps, 56,000 bps, and 64,000 bps. T-services including T-1 lines (1.544 million bps) are other high-capacity digital services available from current carriers.

VIDEO

In the past, customers using video services of the common carriers generally were television stations and television networks. Now many private companies use various kinds of video services including video *teleconferencing* (or videoconferencing). As noted in Chapter 1, videoconferences are real-time electronic meetings of groups of people through the use of video cameras and audio equipment. Usually people at one location take charge of the conference and direct the discussion. Individuals at other locations can also participate and be seen on video monitors as the meeting progresses.

Dedicated video services are available from common carriers and third-party companies for heavy users. Most heavy video service users such as television stations typically schedule certain hours for utilization that usually are associated with programming such as athletic events, concerts, or conventions. Telephone companies use several types of communication media such as satellite, microwave, and fiber optic communication lines to provide these services.

Video services involve the transmission of very large volumes of data that require high bandwidth. Satellite links have the bandwidth needed to furnish high quality video links from one point to another. Microwave links also provide excellent video transmission capabilities. The highest quality video service is offered by fiber optic cable, which transmits light waves and produces an unusually good video signal. Light waves are not easily disturbed by electrical and weather interference that can adversely affect video transmissions over other communications lines.

◆ COMMUNICATIONS MEDIA

Data can be carried in a variety of modes over several different kinds of media. A communications medium provides a physical path over which the data is transmitted. A *circuit* is an electronic path between two points. Many people use the word circuit and communications line interchangeably. However, telecommunications purists often describe a line as the physical cable (such as copper wire or fiber optic cable) over which data is transmitted; a circuit is described as the flow of electrons along the path between two devices regardless of the physical make-up of the cable.

TWISTED PAIR

Twisted pair is the name given to cables consisting of two or more copper wires covered with insulation and bundled together into one composite cable. An example of twisted pair is found in Figure 2-8. Twisted pair is the most common media used in telephone systems. It is also widely used in LANs. Twisted pair varies in quality and capacity. Both quality and capacity are a function of the number of twists per inch and how much insulation (shielding) it has. Twisted pair seemed to be doomed to extinction until, in recent years, research and development of communication links provided techniques of transmitting larger volumes of data across twisted-pair cables.

By twisting the wires around one another, electromagnetic interference from external sources is minimal. By adding shielding, twisted pair is capable of transmitting data at rates that are well in excess of 10 million bps.

COAXIAL CABLE

As may be observed in Figure 2-8, *coaxial cable* is made in a cylindrical construction with the inner conductor made of copper wire. Cable television companies have made extensive use of coaxial cable to provide the high bandwidths required for video. Coaxial cable has several layers of insulation around the central cable. Bundles of coaxial cable can be used to carry thousands of simultaneous voice and data transmissions. These cables are very efficient and are not as susceptible to distortion, signal weakening, and *crosstalk* as twisted pair. Some coaxial cable is very thin and can be laid under carpet in office settings. Coaxial cable has been popular in LANs because several computers can be linked via a multipoint coaxial cable. With a "T" connector (see Figure 2-9), one computer can be easily disconnected or connected to the cable without interfering with other computers that are connected.

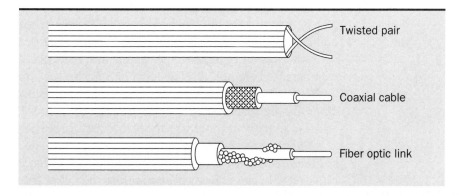

FIGURE 2-8
Cables used in data communications

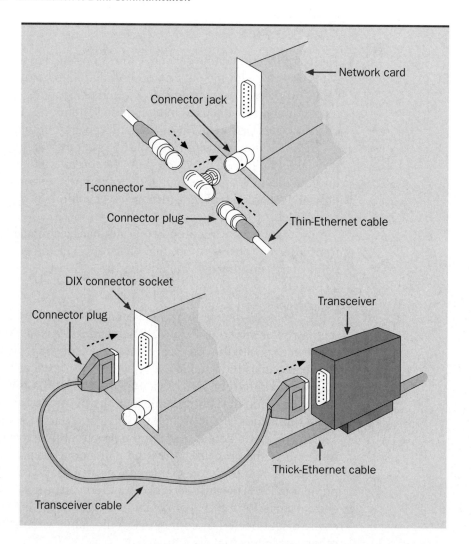

FIGURE 2-9
Twisted pair and
coaxial cable
connections

FIBER OPTICS

One of the most exciting technological advances in communications media has been the development of *fiber optic cable*. This technology uses
light waves to carry data over very thin strands of glass referred to as optical fibers (see Figure 2-8). The terminations on the end of these cables allow electronic equipment to convert the light pulses back into digital data.
The most significant benefit of fiber optic material is its huge bandwidth
and its capacity for rapidly transmitting very large volumes of data.

The materials from which fiber optic cable is made are being improved
continuously. The plastic material used in early fiber optic cable is being
replaced with glass fibers with properties that allow the signal to be carried
farther before being weakened or degraded. However, the drawback to

using these materials is that the light signals must be converted to electronic signals before they can be renewed. This can be a relatively slow procedure that requires expensive equipment. Even so, the distances that light can travel between repeaters (the devices used to renew the signals) are much greater than with any other type of cable currently available. This is why fiber optic LANs can cover much larger geographical areas than LANs using other types of cabling.

Fiber optic cable is much less susceptible to the electronic and magnetic field interference that affects other types of cable. As a result, the error rate for the data transmitted across fiber optic cable is very low. Errors in other physical media are counted in the one per million range while fiber optics error rates are most often counted in the one per billion range. In networks in hostile environments such as chemical factories, petroleum refineries, and other industrial environments that are potentially dangerous for electrical currents, fiber optics can be a much safer alternative.

In spite of its benefits, fiber optic cable does present a challenge for network managers. While common carriers have specialized equipment and trained technicians to install and manage fiber optic systems, few companies can afford to purchase the equipment and expertise required for LANs. While fiber optic cable is very difficult to tap without detection and therefore presents an excellent security tool, it is much more difficult to splice (as one would have to do to add a new workstation to the LAN) than copper wire.

Still, fiber optic cabling is being more widely used in LANs. The Fiber Distributed Data Interface (FDDI) has been defined by the American National Standards Institute and provides hardware and software vendors with guidelines for fiber optic interface. The capacity of fiber optic cables as well as their security and safety make this an ideal medium to carry large amounts of data communication signals from one location to another.

WIRELESS ALTERNATIVES

An increasing variety of wireless alternatives give network managers new options for transmitting data between computers. Some of the most common wireless alternatives are discussed in the following sections.

Microwave Transmission

The term *microwave* is used to describe very high frequency signals directed over a line-of-sight path between two specially designed towers (see Figure 2-10). Microwave signals have a short wavelength, typically above 1,000 megahertz. These have been in general commercial use in the United States for several decades. Because of the line-of-sight requirement, com-

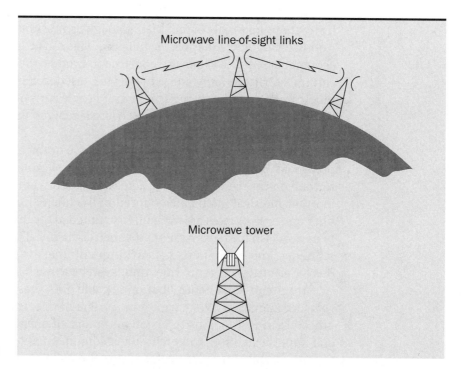

FIGURE 2-10
Microwave
towers and relays

mon carriers space microwave towers every 30 or so miles. Microwave services provide the bandwidth required for video transmission. Although microwave transmission services are capable of higher transmission speeds than is copper-based cabling, their popularity has decreased since the development of fiber optic cabling. Microwave disadvantages include congestion and signal interference from electrical storms, sun spots, and cosmic radiation. Also, because signals can be intercepted between microwave towers, these may be security problems.

Satellite Transmission

In this book, the discussion of satellite communication services is limited to those services available through the common carriers. Because of the huge expense of launching satellites, NASA has put the majority of existing communications satellites in orbit. The initial euphoria created by satellite transmission's high bandwidth and fast speeds subsided somewhat when the problems of dealing with *propagation delay* were better understood. Since the satellite must be in synchronous orbit approximately 22,300 miles above the equator, the sheer distance to and from the satellite presents some unique communications challenges. The propagation delay is the amount of time required for a signal to travel from an earth station to the satellite and back to another earth station. Since the signal must travel the uplink (the path from earth station to satellite) and the downlink (the path from satellite back to earth), this journey introduces almost a second in total delay for one trip. This short time period sounds insignificant, but in today's world, where transmission rates are high, one second is a significant delay. Additionally, many of the error detection and error correction mechanisms in today's sophisticated modems and communication protocols require that blocks of data containing errors be retransmitted. The time for the signal to be sent from the destination station to the sender to retransmit the block and the actual time to retransmit the block can further slow file transfer time and reduce transmission efficiency. In addition, once the signal reaches the terrestrial equipment, processing the signal involves other delays. The use of full duplex circuits coupled with hardware and software programmed to compensate for propagation delay has provided some solutions for these time and distance problems.

Common carriers offer a variety of services over satellite links, including single circuits, master groups, super groups, and wideband circuits. Some companies also offer the lease of a ***transponder*** either for dedicated use or occasional use. A transponder is a component of the satellite that receives the uplink transmission, converts it to a different frequency, and sends the message back to earth over the downlink.

The strongest demand for satellite services commercially has been for audio and video services. Satellite technology provides for an unusually clear signal. The fact that a satellite transmission generally must be relayed only twice between sender and receiver contributes to the data clarity. However, voice transmissions can be problematic because of a variety of factors. First, some transmissions require a "double-hop," meaning that the transmission must be relayed off the satellite twice or relayed off more than one satellite. This causes a perceptible delay which is bothersome in a voice conversation. Second, the echo suppressors used in early satellite transmission tended to lock when the parties on the telephone line began to speak concurrently. So the conversation was stilted because each person had to wait for the other to finish before he or she began speaking. Some customers reported that a conversation over satellite links sounded as though one were speaking in a barrel, complete with the resounding echoes.

There are a number of unique applications that use satellite transmission. One large publisher uses satellite and digital facsimile to publish four regional editions of a daily financial publication. The satellite links enable the regional editions to be printed at locations near the delivery destinations, saving transportation costs. The satellite's large bandwidth enables the publisher to deliver huge volumes of copy in a short period of time to four presses. A number of daily papers such as *USA Today* and *The Wall Street Journal* use similar satellite systems to get their publications to subscribers across the U.S.

A significant problem for many companies is the security exposure of satellite transmission. Just as any consumer can purchase the equipment to receive television programs relayed by satellite, there are places to procure equipment to intercept satellite transmissions of sensitive data as well. As a result, data encryption or scrambling is a common security method used in satellite transmissions.

Cellular Telephone Service

The cellular telephone has rapidly become a mainstay in business communications. The problems of the older mobile communications technology competing for limited bandwidth are overcome by cellular systems. The geography of a city is divided into sections called cells and each cell is equipped with its own antenna. When a subscriber dials into the system the signal travels to the closest antenna. If the subscriber is moving, when the location of the cellular telephone reaches the point where another antenna is closer or stronger in signal, the system automatically routes the signal to the stronger signal source.

The cellular system is becoming pervasive. Almost all of the interstate highway system in the U.S. has coverage, so that when you are traveling across the country by automobile you can remain in touch with your office or customers. For a period of time there was difficulty in receiving calls when you traveled out of your home area, but now most carriers offer a service whereby your calls are routed to you wherever you are within a region. You can even select whether or not you want your calls routed to you.

Data services are also available over cellular systems. Many cellular phone units for automobiles provide interfaces for portable microcomputers. These enable mobile workers to communicate with home offices from just about anywhere. In addition, cellular data networks such as ARDIS have been implemented and standards such as CDPD (cellular digital packet data) have been developed to allow data transmission at speeds of 19,200 bps and higher over cellular networks.

Many vendors have PDAs (personal digital assistants) to provide not only voice, but data and facsimile transmissions over cellular systems using handheld units. And, as noted in Chapter 1, personal communication systems (PCSs) are becoming more popular and many of these take advantage of cellular networks.

Security has become a major issue with the proliferation of so many cellular telephones. The radio frequencies used for cellular phones can easily be accessed by scanners allowing your cellular conversation to be monitored by anyone scanning the cellular radio spectrum in your location. To date, good and reasonably priced encryption mechanisms for cellular data systems are hard to find.

Fraud is a serious problem. The relative ease with which the EPROM (erasable programmable read only memory)—a key component used in uniquely identifying each cellular device—can be altered has invited a host of criminals who change the number and sell the telephone to other people. A large volume of calls can be made before anyone detects that the calls are fraudulent. Many consumers have received bills for thousands of dollars worth of fraudulent calls.

With cellular technologies, you pay for both outgoing and incoming transmissions. This makes such systems more expensive than other data communication services. However, because cellular technologies enable mobile workers to stay in close contact with the home office, the additional expense may be worth it.

The possibility of using the cellular service to connect to LANs and WANs is being explored by more organizations. Systems are available that include software to automatically reestablish connections if disconnected, and to begin transmitting data at the point where the break occurred rather than having to retransmit the entire message. Such convenience could be a good thing, but there are still security concerns. With cellular access, you expose your entire network to new threats.

◆ A BRIEF OVERVIEW OF DATA COMMUNICATION HARDWARE

As described in Chapter 1, in the past the central component of most computer networks was the host computer, usually a mainframe (see Figure 2-11). With the introduction of more powerful minicomputers and personal computers, structures are now primarily determined more by their functions. In other words, a large number of personal computers connected

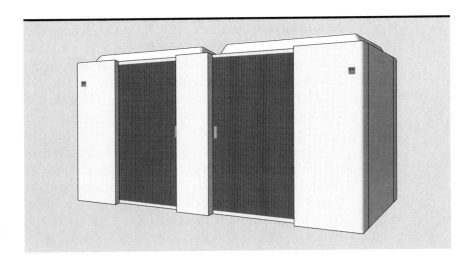

FIGURE 2-11
Mainframe host
processor

by LANs and WANs may now execute tasks that previously could only be performed on mainframes. While mainframes can often process large loads of data more efficiently than LANs, many other tasks can be run as efficiently on LANs. In many host-based networks, mainframe computers use *front end processors (FEPs)*. The FEP is a smaller computer that is used to handle communications tasks for the host. The host is then free to handle the processing tasks requested by users. FEPs are discussed further in Chapter 16.

A key determinant to what kind of host computer is needed is whether or not the system is centralized. If a large database is needed in one place for universal access, a mainframe host may be the only solution. In such cases, the mainframe host will usually require a trained operations staff to maintain the system and keep it running. The system is likely to be equipped with devices such as channel adapters, ports, and **modems**. In the case of analog communications, lines are used to connect user workstations to the host.

Modems have become quite common in home, portable, and office microcomputer systems. When these are used to establish connections to the host, modems and related equipment are often mounted on racks near the mainframe to accommodate incoming data transmissions. Usually this equipment is a small part of a large data processing facility complete with raised floors, special environmental controls, and uninterruptible power supplies. Special modems may be in place for high-speed transmissions of large volumes of data from one mainframe host to another. Such modems may be able to transmit at speeds of 56,000 to 1,544,000 bps or more. Lower speed modems capable of transmitting in the range of 2400 to 14,400 bps are used to communicate with terminals, personal computers, or other end-user devices.

Some offices still have acoustical modems (or acoustic couplers) that enable users to communicate with hosts through telephone handsets. **Acoustical couplers** allow you to place the handset of the telephone in a rubber cradle, and the audible tones are translated into electrical pulses by the modem. For an occasional user or a mobile worker, this arrangement can work well. However, many offices now have personal computers that have internal modems or are attached to a LAN that can be used to establish connections with the host.

The equipment connected to the modem can range from a sophisticated high-end workstation (or clone) to a simple handheld dumb terminal. Today the most common terminals include intelligent workstations or personal computers. In such environments, the personal computer can be used for applications such as word processing, spreadsheets, and small database processing without having to be connected to the host. Through terminal emulation, it can also be used as another terminal. In fact, microcomputer operating systems such as Microsoft Windows, Windows NT, and OS/2 have multitasking capabilities that enable users to simultaneously work on several applications while the microcomputer is connected to a host. Previously, multitasking was available only on larger systems.

Today most terminals, regardless of other attachments, have a video screen. Examples include dumb terminals, intelligent workstations, personal computers, and special purpose workstations such as point of sale (POS) registers. A video enables the creation of easy-to-use data entry screens and also provides a convenient mechanism for the host to respond visually.

Printers are also an important part of data communication networks. Printers can take the form of laser, ink jet, dot matrix, thermal transfer, or daisy-wheel technology. Laser and ink jet technologies are among the most popular today because both allow high-quality character and graphics output on the same page. Color printers are also becoming more popular.

As noted previously, cellular technology has allowed portable computers and handheld devices to interact with a host from remote locations. Some of the other types of technologies that have become popular in networks include:

◆ *Touch screens* where the input is created by users making physical contact with the screen.

◆ *Pointing devices* such as the mouse and trackball. These have become standard equipment in most microcomputer systems.

◆ *Scanning equipment* that can scan a page for input into electronic document management systems, word processing files, and so on. This technology allows the user to convert printed text to a word processing file or document image that can be edited or modified.

◆ *Numeric keypads* for the convenience of those entering large volumes of numeric data.

◆ A BRIEF OVERVIEW OF DATA COMMUNICATION SOFTWARE

Software plays a critical role in successful communications systems. In networking, most communications controls sent across the network are software generated.

Software may be located in many places on the network. First, there is the NOS (Network Operating System), the operating system that controls interactions between networked devices. In large centralized systems, a large amount of communications software is often loaded in the front end processor, including telecommunications access programs and teleprocessing monitors.

Communications software has closer kinship to operating system software than to applications software. Just as for operating system software, if the communications software fails, it can cause the failure of most applications. The primary purpose of communications software is to assure the integrity of the communications link for the duration of the connection between the computers.

The most significant increase in communications software has been for the software used to assist the personal computer to communicate over dial-up lines to hosts, bulletin boards, information services such as CompuServe, and online databases over the Internet. The mushrooming communication activity between personal computers and the various hosts all over the world has triggered a flurry of development of communications software and the associated hardware. Now you can buy software that you just point and click to perform the tasks that required the creation of a whole list of manual commands only a few years ago. A growing list of computer programs can also assist you to connect to a remote personal computer and control it from your location. Many people carry laptop computers home from the office and do work in the comfort of their homes; others telecommute from home rather than fight the traffic to get to the office.

◆ DATA COMMUNICATION PROTOCOLS

Communications software is closely related to the *protocols* used in data communications. Data communication protocols are the steps and procedures that govern the communication between networked computers. Protocols control the order, timing, and execution of the commands needed for two machines to communicate.

One of the most notable movements of the last decade has been the attention received by the *Open Systems Interconnection (OSI)* model (see Figure 2-12). This model has been created by the International Standards Organization (ISO) to standardize communication protocols so that regardless of the hardware or software brand that is used, computers will be able to communicate with one another.

OSI helps overcome incompatibility and interoperability problems. The business community has been plagued for decades with companies entering and then leaving the data processing and telecommunications

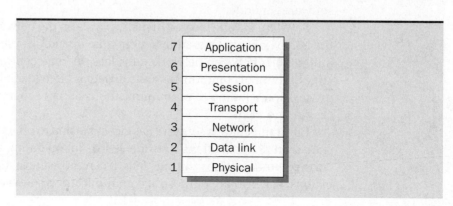

FIGURE 2-12
OSI reference model

manufacturing business. When companies abandon the business or fail completely, users are left with equipment that can only function with compatible equipment no longer in production.

Fortunately, some protocols have emerged as standards for LANs and other networks. Some examples of these efforts are TCP/IP and FDDI. TCP/IP is one of the most widely used protocols in the Internet and client/server networks, and FDDI is the dominant network specification for fiber optic networks.

There are a number of other important data communication protocols. Some of these will be discussed more fully in Chapter 5.

◆ INTERNETWORKING TECHNOLOGIES

For most organizations, it is no longer sufficient to have stand-alone LANs. The trend is definitely toward interconnecting LANs (see Figure 2-13) and interconnecting LANs and WANs. By the year 2000, experts predict that stand-alone LANs will be rare.

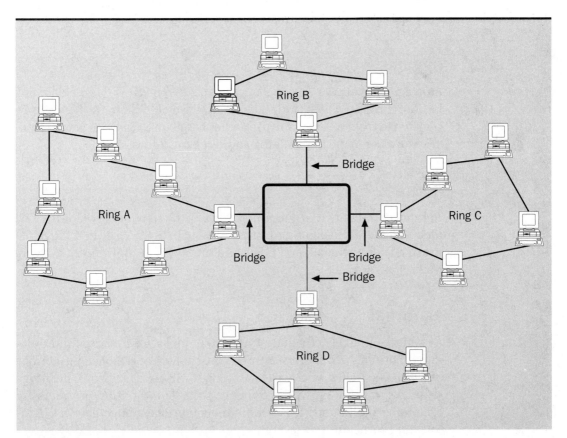

FIGURE 2-13   Interconnected LANs

There is also an increasing interest among organizations in communicating on the Internet, the network created by an international consortium of universities, institutions, governments, and businesses. The Internet is a crucial component of the information superhighway that is being created.

The following is a brief introduction to the primary interconnection technologies used in today's networks. Because of their importance, these are discussed more fully in Chapter 15.

REPEATERS

There is a limit to the distance that either digital or analog signals can travel before *attenuation* (weakening) or noise causes them to degrade. Attenuation is a technical term referring to the decrease in strength of electrical current during data transmission. As mentioned previously, amplification may not be a solution to attenuation because the noise or distortion becomes amplified to the same extent as the signal. A better solution is offered by repeaters. These can extend the normal length of the network and link different network segments to one another. A repeater operates at the physical layer of the OSI model.

BRIDGES

A *bridge* is a network device that connects two networks at the data link layer of the OSI model. Bridges determine which data traffic needs to be transferred from one network to another based on knowledge of the physical addresses of the transmitting and receiving nodes.

ROUTERS

Routers connect two or more networks at the network level of the OSI model. Routers also determine the optimal path to the receiver through the use of an address database known as a routing table and current network traffic statistics.

BROUTERS

As you would likely conclude from the name, a *brouter* combines features of bridges and routers. Brouters are capable of establishing a bridge between two networks as well as routing some messages from the bridge networks to other networks. Some of these combination products are called smart hubs and are a combination of bridge/router/gateway hardware and software.

GATEWAYS

Gateways connect two networks above the network layer of the OSI model. They are capable of converting data frames and network protocols into the format needed by another network.

◆ NETWORK TOPOLOGIES

Topologies can be defined by the physical description of the network. The topology can also be defined by the logical description of the flow of data around a network. The main topologies found in today's networks are illustrated in Figures 2-14 through 2-18.

The *mesh* topology is characterized by physical connections between every node in the network (see Figure 2-14). This topology is seldom used in its purest form, but portions of a network may follow this pattern.

In the case of a *star* network, all devices are attached to a central point, or *hub* (see Figure 2-15). The hub can be described as active or passive. The

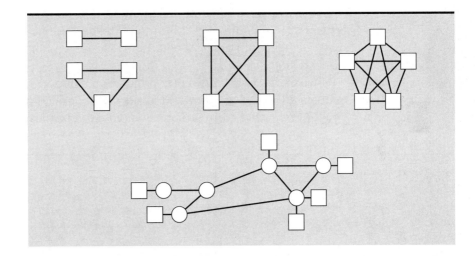

FIGURE 2-14
Mesh topology

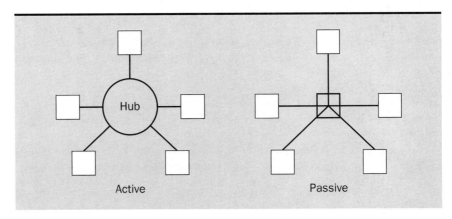

FIGURE 2-15
Star topology

active hub actually initiates the creation of some transmissions, while the passive hub merely provides a connection between networked devices. Since every transmission must pass through the hub, a hub failure can cause the entire network to crash.

A *ring* topology is described as a closed loop of workstations or nodes with messages passing to each station in succession (see Figure 2-16). In this configuration the data travels the loop in one direction.

The *bus* topology uses a straight segment of the media to which all the nodes or workstations are directly connected (see Figure 2-17). This topology is efficient in its use of cable.

There are also implementations which use some combination of the aforementioned topologies. These more complex networks are called *hybrid* topologies (see Figure 2-18).

◆ NETWORK ARCHITECTURES

Network architectures are blueprints (or plans) for how the hardware and software in a network should be deployed or configured. IBM's System Network Architecture (SNA) is among the best known network architectures. Digital Equipment Corporation's DECnet is another well-known network architecture.

Two of the basic hardware configurations supported by network architectures are *point-to-point* and *multipoint* connections. The former is a simple dedicated connection between two computers. In multipoint

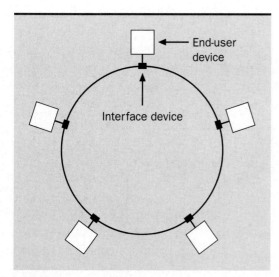

FIGURE 2-16   Ring topology

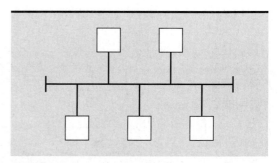

FIGURE 2-17   Bus topology

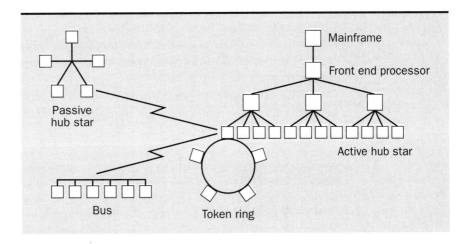

FIGURE 2-18
Hybrid topology

connections, two or more devices may share the same communications medium. Both of these types of connections may be found in both large and small networks.

◆ SUMMARY

Students usually have one of two responses to the introductory chapters of information systems texts. One is excitement about the products being developed and sold. The other is bewilderment at the unfamiliar vocabulary of the technology. The authors hope you will view the introduction provided by this chapter as an opportunity to become involved in the most technologically advanced age the world has ever known.

The world is becoming smaller and smaller with the creation and interconnection of networks. As a student, you can access the Internet and let your fingers do the globe-trotting into libraries and databases around the world.

Your general knowledge of networking starts with understanding the differences between analog and digital signals and how these are transmitted (for example, simplex, half-duplex, and full duplex; synchronous vs. asynchronous; serial vs. parallel, and so on). It is also important for you to know the differences between the major types of cabling used in networks (twisted pair, coaxial cable, fiber optic) and the various types of wireless services that may be used (such as microwaves, satellites, and cellular networks).

Some of the other important fundamental concepts covered in this chapter include major hardware components found in networks, including hosts, front end processors, modems, terminals, printers, and specialized input devices such as touch screens or scanners. This chapter has also provided a brief overview of communications software, communications protocols, internetworking technologies, network topologies, and network architectures.

As you progress through the book, you will continue to build on these fundamental networking concepts. Many will be explored more fully later in the text.

✳ KEY TERMS

Acoustical coupler
Analog
Asynchronous
Attenuation
Audio transmission
Bandwidth
Bits
Bridge
Brouter
Bus topology
Circuit
Coaxial cable
Dedicated
Digital
Encryption
Fiber optic cable
Frequency
Full duplex
Gateway
Half-duplex
Hertz
Hub
Hybrid topology

Mesh topology
Microwave transmission
Modem
Multipoint
Multiplexing
Open Systems Interconnection (OSI)
Parallel transmission
Phase
Point-to-point
Port
Propagation delay
Protocol
Ring topology
Router
Serial transmission
Simplex
Star topology
Synchronous
Teleconferencing
Transponder
Twisted pair

✳ REVIEW QUESTIONS

1. What was the first electronic data communication application in the United States? What protocol did it use?
2. What are the characteristics of EFT systems?
3. What are the differences between analog and digital signals?
4. What is the difference between an analog wave's amplitude and wavelength? How are amplitude and wavelength differences used in data communications?

5. What is crosstalk? Why does this occur?
6. What is a repeater? What functions does it perform? What is encryption? Why is it used?
7. What are the differences among simplex, half-duplex, and full duplex transmissions?
8. What are the differences between asynchronous and synchronous transmission modes?
9. What are the differences between serial and parallel transmission?
10. What is bandwidth and why is it important?
11. Describe the differences among twisted pair, coaxial cable, and fiber optic cable.
12. What is FDDI and why is it important?
13. What types of wideband services are available from common carriers?
14. What are the differences among microwave, satellite, and cellular services?
15. Identify several important types of communication software found in networks.
16. What is a communications protocol? Why are protocols important?
17. What are the differences among bridges, routers, brouters, and gateways?
18. What are the differences among mesh, star, ring, and bus topologies?
19. What is a network architecture?
20. What are the differences between active and passive hubs?
21. What are the differences between point-to-point connections and multipoint connections?

✳ DISCUSSION QUESTIONS

1. If they had it to do over again, why would telephone companies build digital networks instead of analog networks?
2. What network topology would you recommend for a bank with 40 branches? What criteria would you use to pick a topology if you were a manager in charge of installing a new network between the branches?
3. From the discussion in this chapter, what trends seem evident in data communication media use, media signaling, common carrier service, and internetworking?

✳ PROBLEMS AND EXERCISES

1. If you assume an asynchronous transmission speed is approximately 1200 bps, and the data to be transmitted is 24 kilobytes (24,000 bytes) in length, approximately how long would the transmission last? Assume that each byte contains 8 bits.

2. If the transmission speed in problem 1 is increased to 9600 bps how long will the transmission last?

3. After the transmission mentioned in problem 1 you need to print this data. If your printer prints 20 characters per second, how long will it take to print all the data that has been transmitted? (Ignore stop/start bits for printing.)

4. Find a recent article on wideband services offered by common carriers. Summarize the article's contents in a paper or presentation.

5. Find a recent article on high-speed digital services available from common carriers. Summarize the contents in a paper or presentation.

6. Ask a local network manager to come to your class to discuss the types of transmission modes (for example, asynchronous vs. synchronous; half-duplex vs. full duplex, and so on), communication media, and topologies used in the LANs and WANs of his or her organization. Be sure to ask why each was selected.

✳ REFERENCES

Addis, K. and Arbetter, L. et al. "Security Works: Managing Toll Fraud." *Security Management* (September 1993): p. 23.

Anthes, G. H. "VSAT System Offers Bridging, Routing." *Computerworld* (August 16, 1993): p. 63.

Augerson, S. "The Case for Computer Telephony Integration." *LAN TIMES* (March 28, 1994): pp. 43, 46.

Bianchi, A. "World Without Wires." *Inc.* (October 1993): p. 59.

Boudette, N. "Intel to Launch Windows' Personal Conferencing Software Line; Upgrade Modules to Provide Voice, Video." *PC Week* (November 29, 1993): p. 8.

Caron, J. and L. Didio. "Industry's Fast Ethernet Foes Speak Out: 100Mbps is Coming, But 'Town Meeting' Shows Clear Differences." *LAN TIMES* (June 28, 1993): pp. 1, 102.

Collier, A. "High-End Routers Run Close Race." *Communications Week* (April 26, 1993): p. 21.

Comaford, C. "Client-Server Lip Service Is Becoming Reality." *PC Week* (February 14, 1994): p. 55.

Derfler, F. J. "Plugging Into the LAN from the Road." *PC Magazine* (August 1993): p. NE1.

Dryden, P. "Full-Duplex Products are on the Rise." *LAN TIMES* (December 20, 1993): p. 62.

Duffy, C. A. "Peace of Mind: LAN Administrators Juggle Long-Distance Data Security." *PC Week* (July 26, 1993): p. 98.

Feldman, R. "LANs, WANs: Strategies Are on Course: Downsizing, Market Maturation are Prompting Next Wave of Internetworking." *LAN TIMES* (June 28, 1993): pp. 40, 41.

———. "Special Report: Ramping Onto the ATM Superhighway; LAN Hitchhikers Begin to Hit the Road." *LAN TIMES* (January 24, 1994): pp. 61, 62, 66.

———. "ATM Could Tear Down the LAN-WAN Wall." *LAN TIMES* (January 24, 1994): p. 72.

Geier, J. "Critical Access for Mobil Users: Wireless-LAN Implementation Takes Over Where Cable Leaves Off, But Be Aware of Its Pros and Cons." *LAN TIMES* (June 26, 1993): pp. 77, 80.

Howard, B. "The Trouble with Communications." *PC Magazine* (June 29, 1993): p. 105.

Joffe, D. "Have Analog Modems Reached the End of the Road?" *LAN TIMES* (June 27, 1994): p. 24.

Koontz, C. P. "Copper and Optical Fiber Duke It Out on the Desktop." *LAN TIMES* (October 18, 1993): pp. 47, 50.

Kramer, M. "Remote Access to Stand-Alones on the Upswing." *PC Week* (September 6, 1993): p. 96.

————. "Office Wants to Talk to You: Phone Access Server to WordPerfect." *PC Week* (January 24, 1994): p. 13.

Lindstrom, A. "Users Harvest a Variety of Applications." *Telephony* (October 25, 1992): p. 513.

Loudermilk, S. "Dialing for Data." *LAN TIMES* (August 8, 1994): pp. 57, 64.

————. "New Wireless Systems Offer More Distance, Cost Savings." *LAN TIMES* (March 14, 1994): p. 28

————. "Motorola's 'Monet' to Link Wireless Nets." *PC Week* (May 3, 1993): p. 8.

McGuire, M. "Apple Remote-Access Line Delivers Enhanced Security." *PC Week* (November 15, 1993): p. 20.

Miller, M. J. "In Search of Great Communications." *PC Magazine* (November 23, 1993): p. 79.

Musich, P. "Tools Link Microsoft Apps to Host." *PC Week* (September 27, 1993): p. 47.

Rooney, P. "WildSoft Helps Mobile PC Users Stay in Sync." *PC Week* (January 24, 1994): p. 1.

————. "Remote-Control Package Broadens Users' Video Options." *PC Week* (March 29, 1993): p. 33.

Schroeder, E. "VoiceView Allows Data Transfers During Phone Calls." *PC Week* (February 21, 1994): p. 12.

Smalley, E. "Developers Back Novell, AT & T Telephony Services; Apps Range From Voice Processing to Desktop Productivity." *PC Week* (October 11, 1993): p. 16.

————. "NetWare Connect Will Pave the Way For Remote Access." *PC Week* (September 13, 1993): p. 53.

Tam, T. "Remote-Access Software Supports Up to 15 Simultaneous Connections." *PC Week* (November 22, 1993): p. N.9.

VandenBerg, C. "Dial-Up Access to the Internet." *Telecommunications* (December 1993): p. 65.

Yegyazarian, A. "Conferencing via Modem: Sharing Information in Real Time." *PC Magazine* (February 22, 1994): p. 51.

Case Study:
Southeastern State University

Like most other university campuses, the buildings at Southeastern State University are spread across hundreds of acres of property. In SSU's case, the campus spans nearly 1000 acres.

The computer center with its VAX minicomputers is on the bottom floor of SSU's Graduate Studies Center, which places it roughly in the geographic center of the campus. Several of the academic buildings in which LANs have been installed are located more than three-quarters of a mile away from the computer center and almost every LAN on campus has a direct line to the computers in the computer center. In addition, the computer center continues to support remote access to the academic computing resources available on the VAX minicomputers. Remote users typically "modem into" the VAX through 24 available ports.

One of the farthest buildings from the computer center houses SSU's College of Business Administration. The first LAN located in the business building was installed in 1991. It is an Ethernet LAN using Novell NetWare as the network operating system. When first installed, it consisted of 48 IBM-compatible workstations with Intel 386 processors, a file server, and several printers. It is located in one of the corner rooms on the bottom floor of the building. In 1993, a second 24-workstation LAN was created in a room across the hall from the first LAN. This second LAN was essentially constructed to interconnect older IBM microcomputers with Intel 286 processors and limited disk storage (10 megabyte hard drives); now each of these computers has access to a much wider range of software that is stored on the LAN's file server. Both LANs have dedicated communications links with the VAX minicomputers in the computer center.

Prior to the development of these LANs, the student computer lab in the business building consisted of 22 dumb terminals and a high-speed printer connected to the minicomputers in the computer center via multiplexers. The multiplexers (one in the student lab and one in the computer center) allowed the terminals to share a 9600 bps communication line. These terminals were primarily used by information systems majors to interactively develop COBOL programs that would be compiled and run on either the minicomputers on campus or on the IBM mainframe on the campus of the state's flagship university. The high-speed printer allowed program output to be printed out in the lab instead of the computer center. The other main terminal users were students taking marketing research and faculty members completing research projects; the terminals enabled both types of users to access and run statistical packages such as SPSS and SAS that were available either on campus or on the mainframes at the flagship university.

In 1994, the larger LAN in the College of Business Administration was expanded to incorporate the rest of the microcomputers in the building. This meant that all the faculty and clerical offices in the business building were now connected to the server in the larger of the LANs. The cable to which all the workstations are connected snakes through the ceiling of each of the three floors of the 350-foot-long building.

Also in 1994, SSU turned one of the classrooms in the business building into a distance learning classroom for sending and receiving audio and video images between SSU and comparably equipped classrooms on the campus of two-year colleges that are part of the state's university system. These images are transmitted over terrestrial cables available from common carriers. Another change to the business building that occurred in 1994 was the installation of a satellite dish on the roof of the business building; this enables instructors to incorporate publicly available business-oriented videoconferences into the business curriculum.

CASE STUDY QUESTIONS AND PROBLEMS

1. Judging from the distance between the business building and the computer center, should one or more repeaters be included along the communication line? Why or why not?
2. Do you think that repeaters were used when the larger LAN in the College of Business Administration was expanded to incorporate the rest of the microcomputers in the building? Why or why not?
3. What type of interconnection technologies could be used to connect the two LANs in the business building? Which do you feel is most probable? Why?
4. Judging from the description provided, how would you describe the current topology of the larger LAN in the business building?
5. Based on the description provided, how would you describe the topology used to interconnect the minicomputers in the computer center with the LANs in other campus buildings?
6. Explain why a multiplexed 9600 bps line was sufficient to handle the interactive sessions of 22 simultaneous users in the student lab of the business building. (Assume asynchronous communications.)
7. What type(s) of terrestrial cables or services do you think are needed for distance learning transmissions between SSU and two-year colleges in the region? Justify your selection(s).

8. Based on the information provided in this segment of the case, what types of wideband services do you think that SSU is most interested in?

9. As noted in this section of the case, up to 24 remote users can access the VAX minicomputers at one time. How do you think network managers determine the number of remote ports that should be available to remote users?

PART TWO

LAN Fundamentals

LAN Alternatives

CHAPTER OBJECTIVES

After completing this chapter, you will be able to:

◆ Briefly describe the evolution of LANs in organizations.

◆ Briefly describe the differences between LANs and host-based networks.

◆ Identify the benefits of LANs.

◆ Describe the characteristics of, and differences among, sub-LANs, zero-slot LANs, and multiuser DOS LANs.

◆ Identify the major components and characteristics of client/server LANs.

◆ Describe the characteristics of peer-to-peer LANs.

◆ Describe the characteristics of wireless LANs.

◆ Describe the characteristics of PBX- and Centrex-based LANs.

◆ Identify and discuss the criteria used to select a LAN.

The previous chapter covered the major hardware and software components found in LANs as well as LAN topologies and architectures. This chapter focuses on the emergence of LANs as alternatives to host-based networks in organizations. The organizational benefits of LANs are discussed, and an overview of the various types of LANs found in organizations is provided. After reading this chapter, you should have a clearer understanding about how and why LANs are being implemented in organizations.

◆ THE EVOLUTION OF LANs

The emergence of LANs is an offshoot of the introduction and use of microcomputers in businesses. Today, microcomputers are found in most large and small businesses; some companies have purchased thousands of PCs. As of 1993, about half the microcomputers being used in organizations were attached to a LAN or a WAN. The percentage of business PCs included in networks has been steadily increasing since the mid-1980s, and experts expect that almost all microcomputers used by businesses in the year 2000 will be part of one or more networks.

While microcomputers are very common in today's organizations, this has not always been the case. Microcomputers were first marketed during the mid-1970s; the Altaire 8800, Tandy's TRS-80, and the Apple II were among the earliest microcomputers sold on the open market. However, these early machines did not attract the attention of major corporations. This lack of interest was primarily due to a very limited selection of business-oriented software suited to these smaller machines. Only after the development of useful, business-oriented software packages for microcomputers (such as Visicalc—one of the first spreadsheet packages) did businesses start to realize the potential of these smaller computers.

The decision by IBM to develop and market a PC (its first model appeared in 1981) also helped to stimulate business interest in microcomputers. Today there are hundreds of business-oriented software packages for microcomputers, as well as hundreds of models of desktop, laptop, notebook, and handheld PCs.

CENTRALIZED COMPUTING SYSTEMS

Before microcomputers became commonplace, most computer networks in organizations were centralized, host-based systems such as that depicted in Figure 3-1. In these types of systems, the processing power for an entire company was concentrated in a single large processor (the host). In centralized systems, the host is usually a mainframe or a minicomputer.

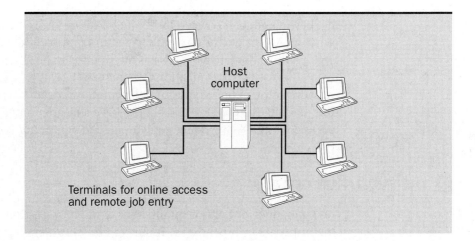

FIGURE 3-1

A centralized computer network

Terminals for online access
and remote job entry

In the 1950s and early 1960s, centralized systems typically operated in a batch-processing mode and commonly used punched cards and tapes for data input and storage. With *batch processing*, data are collected over predefined time periods, then processed all at once. Batch processing contrasts with *immediate processing*—processing data as it is entered. Immediate processing is also called called *realtime processing* or *online processing*.

Online access to the central computer is one of the key characteristics of immediate processing systems. Operating systems with multitasking capabilities (which make it possible for the CPU to concurrently work on multiple tasks for multiple users) became available during the mid-1960s. Multitasking made it possible to create online access and online processing systems that enabled users to carry out processing tasks from distant terminals. Online access systems, remote job entry systems, and online processing systems made it possible for users in all parts of the organization to harness the power of large computers. These developments helped solidify business interest in computing networks and also paved the way for development of distributed processing systems, client/server networks, and LANs.

With online access, all input goes directly from the users' terminals to the central computer for processing. After processing, all outputs to users come directly from the host.

The first terminals used in these systems were called *dumb terminals* because they could do no processing on their own, and only served as an input/output mechanism linking users to the central CPU.

Over time, centralized systems and their support equipment have grown more efficient and sophisticated. In many instances, dumb terminals have given way to smart terminals and intelligent terminals that can relieve the central CPU of some of its processing tasks.

It has also become quite common to use microcomputers in place of terminals in centralized, host-based systems. As noted in the previous chapters, communications software packages such as ProComm and Crosstalk include *terminal emulation* modules that cause the host to recognize the microcomputer as a terminal. Terminal emulation can also be done by installing a terminal emulation board in one of the microcomputer's expansion slots. Both of these options make it possible for the microcomputer user to interact with the host just as from a regular terminal.

In the past, the price differential between terminals and microcomputers dissuaded organizations from choosing PCs over terminals, especially over dumb terminals. However, microcomputer prices have greatly declined. Many organizations now routinely replace malfunctioning terminals with PCs. The PC gives the organization the same interactive capability that it had with the terminal, as well as a machine able to run a variety of other software when it is not being used as a terminal.

Because centralized, host-based systems permit multiple users to simultaneously share a large central computer, these systems are still commonly found in businesses. They are especially likely to be used in organizations that rely on large central databases for order processing, inventory management, and other transaction processing operations.

However, such systems also have shortcomings. For example, using a centralized, host-based computing arrangement puts all the organization's eggs in one basket—the organization relies heavily on the central computer and the lines between it and the terminals. If either is damaged or malfunctions, users' work cannot be done. In addition, the central computer can get bogged down when too many users need to use it simultaneously during peak business periods. Such shortcomings are among the reasons why many organizations have supplemented or replaced their centralized systems with PCs and network computing arrangements.

PERSONAL COMPUTING SYSTEMS

Personal computing systems proliferated in the 1980s. A PC system found in many organizations is depicted in Figure 3-2. Since the late 1980s, worldwide shipments of PCs have consistently exceeded those for the centralized systems that had previously dominated the computer industry.

The philosophy behind personal computing is that a computer should be readily available to help a person work. Many tasks that managers and users perform are inherently individual—for example, word processing, developing graphics for presentations, working with spreadsheets, and keeping personal calendars. Over time, the range of tasks that workers can use PCs for has steadily increased, largely because of the increasing availability of business-oriented applications software for PCs. Some smaller

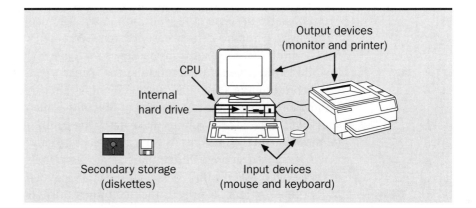

FIGURE 3-2

A stand-alone
PC system

businesses have found that a single PC is all that is needed to handle customer databases as well as accounting, inventory management, and transaction processing operations.

As previously noted, PCs can be connected to one another, or to larger computers, to form networks. However, they need not be connected to other PCs, and in many organizations microcomputers are used as ***stand-alone systems***. A stand-alone system is one not connected to, or not able to communicate with, another computer system (or network). Such a system has its own input, output, and storage devices, as well as its own set of systems and applications software.

Most PCs used in homes are stand-alone systems, as were the first microcomputers introduced to business organizations in the 1980s. This meant, unfortunately, that if a company purchased 50 microcomputers and wished to outfit each with the same word processing package and printer, it had to buy 50 copies of the word processing package and 50 printers.

NETWORKING MICROCOMPUTERS

With modems and communications software, PCs can be connected to a centralized computer via telephone lines. Alternatively, cabling or wireless media may be used to connect multiple PCs (as well as such devices as printers and fax machines) to form a LAN. At present, twisted pair and coaxial cable are most commonly used as the communication media in PC networks. However, fiber optic cabling and wireless media are increasing in popularity.

As shown in Figure 3-3, when microcomputers are networked, workstations consisting of system units, disk drives, keyboards, monitors, and so on, are linked together. Other computing equipment, such as printers and disk storage, are also common in PC networks. For example, in the network in Figure 3-3, users at the different workstations have access to both a laser printer and a large, mass-storage area.

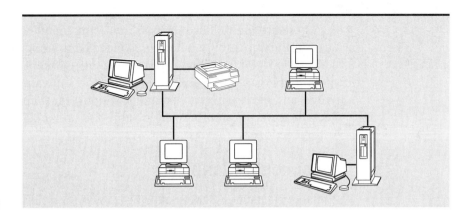

FIGURE 3-3
A PC-based LAN

Local networks do not use host computers; processors within the network itself manage the devices (such as printers) connected to it. In many LANs, the processor that manages the large, mass-storage area used to store applications software for general LAN use is called a file server. The file server is often a high-end microcomputer that controls access to shared disk space, programs and data, and generally controls the LAN. Similarly, a microcomputer that manages printing operations, such as spooling and the print queue, is called a print server. Among the other types of servers found in LANs are database servers, communication servers, and scanner servers.

By networking PCs, users work on their own microcomputers (workstations) but are linked to a network, giving them access to information and computing resources not directly connected to their workstations. Users in these environments can share databases, access needed applications, send e-mail to one another, and work cooperatively on tasks such as writing a technical report. Networks can also provide users with electronic links to peripheral equipment such as printers, plotters, and fax machines, as well as to other networks and external data sources. Providing these resources for each group member's workstation is too expensive for most organizations. It is usually much more cost effective to create a LAN; this allows users to share expensive computing resources. You will see that these are just some of the numerous compelling reasons for creating LANs.

◆ REASONS THAT MICROCOMPUTER-BASED NETWORKS ARE POPULAR

LANs *are* the dominant computing platform of the 1990s. LANs are now performing computing tasks that previously could be done only on mainframes and minicomputers. Today, more LANs are being installed than any other type of computing platform. LANs are the 1990s' most popular network configuration.

There are many reasons LANs are being installed by organizations. Reasons include resource sharing, reduced computing costs, communications, support for workgroup computing, the maturation of LAN operating systems, increased availability of application software, improved maintenance, and decreasing management support for mainframe computing. These reasons are summarized in Table 3-1, and are discussed in the following sections.

RESOURCE SHARING

Probably the greatest benefit of using LANs is the ability to share computer resources, especially expensive equipment and software. Some of the computing resources that can be shared by users in LAN environments are summarized in Table 3-2.

As noted, resources such as programs and files can be made available for common use in LAN environments. In LANs, two or more users can share the same program or file without having to resort to "sneaker-net" (sometimes "floppy-swappy")—walking from one workstation to another with a diskette.

◆ To take advantage of technological advancements in microcomputers (for example, faster microprocessors).

◆ To enable users to share hardware (devices) and software (general-purpose and special-purpose application software).

◆ To reduce the organization's total computing costs. LANs are often less expensive computing platforms to operate and maintain than are either multiple stand-alone systems or mainframe-based systems.

◆ To enable users to electronically exchange messages, files, e-mail, and faxes.

◆ To provide the computing platform needed to support workgroup computing applications, groupware, and group support systems (GSSs).

◆ The maturation of LAN operating systems and LAN management software has made it easier for organizations that have never installed a LAN to do so.

◆ There is an increasing variety of both general-purpose and special-purpose application software packages designed to run in LAN environments. Organizations are able to satisfy many user's computing needs because such software is readily available.

◆ LANs often help organizations standardize their hardware and software, and this can also enhance system maintenance.

TABLE 3-1

Reasons that organizations install LANs

◆ There is decreasing support for centralized mainframe-based systems in many organizations. Many firms are finding it much more cost effective to outsource mainframe applications or to downsize to LAN platforms.

Hardware		Software	
Input	Document scanners and other scanning technologies Video input systems	General-purpose software	Word processing Spreadsheet Database management Desktop publishing Presentation graphics
Output	Printers (especially more expensive color and laser printers) Specialized output devices (for example, plotters)	Special-purpose software	Accounting Inventory management Groupware packages
Storage	High capacity hard disks Disk array systems (redundant arrays of independent disks [RAID]) Optical disk storage systems (for example, jukebox systems) Tape storage and backup systems	Communications-oriented software	E-mail Voice mail Computer conferencing Videoconferencing
Communications	Modems Internetworking technologies (gateways, bridges, routers, brouters, switches) Fax machines		

TABLE 3-2   Some computing resources shared in LANs

Computing equipment can also be shared in LANs. LANs permit users to share printers and other output devices, high-volume direct access storage devices (DASD) and other storage technologies, fax machines, high-speed modems and other communication equipment, and high-resolution scanning equipment and other expensive input technologies.

As LANs continue to develop, the variety of hardware and software available for LAN users to share is steadily increasing.

REDUCED COMPUTING COSTS

Because computing resources can be shared by PCs in LAN environments, an organization's hardware and software costs are typically lower than those associated with supporting multiple stand-alone systems or large centralized systems. Because printers and other peripheral equipment (other input, storage, and output devices) can be shared, less equipment is needed and total equipment costs are lower. The savings generated by resource sharing can be used by network managers to justify the cost of acquiring additional expensive, high-end peripheral devices and making these available to users.

Software costs can also be lowered by installing a LAN. The costs of installing a LAN version of an applications software package (such as Word-Perfect) with 50 site licenses is typically less than what it would cost to purchase 50 copies of the software package for stand-alone systems.

Relative to mainframes and minicomputers, computing costs associated with LANs are also likely to be lower. Microcomputer networks can offer computing power comparable to that of mainframes and minicomputers at a lower cost. Despite dramatic decreases in the costs of computing equipment and storage for larger systems, mainframe and minicomputer computing resources remain expensive. The most dramatic difference in costs is typically found for software. The e-mail, database, and spreadsheet packages for mainframes and minicomputers are almost always more expensive than the equivalent software for microcomputers and LANs. In addition, software maintenance contracts, leasing, and site licenses typically cost more for mainframes and minicomputers than for LANs. Such price and cost differentials have caused many organizations to replace their centralized, host-based systems with LANs; in organizations that have made such changes, reports of better data processing services at lower hardware, software, and personnel costs are quite common. Some companies have reported multimillion dollar annual savings from downsizing to LANs; several examples are discussed in Chapter 11.

In short, LANs are often more cost effective to install, operate, and maintain than are either stand-alone microcomputer systems or centralized, host-based systems. As long as this is the case, LANs will continue to be the most popular computing platform in business organizations.

COMMUNICATIONS

LANs provide users an electronic means of communicating with one another. E-mail systems, such as those described in Chapter 1, are among the most common communications applications found in LANs. These enable users to send messages to one another. It is also possible to establish electronic bulletin board systems (BBSs) and computer conferencing systems in LAN environments.

Some of the message exchanges between LAN users may be in real time, making it possible for two or more users to carry on an electronic conversation. For example, after a user types in a message at the workstation and hits (Enter), the message created displays on the recipient's screen (or on those of multiple other users).

Querying a LAN database to check on customer orders and transferring files from one LAN workstation to another are other examples of communications applications possible in LANs. Many groupware products have built-in communications modules to facilitate message exchanges and collaborative work activities.

SUPPORT FOR WORKGROUP COMPUTING

Most organizations expect their employees to work together. Workers are grouped into subunits, departments, and other hierarchies to facilitate their interaction and to coordinate their efforts toward the achievement of strategic goals.

Managers typically want their organization's computer system to support the interaction of workgroup members and to assist in coordinating their work. LANs are considered by many managers to be the best (and most cost-effective) computing platform for promoting such interaction and coordination.

Managers' interest in teamwork, empowerment, self-managing work teams, and total quality management (TQM) has also contributed to increases in the numbers of teams or workgroups within organizations that need to share tasks and resources to accomplish their mission. The increased emphasis on workgroups and teamwork has also contributed to the widespread acceptance of LANs as important computing platforms. Hence, it should not be surprising to learn that organizations that formerly tried to get a PC on every manager's desk are now deciding to find ways for managers to share information and computing resources.

The development and marketing of groupware and group support system (GSS) packages is one of the offshoots of workgroups and collaborative computing. As noted in Chapter 1, groupware is an evolving concept that may encompass e-mail, computer conferencing, electronic meeting systems, and electronic calendaring systems. In addition, packages such as Lotus Notes make it possible for two or more users to simultaneously work on the same task, such as updating a spreadsheet or writing a technical report. More and more groupware products are being developed and marketed by vendors, and almost all of them are designed for LAN or other network environments.

THE MATURATION OF LAN OPERATING SYSTEMS

LAN operating systems are the systems software in LANs that perform the housekeeping and communication tasks that enable users to share resources, exchange messages or files, and utilize groupware. LAN operating systems are a special type of network operating system (NOS). NOSs have become both more sophisticated and user-friendly over the last decade, and the maturation of the NOSs used in LANs has helped promote the popularity of LANs. As has been the case with other operating systems, graphical user interfaces (GUIs) have become common in NOSs. Examples of the leading LAN NOSs are listed in Table 3-3.

NetWare—a NOS developed by Novell, Inc.—is the most widely used NOS found in PC-based networks. It commands the majority share of the LAN NOS market (with an annual market share of approximately 60 percent). Like other NOSs, users interact with NetWare when they are logging on to the network or when they are dealing with a printer or hard disk that is not part of the microcomputer on which they are working. Because NetWare runs on top of their microcomputer operating system (usually MS-DOS or Windows), users can retrieve or save files on a shared hard disk and can print files on a shared printer.

Since it was first introduced, NetWare has been through several versions. Many of these version changes followed the introduction of new microprocessors (such as Intel's 80386, 80486, and Pentium chips) and were designed to take advantage of the new processors' capabilities. With each new version of NetWare, new capabilities have been added. For example, many of NetWare Version 4.0's capabilities for internetworking LANs were not included in the previous versions.

Each of the other NOSs for LANs listed in Table 3-3 has a definite presence in the business world. Many experts feel that UNIX and Windows NT have the greatest potential to reduce NetWare's market share. Some have predicted that UNIX will have 25 percent of the LAN operating systems market by 1996. However, this won't affect Novell, because it acquired a controlling interest in UNIX in 1993. Windows NT may also become a major player in the NOS market because of the large user base developed by Windows 3.0 and 3.1. It appears that Novell and Microsoft will be the dominant NOS vendors in LAN environments for the rest of this century.

TABLE 3-3
Leading LAN operating systems

LAN OS Type	Product	Vendor
Client/server	LAN Manager	Microsoft
	LAN Server	IBM
	NetWare	Novell
	UNIX	Univell
	Vines	Banyan Systems
	Windows NT	Microsoft
Peer-to-peer	AppleTalk	Apple
	LanTastic	Artisoft
	Windows for Workgroups	Microsoft

INCREASED AVAILABILITY OF APPLICATION SOFTWARE

LAN versions of general-purpose application software (such as word processing, spreadsheet, database, and presentation graphics packages) are becoming more common. Examples of each of these were identified in Chapter 1.

Network versions of special-purpose application software (such as accounting packages, computer assisted design (CAD) packages, and computer assisted software engineering (CASE) tools are also being developed and marketed. Many groupware products would also be classified as special-purpose application software for LANs. The increasing availability of such products helps to stimulate and maintain interest in LANs.

IMPROVED MAINTENANCE

The installation of a LAN can make it easier to maintain end-user computing systems. This is especially true for the application software found in LANs. Vendors regularly release new and improved versions of application software packages, usually after the release of operating system updates.

Upgrading for PC OSs and application software packages is likely to take more time in stand-alone computing environments than in LAN environments. If a company owns only stand-alone microcomputers and wants each user to stay current with the latest operating systems and application software, software upgrades would have to be installed on each workstation. However, if the company's workstations are connected to a LAN, the upgrade will have to be performed only once. Organizations with interconnected LANs, distributed processing systems, or client/server networks can also take advantage of software that automatically upgrades versions of software packages on workstations in every subnetwork.

LAN implementation can also simplify the maintenance of end-user computing systems by standardizing hardware and software within organizations. Standardizing on certain hardware and application packages (that is, purchasing and using some types of hardware and application packages, but not others) makes it easier for LAN managers and the MIS staff to assist end users.

When microcomputers first became popular in organizations, hardware and software purchases were often left to individuals or to departments. This often led to a proliferation of microcomputer types (Apple, IBM, and IBM-compatible), software packages, and a number of problems, including the inability to easily transfer files and data from one system to another. Variety in hardware and software types can make it more difficult to maintain the systems and train users.

LANs can help organizations standardize their microcomputer environments. By supporting particular application packages and operating systems, an organization can help ensure that users will be able to exchange files and data. In addition, standardizing on particular PC models can make it easier for technicians to upgrade machines and make repairs. Standardizing on hardware and software is also likely to make it easier to train users.

DECREASING MANAGEMENT SUPPORT FOR LARGE SYSTEM COMPUTING

Primarily for cost reasons, management interest in centralized, mainframe-oriented computing is waning. As we have already noted, software costs are often significantly lower in LAN environments than in mainframe and minicomputer environments. In addition, mainframe computing resources remain expensive.

Many organizations have learned that LANs can provide cost-effective alternative platforms for many of the applications traditionally run only on mainframes and minicomputers. This has allowed many organizations to downsize some of their key applications—that is, to move them from mainframes and minicomputers to LANs. As mentioned, considerable cost savings have been reported by many companies after downsizing.

Management interest in outsourcing mainframe-based applications has also caused many organizations to look for suitable replacements for large computers. With outsourcing, the organization contracts with another company to perform some of its information processing tasks (for example, transaction processing, payroll, or inventory management). By outsourcing these data processing tasks to another company, an organization may no longer need its mainframe(s) or minicomputer(s). Because LANs can provide users needed processing capabilities, they are often chosen as replacements for larger systems.

Organizations have generally become less mainframe-oriented and more LAN-oriented. In some cases, organizations have "pulled the plug" on their big machines after downsizing or outsourcing. It other cases, the mainframe has been kept but its role redefined within the computer networks that are subsequently created. Often, the mainframe, once the centerpiece of the organization's information system, is used as a server for a particular type of application. Decreased management interest in (and budgetary support for) mainframes has relegated the mainframe to being just another node in the network.

◆ LAN ALTERNATIVES

The wide variety of LAN alternatives ranges from some very basic systems to options that combine data communications with voice (and even video) communications, and provide user interfaces with a variety of links to WANs and value added networks (VANs).

BASIC MICRO-TO-MICRO CONNECTIONS

Essentially, a network is created by connecting two computers to one another in a manner that allows them to exchange messages or files (see Figure 3-4). Another very basic network is created when two PCs share a peripheral device (such as a printer) through a data switch. This is shown in Figure 3-5. These and other basic approaches that connect two or more PCs are described in the following sections.

Null Modem Cables

Null modem cables connect two microcomputers near each other (within 8 to 10 feet) so they can communicate without using modems. Null modem cables can be purchased in computer stores or built by computer technicians. A null modem cable connects the serial communication ports of the two machines. Using communications software, one of the machines is put in the answer mode, and the other machine calls it; when the receiving computer answers the call, data can be exchanged in the same manner as using a dial-up connection.

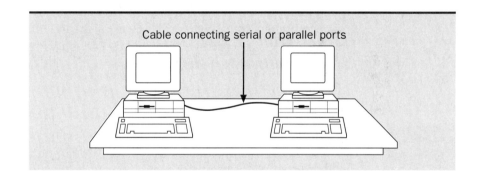

FIGURE 3-4
Two interconnected PCs

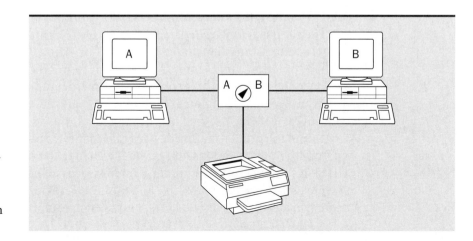

FIGURE 3-5
A basic form of resource sharing—sharing a printer through an A/B switch

Micro-to-Micro Communications Software

Several *micro-to-micro software* packages are available that enable the transfer of data between two microcomputers within a few feet of one another. Most are relatively inexpensive, costing $100 or less, and include required connection cabling. The two PCs are connected via their serial or parallel communication ports; the software causes them to operate as they would when connected by a null modem cable. However, these packages offer several advantages over null cable connections. For example, when using these packages: the data transfer rates are usually higher than those possible through null modem cables; error checking is performed during file transfer; either machine can control the sending and receiving of files; and more than one file can be sent at a time. In addition, some micro-to-micro communication packages allow for printer sharing.

LapLink (a product from Traveling Software) is the best-known micro-to-micro package for communication between two IBM-compatible machines. MacLink Plus and LapLink Mac are two packages permitting file transfer between an Apple Macintosh and an IBM-compatible. Communication between a Macintosh and an IBM-compatible can also be accomplished through null modem cables.

Later versions of MS-DOS (including MS-DOS 6.0) include modules with functionality comparable to that found in LapLink. Once connected, two microcomputers can exchange files by using the INTERLINK or INTERSVR modules. As these come free with the MS-DOS upgrade, they may be used as no-cost alternatives to LapLink.

Shared Printer Connections

In some instances, it is not necessary for two (or more) microcomputers to be able to exchange files, yet it is desirable to have them share a peripheral device such as a printer. The easiest way for two or more microcomputers to share a single printer is to install a print sharing device. A manually operated mechanical switch called an *A/B switchbox* traditionally has been used to connect the microcomputers to the printer.

The simplest *data switch* makes it possible to connect two microcomputers to one printer (illustrated in Figure 3-5). When the switch is turned to the A position, the printer is connected to microcomputer A; when turned to the B position, it is connected to microcomputer B. The primary disadvantage of a mechanical switchbox is that it must be close (within convenient walking distance) to both microcomputers.

Automatic printer sharing devices make it unnecessary for users to get up from their workstations and manually change the data switch setting. These automatic devices contain a microprocessor and program chips, and are usually available in 2-, 4-, and 8-plug capacities. Available for connecting to a PC's serial or parallel ports, they range in price from $200–$600 (significantly more than manual switches). The chips in these automatic devices scan the inputs coming from the various microcomputers; when the chips detect data coming from one of the computers, they stop scanning and pass the data to the printer. When the incoming data stream ends, the chips resume scanning.

Bulletin Board Systems

A data file can also be transferred from one microcomputer to another by first uploading it to a host (using a microcomputer-to-mainframe connection), then downloading it to the other microcomputer. Files can be exchanged in a similar fashion through a *bulletin board system (BBS)*.

A BBS is essentially a type of dial-up connection that can be accessed by communications software and modems. BBSs run on a microcomputer that serves as a host. Users dial-in to BBSs via telephone lines, and once they are logged on, they can leave public messages or questionnaires for other users to complete, upload files to the BBS, download files from the BBS, and run specially written programs (such as an e-mail or database application) external to the BBS through what is called a *door*. (A door is essentially a BBS menu option for users to access an application that is not part of the BBS itself.)

Bulletin board systems are growing more common. Experts estimate that there are at least 50,000 BBSs worldwide and more than 35,000 in the U.S. BBSs are also being used more frequently by corporations both for external and internal communications. Corporations have set up BBSs to collect data from customers, make newsletters and other reports available to the public, and as an alternative to telephone help-line services.

Most BBSs operate at slow speeds—1200 or 300 bps transmission speeds are common. In addition, users should be cautious of the files and programs downloaded from BBSs because of the possibility of virus infections. Network managers should ensure that users are required to run a virus check on all programs or files acquired from BBSs.

SUB-LANs

Other configurations provide users with a subset of the capabilities found in full-fledged LANs. Included in this set of alternatives are sub-LANs. The two major services available in such systems are file transfer and peripheral device sharing.

Sub-LANs are typically implemented with data switches, which provide a connection between microcomputers (and between microcomputers and peripheral devices such as printers). Figure 3-6 shows an example of a sub-LAN. The data switch shown in Figure 3-6 is used to establish a physical connection between devices in much the same way that an A/B

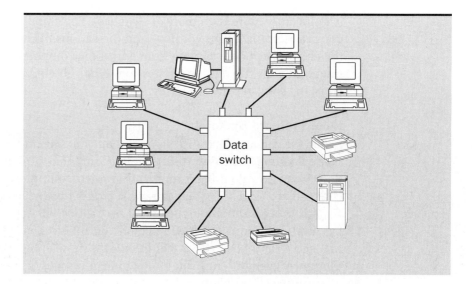

FIGURE 3-6
A sub-LAN

switch is used to allow two or more microcomputers to share a printer or peripheral device.

Sub-LANs are relatively inexpensive to implement; the cost per workstation is often less than $50. The microcomputers and peripheral equipment are connected to the data switch via their serial or parallel ports. Like A/B switches, sub-LAN data switches can be either manual or software-controlled.

Usually, only low-speed data transfer rates (19.2 Kbps or less) are possible in sub-LANs; these speeds are much lower than those found in most true LANs. One of the problems associated with lower speeds is that the switch can be tied up for considerable periods of time when large files are being transferred between connected devices. Some of the more expensive "intelligent" data switches have built-in buffers to alleviate such potential bottlenecks.

ZERO-SLOT LANs

Zero-slot LANs link PCs and printers via cables connected to the PCs' standard serial or parallel ports (see Figure 3-7). Software controls the data transfer between the networked machines. Zero-slot LANs get the name from the fact they do not require an additional slot on the workstation's motherboard for a network interface card (NIC). The adapter plugs used in these LANs can be plugged into a serial or parallel port; an expansion slot on the motherboard is not used.

The operating system for a zero-slot LAN works in conjunction with DOS. Some vendors of zero-slot LAN systems include Artisoft, Brown Bag Software, Fifth Generation Systems, Grapevine LAN Products, and Traveling

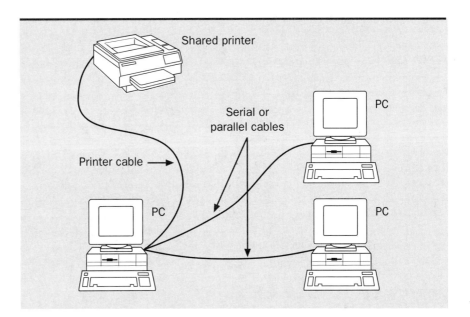

FIGURE 3-7
A zero-slot LAN

Software. Sub-LAN functionality can also be used via the INTERLINK and INTERSRV modules of MS-DOS 6.0.

Zero-slot LANs are among the least expensive of the LAN alternatives; they usually cost from $50–$100 per workstation. However, these LANs also provide users limited network services. All zero-slot LAN systems provide file sharing and file transfer capabilities, along with printer sharing, and some products also provide e-mail and printer spooling functions. Although the range of networking services available to users is limited, zero-slot LANs are an inexpensive option when all that a few of users need are file and print sharing capabilities.

MULTIUSER (D)OS LANs

Another relatively inexpensive LAN uses a multiuser operating system (OS) to carry out network functions. *Multiuser OSs* enable multiple users to concurrently run tasks on the same microcomputer. The microcomputer on which user tasks are performed is typically a high-end machine equipped with a high-speed 486- or Pentium-based microprocessor. Like many server-based LANs (discussed next), multiuser OS LANs are centrally managed (see Figure 3-8). While best suited to use by fewer than 25 users, many multiuser OS systems connect to other multiuser OS LANs, and some also connect to server-based LANs and mainframes.

The most common OS used in this type of LAN is Multiuser DOS, an operating system that runs MS-DOS software, uses MS-DOS commands, and follows MS-DOS conventions. Some of the vendors that have developed

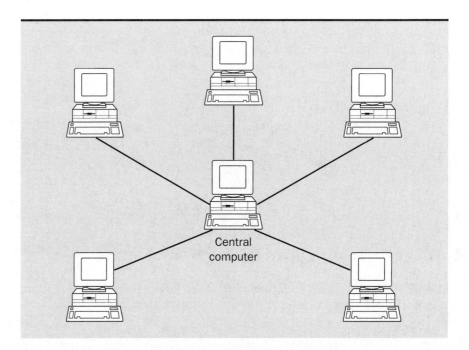

FIGURE 3-8
A multiuser
OS LAN

multiuser DOS products include Concurrent Controls, Microbase, The Software Link, and THEOS Software Corporation. Novell DOS, a product originally developed by Digital Research, is another multiuser DOS system.

Because all the processing in a multiuser OS LAN is performed on a central, high-performance microcomputer, the power of the central PC's microprocessor is available to each of the linked workstations. The workstations in these LANs do not have to have state-of-the-art PCs; they can be older, less expensive microcomputers, or even dumb terminals. Diskless workstations (microcomputers lacking disk drives and secondary storage capabilities) can also be used in these LANs because user files can be stored on high-capacity hard drives attached to the central computer.

Most multiuser OS LANs offer user workstations multitasking capabilities. Multitasking allows each user to concurrently run two or more applications. For example, a user could begin recalculating a large spreadsheet (often a time-consuming task), then return to working on a word processing document while the recalculation is performed in the background. While multitasking is a common feature in many of the newer 32-bit versions of microcomputer operating systems (including Windows NT and Macintosh System 7), such operating systems typically cannot be run on older, slower microprocessors. Multitasking versions of DOS provide older microcomputers with computing capabilities previously possible only on machines that possess newer processors.

Because newer, 32-bit operating systems are not used, the applications available to users in such LANs are usually non-graphical, DOS-based applications. Software with graphical user interfaces (GUIs) is typically not found in such systems, especially if dumb terminals and pre-80286 machines are used as workstations.

CLIENT/SERVER LANs

Client/server networks are emerging as the dominant computing platform. Within client/server systems, shared resources are placed at dedicated servers that manage the resources on behalf of client workstations. As noted in Chapter 1, *client/server computing* occurs when two or more computers cooperate to perform a processing task. The hardware, software, and data resources of two or more computers are combined to solve a problem or to process a single application. A *client/server network* is a computer network that enables client/server computing to occur. A client/server network can be viewed as an information system in which the entire network becomes the computer to solve a particular problem or to process an application.

Client/server LANs are one of many types of client/server networks. An organization's client/server network may consist of a single LAN, interconnected LANs, a WAN, interconnected WANs, or interconnected LANs and WANs. The computers in client/server networks can range from microcomputers to mainframes (or even supercomputers). In this section, however, we will be focusing on microcomputer-based client/server LANs.

In client/server LANs, the software and data resources to be shared are placed on servers, which manage these resources on behalf of the other workstations (clients). Servers also enable users to share special-purpose hardware devices attached to the LAN such as printers, scanners, and optical storage units. In a client/server LAN, a microcomputer can be a server for one type of application and a client for other types of applications.

A wide variety of servers may be found in client/server LANs; some are summarized in Table 3-4. Of these, file servers, printer servers, network servers, and database servers are the most common (and will be most fully discussed in this text).

The level of interaction between clients and servers when solving a problem or processing an application can vary. The major types of client/server computing arrangements used in LANs are summarized in Table 3-5 on page 97. In the 1990s, client-driven and peer-to-peer client/server computing arrangements have received the greatest amount of interest.

Client-driven client/server computing is the norm for LANs with database servers. However, for enterprise networks, peer-to-peer client/server computing is considered to be the wave of the future. In such systems, clients and servers share the work needed to complete a particular processing task; in addition, multiple servers may help a single client complete a processing task.

Type of Server	Function/Purpose
File	Stores application software and sends it to user workstations when requested; can also store user files.
Printer	Enables users to share a printer. Also, manages printing operations such as spooling and print queues.
Database	Provides users with access to data in shared databases. Some database servers process user queries and carry out data manipulation functions for users.
Communications	Manages communications between workstations in the LAN and other networks. Provides users with access to other LANs and networks (both internal and external). May also provide remote users with access to the LAN.
Fax	Enables users to share fax capabilities. Makes it possible for users to both send and receive documents through their workstations. Manages both incoming and outgoing spooling operations and fax queues.
Scanner	Manages the operation of document scanners or other scanning technologies. Commonly found in image processing systems.
E-mail	Manages mail functions for LAN users. Most commonly found in larger LANs with heavy e-mail traffic.
Optical disk	Carries out many of the same functions as a database server for data, files, and images stored in optical disk storage units (jukebox systems). May also manage the recording (writing) of data and document images in image processing systems.
Special-purpose	This category of servers may include video servers (which enable the storage and retrieval of video clips), voice mail servers, expert system servers, and so on, depending on the functions that LAN users perform.

TABLE 3-4
Types of servers
in client/server
LANs

Client/server LANs are relatively easy to expand. As new hardware and software becomes available, it can be added to the LAN to enable users to share and take full advantage of what these new technologies have to offer. Growth and expansion of the network's processing capabilities can take place in planned, manageable increments.

LANs WITH NONDEDICATED SERVERS

Client/server LANs with dedicated servers are the most widely used type of configuration, especially for large LANs. The microcomputers used as

Approach	Description
Host-driven terminal emulation	Occurs when the client connects to the server in the same manner that a terminal connects to a host. Virtually all processing is done by the server. Terminals are used in place of microcomputers in some of these LANs.
Host-driven front ending	Occurs when the client converts messages transmitted by the server (host) to a more user-friendly interface than the standard interface provided by the server. Also, provides users with a friendly interface for sending messages to the server. All important processing is done at the server, and the creation of a user-friendly interface is the only real work done by the client.
Host-driven client/server	Occurs when the client carries out specific actions for the server (host) such as displaying a menu, window, or dialog box. Typically, some of the displayed options are carried out by the client's computer under the direction of the server (host).
Client-driven client/server	Occurs when the server carries out processing as requested by the client. The client essentially requests the server to perform a processing operation; the server then carries out the operation and sends the results to the client.
Peer-to-peer client/server	The most cooperative and sophisticated type of client/server computing, occurs when the client and server cooperate in performing a processing operation. The roles of clients and servers may switch several times during the actual processing operation; both clients and servers may request services from one another. In many peer-to-peer arrangements, host-driven and client-driven client/server computing structures are combined.

TABLE 3-5 Client/server computing in LANs

dedicated servers are often more expensive, high-end machines utilizing faster and more powerful microprocessors, and often have higher-capacity disk systems and faster disk access times than those found on user workstations.

In smaller LANs (such as those found in small offices with 20 or fewer workers), dedicating such machines to server functions may make the network less cost-effective than it could be. This is because the dedicated machine is likely to be underutilized (too often idle) when there is a relatively small number of concurrent users (8 or fewer user workstations). In such situations, network managers may consider configuring the faster and more powerful microcomputer as a nondedicated server.

A ***nondedicated server*** operates both as a server and a workstation. Several NOSs for LANs provide this option. Typically, a nondedicated file server is a high-end microcomputer at which most of the network resources are stored. However, to take full advantage of its processing capabilities, a nondedicated server can be used as a workstation in addition to carrying out server functions for the other workstations in the LAN. For smaller LANs, the primary advantage of nondedicated servers is more cost-effective use of the server's resources.

Some of the potential disadvantages of nondedicated servers include the possible reduction in service to users during peak work volumes and greater vulnerability to server failure. Because a nondedicated server divides its work between server functions and the application(s) being run by the user utilizing the machine as a workstation, conflicts may occur when there are heavy workloads. If too many conflicts occur, service to both the user at the nondedicated server and the other users in the LAN may suffer. If this happens too frequently, it may be necessary to reconfigure the server to make it a dedicated machine.

Simultaneously carrying out both server functions and running user applications also increases the likelihood of server failure. If the user's application locks up, server functions may also be disrupted. Inexperienced users working at the nondedicated server can also cause LAN problems by turning the machine off or by inadvertently reformatting its hard drive. These potential disadvantages should be taken into account by network designers prior to LAN implementation.

PEER-TO-PEER LANs

Peer-to-peer LANs feature distributed rather than server-managed systems. In effect, any workstation in a peer-to-peer LAN can function as a server by making the data on its hard drive or peripheral equipment (for example, its printer) available to the other workstations in the LAN. Any of the microcomputers in peer-to-peer LANs can function both as a server and as a user workstation. In addition, such LANs can be configured so that every microcomputer attached to the LAN can access all of the network's resources. An example of a peer-to-peer LAN is depicted in Figure 3-9.

Peer-to-peer networks can often be used to connect stand-alone microcomputers without major reconfiguration of the stand-alone systems; the printers and files can remain on the computers to which they were originally assigned. After installation, these resources can be shared with the other workstations in the LAN.

Peer-to-peer LANs are often feasible only in LANs with 25 or fewer workstations, but some vendors have released products capable of supporting LANs consisting of several hundred workstations. In addition, most peer-to-peer LANs are quite easy and inexpensive to set up, often at a cost of under $200 per workstation. If connections are made through the workstations' serial or parallel ports (as is normally the case for lower-speed, peer-to-peer LANs), the cost per workstation may be less than $100.

Two of the most popular operating systems for peer-to-peer LANs are LANtastic and Microsoft's Windows for Workgroups. Novell's NetWare Lite,

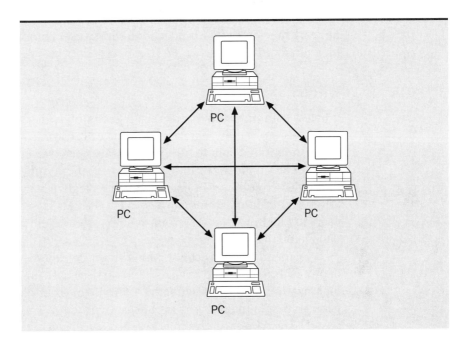

FIGURE 3-9
A peer-to-peer
LAN

another peer-to-peer network operating system, has been popular in the past. However, LANtastic and Windows for Workgroups have become the dominant peer-to-peer LAN products of the 1990s.

Most peer-to-peer operating systems provide such user-oriented network services as logon security, the ability to restrict user access to available resources, disk and directory mapping (to help users find resources stored at the various workstations), message distribution, and print spooling. Some products also provide user interfaces to other LANs, including client/server LANs. Several peer-to-peer operating system products such as LANtastic also enable network managers to perform network audits of user logons and logoffs and to backup system files.

In general, security is a bigger problem in peer-to-peer LANs than in client/server LANs. Peer-to-peer LAN software often does not include mechanisms for protecting confidential files and directories. If the disk(s) at each workstation are accessible from each of the other workstations, protecting the confidentiality of files and directories can present some real challenges to network managers.

Another often-reported shortcoming is the slowness of these systems in transferring and printing out graphically-oriented files. Such file transfers and print operations (especially when the output is directed to a printer attached to someone else's workstation) can temporarily bog down these types of systems.

In spite of these shortcomings, peer-to-peer systems are relatively inexpensive and easy to install. In many work settings, a peer-to-peer network may be all that is needed to satisfy user computing and communication requirements.

WIRELESS LANs

Wireless communications are already common and are increasing in both popularity and variety. Wireless LANs typically use microwave signals, radio waves, or infrared light to transmit messages between computing devices.

Wireless data communications are often an option for applications that require mobility such as the need to regularly move a workstation. Wireless communications can also expand a LAN by making it possible for users to dial in to the LAN from remote (or mobile) locations via cellular data networks.

In general, wireless connections are more expensive than most types of physical cabling used in LANs. Still, wireless LANs may sometimes be an organization's most cost-effective option. For example, in old buildings, wireless media may be less expensive than renovating to install a cable-based LAN. Wireless LANs may also be cost-effective options in buildings with full wiring ducts. These LANs may also be desirable alternatives in hazardous environments, temporary operating locations, and for disaster recovery.

The two basic types of wireless LANs are controller-based wireless LANs and peer-level systems. In controller-based wireless LANs, a dedicated server (controller) manages message transmissions between the workstations. The controller handles all communications between devices in the LAN. Figure 3-10 depicts an example of a controller-based wireless LAN.

Peer-level systems allow each computing device to communicate with every other computing device in the LAN. This is less expensive to implement than central controller-based systems because master control devices (and software) are not needed. However, these slow down when there are many users and with heavy message traffic. They are also less secure than controller-based systems. An example of a peer-level wireless LAN is depicted in Figure 3-11.

The actual communications between workstations may be channelled through special devices that divide the total LAN into cells, illustrated in Figure 3-12 on page 102. In most instances, these devices handle communications between devices within each cell, as well as communications between devices in different cells. Figure 3-12 also illustrates that wireless LANs can be connected to an organization's cable-based LANs and extend an organization's backbone network.

Wireless technologies are still relatively new but are expected to become more commonplace in the 1990s. Some of the major vendors of wireless LANs and their products are listed in Table 3-6 on page 102. As Table 3-6 indicates, radio waves and infrared light are wireless LANs' most common

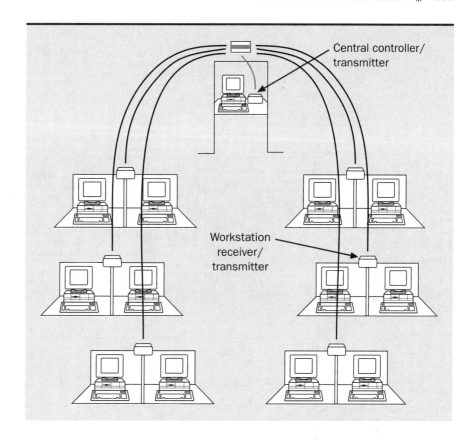

FIGURE 3-10
A controller-based wireless LAN

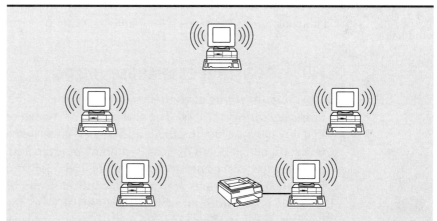

FIGURE 3-11
A peer-level wireless LAN

communications media. Spread spectrum radio systems (such as NCR's Wavelan) that use unregulated radio frequency bands for data transmissions are rapidly growing in popularity.

Wireless LAN technologies are expensive, but like most computing and communications technologies, are expected to continue to drop in price.

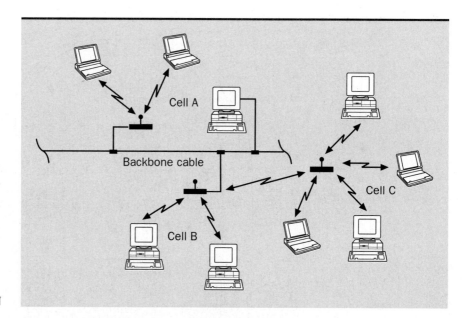

FIGURE 3-12
Wireless LAN
intercon-
nected to a
cable-based LAN

Product	Vendor	Media
Arlan	Telesystems	Radiowave
Altair	Motorola	Radiowave
Wavelan	NCR	Radiowave
BICC	BICCO	Infrared light
Photolink	Photonics	Infrared light

TABLE 3-6
Some leading
wireless LAN
products

PRIVATE BRANCH EXCHANGES (PBXs)

Most organizations have their own telephone switchboards, or *private branch exchanges (PBXs)*. The earliest PBXs—which one can still observe in many movies from the 1930s and 1940s—consisted of telephone operators stationed at board devices, switching plugs in and out of the board to provide physical connections for callers. Today, of course, the vast majority of such connections are software-controlled and handled by computers. This is true both for calls made between workers in the same office complex whose connections are handled by the organization's PBX as well as for long-distance connections handled by telephone companies. Sometimes the computer-based private branch exchanges found in businesses are referred to as CBXs (for computerized branch exchanges) or PABXs (for private automatic branch exchanges).

Besides phones, PBXs can be used to interconnect other devices such as computer terminals, microcomputers, printers, and facsimile machines.

Virtually all of the newer PBXs use digital signals for both voice and data communications, and most provide *voice-over-data* capabilities that allow users to simultaneously engage in voice and data communications over the same medium. With voice-over-data, a user can be transferring a large computer file while talking to a coworker on the phone.

Because a PBX can interconnect computing technologies used within an organization, it can be used to create a LAN. As it does for voice communications, the PBX can serve as a sophisticated (data) switch capable of supporting numerous concurrent connections between workstations. For example, in addition to supporting conference calls, most PBXs can provide three or more users with realtime computer conferencing capabilities.

A PBX can also interconnect LANs, connect LANs with host-based systems, and interconnect LANs and WANs. Such interconnections are illustrated in Figure 3-13.

Because most PBXs use digital signals, they can also provide the media needed for modem-less micro-to-mainframe links. Some PBXs are also being used to support desktop-to-desktop videoconferencing.

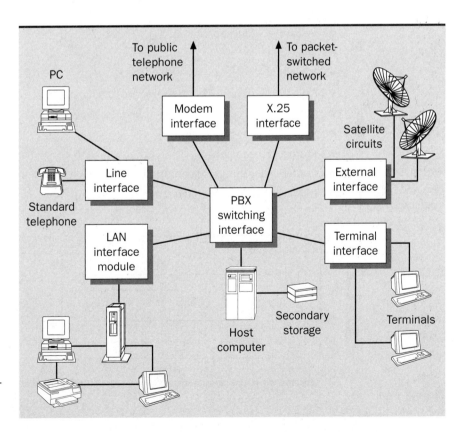

FIGURE 3-13

PBX interconnections to LANs and WANs

CENTREX SERVICES

For companies that desire PBX-like capabilities, but do not wish to bear the equipment and operational costs associated with the acquisition of a PBX, Centrex services may fit the bill. *Centrex services* can be leased from local telephone companies and can provide subscribing organizations with most of the voice and functions found in a PBX (see Table 3-7). The primary difference between a PBX and Centrex is that with Centrex, the switching is done at a local telephone company's central office rather than at a switch located on the customer's premises—see Figure 3-14. The telephone company is paid a monthly fee for these services, and it (not the subscribing organization) is responsible for managing and operating the switch.

Centrex offers the organization lower-cost voice and data communication services than possible with a PBX. However, acquiring a PBX gives a company greater control over the use of its telephone system, provides higher data transmission speeds, and offers additional features and services not available with Centrex. Also, because the line costs associated with a PBX can be amortized over its service life (rather than being a recurring monthly

TABLE 3-7
PBX and Centrex systems services

♦ Voice-over-data capabilities

♦ Voice mail

♦ E-mail

♦ Asynchronous and synchronous data transmission

♦ Authorization codes to restrict access (for security)

♦ Store-and-forward capabilities

♦ Protocol conversion

♦ Conference calling and computer conferencing

♦ LAN capabilities controlled at the switchboard

FIGURE 3-14
A Centrex-based LAN

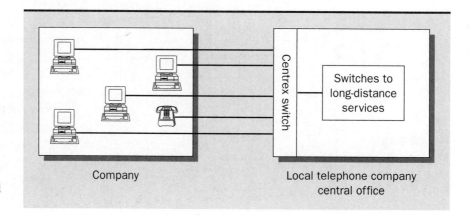

lease charge), the ultimate costs of a PBX may be lower than those for Centrex services. These are a few of the trade-offs that an organization must consider when choosing between a PBX and Centrex services.

SERVICE BUREAUS

Another external option to interconnect microcomputers within an organization is a service bureau. A *service bureau* is an organization providing information processing services to subscriber organizations. The processing services are similar to those available on a large, central host. However, by using a service bureau, the organization would not have to purchase, operate, or maintain the mainframe or minicomputer. The subscribing organization gains access to the service bureau's computers and software via a leased (typically dedicated) telephone line.

Subscribing organizations are billed monthly, and the fee typically consists of a standard monthly connection charge (which covers the cost of the leased line) as well as a second charge based on the volume of processing done for the organization. When first subscribing, an organization may have to pay for all the computers (or terminals) it needs and installation to connect them to the leased line.

The service offered by a service bureau is usually specialized and focused on a particular industry. For example, service bureaus for credit unions (such as FiTech and USERS) provide online processing for all the major services that credit unions provide (savings, checking, loans, ATM transactions, credit card transactions, certificates of deposit, savings bonds, and so on). Credit union service bureaus also perform all accounting, bookkeeping, and financial reporting operations for their clients. This makes it unnecessary for a credit union to have to own or operate an in-house computer system. For smaller credit unions, service bureaus are often the most cost-effective way to process their information. Other service bureaus are available for a variety of other types of organizations, including truck terminals, financial services firms, independent insurance agencies, and car rental companies.

In general, if a service bureau is available for a certain type of organization, it should not be overlooked as a LAN alternative. This is particularly true if all that users need are basic LAN services such as file and printer sharing.

Service bureaus often enable users within organizations to exchange files and share peripheral devices even though the actual communications functions and information processing may be done hundreds (or thousands) of miles away. However, the communications speeds available from service bureaus are often lower than most of the other LAN alternatives that we have discussed. In addition, response times may slow down during peak processing volumes, and subscribing organizations may be without computing capabilities when the service bureau's computers are down.

◆ CRITERIA FOR COMPARING LAN ALTERNATIVES

Many criteria are used to compare LAN alternatives, and it would be impossible to present an exhaustive list covering all situations that LAN managers may encounter. Some of the more important ones are summarized in Table 3-8 and are discussed in this section. As you will see, most of these center around cost considerations, the ability to satisfy users' needs, and network management issues.

NUMBER OF USERS AND WORKSTATIONS

One of the primary considerations in selecting from among the various LAN alternatives is the number of users and workstations that must be supported. Some of the alternatives discussed in the previous section of this chapter are best suited for smaller LANs.

For LANs that will have few workstations and peripheral devices (ten or fewer), zero-slot, multiuser DOS, and peer-to-peer LANs may be most cost effective. It may also be feasible to build the LAN around a nondedicated server or to subscribe to a service bureau when the number of workstations and shared peripherals is limited.

◆ Number of users and workstations to be supported

◆ The availability of application software

◆ Type(s) of workstations supported

◆ Printer support options and limitations

◆ Expandability

◆ Communications speeds

◆ Available media options

◆ Geographic limitations

◆ Interconnectivity with other networks

◆ Adherence to widely recognized standards

◆ Availability of multiple sources of LAN hardware and software

◆ Vendor support

TABLE 3-8　◆ Availability of LAN management software
Criteria to
compare LAN　◆ LAN security
alternatives
◆ Overall costs

If the LAN must support more than 10 but fewer than 100 users, both client/server and peer-to-peer networks may be options. LANs built around PBX or Centrex services may be appropriate in these situations, but these can also be used to support thousands of users. Client/server computing is the most common option used for very large LANs (consisting of hundreds or thousands of users). However, some peer-to-peer products, such as LANtastic, are capable of supporting several hundred users.

When determining the number of workstations to be supported, the network manager should not overlook growth projections of the number of users in the organization. If the organization is small but growing, it may be wisest to choose an option that will support more than the current number of users. While this may mean greater cost in the short run, it can avoid having to switch from one type of LAN to another (say from zero-slot to client/server) in the future.

AVAILABILITY OF APPLICATION SOFTWARE

Another important consideration in determining the type of LAN to implement is the availability of the application software needed by LAN users. Many experts consider the ability to support both current and projected user application needs to be the most important criterion in LAN selection. If necessary application support is available only for certain types of LANs, the choices may be quite limited.

There is an abundance of general-purpose application software for stand-alone microcomputers; many of these packages have been adapted for use in LANs. However, current versions of many application packages may be feasible for only a subset of LAN alternatives. For example, Windows-based applications may not be possible on zero-slot LANs and, because of their tendency to bog down peer-to-peer LANs, they may not be desirable on these either. In addition, the performance of application packages may be better on some types of LANs than on others. For example, a particular spreadsheet package may perform almost flawlessly in a client/server environment, but may misbehave in a peer-to-peer LAN.

The availability of general-purpose application software is closely linked to the network operating system (NOS) used in the LAN. Application software communicates with the NOS (and hence, the network) via an application program interface (API); APIs differ from NOS to NOS. Such API differences may cause variations in the performance of a particular application package. For example, a database management package that runs correctly on LANs using Novell NetWare 3.11 may not perform as well on a LAN using Banyan Vines as its NOS.

A general-purpose application software package that runs correctly in a LAN environment when used by only one user at a time may misbehave when two or more users try to use the application at the same time. Hence, it is important for network managers to consider the application's ability to handle multiple concurrent users. If the application software is needed by users but does not run correctly when multiple users access it, LAN managers may have to consider installing copies of the software on the hard disks of each of the workstations in the LAN rather than on the file server. Some of the problems that result from multiple concurrent users may also be controllable through site license agreements and user profiles that restrict all but essential users from accessing the particular package (both site licenses and user profiles are discussed later in the book).

As has been noted, one of the main reasons for creating a LAN is to allow users to share computing resources such as application packages and peripheral devices. As a result, the ability of a LAN to run particular applications cannot be taken lightly. Because of this, LAN managers should require potential network vendors to demonstrate the ability of particular packages to support multiple users under both normal and peak work volume conditions. Of course, LAN managers should also consider the LAN alternative's ability to support both current and planned applications.

TYPE(S) OF WORKSTATIONS

The ability of an application package to support a variety of workstation types and configurations may also be an important criterion when selecting a LAN alternative. For example, if an application package must be available to users in a LAN consisting of a mix of old and new PCs, the network manager's choices among LAN configurations may be limited.

The multiuser DOS LAN is quite good at supporting wide variations in microcomputers. A multiuser DOS LAN can even utilize dumb terminals as workstations. However, this is really an option only when the number of users is quite small and when users are content with DOS-based applications lacking GUIs. Some type of client/server configuration is likely to be used when a relatively large number of users with various types of workstations needs to be supported.

Special challenges exist when the LAN will use both IBM-compatible and Apple computers; for some LANs, this is possible, but others can support only IBM-compatible (or only Apple) computers, but not both brands. Once again, this may be due to the NOS used as well as to the application packages' versatility.

When comparing various LAN alternatives, it may be important to consider the ability to support a variety of workstation configurations (and brands). Network vendors should be required to demonstrate a product's ability to enable users to simultaneously run particular applications on the various types of workstations that will be included in the LAN.

PRINTER SUPPORT

Printer needs are often an important issue in LAN selection decisions. Printing can be a crucial part of the work that users perform. Because of this, network managers may need to pay particular attention to a LAN alternative's ability to support user printing needs and printing volumes.

Some LAN alternatives are more restrictive than others in the number of printers supported per server and in the variety of ways in which printers can be deployed. This is often determined by the NOS that will be used. Some NOSs for LANs require printers to be attached to a server—either a file server or a specially configured printer server. Furthermore, there is usually a set limit on the number of printers that can be attached to one server; this means that multiple file or printer servers may be needed to support a large number of users. Other network operating systems for LANs enable a user to access printers attached to other workstations in the LAN; many peer-to-peer LAN products offer this capability.

LAN managers must also consider the types of printers that will be used in the LAN and ensure that the printing and spooling drivers needed to support them are available. This is an especially important criterion when printer types vary or when support for color printers is needed.

EXPANDABILITY

The ability to add workstations and specialized peripheral devices is often important, especially in organizations that are increasing in size. If an organization's number of employees is increasing, it is likely to be regularly adding new workstations to its LANs.

As noted, some LAN alternatives (and NOSs) can support more workstations than others. Hence, network managers should consider user growth projections when selecting from among the various LAN alternatives. Growth in the number of users or in the variety of specialized peripheral devices is likely to be most easily accommodated in client/server networks.

The ease of adding workstations is often related to the type of communication media that will be used. For example, the task of adding new workstations may be quite easy when twisted pair or coaxial cabling is used, but it can be much more difficult to attach additional microcomputers in LANs that use fiber optic cable.

The ability to expand the LAN by adding specialized peripheral devices (for example, scanner or optical disk storage technologies) may also limit the number of feasible LAN alternatives. Once again, this is often determined by whether or not appropriate device drivers for the specialized devices are available.

COMMUNICATIONS SPEED

The ability of LAN users to quickly exchange files or data is directly related to the communications speeds that the LAN can support. Lower-cost LAN alternatives (such as zero-slot LANs) are generally also lower-speed LANs. LANs built around PBX or Centrex services also provide significantly lower speeds than most server-based LANs; high-end systems typically provide a maximum of 64,000 bps and many can provide only 19.2 Kbps or 9600 bps data communication speeds. Many of the wireless LAN alternatives discussed are also limited to 14,400 bps or less, but as these evolve, significantly higher transmission speeds are expected.

Some peer-to-peer LANs offer speeds up to 10 million bps, however, speeds of 2 million bps or less are common. Some of the lowest speeds for peer-to-peer LANs are found in LocalTalk LANs; the transmission rates for such LANs are usually less than 250,000 bps.

Client/server LANs typically offer the highest transmission speeds. Most client/server LANs support speeds between 4 Mbps and 16 Mbps. However, client/server LANs with speeds of up to 100 Mbps (and more) have been implemented.

LAN speeds will continue to increase in the years ahead. By the late 1990s, speeds of up to 100 Mbps are expected to be available for copper-based cabling (twisted pair and coaxial); currently, such speeds are possible only for LANs using fiber optic cabling. For speeds in excess of 100 Mbps, fiber optic cabling is likely to remain the only feasible option.

When determining the speeds needed in a LAN, network managers should consider the type of work performed by users and the size (and nature) of the files and data that they need to exchange or transmit across the LAN. If the files that will be exchanged are large, transfer rates will be much faster if higher-speed LAN options are selected. Also, if graphics files are to be exchanged or transmitted to peripheral devices for storage or output, high-speed LANs may be the only feasible option.

MEDIA OPTIONS

In Chapter 2, twisted pair wires, coaxial cable, and fiber optic cable were identified as the three most common types of communication media found in cable-based LANs. In this chapter, radio waves and infrared light have been identified as the most common media used in wireless LANs.

Rather than installing new cabling, it is often tempting for network managers to try to take advantage of existing wiring. Many PBX and Centrex-based LANs can be implemented using existing unshielded twisted pair telephone wires. However, if high speeds are desired and if existing telephone lines pass through areas where electromagnetic interference (EMI) or radio frequency interference (RFI) cause data transmission errors, new cable that is more resistant to such interference (such as shielded twisted pair, coaxial cable, or fiber optic cable) may have to be installed. Such installation may mean pulling new cabling through already-crowded conduits. If existing conduits are already full or if it would be too expensive to install new ones, a wireless LAN may be the only feasible option.

Standards have been established for LAN cabling options. The most widely used set for unshielded and shielded twisted pair cabling (consisting of five levels) is summarized in Table 3-9. Twisted pair cabling is the currently dominant media used in LANs. IBM has also developed standards for twisted pair cabling, but the five-level set of standards is the most widely known. As a general rule, cabling able to support higher communication speeds is more expensive per foot as well as more resistant to EMI and RFI interference.

The type of work performed by users and their file and data communication needs should be considered by LAN managers when selecting communications media. If graphics or large files are to be transmitted across the LAN, cabling capable of supporting higher speeds should be selected. In addition, since it is less expensive to install high-capacity wire the first time than to rewire later, it generally makes sense to use the better cable grades from the beginning.

GEOGRAPHIC LIMITATIONS

At present, fiber optic LANs are able to span the greatest geographic area, but these are also among the most expensive of the LAN options. In general, higher-speed LANs serve a limited geographic area. This is because it can be very expensive to maintain high speeds and high-quality, error-free signals over long distances. When selecting from among LAN alternatives, network managers should be aware of the geographic limits of the various LAN alternatives.

Many lower-cost LANs are designed for microcomputers and workstations that are physically close to one another. This is true for zero-slot LANs, multiuser DOS LANs, and most of the lower-end peer-to-peer options.

TABLE 3-9
Standards for twisted pair cabling

Level	Communications Speeds Supported
1	Less than 1 million bps
2	Up to 4 million bps
3	Up to 10 million bps
4	Up to 16 million bps
5	Up to 100 million bps

Client/server LANs may cover a larger geographic area, but (as you will see in Chapter 5) many of these have maximum cable lengths and may be restricted to a single room, building, or campus. PBX and Centrex services can provide LAN connections over relatively large geographic areas, but at much lower communications speeds than those available for some of the other options.

INTERCONNECTIVITY WITH OTHER NETWORKS

The ability to interconnect two or more LANs (or a LAN to a WAN) may be an important issue in LAN selection. As LANs have emerged as key computing platforms, it has been common for different types of LANs to be implemented in different parts of the same organization. For example, a small peer-to-peer LAN may be used in one department, a zero-slot LAN in another, and server-based LANs in several others. Interconnecting a variety of LANs or giving users in these LANs access to a central mainframe can be a considerable challenge.

Although there are a variety of connection options (typically involving bridges, routers, gateways, and other interconnection technologies) that enable one type of LAN to communicate with another type of LAN or a WAN, it is often best to choose a LAN option (and NOS) that supports the communications protocols already in place. In fact, the wisest choice is often the LAN option (or NOS) already most common in the organization. This choice makes it possible to capitalize on the organization's experience with this type of LAN and generally facilitates the process of interconnecting the new LAN with existing LANs of the same type. Interconnecting LANs that adhere to widely accepted standards is also easier than interconnecting LANs that do not.

ADHERENCE TO WIDELY ACCEPTED STANDARDS

In Chapter 5, several internationally recognized standards for LANs are discussed. De facto standards (standards not internationally recognized but so widely used that they have become major factors in the marketplace) and proprietary standards (standards created by a particular hardware or software vendor for the technologies it markets) are also discussed.

Choosing a LAN that conforms to recognized standards can be beneficial, and doing so is often considered to be a relatively safe decision. Selecting from among options that conform to standards enables network managers to capitalize on the experience and expertise that is already present among users of such systems. In addition, the existence of standards generally means that there are multiple hardware and software vendors with competing products for this type of LAN. Competition among vendors usually helps to hold down the costs of both equipment and software applications.

Network managers are likely to find new or innovative types of LAN options are available that are not covered by standards. Standards usually emerge only after a particular type of LAN has survived a while in the marketplace. If a LAN alternative does not conform to existing standards, it may still be worthy of serious consideration, especially if it is capable of satisfying users' information processing needs more fully than other LAN alternatives.

AVAILABILITY OF HARDWARE AND SOFTWARE

The availability of the information technologies (both hardware and software) that enable users to run the applications they need is often an important criterion in the LAN selection process. For NOSs or LAN types that have captured large segments of the market, there are often multiple vendors for LAN hardware and software. The presence of multiple hardware vendors creates competition and helps keep the cost of LAN components such as network interface cards (NICs), servers, and shared peripherals at reasonable levels. Competition among vendors is also likely keep the cost of LAN software (such as LAN management software, electronic mail, and network utilities) low.

There are usually more end-user application software options available for the LAN operating systems that have captured large market shares. Software vendors work to ensure that their products function properly in the most popular LAN operating systems. There are likely to be significantly fewer sources of application software for less popular NOSs. Because the availability of multiple vendors of application software can help hold application software costs down, network managers often limit their selections to LAN options that run the most popular NOSs.

VENDOR SUPPORT

The extent and quality of vendor support of LAN products is often a very important criterion in LAN selection decisions. It is reassuring to know that vendors will help solve problems, improve the ability of the LAN to support user applications, enhance the overall performance and throughput of the LAN itself, and generally maintain the LAN. It is also comforting to know that vendors can provide replacement parts, new equipment, and hardware (or software) upgrades.

Objective information about vendor support can be hard to come by if the only information available is from the vendors themselves, but information can be found in trade periodicals. Comparisons of user (or LAN manager) views of the extent and quality of support provided by competing vendors of LAN products are commonly published in periodicals such as *Computerworld, Communications Week, Network World,* and *LAN Magazine.* These comparisons often cover a fairly wide range of support types and are almost always more objective than the information received directly from vendors or from the list of references they provide.

It is important to note that more vendors are moving toward paid support; free support is fast becoming a thing of the past. Vendors have found that supporting product users can be expensive and are now requiring users to pay fees for many of the services that were previously free. The support vendors provide for large users of their products (such as Fortune 1000 companies) is often more extensive and superior to that provided to smaller user organizations.

AVAILABILITY OF LAN MANAGEMENT SOFTWARE

The availability of software that will assist the network manager in monitoring and managing the day-to-day operations of the LAN can be an important consideration when selecting from among LAN alternatives. LAN management software (which will be discussed further in Chapters 13 and 15) is often available from NOS vendors or from third-party sources. The actual tools needed are determined by the size of the LAN (the number of user workstations) and the complexity of user applications. However, most LAN management software packages enable managers to set user security levels, manage user access to printers, add or delete workstations from the LAN, and monitor user logons/logoffs as well as the use of particular applications.

Similar to vendor support comparisons, LAN management software comparisons are often published in trade periodicals, at least for the leading products. Such comparisons can facilitate the selection of LAN management software.

The general manageability of the LAN itself can also be an important LAN selection issue. The ease in troubleshooting network problems, installing hardware and software upgrades, adding workstations, and rectifying printer problems may be important considerations when determining the type of LAN to install. These issues should be thoroughly explored before making a selection, especially in organizations that desire to keep their LAN management staffs as small as possible.

LAN SECURITY

Much of the data stored in LANs is shared. Other data is private. Of course, users at the various workstations in the LAN should be able to access shared data. For private data, on the other hand, it is often important to restrict access to single users (or to a limited number of users).

The ability to restrict user access to private data is much more easily done in some types of LANs than in others. For example, it is often easier to create software controls for data access in client/server LANs than in peer-to-peer LANs. Password systems and user profiles are among the most commonly used mechanisms to control access to private data.

Security and user access become even more crucial in organizations that wish to interconnect their LANs to provide dial-in access to LANs from remote locations.

OVERALL COSTS

It should be apparent that some types of LANs are less costly than others. Some of the LANs discussed were explicitly identified as lower-cost alternatives (zero-slot, multiuser DOS LANs, and some of the peer-to-peer options). Many of these are less expensive because they are relatively easy to install and can take advantage of existing microcomputers and peripheral equipment. Using PBX or Centrex services to create a LAN may also be relatively inexpensive to install if existing telephone wiring can be used; up to 80 percent of the costs of implementing a LAN can come from installing the needed cabling. Although wireless LANs obviate the need to install cabling, they are still more expensive (per workstation) to implement than most of the cable-based systems.

The faster a LAN is, the more it will cost. This is due to the fact that NICs and LAN adapters are usually more expensive for high-speed LANs than for low-speed LANs. The cabling needed to support high LAN speeds can also be significantly more expensive (per foot) than that needed for lower speeds. In addition, the LAN software in high-speed environments tends to be more expensive than that needed for low-speed LANs.

There are personnel costs associated with all types of LANs, both large and small, fast and slow. It is usually necessary to have one (or more) full-time LAN manager for each large LAN or high-speed LAN. For LANs that operate continuously, many individuals are likely to be needed. However, for smaller or slower LANs, LAN management may take up only part of an employee's day. Such personnel costs are just some of the recurring costs associated with keeping a LAN up and running.

Some of the major costs associated with implementing and operating a LAN are summarized in Table 3-10. Since cost and budgetary issues can be paramount among the criteria used to choose the type of LAN to be installed, they will be addressed throughout this book.

◆ SUMMARY

LANs have been quite common since the mid-1980s. Prior to the introduction of microcomputers, centralized host-based computing was the most common type of computer network.

◆ Backup systems (disk mirroring and server duplexing)

◆ Cabling and cable installation

◆ Hardware and software upgrades

◆ Replacement hardware (NICs, adapters, and so on)

◆ Site preparation (renovations, installation of wiring closets)

◆ Staffing costs (LAN manager, technical support personnel)

◆ Supplies (printer paper and cartridges, diskettes)

TABLE 3-10

Costs associated with LANs

◆ Training costs (for users, LAN management staff, technicians)

◆ Site licenses for LAN NOSs, software packages, and so on

Mainframes and minicomputers are the typical hosts in centralized computing systems, and terminals (dumb, smart, or intelligent) can be used as workstations. Terminal emulation software also makes it possible to use microcomputers as workstations in centralized computing networks.

Stand-alone microcomputer systems are not connected to other computers or networks; these are self-contained systems that do not electronically communicate with other computers. A stand-alone system has its own input, output, and storage devices, as well as its own set of systems and application software.

When microcomputers are networked, workstations are linked together. At a minimum, this enables users to exchange data and share peripheral devices. This can help users be more productive and can also help to contain an organization's cost of information technology.

Resource sharing, reduced computing costs, communication, support for workgroup computing, the maturation of LAN operating systems, an increasing variety of application software, improved maintenance, and decreasing management support for mainframe computing are some of the major reasons why organizations are implementing LANs.

One of the greatest benefits derived from installing a LAN is the ability to share computer resources, especially expensive equipment and software. Because such resource sharing is possible, it may be less expensive to implement a LAN than to support multiple stand-alone systems. In addition, computing costs for LANs are often less than those for mainframe- or minicomputer-based systems.

E-mail systems are among the most common communications applications found in LANs. LANs can also be used to satisfy workgroup computing needs. Groupware and group support systems (GSSs) are becoming more common; these are designed to run in LAN environments.

Network operating systems (NOSs) for LANs are the system software that performs the housekeeping and communication tasks that enable users to share resources, exchange messages or files, and utilize groupware.

As NOSs have become more sophisticated, user-friendly, and mature, they have helped to promote the popularity of LANs.

LAN versions of general-purpose application software are becoming more common. Network versions of special-purpose application software such as accounting software, computer assisted design (CAD), and computer assisted software engineering (CASE) tools are also available.

The maintenance of end-user computing tools is often easier in LAN environments. Software upgrades may have to be installed only once; with multiple stand-alone systems, software upgrades are needed for each computer. LANs can also enhance the maintenance of end-user computing systems by promoting the standardization of hardware and software within organizations.

Mainframe computing resources have generally remained expensive. Many organizations have found that LANs can provide cost-effective alternatives for many of the applications that they traditionally ran only on mainframes and minicomputers. Because of this, numerous organizations are downsizing to LANs.

There are many LAN alternatives for organizations to choose from. These range from very basic systems that offer rudimentary resource and communication capabilities to very sophisticated options, including some that combine data and voice communications and provide user interfaces to WANs.

The most basic configurations may consist of: null modem cables that allow two contiguous microcomputers to exchange files without using a modem; microcomputer-to-microcomputer communication software packages that include both the software and cabling needed to transfer files from one microcomputer to another; or shared peripheral device mechanisms that usually include manual or automated switches. A data file can also be transferred from one microcomputer to another by first uploading it to a host, then downloading it to the other microcomputer. Files can be exchanged in a similar fashion through a bulletin board system (BBS).

Sub-LANs provide users with a subset of the capabilities found in full-fledged LANs. The two major services available in such systems are file transfer and peripheral device sharing. These are typically implemented with data switches that connect microcomputers or microcomputers and peripheral devices.

Zero-slot LANs can also link PCs with one another and with peripheral devices such as printers. These get their name from the fact that they do not require a slot on the computer's motherboard for a network interface card (NIC). Instead, an adapter plugged directly into a serial or parallel port of the microcomputer is used.

Multiuser OSs enable multiple users to concurrently run tasks on the same microcomputer. The microcomputer on which user tasks are performed is typically a high-end machine. The power of the central computer's microprocessor is made available to each of the linked workstations. The workstations in these LANs do not have to be state-of-the art microcomputers; older microcomputers, dumb terminals, and diskless workstations can be used. Most multiuser OS LANs provide user workstations with multitasking capabilities.

Client/server networks are emerging as the dominant computing platform in today's organizations. Within client/server systems, shared resources are placed at dedicated servers which manage the resources on behalf of client workstations. A wide variety of servers may be found in client/server LANs; among these, file servers, printer servers, network servers, and database servers are the most common.

In smaller LANs, dedicated servers may not be the most cost-effective option. Because a dedicated server may be underutilized in smaller LANs, it is often desirable to build the LAN around a nondedicated server that can also be used as a workstation. Nondedicated servers are more prone to failure than are dedicated servers, but can be less expensive to implement than dedicated server-based LANs.

In peer-to-peer LANs, workstations can function as both servers and clients. Any workstation in this type of LAN can function as a server by making the data on its hard drive or its peripheral equipment (for example, its printer) available to the other workstations in the LAN. In some peer-to-peer LANs, users can access all of the network's resources from their workstations. Peer-to-peer LANs generally aren't as large as client/server LANs; in addition, security is a major issue.

Wireless communications are increasing in popularity and variety. These typically use microwave signals, radio waves, or infrared light to transmit messages between computing devices. While wireless connections are more expensive than most types of physical cabling, there are numerous situations in which the implementation of a wireless LAN may be an organization's most cost-effective option.

In addition to providing connections among a company's telephones, a private branch exchange (PBX) can be used to connect computer terminals, microcomputers, printers, fax machines, and other devices. Most PBXs use digital signals for both voice and data communications and provide voice-over-data capabilities that allow users to simultaneously be engaged in voice and data communications over the same medium.

For companies that desire PBX-like capabilities but do not want the associated equipment and operations, Centrex services can be leased from local telephone companies. The main difference between a PBX and Centrex is that with Centrex, the switching is done at a local telephone company's central office rather than at a switch at the customer's location.

Another option to interconnect an organization's microcomputers is a service bureau. A service bureau is an organization that provides subscribers with information processing services similar to those available on a large, central host. A subscriber's organization gains access to the service bureau's computers and software via leased telephone lines.

There are many criteria that network managers may use to compare LAN alternatives. Most of these center around cost considerations, the ability to satisfy users' needs, and network management issues.

One of the primary considerations in choosing from among the various LAN alternatives is the number of users and workstations that must be supported, both now and in the future. Another important consideration is the availability of the application software needed by LAN users; a greater variety of both general-purpose and special-purpose application software is available for some types of LANs than for others.

The ability of an application package to support a variety of workstation types and configurations may also be an important criterion. Because access to printers or specialized peripheral devices can be a crucial part of users' work , a LAN's ability to support such access may be a key consideration. The ability to expand the network by adding new workstations, workstation configurations, and devices may also be an important LAN selection issue.

Users' ability to quickly exchange files or data is directly related to LAN communications speeds. Speed may also be affected by the type of media used. In general, higher-speed LANs and LAN media are more expensive to implement than are lower-speed alternatives.

Some types of LANs can cover larger geographic areas than others. This can be an important consideration for some organizations, especially for those that consist of multiple buildings or operating locations.

The ability to interconnect two or more LANs or a LAN to a WAN can be an important selection issue. LANs that adhere to widely accepted standards are generally easier to interconnect.

For LANs that conform to existing standards, there is often a larger number of competing hardware and software vendors. Competition among vendors encourages multiple sources for the same type of equipment or software; such competition also helps to keep hardware and software prices down.

The extent and quality of vendor support is often an important criterion in LAN selection decisions. Objective comparisons of vendor products and support can be found in trade periodicals.

The availability of software that will help the network manager monitor and manage the day-to-day operations of the LAN can also be an important consideration when selecting from among LAN alternatives. Most LAN management software packages enable managers to set user security levels, manage user access to printers, add or delete workstations from the LAN, and monitor user logons/logoffs as well as the use of particular applications. Security and user access to private data have become particularly important issues in organizations that wish to interconnect LANs or to provide dial-in access from remote locations.

Cost considerations are almost always very important issues in LAN selection decisions. LAN costs can include installation, operating costs, personnel costs, and maintenance costs. Overall costs vary greatly among the LAN alternatives discussed in this chapter.

✳ KEY TERMS

A/B switchbox	Peer-to-peer LAN
Bulletin board system (BBS)	Private branch exchange (PBX)
Centrex services	Service bureau
Data switch	Stand-alone system
Micro-to-micro software	Sub-LAN
Multiuser OS	Voice-over-data
Nondedicated server	Zero-slot LAN
Null modem cable	

✳ REVIEW QUESTIONS

1. When did microcomputers first become popular in business organizations? What caused microcomputers to become popular among businesses?
2. Describe the characteristics of centralized computing systems.
3. What does online processing mean in host-based computing systems? Why was the development of online processing capabilities significant in the evolution of computer networks?
4. What is terminal emulation? How does it allow microcomputers to communicate with mainframes and minicomputers?
5. Why are many organizations substituting microcomputers for terminals in centralized computing systems?
6. What are the characteristics of stand-alone microcomputer systems?
7. Identify the major reasons why LANs are being implemented more often than any other type of computing platform.
8. What types of computing resources may be shared by users in LANs? Why is the ability to share resources important?
9. Why are LANs typically more cost effective than having multiple stand-alone systems? Why may LANs be more cost effective than centralized computing systems?

10. What are the major types of communication capabilities available to LAN users?
11. How are LANs used to support workgroup computing needs?
12. What are the major characteristics of LAN operating systems? How do LAN OS characteristics change as new versions are released?
13. What is the difference between general-purpose and special-purpose application software for LANs? Provide examples of each.
14. How may maintenance be improved by implementing a LAN?
15. How may the development of LANs help an organization standardize its hardware and software?
16. Describe the reason for the decreasing management support for centralized mainframe-based computer systems in many organizations.
17. What are the characteristics of null modem connections between microcomputers?
18. What are the characteristics of micro-to-micro communication packages? What functions do these carry out?
19. What are A/B switchboxes used for? What different types of these may be found?
20. What are the characteristics of bulletin board systems? How may users exchange files via a BBS?
21. What are the characteristics of sub-LANs? What are the advantages and disadvantages of sub-LANs?
22. What are the characteristics of zero-slot LANs? Why are they so named?
23. What are the characteristics of multiuser OS LANs? What are the advantages and disadvantages of multiuser OS LANs?
24. What are the characteristics of client/server LANs? Why are these becoming the most popular type of LAN?
25. What is a nondedicated server? What are the advantages and disadvantages of building a LAN around a nondedicated server?
26. What are the characteristics of peer-to-peer LANs? What are the major advantages and disadvantages of peer-to-peer LANs?
27. What are the differences between controller-based and peer-level wireless LANs? What are the major advantages and disadvantages of wireless LANs?
28. What are the characteristics of private branch exchanges (PBXs)? What are the advantages and disadvantages of PBXs?
29. What is meant by *voice-over-data* capabilities?
30. What are the similarities and differences between PBXs and Centrex services?
31. What is a service bureau? How may a service bureau be a mechanism for creating a LAN within an organization?
32. Why is the number of users that a LAN is capable of supporting a common LAN selection criterion? Identify several LAN options that are best

suited to a small number of users. Also identify several LAN options that can support large numbers of users.

33. What are some of the factors that determine the availability of application software for a LAN alternative? Identify LAN alternatives for which the range of application software is quite large.
34. How may LANs vary in their ability to support a variety of workstation types? Why may this be an important LAN selection criterion?
35. How may LANs vary in their ability to support user printing needs?
36. Explain why the expandability of a LAN may be an important factor in selecting a LAN alternative. What types of expandability may network managers seek?
37. How much variation is there in LAN communication speeds? Which LAN alternatives offer the highest communication speeds? Which LAN alternatives provide lower communication speeds?
38. Why may media speed be an important issue in selecting a LAN? What factors affect media speed?
39. What types of LANs are most likely to be used when it is important to cover a relatively large geographic area (for example, several buildings or a campus)?
40. What factors determine the ability of a LAN alternative to be connected to other LANs or to WANs?
41. Why may adherence to standards be an important issue in LAN selection decisions?
42. Why is it desirable for LAN managers to have a variety of LAN hardware and software vendors (and products) to choose from?
43. Why is vendor support often an important issue in LAN selection decisions?
44. Why may the availability of LAN management software be a factor in LAN selection decisions?
45. Describe how security could be a factor when selecting from among LAN alternatives.
46. What are some of the major cost factors that may be considered when choosing a LAN?

✳ DISCUSSION QUESTIONS

1. Why may LANs be considered a natural offshoot of the evolution of the use of microcomputers in organizations? Discuss how microcomputers are likely to continue to evolve. Also discuss how these changes are likely to affect LANs.
2. Why is it often more costly to equip, maintain, and upgrade multiple stand-alone systems than it is when the same number of microcomputers

are connected to a LAN? What benefits are available to users in LAN environments that are not available to users with stand-alone systems?

3. Why has support for workgroup computing become important in organizations? Why are LANs among the best available options for supporting workgroup computing?

4. Why is there a greater variety of hardware and software options available for organizations with LANs that use Novell's NetWare? Why does the popularity of NetWare promote competition among vendors? How does the availability of hardware and software help keep NetWare popular?

5. Discuss the issues that you feel are likely to cause an organization to choose a full-fledged LAN over the more basic micro-to-micro connections such as A/B switches, null modem cables, micro-to-micro communication software packages, and sub-LANs.

6. Discuss why security is often an important issue when considering the implementation of a peer-to-peer LAN. Identify some of the mechanisms that could be used to protect the privacy of files in a peer-to-peer LAN.

7. Multimedia in LAN environments is likely to present LAN managers with some major implementation and maintenance problems. Identify the LAN selection criteria that become central issues if an organization intends to include the exchange of multimedia files between LAN users. Discuss why each of these is a central issue in this situation.

8. Some experts claim that overall LAN performance and user satisfaction is likely to be high when there is good vendor support available, even if the LAN is generally mediocre or below average in quality. They even go so far as to claim that performance and user satisfaction may be better in these instances than in cases where poor vendor support is available for what is generally considered a high-quality LAN. Discuss the validity of these claims.

9. The total costs of a LAN are often broken down into two categories: implementation costs and recurring costs. Identify the types of costs that should be included in each category and discuss why each should be classified in this manner. Also discuss why organizations may be more concerned with recurring costs than with implementation costs when selecting a LAN alternative.

✳ PROBLEMS AND EXERCISES

1. Identify a local company that has implemented one (or more) of the LAN alternatives discussed in this chapter. Interview one or more of the LAN users (or the LAN manager) and ask them what they like most

about the LAN and what they like least. Also ask them how the LAN could be improved. Share your results with your classmates.

2. Identify a local company that is in the process of selecting a LAN (or that has recently implemented one) and find out from the LAN manager why they decided to implement a LAN and what criteria are being (or were) used to select the LAN. Compare these to the LAN selection criteria described in this chapter; look for similarities as well as for additional criteria that are not described in the text. Share your findings.

3. Identify a local company that is in the process of downsizing (moving its applications from a mainframe or minicomputer to a LAN or set of LANs). Ask the company's highest-ranking network manager to speak to your class about why it has decided to downsize and why it has decided to move toward a LAN-oriented computing environment.

4. Find a recent trade article comparing vendors on the support that they provide to users. Identify the criteria on which the vendors were compared and describe what each means. Present you findings in class.

✴ REFERENCES

Bates, R. J. *Disaster Recovery for LANs*. New York: McGraw-Hill, 1994.

Berg, L. "The Scoop on Client/Server Costs." *Computerworld* (November 16, 1992): p. 169.

Brown, R. "The Wireless Office Untethers Workers." *Networking Management* (June 1992): p. 42.

Burns, N. and K. Maxwell. "Peering at NOS Costs." *LAN Magazine* (May 1992): p. 69.

Derfler, F. "DOS Based LANs Grow Up." *PC Magazine* (June 25, 1991): p. 167.

———. "Low-Cost LANs Grow in Features and Performance." *PC Magazine* (April 14, 1992): p. 209.

———. "Peer-to-Peer LANs: Teamwork Without Trauma." *PC Magazine* (May 11, 1993): p. 203.

Fitzgerald, J *Business Data Communications: Basic Concepts, Security, and Design*. 4th ed. New York: John Wiley & Sons, 1993.

Frank, A. "Networking Without Wires." *LAN Technology* (March 1992): p. 51.

Frenzel, C. W. *Management of Information Technology*. Boston, MA: Boyd & Fraser, 1992.

Goldman, J. E. *Applied Data Communications: A Business-Oriented Approach*. New York: John Wiley & Sons, 1995.

Keen, P. G. W. and J. M. Cummins. *Networks in Action: Business Choices and Telecommunications Decisions*. Belmont, CA: Wadsworth, 1994.

Panchak, P. L. "Network Alternatives: Making the Right Choice." *Modern Office Technology* (March 1992): p. 22.

Parker, C. and T. Case. *Management Information Systems: Strategy and Action*. 2nd ed. New York: Mitchell/McGraw-Hill, 1993.

Rains, A. L. and M. J. Palmer. *Local Area Networking with NOVELL Software*. 2nd ed. Danvers, MA: Boyd & Fraser, 1994.

Salamone, S. "Test Shows Which NOS Performs Best for You." *Network World* (January 20, 1992): p. 1.

Saunders, S. "PBXs and Data: The Second Time Around." *Data Communications* (June 1993): p. 69.

Schatt, S. *Data Communications for Business*. Englewood Cliffs, NJ: Prentice Hall, 1994.

Stamper, D. A. *Business Data Communications*. 4th ed. Redwood City, CA: Benjamin/
     Cummings, 1994.
———. *Local Area Networks*. Redwood City, CA: Benjamin/Cummings, 1994.
Strauss, P. "Welcome to Client-Server PBX Computing." *Datamation* (June 1, 1994): p. 49.
Therrien, L. and C. Hawkins. "Wireless Nets Aren't Just for Big Fish Anymore." *Business Week*
     (March 9, 1994): p. 84.
Tissot, A. F. "The Changing Role of the PBX in Today's Office Environment." *Telecommunica-
     tions* (November 1993): p. 51.

Case Study:
Southeastern State University

As is the case for many state-supported colleges and universities, Southeastern has limited funds available for purchasing computer hardware and software. In the 1980s, budget limitations constrained SSU's ability to purchase microcomputers for its faculty and staff and forced the university to slowly install these in campus offices. Because limited numbers of microcomputers and microcomputer software were purchased each year, faculty members and some clerical workers had to wait several years longer than did their colleagues to get their first microcomputers. However, the machines installed in the offices of SSU employees who had to wait the longest were invariably faster and more powerful than the machines of employees who received their computers early on.

The slow proliferation of microcomputers in the offices of SSU is still apparent. You can walk down the halls of most buildings on campus and find several generations of microcomputer hardware in the offices. In colleges that purchased IBM and IBM-compatible computers, the machines in use may range from those with 8086 processors and two 5¼" low density disk drives (no hard drives) to 90+ megahertz machines with Pentium processors, CD-ROM drives, and 500+ megabyte hard drives. In colleges that purchased Apple computers (primarily the College of Education) several generations of Macintoshes and a few Apple IIs can be found.

Several generations of processors as well as vendor brands are found in most of the student computer labs at SSU, including those that are LANs. In fact, the student labs have often served as a dumping ground for the older machines that were once found in faculty and staff offices. As the colleges have found the funds to replace the older machines, they are often moved to the student labs in order to get the most out of the university's investment in these systems. Department chairs often justify purchases of new computers for faculty offices by arguing that this is the most cost-effective way to satisfy the computing needs of faculty members and to expand the computing resources available in student labs at the same time.

For most of the colleges at the university, it took four or more years to purchase computers for all their faculty and clerical staff. Now, it takes just about as long to replace existing computers with newer, more powerful, and more sophisticated machines. By 1994, most colleges had replaced almost all of their 8086 and 8088 machines (and Apple IIs). In addition, the number of 80286 machines found in faculty and administrative offices was decreasing. At this time, most of the computers in use were 80386 machines; however, significant numbers of 80486 processors were in place (and increasing in number), and a few machines with Pentium processors could be found.

As a repository for the replaced faculty and staff computers, the student labs contain a much higher percentage of older, less powerful machines than state-of-the-art microcomputers. 8086, 8088, and 80286 machines with monochrome monitors are still quite common. However, the labs also contain some 80386 computers and a limited number of 80486 machines that have been purchased from a student computer lab budget created by the university's administration in 1990. In 1994, the student labs primarily received 80286 machines from the colleges. However, some 80386 machines were also filtering in.

As microcomputers became more common on the SSU campus, the university realized that some type of standardization was needed. The different colleges were purchasing a variety of computer models from a variety of vendors. In addition, the colleges were spending their money on a wide range of application software packages. An inventory taken of campus-wide computing resources in 1988 found 11 different word processing programs, 7 different spreadsheet programs, 6 different file/database management programs, and 5 different LAN operating systems in use. The committee also found that more than 50 different models of microcomputers had been purchased by university subunits since 1983.

By the late 1980s, SSU's Computer Services division was being pressured to assist end users in maintaining the diversity of systems that had been installed. They were being asked to develop training programs for the software purchased and were called upon daily to help users with problems they encountered when trying to run the different programs. With a limited budget of its own, Computer Services was overwhelmed by these requests.

In 1989, a university committee was formed to address the hardware and software diversity problems that had emerged. After studying the problem and how similar problems were being addressed by other universities, the committee recommended that only IBM and IBM-compatible microcomputers be purchased for clerical and administrative offices. It also recommended standardizing on a word processing package (WordPerfect), a spreadsheet package (Supercalc), and a database package (dBase III) campus-wide. The committee reasoned that such standardization would make it easier for Computer Services to help end users. It also reasoned that standardization would make it possible for either Computer Services or the Personnel Division to develop training programs for the three software packages.

Standardizing on hardware and software did have the effects that the committee expected it to have. For example, training programs on the three packages were initiated and the Computer Services staff was able to say, "No, we're sorry, we can only provide support for the standard packages and equipment." However, the effects were not as positive as many thought they would be. This was partly due to the fact that the software packages chosen were selected primarily because they were the most widely-used ones on the SSU campus. Some subunits that had grown comfortable using other brands expressed their dissatisfaction with the selections and their intention to continue using what they already had.

Within 24 months, the standards changed in response to the state university system's decision to acquire state-wide (university system-wide) site licenses for WordPerfect (at $25 per workstation, networked or not) and for Borland International's software (including Quattro, a spreadsheet program, and Paradox, a database package). The university system's agreement with Borland was financially significant for most of the state's colleges and universities, including SSU; it essentially meant that Borland products could be installed on all of SSU's microcomputers for free.

These incentives were too good for SSU to pass up, and in early 1991, the decision was made to end the university's site license agreements for Supercalc and dBase III and to restandardize on Quattro and Paradox. As SSU had already standardized on WordPerfect, and as the $25 per workstation fee was a better deal than SSU's existing site license agreement with WordPerfect Corporation, the university did not change its standard for word processing.

This restandardization created the need for new rounds of training. Actually, the university found that even with standards, software training must be an ongoing effort. Like hardware, software packages have progressed through several generations (or versions) since the late 1980s. For example, the DOS versions of WordPerfect have progressed from WordPerfect 4.2 to 5.0, and then to 5.1. WordPerfect 5.2 was the first version for Windows, and this was later upgraded to the 6.0 version. As a result of these version changes and upgrades and the fact that the latest versions are not found in all administrative and clerical offices (many still lack machines capable of running Windows and Windows applications), the university has routinely offered training for several WordPerfect versions since 1991. This has also become necessary with the spreadsheet and database packages; as Windows and Windows-capable machines have become more common in the university's administrative and clerical offices, the Windows versions of Quattro and Paradox are also being more widely used.

As you can see, the microcomputer hardware and software used at SSU have changed significantly over the past decade. Further changes are expected. Everyone expects continued microprocessor enhancements, operating system upgrades, and new versions of application software packages. SSU's administration knows that it will have to continue to upgrade its microcomputer hardware and software and add new technologies. As noted previously, internetworking, multimedia applications, distance learning applications, and videoconferencing are looming larger on SSU's computing horizon. All of the changes are expected to present SSU with ongoing managerial and budgetary challenges.

CASE STUDY QUESTIONS AND PROBLEMS

1. In light of the information given in the case, which two types of LANs would seem to be good alternatives for an academic department at SSU? Justify your selections. (Assume a 25-person department outside the College of Education that is able to replace approximately 25 percent of the machines in faculty offices per year.)

2. From the information provided, which types of LANs seem to be good options for the student labs at SSU? Justify you selections.

3. If you were asked to form a LAN within one of the colleges at SSU (but not at the College of Education), what do you feel should be the five most important selection criteria? Justify using each of the selection criteria on your list.

4. What types of problems could restandardizing cause SSU? What factors could cause an organization to restandardize? What factors do you consider important in determining whether or not to restandardize? Justify your selections.

LAN Hardware and Media

CHAPTER OBJECTIVES

After completing this chapter, you will be able to:

- Describe some of the major types of terminal equipment found in LANs.

- Identify some of the microprocessors used in today's networks.

- Describe the major storage technologies used in LANs.

- Identify some of the major output technologies found in LANs.

- Describe how queuing affects network performance.

- Describe the characteristics of network interface cards.

- Describe the major cabling and adapter technologies used in LANs.

- Discuss the key role of servers in LAN environments.

- Identify some of the major trends in LAN technologies.

The manner in which data and information are input into a network depends on the terminal device(s) used. A *terminal* can be defined as an input device connected to a computer or host. A terminal may be dependent on the host computer for all (or some) processing operations or it may perform its tasks independently.

◆ TERMINAL DEVICES

A wide variety of terminal devices is used in LANs; the most popular is the personal computer. A PC (microcomputer) attached to a LAN can serve as a host computer or as a terminal for another host. In some networks, a PC may perform both functions. For example, a PC in a LAN may have an IBM 3270 terminal card; in addition, it may control access to disk files requested by other microcomputers in the network.

Although some information technology purists argue that there are differences between PCs and workstations (that is, that workstations are more powerful and can perform more sophisticated tasks), most often the term *workstation* is used to describe any microcomputer attached to a network. That is, workstations are the networked microcomputers at which users work.

In this text, *workstation* has this more general sense. However, be aware that some IT professionals use the term *workstation* to refer to only very sophisticated and powerful microcomputers used for engineering and scientific applications.

◆ TERMINAL EQUIPMENT SELECTION ISSUES

Network managers must consider many factors when determining which type(s) of terminal equipment to use in a network. Four considerations are briefly summarized in the following paragraphs.

First, what transmission and processing speeds are needed? Which terminal options have adequate speeds for the task to be performed on the network? Speed is often related to the cabling and transmission mode used (asynchronous vs. synchronous), so these factors should also be considered.

The second consideration is the nature of the user input interfaces. What type(s) of terminal equipment can accommodate the volume and nature of the data to be input? Because the input may range from bar codes or alphanumeric text to sophisticated graphics and multimedia, this is an important issue.

Third, what terminal equipment option can handle the output that users need or want? Output can range from text-based output to color graphical output, which is rapidly becoming the norm.

A fourth criterion is cost. Network managers must determine if the terminal option's cost is proportional to the value of the tasks that it will perform. An expensive, bells-and-whistles workstation is likely to be unnecessary for relatively straightforward text and number-based transaction processing operations; in fact, a dumb terminal may be all that is needed. Also, regular microcomputers fare poorly in dirty working environments; protection from contaminants is needed.

◆ TYPES OF TERMINALS

Many network managers use *terminal* to refer to a device that consists of a video screen and keyboard. In the days of highly centralized systems, such a terminal—typically a dumb terminal—was economical and adequately captured character-based data and transferred it to the host computer. This type of terminal is still able to perform some tasks well, depending on the applications.

Currently, there is a substantial number of LANS and distributed processing systems. There is also a greater need for terminals to be able to carry out some processing activities on their own. By definition, distributed processing requires processing capability at points in the network other than the central host; ideally, processing should start at the terminal level. In distributed processing networks, dumb terminals will not serve, and even smart terminals with limited processing capabilities may be insufficient. In such situations, intelligent terminals and microcomputer workstations may be the only option.

MICROCOMPUTER WORKSTATIONS

In the early 1980s, a dumb terminal cost $450 (or less) and an intelligent terminal cost as little as $850. Microcomputers ranged in cost from about $3,000 to $10,000. Then, sophisticated microcomputers could cost over $10,000. Because of the significant price differences, managers needed truly compelling reasons to buy microcomputers for use as network terminals. Economics alone presented a very persuasive argument against doing so.

In the mid-1990s, dumb terminals can cost less than $200, intelligent terminals as little as $300, and microcomputers can be purchased for under $800. Of course, high-end microcomputers, such as those needed for engineering workstations, are more expensive; these range in cost from around $5,000 to $40,000.

While price differentials still seem to favor terminals over microcomputers, managers now have a number of compelling reasons to purchase microcomputers. For example, an increasing number of applications involve as much input of graphically-oriented data as character data. Most terminals, especially dumb terminals, cannot handle such inputs. Also, because microcomputers can perform processing operations on their own, they are better suited to the distributed processing networks emerging in many organizations.

WORKSTATION INPUT TECHNOLOGIES

The most prevalent input hardware is still the keyboard. Keyboard layouts have changed and keyboard supplements (such as the mouse) have become standard microcomputer equipment. These changes have made microcomputers easier to use.

The keys on a keyboard are organized as they are on typewriters. Of course, microcomputer keyboards also include from 10 to 15 function keys, 10-key numerical keypads, and auxiliary keys such as the microcomputer (Alt), (Ctrl), (Ins), (Del), (Home), (PgUp), and (PgDn) keys. The function keys are programmed to perform specific functions depending on the software being used; typically, these keys perform a specific macro or small program operation.

Some computer keyboards now substitute for older kinds of office equipment. Accountants who continue to use exclusively a 10-key calculator, instead of the 10-key numeric keypad on a keyboard, are rare. The keyboard's keypad enables the accountant to enter data by touch, but to also have the advantage of the microcomputer's memory.

Custom keyboards have been designed for use in particular industries. For years, airlines have used terminals that function as typewriters and also include keys for performing specific functions unique to the ticketing requirements of the airline industry. Ticket agents are specially trained on these keyboards.

For international businesses, specialized keyboards to accommodate language differences have been developed. Virtually every DOS-compatible microcomputer loads a keyboard driver designed to function in the language environment of the country in which it is sold. Language divisions, such as American English and British English, are accommodated by separate drivers.

The *mouse*—not the rodent, but the small plastic holder with a rubber sphere that rolls on a table—has become a standard peripheral for most microcomputer systems. Software enables a user to direct a mouse pointer onscreen to display menu commands, making it unnecessary for users to memorize command strings. Software has become more user-friendly; mouse users can access pull-down menus, which almost intuitively guide the user to completion of a task.

A mouse variation is the trackball. The sphere is on the top of the plastic cover; the thumb or finger manipulates the pointer onscreen. Although used more for games than for serious commercial programs, the joystick is another cursor-control device. The joystick is best suited to specialized applications requiring extensive interactive use. Another type of mouse looks very much like a large pen with a roller on the lower end. This device is often used with drawing packages—it permits hand-drawn input—and is called a stylus or digitizer.

Reader/scanner devices are among the special input technologies successfully used with a variety of industry applications. For example, bar code technology has become common in retailing and warehousing. Portable terminals help retailers and warehouse workers perform physical counts of inventory. Some of these terminals are actually specialized microcomputers that automatically determine the value of on-hand inventory by multiplying the unit prices by the inventory count.

Reader/scanner devices are also used in the validation of credit cards at point-of-sale terminals in retail operations, grocery stores, and restaurants. Reader/scanner devices have become very common in LAN-based point-of-sale systems.

In general, scanning devices used to input graphics are becoming common in the industrial world. Scanning devices include the flatbed scanner. The user lays a hardcopy page on a glass plate; a light bar that looks like a fluorescent bulb mechanically scans the page, converting the image to machine- (computer-) readable form. The scanned image can be then stored in RAM or on disk in a format recognizable by other applications. Special software is used to control the scanning processes.

Hardcopy images can also be scanned to compare text to a template of characters. By so doing, the processor can select which of the ASCII characters is contained in the image. The scan of a page of text can then be transformed into an ASCII file that can be edited by a word processor. This process is generically called *optical character recognition (OCR)*; special OCR software (not the scanning software used for scanning graphics) is used for such conversions.

Digital cameras are now available that offer many features you would expect of high-quality cameras. The "photographic" image is stored in the internal memory of the camera until it is uploaded to a computer. Once in the computer's memory, the image can be stored in formats permitting it to be imported into a database, document, or other user application.

Microcomputers and other terminals can also receive graphic, facsimile input. The image can be automatically converted into a text file using a process similar to that used in OCR. When these images are received by microcomputers attached to LANs, they can be made accessible to other LAN users.

The list of input technologies used in LANs and other networks continues to grow. Light pens, touch screen technologies, and wireless personal digital assistants (PDAs) are some of the other input technologies available. These and all the other input technologies continue to become both more user-friendly and sophisticated. In the following sections, you will find that this is also happening with many other LAN technologies.

◆ PROCESSOR OPTIONS

For some time, the Intel 80X86 series of chips has dominated the central processing unit choices for many network workstations. As the development of processor chips continues, network managers are likely to have more options. For example, Hewlett-Packard is developing a chip that will function with Novell NetWare to provide more advanced redundancy (backup and fault tolerance) features in the LAN. In addition, Digital Equipment's Alpha chip promises a six-fold increase in brute processing power over the Intel 80486 chip. And, for engineering applications at Sun workstations, the SPARC chip is typically used to process graphics and for other processor-intensive tasks.

The Pentium processor, another Intel product, has also found its way into LANs. The PowerPC chip (developed by a consortium of companies including Apple, Motorola, and IBM) is found in recent generations of Apple Macintosh and NeXT computers.

Due to the wide variety of processors in use, it is important to ensure that networked machines using different processors are able to communicate with one another and share network resources. This boils down to overcoming not only the differences among processor technologies, but also differences in the operating systems and software used on the different machines. At present, being able to accommodate a wide variety of processor brands and generations with the same LAN can be a networking challenge. Indeed, it is often easier to segregate different processor brands into separate LANs and use internetworking technologies (routers and gateways) to connect them than it is to achieve effective communication between machines with different processors within the same LAN.

A number of software solutions have been developed to address the challenges posed by the diversity of processors. For example, Novell NetWare enables Macintosh computers to connect to LANs that primarily consist of IBM and IBM-compatible machines; this NOS contains the modules needed to enable Macintosh users to access data that would otherwise be unreadable. In addition, there is a general thrust to have DOS machines with either Intel or Motorola chips be able to exchange files with UNIX systems. In 1993, Novell bought UNIX; Novell has announced its intention to work toward effective connectivity between DOS and UNIX machines.

◆ STORAGE SYSTEMS

In the days when personal computers stored most applications and data on a few floppy disks, few people gave a thought to the day that gigabytes of disk storage would be required for LANs. The average personal computer in commercial use in 1994 had a hard disk holding 100 megabytes of information. As you might expect, the average size of hard disks has been increasing and is expected to continue to increase.

As more companies review their information systems in light of cost efficiency, it is clear that larger storage systems will proliferate. One of the computers used to write this book is a laptop 80486 with over 1.5 gigabytes of disk storage available. As more users take their portable computers on the road and communicate back to the desktop or mainframe over telephone lines or wireless media, the need for larger storage systems becomes greater.

The popularity of multimedia applications is also driving the need for larger and larger secondary storage systems. Multimedia files can consume great amounts of storage space; this is one of the reasons that high-capacity CD-ROM systems are commonly used to store multimedia applications.

Optical storage technologies are generally viewed as the best answer to increasing demands for storage. Removable **optical read/write disks** can hold more than 100 megabytes of information; some can store nearly a gigabyte of data on a single disk. While these disk systems are still considerably more expensive than comparable magnetic disk drives, as the number of optical storage users grows, prices will gradually move toward those of magnetic disk systems.

Large appetites for disk storage are not limited to desktop units. Many portable units are being introduced that enable high-capacity disks (both optical and magnetic) to be used. Trends indicate that it won't be long before the capacities of the average secondary storage systems for all microcomputers are measured in gigabytes.

Many libraries at academic institutions have made large databases available to online users. Often, this is done through the use of banks of disk drives using CD-ROM technology (such systems are sometimes called *optical jukeboxes*).

The disk systems used in LANs are no longer simple stand-alone, multiple-platter disk drives. Many of the storage schemes now in use allow for storage across an array of disk drives allowing read/write operations to several (physical) disks simultaneously. A disk array—a cluster of hard disks—allows for faster read/write processes than does reading/writing in a serial fashion, sector by sector (the process typically used in a single multiple-platter disk drive).

Another emerging secondary storage technology for LANs is **redundant array of independent disks (RAID)** systems. RAID systems replace large, expensive disk storage systems with groups of smaller, less expensive disk drives. As a result these are sometimes called redundant arrays of inexpensive disks. These systems can provide high-capacity storage at relatively low cost; they can also increase access speeds and bolster reliability. Figure 4-1 illustrates how Novell NetWare can monitor RAID drives.

RAID reads and writes data in parallel. This means dividing data between disk drives at the block, byte, and bit levels. This procedure is called *data striping,* and it can boost data transfer rates by a factor equal to the number of drives in the array. RAID can also increase the fault tolerance of a network. Because of this, RAID systems will be described more fully in Chapter 14.

Large disk storage units intended for LAN use can measure their capacity in terabytes (trillions of bytes). For years, networking software has allowed for such high capacities. As LAN customers began to demand very large units, the manufacturers responded. Some state-of-the-art data processing installations have disk systems supporting LANs having capacities as large as those used in mainframe computers.

Archiving, or *backing up,* has become a major security concern for LAN managers. Early in the evolution of LANs, making duplicate copies of user data and applications was largely ignored by network managers because these files were perceived as unimportant. However, when users in financial institutions, research consortiums, academic institutions, and other large organizations began to store sensitive data on network drives, the need for backup and security became apparent.

In many LAN environments, archiving is done through the use of magnetic tape. Some tape backup systems are more automated than others. For some, the NOS or network management software is used to begin the

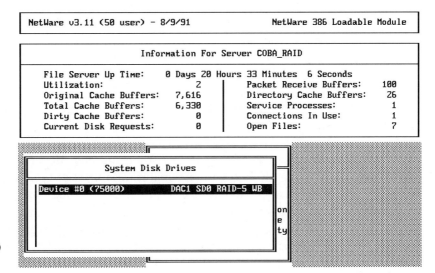

FIGURE 4-1
Monitoring RAID drives

backup automatically at a preset time. However, as the backup data may fill up many tapes, unless an automatic tape carousel is used, changing tapes must be done by hand.

Tape backup speeds have improved dramatically over the years. Not long ago, many companies faced the problem of having insufficient time to back up the system between the end of one business day and the start of the next. Now, faster tape backup systems have alleviated many of these problems.

Ideally, storage systems should be able to retain data through power outages, power fluctuations, and variable environmental conditions. However, network managers should regularly test LAN backup mechanisms to ensure that they are working properly. Backup systems that do not function properly can cause major problems. For example, a hospital in an eastern state failed to test the restoration process for a system that archived $25 million in accounts receivable data. After the system crashed, the restore option also failed. The hospital was able to reconstruct only $17 million of the receivables. Failure to test that system cost the hospital at least $8 million and the goodwill of many patients whose records were lost.

◆ LAN OUTPUT TECHNOLOGIES

LAN outputs can vary significantly from those of stand-alone PC systems. This is especially apparent for printing operations, arguably the most important output processes in LANs. Many new LAN managers, having attached a printer that has worked well on a stand-alone system to the LAN, find that it is unable to handle the large volumes of output required by LAN users.

Printing becomes a complex subject in a LAN when user applications include a wide variety of software with unique requirements for the printed output. For example, it is not unusual even in smaller LANs to have some users who require no more than simple, character-based output, while others need sophisticated graphics output. In other settings, continuous-feed preprinted forms may be required for output.

It is rarely possible to meet all of these diverse requirements with one central printer. Most LANs have a *print queue* facility that allows the individual workstation to complete the printing task quickly so that it can move on to the next task. The network's queuing mechanism ensures that user print jobs are output in a timely manner. Most networks have the capability of directing the normal print output to the LAN queue. The network must then determine which printer can successfully complete the job in the correct format.

In Novell NetWare LANs, most of these tasks are accomplished through the PRINTCON (printer configuration) and the PRINTDEF (printer definition) utilities. Figure 4-2 provides an example of a NetWare printer server set-up and Figure 4-3 illustrates some of the print queue settings for a particular LAN.

The print queue in a LAN works as does the Print Manager in Microsoft Windows on a stand-alone personal computer. In Windows, a print job is initiated by the application and is directed to a file on the disk. From this file, the Print Manager determines to which printer and which port the printing should be directed. It tests to make sure that the printer is online and, as soon as sufficient system resources are available to print the job, it releases the data in the temporary file to the print queue. In the LAN, the direction of the print jobs to network printers is just a little more complex, because the print jobs may be initiated by any user application that has the

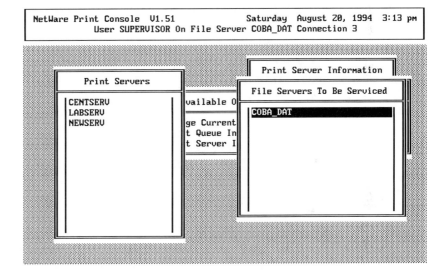

FIGURE 4-2

Printer server set-up

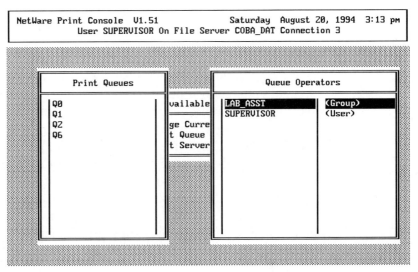

FIGURE 4-3

Print queue settings

proper configuration to allow printing. Once initiated, the print job is directed to the appropriate print queue by the network operating system.

In some LANs, the physical location of the user initiating the print job can be important. If the LAN stretches over a large geographic area or if certain workstations are located at a remote location, it may be difficult for the user to verify that the output is valid. Even though some application software permits users to see a simulated printout on the screen, the simulation is not a guarantee of perfect results on paper. To address such problems, LAN printers may be geographically scattered so that users in all LAN segments can easily access network printers.

While the financial justification for implementing a LAN may include the elimination of future printer purchases for individual departments, this seldom happens; departments often need different types of printing. Most LANs can, however, distribute printers to particular locations or attach printers to specific computers.

When a printer is attached to a particular workstation, the computer essentially assumes the role of ***print server***. Novell NetWare enables the file server to also be the printer server for the network; however, it also makes it possible for another computer in the network to function as either a dedicated or nondedicated print server. A ***dedicated server*** performs only network printing operations; a nondedicated print server can be used as a workstation in addition to handling network printing chores.

In some situations, using the file server as a network print server can negatively impact overall file server performance. However, many smaller LANs use the file server for this purpose. A dedicated print server requires a powerful computer with adequate RAM and disk space assignable exclusively to this task. In most instances, the printers are physically attached to the print server; however, it is also possible to attach them to a workstation that is logically attached to (and controlled by) the machine that functions as print server. In Novell LANs, print queues can be monitored onscreen using the PCONSOLE utility print. This utility enables a single member of the network staff to monitor a large number of print jobs.

Laser printers are among the most popular printers found in LANs. Part of this popularity is due to the outstanding quality of laser printer output, their quiet operation, and their speed. Many laser printer vendors have models designed to be attached to a network; these printers are generally more expensive than regular desktop laser printers. Laser printers for networks are designed for almost continuous duty and can achieve peak output volumes higher than many other types of printers. Many LAN managers have found that less expensive desktop laser printers are insufficient for meeting LAN user printing demands even in small networks. LAN managers have also learned that backup printers should be readily available because printer failures are inevitable.

Dot matrix and inkjet printers can also perform satisfactorily in LAN environments, but they are generally slower than laser printers. Slow printer output speeds can affect the entire network. If a print job is delayed because of a slow printer, other print jobs in the same queue will be delayed, and other network resources could be adversely affected as well. It is generally better to attach dot matrix and inkjet printers to individual LAN workstations than to the print server; this arrangement also provides a relatively inexpensive solution to the problem of users insisting that they see printouts immediately.

Networks are capable of other types of output as well. For example, networks commonly use fax servers to send and receive fax transmissions for network users. Using a fax server avoids some expense in providing a text-file-to-fax converter or fax modem for every workstation on the network. Voice mail applications (and voice mail servers) are also becoming more common in LAN environments.

◆ NETWORK INTERFACE TECHNOLOGIES

To attach a microcomputer to a LAN, there must be an electronic connection between the workstation's I/O bus and the network communication medium. Today, both physical and wireless connections are possible. However, no matter what option is selected, there must be a means of connecting the microcomputer's bus to the network medium.

Some of the many important network interface technologies found in LANs include the network interface cards, connectors, hubs, adapters, and PCMCIA technologies depicted in Figures 4-4 and 4-5. These and other important network interface technologies are discussed next.

NETWORK INTERFACE CARDS

Network interface cards (NICs), also known as network adapter cards, establish the physical link between the workstation (or server) and the network's communication media. If the microcomputer is to communicate with other workstations attached to the LAN, it must be equipped with an NIC that adheres to the LAN's media access control method (CSMA/CD or token passing) and network architecture (Ethernet, Token Ring, Arcnet, FDDI, and so on). Both media access control methods and network architectures are explained more fully in Chapter 5. The software that controls access to the network's communication media is implemented on the NIC.

The NIC essentially acts as a mediator or translator between the workstation and the network. The main job of a network adapter card is to access the network resources that the workstation's user needs and to adhere to the network architecture's media access control methodology.

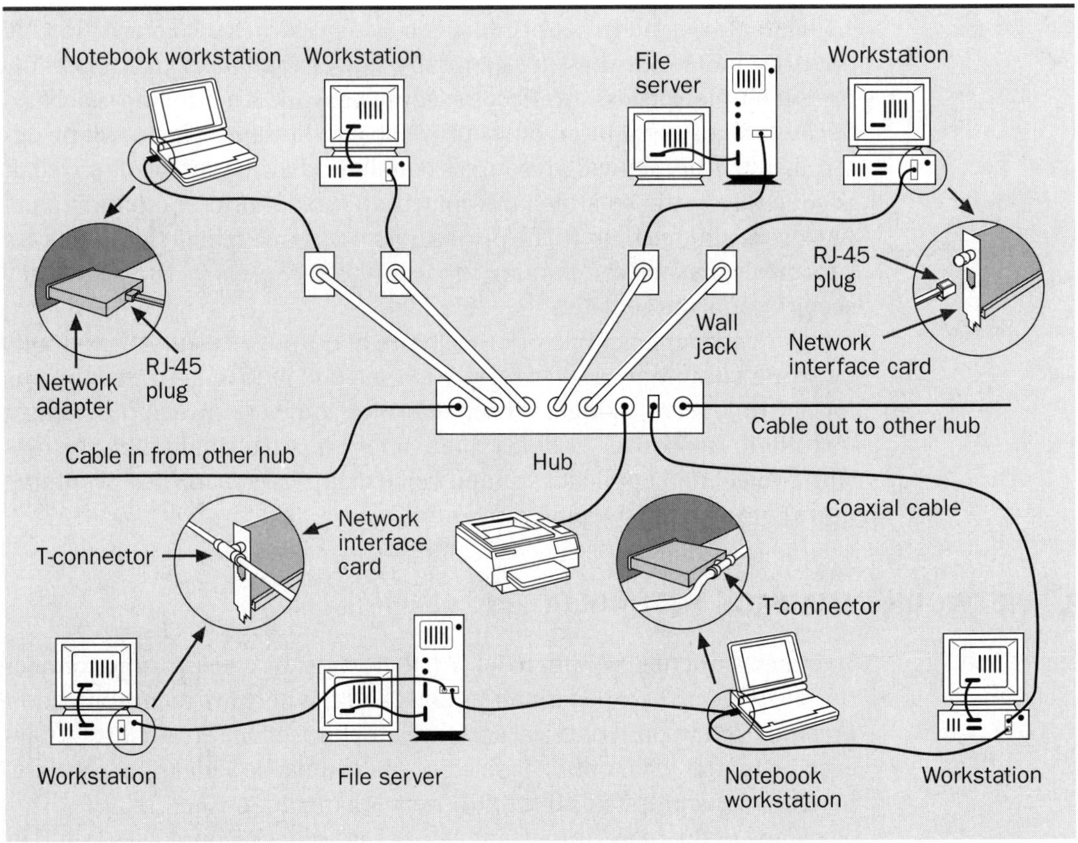

FIGURE 4-4 Network cabling

The bus in the workstation to which the NIC is attached is generically known as the *I/O* (input/output) *bus*, or sometimes the *data bus* or *expansion bus*. This bus is the main data transfer channel to the workstation's CPU and RAM.

The speed and efficiency of data transfers between the NIC and the workstation's RAM are related to the width of the bus—the bus' width determines how many bits can be simultaneously transferred between system components. The width of the I/O bus is a key factor in selecting the NIC needed for a particular workstation. For example, for microcomputers with *ISA (Industry Standard Architecture)* buses, the bus width is either 8 or 16 bits. A machine with an *EISA (Extended Industry Standard Architecture)* bus, however, has a bus width of either 16 or 32 bits. A machine with an *MCA (microchannel architecture)* bus also has a bus width of 16 or 32 bits, as does an Apple Macintosh with its *NUBUS* architecture. In short, when selecting an NIC for a particular workstation, the network manager must know the microcomputer's bus architecture; of course, he or she must also know the network's architecture (Ethernet or Token Ring) and media access control method.

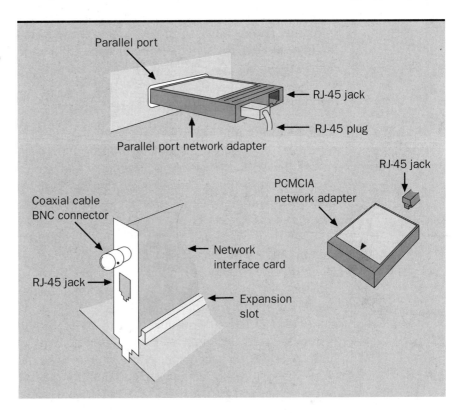

FIGURE 4-5
Network interface cards and adapters

A network interface card may perform several functions. Some of these include:

◆ monitoring activity on the communications media

◆ providing each workstation a unique identification address

◆ recognizing and receiving data destined for the workstation

◆ transmitting data from the workstation

As noted, to carry out these functions, the workstation's NIC must include the electronic logic needed to implement the media access control method used in the network. For example, if token passing is the media access method, the workstation must be equipped with an NIC that supports token passing. If a contention-based media access control method such as CSMA/CD is used, the workstation's network adapter card must be able to support this media access control methodology.

The connections on the NIC must match the physical communication media. For example, if the media uses thin-net coaxial cable, there must be a ***Bayonet-Neill-Corcelman (BNC)*** coaxial cable connection on the NIC; if thick coaxial cable is used, the NIC may need an ***attachment unit interface (AUI)*** connector. When twisted pair cabling is used, usually an ***RJ-45 jack*** (which is similar to an ordinary telephone jack) must be used to connect the cabling. Several NIC manufacturers are offering cards with BNC, RJ-45, and other connections, so that if you move the workstation or decide to change your cabling scheme, you have multiple options for the physical connection.

NETWORK ADAPTER CARD DRIVERS

Another important factor in effective network interfaces is the driver software used for the NIC. This ***network adapter card driver*** software must be compatible with the network adapter card hardware. Also, since the network adapter card may have to be compatible with two or more different network operating systems, most NIC vendors ship multiple drivers with their products.

If an inappropriate NIC driver is used, the microcomputer cannot communicate with the network adapter card, which means that it cannot communicate with the other machines in the network. In effect, an inappropriate NIC driver prevents the workstation from being connected to the network. Network managers must be sure that the NIC they purchase includes appropriate network adapter card drivers for the network operating system(s) used in their LANs. Examples of network card drivers and the network operating systems they work with are summarized in Table 4-1.

Some systems make it possible to monitor the operations of network cards and network card drivers. Novell's NetWare is one of these. Figure 4-6 illustrates how NetWare monitors a LAN driver named TCE32ESH.

◆ OTHER NETWORK INTERFACE TECHNOLOGIES

Other important network interface technologies include transceivers, connectors, resistors, terminators, and alternatives to NICs such as pocket adapters and PCMCIA technologies. These are briefly described in the following sections.

TABLE 4-1
Some network adapter card drivers and NOSs with which they work

Driver Name	Network Operating System(s)
AppleTalk Phase 2	Macintosh System 7, AppleTalk
Internet Packet Exchange(IPX)	Novell NetWare
Network Driver Interface Specification (NDIS)	Banyan Vines, Microsoft's LAN Manager, IBM's LAN Server
NETBIOS	NOSs that understand NETBIOS
Open Data Link Interface (ODI)	Novell NetWare
TCP/IP	UNIX and other NOSs that understand TCP/IP

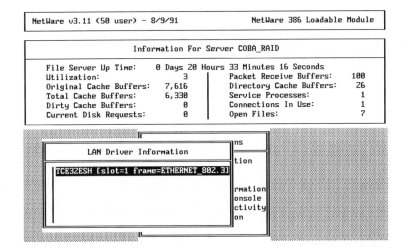

FIGURE 4-6

Monitoring a
LAN driver

TRANSCEIVERS

A *transceiver* is the NIC component that establishes the connection to the communications medium and implements the transmit/receive functions of the media access control method. The term *transceiver* is an abbreviation for the term *transmitter-receiver*; the abbreviation concisely summarizes the primary role played by this network interface technology.

The transceiver's main function is to determine how rapidly data can be transmitted and received by the workstation. This is determined by the speed of the media access control method used, as well as by the network cabling. For example, Ethernet LANs may operate at different speeds and utilize different types of media—there is no single Ethernet transceiver. Rather, different transceivers are needed for different implementations of Ethernet such as 10BaseT, 10Base5, and 10Base2 Ethernet LANs; the distinctions between these are explored more fully in Chapter 5.

In some Ethernet LANs, the transceiver is directly connected to the communications medium. However, in most Ethernet LANs and in most other types of LANs, the transceiver is physically located on the network interface card.

CONNECTORS, RESISTORS, AND TERMINATORS

Some peripheral hardware items must be used with NICs in certain instances. For example, a BNC connector is attached to the network interface card by means of a *T-connector* (see Figures 4-4 and 4-5). The T-connector permits the connection with the computer to be severed without interfering with the network circuit. The circuit continues to run through the top of the T whether or not the NIC is connected to the lower portion of the T-connector.

If the computer is the last in a row of computers connected to the network, a terminator with electrical resistance prescribed by the protocol chosen must be used. In the case of Ethernet, a 50-ohm resistor must be used on the side of the T-connector that is opposite the source cable. In 10Base2 Ethernet networks, resistors must be placed on both ends of the network, and one of these must be grounded.

It is important for network managers to realize that NICs, adapters (such as BNC adapters), and other NIC peripherals can fail. When they do, they may cause significant network problems. For example, if you are investigating a row of computers that cannot communicate with the network and you have exhausted all other possible sources of failure, test the terminator. Replacing the terminator may bring the row of workstations back online. Coaxial terminations, whether thick-net or thin-net, are especially fragile.

If installed properly, twisted pair media is relatively trouble free in operation. However, network managers should always be on the lookout for prefabricated cables with improperly constructed terminations. The wires should be checked to ensure that they are consistent on each end. They are color coded, so comparison is easy.

Initially, LANs required one type of communication media to be used throughout the LAN. For example, if you started using thin-net (thin coaxial cable) in one segment of the LAN, it had to be used throughout the entire network. This is no longer true. Several manufacturers produce repeaters and other interconnection technologies (including bridges, routers, and brouters) that can make it possible to use different types of media in different parts of the network. In addition, wireless LAN segments can be added to cable-based networks.

ALTERNATIVES TO NETWORK ADAPTER CARDS

In the past, NICs were almost always installed in an expansion slot on the motherboard of the workstation. Now this is no longer required. Many portable computers have no space for such cards and, as a result, two alternative means of connection have emerged—these are illustrated in Figure 4-5. There are likely to be more in the future. One popular item is the pocket parallel port adapter—the pocket being where most people carry it. This type of network interface is designed for portable computers; it is used by technicians and administrators who need to quickly hook up a workstation to the network or troubleshoot the functionality of an NIC by providing a bypass around the regular interface.

For the most part, the pocket parallel port adapter installs in a plug-and-play mode. These adapters usually employ a slower rate of transmission than does a regular NIC, but the bidirectional parallel ports available on newer machines can provide extra speed through these connectors.

A second method of connection for machines not capable of using expansion cards are the PCMCIA (Personal Computer Memory Card International Association) slots found on newer portable computers. These connectors have been dubbed "credit card slots," reflecting the shape of the connector used. Most experts predict that even desktop machines will soon incorporate PCMCIA-compatible components for network attachments.

◆ HUBS AND WIRING CLOSETS

In some networks, once the computer is connected to the medium, it becomes an active node on the network. However, in other LANs, wiring **hubs** such as the one shown in Figure 4-4 are needed to establish workstation-to-workstation communications. For example, in IBM Token Ring LANs, the individual workstations are connected to wiring hubs called *multiuser access units (MAUs)*. In addition, some networks employ different types of hubs. Arcnet, for example, utilizes both passive and active hubs. In Arcnet LANs, **passive hubs** provide a limited number of connections and offer no signal regeneration; this limits workstations to being no more than 100 feet from the hub. However, since **active hubs** have signal regeneration capabilities, individual workstations can be located up to 2,000 feet away. In addition, active hubs can handle a significantly larger number of connections. Larger Arcnet networks and those covering longer distances almost always require active hubs.

Hubs are often located in *wiring closets* that may also house telephone equipment and heating/air conditioning equipment. The installation of hubs and wiring closets usually requires expertise in network implementation. The success of the network is often dependent on the types of hubs used, the number of workstations and peripherals connected to the hubs, the length of the cables going to individual workstations, the total cable length, and the type of communication medium used. Because of this, network managers should always review hub vendor recommendations about cable lengths, number of cable connections, and so on. They should also be wary about pushing these limits; staying within vendor specifications is usually the wisest course of action.

◆ SERVERS

In addition to terminal equipment and network interface technologies, servers play a very important role in most of the networks found in organizations today. The characteristics of servers have changed. The trends observable in server technologies are briefly discussed in this section.

FILE SERVERS

In many of the early LANs, server functions were performed by one of the microcomputers attached to the network. This arrangement essentially fits the description of non-dedicated servers provided in Chapter 3. While this configuration may be adequate for smaller LANs, as networks grow (in terms of the number of attached workstations), it quickly becomes cumbersome and begins to impair network performance. Because of the detrimental

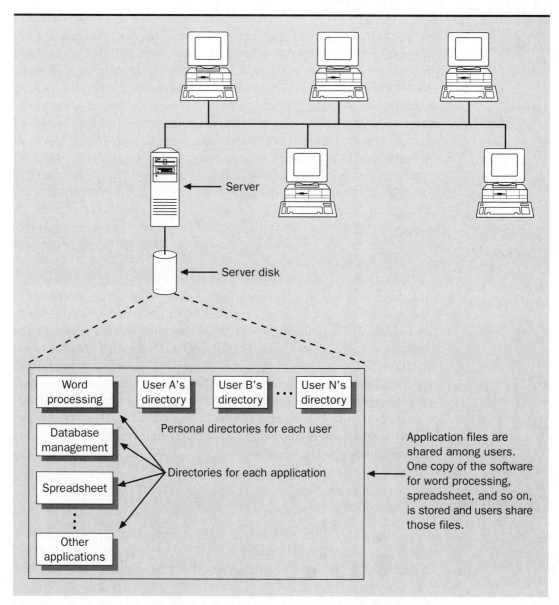

FIGURE 4-7 File server technology

effects that can occur, nondedicated file servers are most appropriate for very small LANs, and some NOS vendors have ended their support for this approach. For example, Novell, the vendor with the largest market share of NOSs for LANs, stopped supporting the concept of a shared workstation/file server when it released NetWare version 3.xx.

The primary function of a file server is to assist the workstation in loading programs and accessing data on the file server disk (see Figure 4-7). However, the server software also functions as a network traffic director by coordinating access to shared files. Several figures have been included to illustrate how server software assists the network management function. For example, Figure 4-8 illustrates how NetWare can be used to access and maintain the list of available servers from a remote console. Figure 4-9 illustrates an entry in Novell NetWare's File Server Information file that

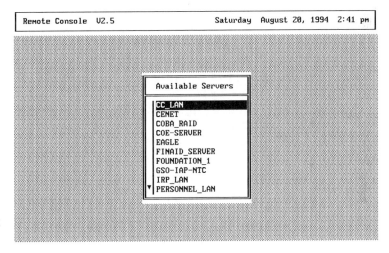

FIGURE 4-8
Monitoring
available network
servers

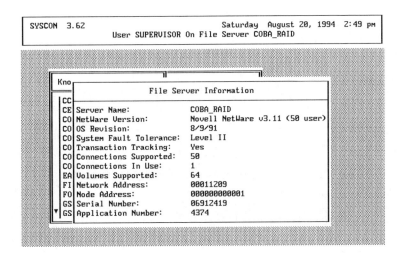

FIGURE 4-9
Network
and server
configurations

stores data on the configuration of servers in the network. Figure 4-10 illustrates the supervisory options that network managers can exercise for a file server in NetWare LANs. And, Figure 4-11 shows a listing of the system modules that have been loaded for a particular file server.

Many software management issues can be facilitated by file server software. One of these is ensuring that the organization adheres to its license agreements with software vendors (license agreements are discussed more fully in Chapter 11). A license essentially dictates how many network users can concurrently (simultaneously) be using the software product. For example, if you have a license for a maximum of 10 concurrent users of WordPerfect, the server software can be configured to reflect this and enforce the license (by not allowing an eleventh user access to WordPerfect if 10 other users are currently using the package).

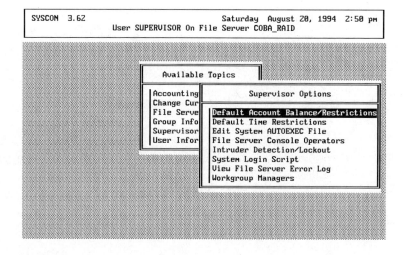

FIGURE 4-10

File server supervisory options

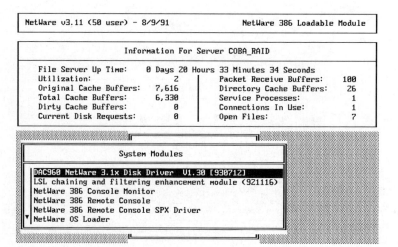

FIGURE 4-11

System modules

Over time, file server software has incorporated increased redundancy and fault tolerance into the server management system to prevent total server and network crashes. For example, even in its early versions, Novell included modules in NetWare for creating duplicate file allocation tables, the index to the files on the disk. Additionally, Novell introduced the *hot fix* concept, whereby when the file server senses a physical deformity on the disk, the data is removed to an area of the disk without blemishes. NetWare's hot fix capability is illustrated in Figure 4-12. Novell has organized its redundancy and hot fix features into fault tolerance levels describing the extent to which the network should resist a total crash.

Currently, most of the major NOSs for LANs support the online duplication of disks and disk controllers (duplexing) and many support server duplexing. With server duplexing, a second machine that duplicates the activities of the primary server is available; if the primary server fails, the secondary can march right on.

DISK SERVERS

A disk server enables a group of workstations to concurrently access a disk or group of disks (Figure 4-13). Disk servers were quite common in some of the early LANs but are becoming less common as file servers have taken their place.

For the most part, disk servers allow a given workstation to treat the disk as a logical peripheral with limited sharing of certain common access areas. Disk servers have both public and private storage areas. All LAN users can access files in the public areas; however, the files in the private areas are accessible only to individual users (at least in theory). Because many disk server systems lack file management tools such as file and

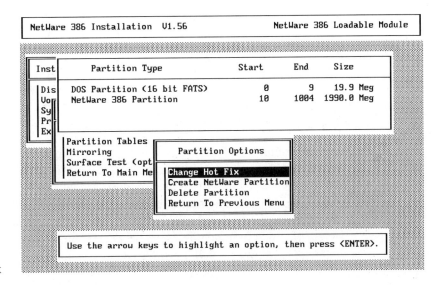

FIGURE 4-12
NetWare's hot fix

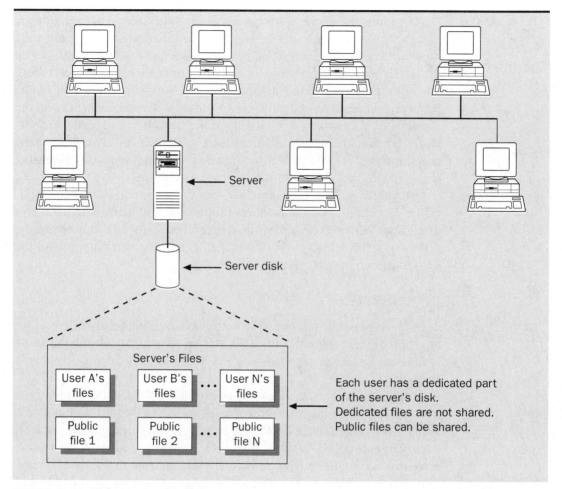

FIGURE 4-13   Disk server technology

record locking, their ability to protect against accidental erasure and over-writing of files is limited. As file servers usually include these and other file management tools, it should not be too surprising to learn that most systems in commercial use now use file servers rather than disk servers.

DATABASE SERVERS

In organizations downsizing their computer platforms from mainframes to LANs, database servers have become more common. The main function of database servers is to process user requests for data, to manipulate data, and to facilitate the storage of new data entered by users (Figure 4-14). In brief, database servers become responsible for many of the functions performed by the database management system (DBMS) on a centralized host. As you will read in Chapter 9, SQL (Structured Query Language) servers—one type

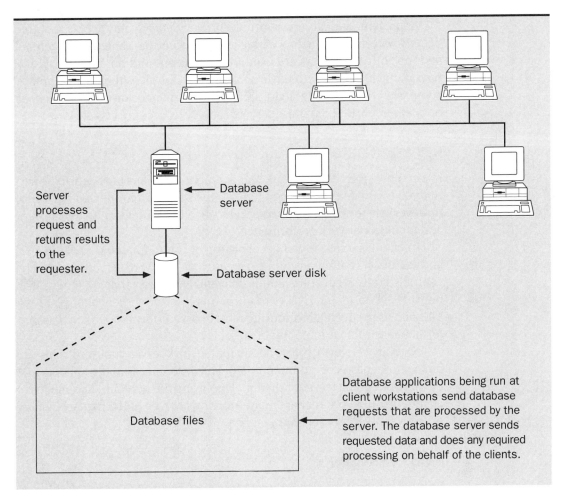

Server processes request and returns results to the requester.

Database server

Database server disk

Database files

Database applications being run at client workstations send database requests that are processed by the server. The database server sends requested data and does any required processing on behalf of the clients.

FIGURE 4-14 Database server technology

of database server—are especially popular in client/server networks in which data (or data requests) can originate in multiple network segments with different NOSs. With the emergence of client/server networks in many organizations, LAN managers should anticipate learning more about database servers.

OPTICAL DISK SERVERS

If data needed is stored on optical disks, an *optical disk server* may be used in the LAN. Optical disks are becoming more popular; many vendors of commercial databases are marketing their information on CD-ROM or other types of optical disks largely because the disks can store very large volumes in several formats (including audio, video, text, and graphics). Once information is stored, optical disk servers can provide an efficient and relatively inexpensive distribution system.

At present, optical servers primarily provide read-only access to stored data. However, in image processing and electronic document management systems, optical servers are used for both input and data manipulation operations in addition to retrieval activities; this gives them functionality comparable to that of database servers. Over time, the distinction between optical servers and database servers is likely to become less clear.

NETWORK SERVERS

In a stand-alone LAN, network management functions are likely to be carried out by the file server. In organizations with interconnected networks, however, the term *network server* refers to a computer that helps monitor and manage network performance.

In client/server networks consisting of interconnected LANs, WANs, and servers, it is often desirable to create a hierarchy of servers that can be centrally controlled. For example, Novell's NetWare version 4.0 permits a network server in one LAN to monitor one or more servers in other LANs. This allows for overall monitoring of network performance as well as performance within various subnetworks.

NetWare version 4.xx also allows for network server duplexing. This can build another level of redundancy and fault tolerance into the network and the network management function. Duplexing the network server helps to ensure that network management functions can be performed even if the primary network server should fail.

PRINT SERVERS

Printing is one of the slowest operations that occurs in any computing system, including LANs. From the very beginning, the sharing of printers has been common in networks and so have printer servers (although they may not have been called such).

Essentially, a print server provides printing services for workstations on the network that do not have attached printers (or lack attached printers of the same type or quality as those controlled by the printer server). This service can be provided at different levels. For example, Windows for Workgroups allows a workstation connected to a network to send output to be printed to a queue serviced by a printer anywhere in the network. This enables the workstation to transmit print output as a background task while the user continues work on another task in the foreground. Once the Print Manager in Windows has released the print job to the print queue, the source workstation has no further responsibility. The network's print server will store the print job on a file server disk until it is ready to actually direct the job to a physical printer.

Even without Print Manager functions such as those found in Windows or OS/2, the network administrator can manage the print server function for the workstation. For example, in some client/server LANs, the printer server is another task handled by the file server. Although this may degrade other file server operations, it may be a cost-effective implementation of print server functions in smaller LANs. In other small- to moderate-sized LANs, a nondedicated print server may be used, that is, a PC used as a workstation also carries out print server operations. In larger networks, however, the print server is almost always a dedicated microprocessor and can control network printing operations on a number of printers. For example, in version 3.xx of Novell's NetWare, a single printer server can serve 12 to 16 printers; version 4.xx allows 25 printers to be attached to a single print server.

If users need a variety of printed outputs (text-based, graphics, and intermingled text and graphics), print functions can be a management challenge. In addition, in many instances, changing the size or color of the paper used in the printer is done by hand. Manual changes are also likely to be needed when organizations use preprinted forms (such as invoices, statements, and billing notices). Hence, in many of today's networks, it is important for the printer server software to prompt the network user to change printer parameters and feeds when necessary.

Other print server management issues follow.

Print Volume

When network printers are expected to produce high volumes of printer output, it is important to ensure that there is an adequate path from the disk that holds the print file to the printer itself. Continuous-duty printing often requires mechanical connections capable of sustaining heavy-duty work. In addition, the print queue within the computer handling printer server operations must be large enough to store all the data waiting to be delivered to the printer. Because of this, network managers typically must have good estimates of peak print volumes when setting the parameters for print queues. Also, in order to enhance print output efficiency, the network manager may configure the printer server to direct some types of jobs to certain printers and other types of jobs to other printers. For example, quick, small jobs may be directed to an inexpensive printer rather than to a more expensive printer. With two or more printers, more jobs can usually be handled in a given time period, even if paper or style changes are required.

Physical Distribution

Some networks' workstations are widely dispersed. In such networks, a print job could be physically executed on a printer located miles away from the user workstation. As you might expect, with this type of arrangement network managers have to be more concerned about coordinating the manual tasks associated with printing operations (such as paper changes, printer jams, and form changes), because after releasing print jobs to the printer server, users are not likely to be aware of problems.

Job Sequencing

In today's world, network users are likely to need both character-based documents and graphics-based documents. And, if there is only a single printer, it may have to be reset between jobs. One of the ways around such a problem is to direct character-based and graphics-based jobs to different printers. Another possible solution is to manage the print queue so that resets between jobs are minimized (that is, several character-based print jobs are printed one after another before the printer server prompts the network staff to reset the printer settings for a batch of graphics-based jobs).

Directing output to specific printers can often be performed with network versions of popular application programs. For example, Lotus Organizer for Windows output can be directed to a specific printer through the application package's interfaces with NetWare (see Figure 4-15).

OTHER TYPES OF SERVERS

A number of other servers are becoming more common in LANs. For example, communications servers can provide dial-in access for remote users; those with dial-out capabilities can allow users to access computers in

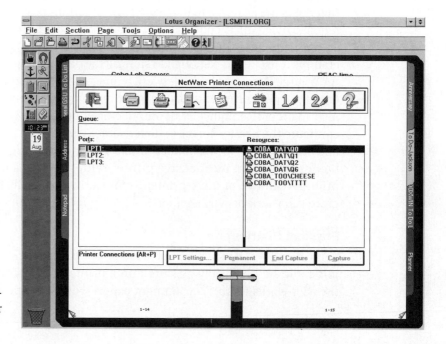

FIGURE 4-15
Printing connection windows for the network

other networks, as well as external databases and information services such as those available from CompuServe or the Dow Jones News Retrieval Service. Also, as noted, fax servers and scanner servers are increasing in popularity.

In general, experts expect the variety of server technologies found in LANs and other networks to increase. Network managers need to stay abreast of these developments to determine if the advances provide cost-effective solutions to computing needs in the networks they administer.

◆ IMPORTANT TRENDS IN NETWORK TECHNOLOGIES

Because networking is one of the hottest topics in most businesses, you can expect to see better, more sophisticated, and more user-friendly network solutions in the years ahead. For example, some networks now provide safeguards for data and files that previously were available only in mainframes. In addition, network security, and fault tolerance are being enhanced through the incorporation of more redundancy (server duplexing, RAID technologies, disk mirroring, and so on) and backup features. Network management systems (the software that network managers use to monitor and enhance network performance) are also becoming more sophisticated and able to manage interconnected networks.

Like LAN workstations, servers are becoming faster and more powerful largely due to advances in processor technologies and the trend toward 32-bit architectures for all computing equipment. In addition, communications speeds in LANs will be increasing in the years ahead with developments like Fast Ethernet (with speeds up to 100 megabits per second), CDDI (copper distributed data interface), a copper cabling cousin to FDDI that should support speeds up to 100 megabits per second), and ATM (asynchronous transfer mode), that will allow the concurrent transmission of multimedia files at speeds over 100 megabits per second. Even wireless transmission speeds are expected to at least double by the turn of the century.

◆ SUMMARY

Terminal equipment is one of the most important components of LANs and other networks. In LANs, most of the workstations are microcomputers. However, video display terminals with or without processing capabilities of their own may also be included in LANs.

The important issues that network managers should consider when selecting terminal equipment include the transmission and processing speeds needed, the types of user input interfaces required, the types of outputs needed by users, and whether the cost of equipment is justifiable.

Because microcomputers and video display terminals are the two most popular types of workstations found in networks, keyboards are the most common type of network input technology. Other important input technologies include the mouse, trackball, reader/scanner devices, and specialty input equipment such as digital cameras, fax technologies, light pens, touch screens, and wireless personal digital assistants (PDAs).

The workstations in LANs and other networks are most likely to use Intel microprocessors such as the 80386, 80486, and Pentium chips. However, microprocessors manufactured by other chip makers such as Motorola—which supplies chips for Apple Macintosh computers—are not uncommon. PowerPC chips have been predicted to show significant growth in network usage in the 1990s. A number of internetworking and software challenges have been posed by the increasing variety of processor technologies found in LANs.

Storage technologies are also changing in network environments. For example, the average capacity of the hard disk systems found in LAN workstations continues to increase. The trend toward multimedia applications is one of the driving forces behind the need for larger storage systems in LANs; it is also one of the reasons why optical storage technologies are becoming more common. RAID (redundant array of independent disks) systems are also increasing in popularity in LANs because of their high storage capacities and because they provide disk striping, disk mirroring, and disk duplexing capabilities that add redundancy and fault tolerance to networks. Tape backup systems are also quite common in LANs; these are generally increasing in both capacity and sophistication.

Video display and printed output are the most common types of LAN outputs. Because of differences in the output needs of users, managing LAN printing operations and print queues can pose special challenges to network administrators. In many LANs, either dedicated or nondedicated print servers handle the automated portions of printing operations. However, in some LANs, the file server is also used as the print server.

Laser printers are the most popular type of printer found in LANs. However, dot matrix and ink jet printers are also quite common. Fax servers and voice mail applications are other increasingly popular output technologies.

Network interface technologies play a crucial role in LANs and other networks. These collectively establish the physical connection between a microcomputer and the communication medium in the network that enables the computer to communicate with other workstations.

A network interface card (NIC) is installed in an expansion slot on the PC's motherboard. NICs must conform to the media access control method used as well as to the network architecture. In addition, the NIC must be appropriate for the microcomputer's I/O bus architecture and must have the physical interface needed to connect it to the network's communications medium.

NIC drivers are the software that enables the microcomputer to utilize the NIC to communicate with other networked workstations. NIC drivers must be compatible with the network operating system used; if the wrong driver is installed, the workstation will not be able to communicate with other networked machines.

Other network interface technologies include transceivers, which establish the physical connection to the communications media and implement the transmit and receive functions of the media access control method used. Connectors (such as BNC connectors, T-connectors, and RJ-45 jacks) are essential for physically interfacing with the communications medium. In addition, resistors and terminators are used to implement the electrical end points of the network.

Alternatives to NICs include parallel pocket adapters and PCMCIA technologies. Both of these are commonly used to connect portable computers to networks; however, they can also be used with desktop systems.

Hubs and wiring closets may also be used to establish workstation-to-workstation connections. In IBM Token Ring LANs, multiuser access units (MAUs) are used as hubs. Arcnet LANs have both passive and active hubs; as active hubs can regenerate signals, workstations can be located much further away than they can be with passive hubs.

Of course, servers are also important network components. There are both dedicated and non-dedicated servers. In addition, in some networks, one machine may carry out more than one type of server operation. For example, the same computer may be used to perform file server, printer server, and network server functions. Database servers, optical disk servers, and specialty servers such as fax servers are all becoming common in LANs.

In general, servers are becoming faster and more powerful. LAN technologies are incorporating more security, redundancy, and fault tolerance, and the network management systems used to monitor and enhance network performance are becoming more sophisticated. Media speeds are increasing, and LAN speeds of 100 megabits per second are expected to be commonplace by the year 2000. These changes guarantee that network managers will be faced with many technological challenges in the years ahead.

✳ KEY TERMS

Active hub	Hot fix
Attachment unit interface (AUI)	Hub
Bayonet-Neill-Corcelman (BNC)	Mouse
Communications server	Network adapter card driver
Dedicated print server	Network interface card (NIC)

Network server
Optical character recognition (OCR)
Optical disk servers
Optical read/write disk
Passive hub
Print server
Reader/scanner device

Redundant array of independent
    disks (RAID)
RJ-45 jack
T-connector
Terminal
Transceiver
Workstation

✳ REVIEW QUESTIONS

1. Describe the criteria that network managers consider when selecting terminal equipment.
2. What may the term *workstation* mean to different network managers?
3. In addition to keyboards, what other types of input technologies may be found in LANs? Briefly describe the characteristics of each.
4. Identify the major processor technologies found in LANs. What approaches may be used to facilitate communications between computers with different processors?
5. How is the trend toward multimedia systems contributing to the trend toward high-capacity storage systems?
6. What are RAID technologies? Why are these becoming more prevalent in LANs?
7. Describe the technologies used in LANs for archiving or backing up data, files, and applications.
8. Identify the optical disk technologies becoming more common in LANs. Why are these popular?
9. Identify the major types of output technologies found in LANs.
10. What is the difference between a dedicated and a nondedicated print server?
11. Why are laser printers common in LANs?
12. What functions are carried out by print servers?
13. What are the advantages and disadvantages of using a file server to carry out print server operations?
14. What are the characteristics of network interface cards? What functions are performed by NICs?
15. What factors must be taken into account when selecting a network interface card for a particular microcomputer?
16. What are network adapter card drivers? Identify several examples. What factors must be considered when selecting network adapter card drivers?
17. What is the role of a transceiver?
18. What are the roles of connectors, resistors, and terminators in LANs?

19. What alternatives to NICs are available? Describe how these differ from one another.
20. What are hubs? Why are hubs important? What is the difference between active and passive hubs?
21. Identify the different types of servers that may be found in LANs. What trends may be observed for each type?
22. What are the print server management issues important to network managers? Briefly describe why each is important.
23. What are some current trends in networking technologies?

✳ DISCUSSION QUESTIONS

1. Assume that a new output peripheral has been invented and that its initial market acceptance is exceptionally good. As manager of a large corporate LAN, you view the new peripheral as necessary, but your budget is insufficient for purchasing one for each user. Describe the approaches you would use to make this peripheral available to LAN users without purchasing one for each workstation.
2. Assume that you are faced with almost 80 percent of your users requesting a fax machine, each user having a station on the LAN. What would you recommend, assuming that almost all the users said that the fax traffic they would originate would largely be documents in their word processing program on the LAN?
3. Assume that you are the Chief Information Officer for a major corporation. You have been requested to supply a five-year forecast for technical support personnel in two categories: software support and hardware support. Discuss what increase or decrease you might project in each category and why you would project an increase or decrease. Also, describe how the roles of hardware and software support personnel are likely to change over the next five years.
4. Why are redundancy and fault tolerance increasingly important in LANs?

✳ PROBLEMS AND EXERCISES

1. Invite a local network manager to come to your class to discuss how workstations are physically connected to the communication media used in the networks in the organization.
2. Invite a network consultant or network manager to speak to your class about the current trends in the networking technologies used in your local area. Determine how these predictions differ from the trends described in this chapter.

3. If network connections are available on your campus, go to the lab and learn how to logon through a workstation. Write down the steps and bring them to class.

4. Find a recent article comparing print servers in a networking or computing periodical. Summarize the contents of the article in a paper or presentation.

5. Find a recent article on NIC drivers in a networking or computing periodical. Summarize the contents of the article in a paper or presentation.

✳ REFERENCES

Ambrosio, J. "Warehouses Cling to Mainframe." *Computerworld* (August 23, 1993): p. 91.

Baker, L. and E. Bowden. "Spotlight on 100MHz Cable Testers: Latest Crop of Cable Testers Let Network Managers Address the Strict Demands of Category 5 Wiring." *LAN TIMES* (March 14, 1994): pp. 73, 74, 81.

Ballou, M. "Client/Server Enters Mission-Critical Game." *Computerworld* (November 15, 1993) p. 112.

Bozman, J. S. "Low-Cost Workstations Flood Market." *Computerworld* (January 17, 1994): p. 51.

Brandel, W. "Rush Is on for Plug-and-Play Leadership." *Computerworld* (May 2, 1994): p. 81.

———. "HP Pursues Network Storage Market." *Computerworld* (April 18, 1994): p. 82.

Csenger, M. "Computerm Acquires McData LAN-Mainframe Gateways." *Communications Week* (April 25, 1994): p. 31.

Dern, D. P. "TCP/IP for Ma and Pa: A Simple Solution at Last." *LAN TIMES* (August 9, 1993): pp. 35, 40, 42.

DiDio, L. "Remote Routers to Make Debut." *Communications Week* (April 25, 1994): p. 41.

———. "3Com Stacks Hub With Flexibility." *Communications Week* (April 25, 1994): p. 1.

———. "Cisco Routers to Be Included in New Stackable Hubs from LanOptics." *Communications Week* (May 2, 1994): p. 53.

———. "Switching Hubs Push Routers to the Edge." *Communications Week* (May 2, 1994): pp. 1, 14.

———. "Vendors are Stacking up Their Hubs." *Communications Week* (April 25, 1994): p. 31.

———. "Racal-Datacom Enhances Its Ethernet Line: New Workgroup Filtering Should Reduce LAN Bottlenecks, Data Collisions." *LAN TIMES* (November 15, 1993): p. 7.

Dostert, M. "Networking Exotic Peripherals Still Difficult." *Computerworld* (June 14, 1993): p. 65.

Dryden, P. "NICs Evolving as New Needs Surface: High-Speed Technologies Race to Be Next Major Desktop Delivery System." *LAN TIMES* (June 26, 1993): pp. 1, 109.

———. "Lexmark Ships Interactive Print-Server Ware." *Communications Week* (April 25, 1994): p. 23.

———. "Intel Enhances Print-Server Line." *Communications Week* (April 18, 1994): p. 25.

Ellis, P. "Matching Array Technology to Application Requirements." *Network World* (May 2, 1994): p. L.16.

Fitzgerald, M. "Apple Pins Server Hopes on PowerPC." *Computerworld* (April 25, 1994): p. 12.

Fogarty, K. "Xerox Offers Print Services Middleware." *Network World* (April 25, 1994): p. 4.

Graziano, C. "Vendors Work to Combine Storage Management and Network Management." *LAN TIMES* (February 14, 1994): pp. 29, 30.

———. "EFI Software to Monitor Mix of UPSes." *LAN TIMES* (August 8, 1994): pp. 47, 51.

———. "Power-Management Software: As Networks Grow, Administrators Get a Handle on Uninterruptible Power Supplies." *LAN TIMES* (June 27, 1994): pp. 47, 50.

———. "Companies Sign Flat-Fee Licensing Agreement." *LAN TIMES* (June 26, 1993): p.73.

Griffith, C. and B. Homer. "Reconsidering RAID: Today's Low Disk-Drive Prices Narrow the Benefits of RAID." *LAN TIMES* (January 24, 1994): p. 75.

Harper, E. "Caching in on Network Performance: An In-Depth Look at How the Disk I/O Channel Affects Performance and What Role Caching Controllers Play." *LAN TIMES* (October 18, 1993): pp. 81, 84.

———. "Finding a Hard-Disk System That Works for You: In the Old Days, It Was a Breeze; Now It's a Battle Among ESDI, IDE, and SCSI." *LAN TIMES* (October 18, 1993): p. 89.

Harrison, P. J. "Optical Storage: Where Is It Now?" *LAN TIMES* (October 18, 1993): p. 50.12.

Klett, S. P. "Remote Router IQ Rises." *Computerworld* (April 11, 1994): p. 12.

———. "Hub Encroaches on Router Turf." *Computerworld* (February 7, 1994): p. 16.

"Magneto-Optical Disc Servers." *LAN Magazine* (November 1994): p. 81.

Margolis, N. "Client/Server Ship Docks in Seattle." *Computerworld* (July 12, 1993): p. 6.

Musich, P. and W. Pickering. "SynOptics Unwraps Fast Ethernet Hubs." *PC Week* (April 25, 1994): p. 8.

Rauen, C. "Standards: Blueprint for the Future or Roadblock to Progress?" *LAN TIMES* (August 23, 1993): pp. 73, 76.

Roberts, E. "Hardware Investments May Dictate Hub Choices." *LAN TIMES* (March 14, 1994): pp. 43, 46.

———. "HP Delivers Fruits of 110VG Labor: New Stackable Hub and Routers Target Central and Remote Sites." *LAN Times* (May 9, 1994): p. 35.

Streeter, A. "Internetwork Printing Made Easier: New Hardware, Software Ease Laser-Printer Hook-Up to Mulitprotocol Environments." *LAN TIMES* (November 15, 1993): p. 50.

Wexler, J. M. "'Superhubs' Invade Router Turf." *Computerworld* (October 4, 1993): p. 12.

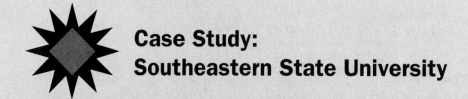

Case Study:
Southeastern State University

As noted in the previous segments of this case discussion, the student computer labs at SSU have a range of microcomputers acquired over the years by the university. As a result, several generations of microprocessors are found in most student labs and LANs. There are pre-80X86 processors (8086 and 8088) still in use in several computer labs, though these are being replaced as quickly as budgets allow. There are also several generations of processors manufactured by Motorola in the Apple II and Macintosh computers found on campus; these tend to be concentrated in the College of Education. At the other end of the spectrum are a handful of Pentium processors, machines with PowerPC chips (primarily high-end Macintosh computers), and SPARC workstations (found in the College of Engineering). At present, there are more 80386 microcomputers in use at the university than any other type.

The university has ceased having its older microcomputers repaired if they break down, largely because university officials can no longer justify the expense of prolonging the lives of what are essentially outmoded computers. This repair policy contributes to the high attrition rates of such machines. However, in most instances, older machines continue to be used until they experience a breakdown that requires them to be taken out of service.

LAN managers at SSU have mixed feelings about not repairing older computers when they break down. While, on the one hand, such breakdowns mean that SSU students utilize computers that are closer to being state of the art, when there is already a general shortage of computers for student use on campus, the loss of any machine hurts. However, most LAN managers are not sorry to see them go. Many complain that the older machines with their 8- and 16-bit bus architectures are difficult to incorporate into the LANs. These network administrators note that most of the older computers are either incapable of running the latest DOS and Windows versions of software packages that the university has standardized on, or run them at speeds that are frustrating to users. They complain that users often can't tell the difference between older and newer machines and cause their workstations to lock up when they attempt to access a version of an application that their machine cannot handle. They also complain that the differences in processor technologies and bus architectures found in workstations have forced them to support multiple DOS and Windows versions of the application packages on which the university has standardized.

Such complaints are not limited to LANs using a particular brand of network operating system. It is a widespread concern heard from managers of Novell LANs (the most common), LAN Manager LANs, Banyan Vines LANs, and even managers of AppleTalk LANs. In addition, the concern exists for both stand-alone LANs and those with TCP/IP connections to VAX computers and the Internet. In brief, the issue is important among all the LAN managers on campus including those overseeing coaxial-cable-based Ethernet LANs, those administering twisted pair-based Token Ring LANs, and those whose LANs employ fiber optics as the communication media.

Because of price differences among network interface technologies, most of the LANs on campus are coaxial-cable Ethernet LANs; most use thin coaxial cable, but thick coaxial cable is used in some LANs. The NICs, connectors, terminators, resistors, hubs, and so on for Ethernet LANs typically cost less than half of the per-workstation connection cost of Token Ring LANs. And, the per-workstation costs for fiber-based LANs are almost invariably higher than for Token Ring LANs.

Alternatives to NICs are finding their way into some of the LANs on campus, primarily the ones connecting faculty and staff computers. For example, SSU's Admissions Department has started to replace its desktop machines with high-end portable computers. These portables can be connected to the LAN in the Admissions Department through parallel pocket adapters at "docking stations" that access normal-sized keyboards and monitors. This arrangement makes it possible for an admissions staff person to take the same machine and software (stored on the portable's hard disk) used in the office either home for use after normal work hours or on recruiting visits to college fairs and graduate school fairs at high schools and universities. The machines that recruiters are equipped with are used for multimedia presentations at recruiting functions and can be used to create and submit fax versions of SSU's application forms. Recruiters at recruiting functions can also access databases in the Admissions Department's LAN in order to check the status of an application submitted by a potential new SSU student.

CASE STUDY QUESTIONS AND PROBLEMS

1. In light of the information provided in the case study and the chapter, identify and briefly describe several ways in which the incompatibilities among processor technologies could be addressed by SSU.
2. Based on the information provided in this segment of the case study, inventory the different types of workstation bus architectures, NICs, NIC drivers, transceivers, and connectors found at SSU.

3. Briefly describe how the NIC drivers used in SSU's stand-alone LANs differ from those used in LANs connected to the university's VAX computers.

4. Based on the description provided, discuss the likely configuration of the portable computers used by recruiters; that is, discuss the characteristics of their storage systems, as well as the input, output, and communication technologies used in these machines. What could be used at the docking stations instead of parallel pocket adapters to connect the portable to the network?

5. Discuss the general advantages and disadvantages of replacing employee desktop systems with portable computers and docking stations.

6. Based on the information provided in this segment of the case study, identify and briefly describe the types of servers that may be found in the LAN in SSU's Admissions Department.

LAN Protocols and Operating Systems

CHAPTER OBJECTIVES

After completing this chapter, you will be able to:

◆ Describe the difference between system software and application software.

◆ Discuss the characteristics of communications protocols.

◆ Describe the differences between polling, contention, and token passing.

◆ Identify some of the major network standards and standard setting organizations.

◆ Briefly describe the OSI reference model.

◆ Briefly discuss the importance of TCP/IP.

◆ Briefly discuss the IEEE.802 standards.

◆ Briefly describe the FDDI standard.

◆ Describe the major functions performed by LAN OSs.

◆ Identify the major OSs used in client/server LANs.

◆ Identify the major OSs used in peer-to-peer LANs.

◆ Describe some of the important trends in LAN protocols and OSs.

In this chapter, we will focus on the software that enables LAN users to communicate with one another and to run the applications that they need to perform their jobs. We will begin by discussing communications protocols—the rules that determine when an individual workstation is allowed to transmit data to another computer or peripheral device in the LAN. We will then discuss the functions that LAN operating systems (LAN OSs) carry out as well as some of the specific OS products found in both client/server and peer-to-peer LANs. The chapter will conclude with a discussion of some important trends that may be observed in LAN protocols and operating systems.

◆ SYSTEM SOFTWARE VERSUS APPLICATION SOFTWARE

The computer programs that are found in LANs can be broken down into two main categories: system software and application software. Examples of the types of software falling into each category are shown in Table 5-1.

Application software includes the user-oriented programs that enable users to carry out their work activities. As we have already noted in previous chapters, there are two major types of application programs: general-purpose and special-purpose. General-purpose application software includes programs and packages that are likely to be used by workers in all (or most) of an organization's subunits—such as word processing, spreadsheets, and database packages. Special-purpose application software includes programs and packages that are likely to support the activities in a particular subunit of an organization; accounting/bookkeeping programs, inventory management programs, payroll software, and production scheduling programs are typically classified as special-purpose application software.

System software consists of "background" programs that enable application software to run smoothly on the computer hardware that is being used. System software provides an interface between application software and the computer hardware and carries out many of the "housekeeping"

System Software	Application Software
LAN Operating Systems (LAN OSs)	General-purpose:
Media interface software	Word processing
Redirector modules	Spreadsheet
	Database management
Server system software	
Workstation software interface	Special-purpose:
	Groupware
Print spoolers	Accounting
Backup utilities	

TABLE 5-1
LAN software

tasks that users need on a regular basis, such as physically storing a file on a disk or transmitting it to a printer for output. The most important piece of system software found in any computer system is the operating system. Utilities and other kinds of system software also play important roles in virtually any type of computer system.

The system software found in LANs is essentially an extension of the system software used for a stand-alone microcomputer. LAN system software provides the interface through which user applications can access shared computing resources without having to worry about how the hardware-oriented details (such as directing a file to the appropriate printer or obtaining an application program from the file server) are actually carried out. As you will see, LAN system software is found in workstations as well as servers.

Later in the chapter we will discuss LAN operating systems in more detail. But before we do, it is important to discuss the communications protocols found in LANs. Because protocols determine how LAN devices are allowed to communicate with one another, they have a major impact on how the network operating systems (NOSs) for LANs operate. In fact, most LAN NOSs are designed to support one or more of the media access control methods that we will discuss in the next section.

◆ WHAT ARE COMMUNICATIONS PROTOCOLS?

The workstations within a LAN are able to communicate with each other, with servers, and with shared peripheral devices because of a protocol. In general a communication **protocol** governs the flow of data between sending and receiving workstations; it is the set of rules that determines how and when workstations are allowed to transmit or receive data, how the data is formatted, and how error checking tasks are performed. Protocols also typically include rules for message lengths, media access, and addressing. In brief, protocols are the communications rules that control the transmission and reception of messages in data communication networks.

The protocols used in data communication networks are often mapped into the layers of the *OSI* (Open Systems Interconnection) *reference model*. Because of this, articles found in networking and computing periodicals often address network layer protocols, data link protocols, and so on, as well as protocols for devices that operate at the different levels of the OSI model (for example, routing protocols for routers). However, for LANs, data link layer protocols are most important. Because the communication rules at this level focus on how workstations gain access to communications media in order to transmit and receive messages, data link protocols are also called *media access control (MAC)* methods. In this text, we will generally use the acronym MAC and the term *media access method* to differentiate data link level protocols from other types of data communication protocols.

Three types of media access methods are used in computer networks: polling, contention, and token passing. Of these, contention and token passing are most commonly found in LANs.

POLLING

Polling is most likely to be used in networks that have a star topology such as the one illustrated in Figure 5-1. When polling is used, the central computer controls the transmission of data between workstations. The central computer is sometimes called the *primary station* or *supervisor* and the other computers in the LAN are *secondary stations*. Secondary stations are only allowed to transmit data when they are given permission to do so by the supervisor. The process that the supervisor uses to ask secondary stations if they have data to transmit is polling.

The supervisor polls the workstations attached to it in a round-robin fashion. Each of the secondary stations is asked, in turn, if it has a message to send. If a workstation has a message to transmit, it sends it to the supervisor. The supervisor then forwards it to the workstation to which the message is addressed.

The supervisor uses a prespecified polling list for the workstations attached to it. In order to give equal attention to each workstation, the supervisor's polling list is likely to be something like 1, 2, 3, 4, 5, 6, 7, 8, 1,

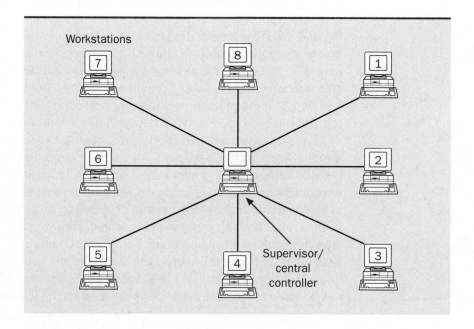

FIGURE 5-1

Polling in a LAN with a star topology

2, 3, 4, 5, 6, 7, 8, 1, and so on. If one workstation is typically busier than others (or should be given higher priority), the polling list can be modified so that it is polled twice as often (or more) as the other workstations (such as 1, 2, 3, 4, 5, 1, 6, 7, 8, 1, 2, 3, 4, 5, 1, 6, 7, 8, 1, and so on).

One of the main drawbacks of polling is the fact that if the supervisor malfunctions and is unable to carry out its polling functions, communications between workstations cannot occur. This problem has been addressed in some networks by having a backup supervisor that automatically takes over the communications chores when the regular supervisor malfunctions.

Polling is not as common in LANs as contention or token passing. However, it is sometimes found in multiuser DOS LANs (discussed in Chapter 3), and in some of the other products for creating smaller LANs that have a star topology.

CONTENTION

Contention is most commonly used in LANs with bus topologies such as the one depicted in Figure 5-2, but like any other media access method, it may be implemented in any one of the fundamental network topologies. Contention is so named because workstations must contend for use of the communications medium to which each has access.

When a pure contention media access method is used, each workstation has equal access to the communications medium. Each workstation monitors ("listens") to the medium to determine if it is in use (that is, if data is being transmitted by another workstation). In such LANs, the workstations are said to have *carrier sense (CS)*—the ability to detect when the medium's carrier signal is being used to transmit data. If a workstation that has a message to send detects that the medium is idle, it can begin to

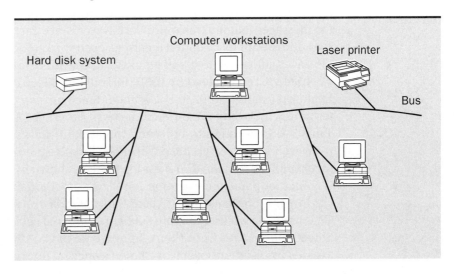

FIGURE 5-2

A bus topology

transmit data. Because any of the workstations with access to the medium can transmit data when they detect that it is not being used, the LAN has *multiple access (MA)*.

When two or more workstations with data to send detect that the medium is idle, they may begin transmitting at the same time. When this happens, the message signals interfere with one another, become garbled, and cannot be transmitted successfully; the result is called *collision*. Because collisions are possible (and actually occur quite frequently), it is essential for the transmitting workstations to be able to detect when one occurs and to attempt to retransmit their data after the collision; this ability is called *collision detection (CD)*. If each transmitting workstation tried to send its data immediately after a collision has occurred, the result would be another collision. To reduce the possibility of a second collision, each workstation waits a randomly selected time interval before it attempts to retransmit its data.

The medium access control method that has just been described is **CSMA/CD** (carrier sense multiple access with collision detection). It is the best known of the contention-based MAC methods and it is the most commonly used medium access control approach used in LANs with bus topologies. With CSMA/CD, workstations can access the network medium whenever it is not in use; in polling or token passing, workstations must wait for a poll or for the token. The collision-detection feature ensures that when two devices transmit messages at the same time, causing the messages to collide and interfere with each other, receivers will ignore the garbled message and senders will retransmit their messages when the line is clear.

CSMA/CD is sometimes called a "listen-before-talk" MAC method, because all workstations monitor the medium and do not attempt to transmit messages until they detect that the medium is not in use. It is known as a broadcast media access method because all workstations receive all the messages that are transmitted. However, since each message includes the addresses of its intended receivers, each workstation accepts only the messages addressed to it and ignores and discards the others.

CSMA/CD is covered by IEEE (Institute for Electrical and Electronics Engineers) standard 802.3. This standard recognizes a variety of media alternatives including the different types of Ethernet LANs summarized in Table 5-2. In addition to Ethernet, Ungermann-Bass's Net/One is a relatively well-known cable-based local network design that uses CSMA/CD. An example of a CSMA/CD message frame is shown in Figure 5-3.

Another popular contention-based MAC method that is a variation on CSMA/CD is **CSMA/CA**. The CA stands for *collision avoidance*. When CSMA/CA is used, the workstations are configured to avoid collisions rather than to detect and recover from them. In one type of CSMA/CA, collisions are avoided by assigning each workstation a specific time interval that it must

Name	Type	Description
1Base5	Baseband	A 1 Mbps medium whose maximum length can be 500 meters before repeaters are needed to amplify the signal.
10Base5	Baseband	A 10 Mbps medium whose maximum length can be 500 meters before repeaters are needed to amplify the signal.
10Base2	Baseband	A 10 Mbps medium whose maximum length can be 200 meters before repeaters are needed to amplify the signal. In Ethernet LANs, such a medium is usually called *Thinnet* or *Cheapernet*.
10BaseT	Baseband	A 10 Mbps medium using unshielded twisted pair wiring. The "T" stands for twisted pair. Repeaters are usually required after 100 meters and such LANs are usually implemented in a star topology.
10Base36	Broadband	A 10 Mbps medium whose maximum length without repeaters is 3,600 meters.

TABLE 5-2   CSMA/CD media options covered by IEEE standard 802.3

Preamble	Start frame delimiter	Destination address	Source address	Length	Data + Pad	Frame check sequence
7 bytes	1	2–6	2–6	2	0–1500	4

FIGURE 5-3   CSMA/CD (Ethernet) packet structure for IEEE 802.3

wait to begin transmitting after detecting that the medium is idle. By assigning each workstation a specific time slot in which to begin transmitting, collisions can be avoided. In a second type of CSMA/CA, a workstation with data to transmit sends out a small message (often only three bytes long) that signals the other workstations that it intends to transmit data over the medium; of course, this small message is sent only if the workstation "hears" that the medium is currently idle. The small message tells the other workstations to wait to transmit data until the signaling workstation has sent its data. In this system, the only collisions that occur are between the preliminary small message packets and not between actual data transmissions.

This second type of CSMA/CA is used in AppleTalk LANs that consist of Apple Macintosh workstations and peripheral devices. Although formerly referred to as AppleTalk, the actual cabling system used in Apple-based LANs is called LocalTalk, which uses a CSMA/CA protocol, the LocalTalk link access protocol (LLAP). LocalTalk is a relatively slow system that supports transmission speeds of 230,400 bits per second. The maximum cable length in LocalTalk LANs is 300 meters and the maximum number of workstations is 32.

TOKEN PASSING

Another media access control method widely used in LANs is ***token passing***. This is used in LANs with both bus and ring topologies (although, as we will discuss shortly, a "true" ring is only found in FDDI LANs). With token passing, a small packet called a "token" is passed around the ring from one device to the next. The general process is summarized in Table 5-3. When a workstation has a message to send, it "captures" the token, puts its message and the address of the workstation that should receive it in a data frame, and then passes the token and message along to the next device in the LAN. As the token and message pass from device to device, each workstation checks to see if the message is addressed to it. When it is, the workstation accepts the message, appends an acknowledgment of receipt for the sender and passes the token along to the next workstation. When the acknowledgment is received by the sender, the token is released and sent to the next workstation. With token passing, devices cannot collide as they vie to use the medium.

Two major types of token-passing LANs are covered by IEEE standards. IEEE standard 802.5 covers token-ring networks, while IEEE 802.4 addresses token-bus networks. FDDI LANs are essentially high-speed token-ring networks that use fiber optic cabling as the communication medium. Each of these is discussed more fully below.

Token-Ring Network

In a token-ring network, the electronic token is passed from workstation to workstation in one direction. An example of a token-ring network is shown in Figure 5-4. When being passed, the status of the token is either free (empty) or busy (in use). A workstation with data to send must wait until it receives a free token. It then changes the status of the token to busy and inserts its data and the address of the workstation that should receive it. The busy token is then passed from workstation to workstation until it arrives at the intended recipient. The receiving workstation accepts (copies) the message and attaches an acknowledgment to the sender. The busy token containing the acknowledgment is passed from device to device until it arrives back at the sender. The sender then changes the status of the token from busy to free (or generates a new free token) and passes it along

---

TABLE 5-3
Process used by token-passing media access control protocols

1. A workstation with data to transmit waits for the token.

2. If the token is not in use, the workstation transmits its message and the address of the workstation that the message is intended for.

3. The workstation receives an acknowledgment from the message recipient.

4. The workstation passes the token to the next workstation in the LAN.

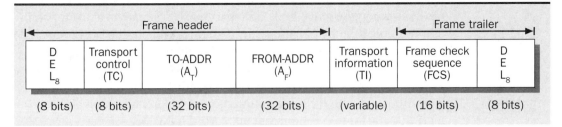

Frame header					Frame trailer	
D E L$_8$	Transport control (TC)	TO-ADDR (A$_T$)	FROM-ADDR (A$_F$)	Transport information (TI)	Frame check sequence (FCS)	D E L$_8$
(8 bits)	(8 bits)	(32 bits)	(32 bits)	(variable)	(16 bits)	(8 bits)

FIGURE 5-4   Packet structure for IBM token-ring

to the next device. Since only one workstation can transmit at a time when this approach is used, messages do not interfere with one another.

Tokens and messages travel from LAN device to LAN device in a ring (sequential) fashion. This is why such LANs are called token-passing rings. However, the cabling used in many token-ring networks is physically a star topology. Most of these are modeled after IBM's Token-Ring Network, with a hardware box (a multistation access unit (MAU)) that provides plug-in interfaces for either four or eight workstations. Usually MAUs provide plugs for unshielded twisted pair wires (see Figure 5-5). Depending on the cabling and network interface technologies used, token-ring networks operate at either 4 or 16 million bps.

One or more of the plug-in interfaces in a MAU can be used to provide connections to other MAUs. Hence it is possible to "daisy chain" MAUs and thus expand the number of workstations in a LAN. The twisted-pair cabling typically used connects the workstation's network interface card (NIC) to the MAU. The token passing that occurs takes place within the MAU's circuitry.

A malfunctioning workstation can be detached from a token-passing ring LAN by unplugging the cabling at either end; this can also be accomplished through LAN management software or network OS commands and

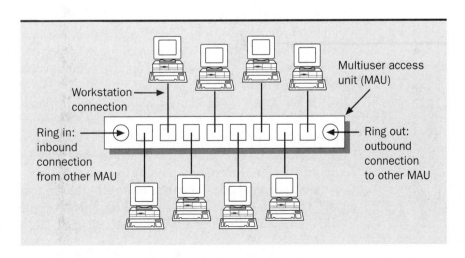

FIGURE 5-5
An 8-port token-ring MAU

via controls on the MAU itself. Adding a new workstation to a token-passing LAN can be done by plugging it into an open plug in the MAU. Moving a workstation from one location to another may mean little more than unplugging it from one MAU and plugging it into an open slot in a MAU at the new location. These LAN changes (moves, adds, and drops) can usually be accomplished with little disruption to normal LAN functions. Token-ring networks are more common than token-bus networks.

Token-Passing Bus

IEEE standard 802.4 defines token-bus networks. In such networks, an unused token is simply a predefined frame and a used token contains both the token (the predefined frame) and a message frame. When a workstation with data to send receives an unused token, it puts its message in a message frame along with the address of the intended receiver. The receiving device copies the message, adds an acknowledgment and then returns the token to the sender, which then passes the token to the next workstation.

In token-passing bus networks, the token is passed from device to device as determined by an address list. This list specifies the order in which each station receives the token. To give some workstations higher priority than others, their addresses can be listed multiple times in the address table. The token is passed in either descending (the most common) or ascending address order. When being passed in descending order, the token is passed to the device with the next lower address in the table; the device with the lowest address forwards the token to the device listed first in the address table.

If a workstation attempts to send the token to an inactive or malfunctioning device, it waits a predetermined time interval for an acknowledgment and then attempts to send the token a second time. After a fixed number of failed attempts (usually two or three) to send the token to the device whose address is next in the list, the sender issues a message that includes the address of the unresponsive device. This message essentially asks, "Who is next?" The successor of the unresponsive device recognizes the address as its predecessor in the list, responds that it is next, and receives the token.

When a new workstation is added to a token-passing bus, the address table is updated to ensure that it receives the token and has access to the communication medium. This is usually accomplished automatically through autoreconfiguration, in which workstations are invited to join the network by the workstation holding the token frame.

Some experts describe token-bus networks as a token-passing contention-based media access control method. This description has some merit, but the rationale behind it is beyond the scope of this book. Some of the better known examples of token-passing bus networks include Datapoint Corporation's Arcnet and LANs found in manufacturing environments using MAP (Manufacturing Automation Protocol) or TOP (Technical and Office Protocol). Some of the wireless LAN products are also token-passing bus networks.

Arcnet is a baseband token-passing bus (or star) architecture that does not conform to the IEEE 802.4 or 802.5 standards. However, its low cost has made Arcnet extremely popular and it is commonly recognized as a de facto LAN standard. This status may change because the American National Standards Institute (ANSI) is expected to designate Arcnet as ANSI standard 878.1. If this happens, there will be pressure on the International Organization for Standardization (ISO) to recognize Arcnet as a worldwide standard.

Arcnet can support approximately 250 workstations in a single LAN. Arcnet can be used with coaxial (the original medium), twisted pair, and fiber optic cable. Arcnet Plus supports transmission speeds of 20 Mbps; this is eight times faster than the speed found in most Arcnet LANs (2.5 million bps). In a bus topology, the maximum cable length without repeaters is 1,000 feet. In a star configuration, the maximum distance between a workstation and the central hub can be 2,000 feet.

MAP and TOP follow IEEE standard 802.4. These are broadband token-passing bus networks that typically employ coaxial cable as the communications medium and support speeds of 1, 5, 10, or 20 Mbps. As noted, these are most commonly found in manufacturing environments and have been designed to optimize factory automation and computer integrated manufacturing (CIM) applications.

FDDI LANs

High-speed LANs using fiber optic cable as the communications medium are becoming more common. *FDDI* (Fiber Distributed Data Interface) is a set of standards created by ANSI (the ANSI X3T9.5 Task Group) to govern such networks.

The FDDI standard is used in LANs (or MANs) with a true ring configuration where fiber optic cable connects adjacent workstations or nodes (see Figure 5-6). The media access control scheme used in FDDI LANs is a variation on the IEEE 802.5 token-passing ring standard that we have already discussed. These LANs use a "timed-token rotation" that limits the amount of time that the token can be captured by a workstation or node before passing it along to the next device in the ring.

FDDI LANs use two fiber optic cables to connect workstations or nodes. In most cases, data travels on the primary ring, and the secondary ring is used as a backup data transmission path if the primary ring fails. In some cases, both the primary and secondary ring are directly involved in

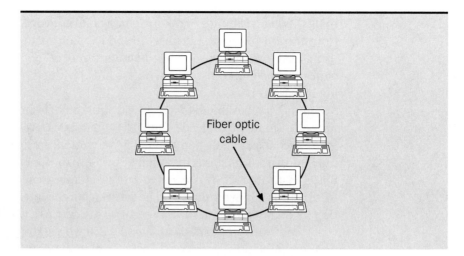

Fiber optic
cable

FIGURE 5-6

A ring topology
where adjacent
workstations are
connected by
fiber optic cable

data transmission: the primary ring carries input data and the secondary
ring carries output data. The presence of two fiber optic cables also means
two types of workstations or nodes can be found in FDDI LANs. Dual-
attachment stations (DASs) are physically connected to both the primary
and secondary cables. Single-attachment stations (SASs) are only connected to the primary cable. The main advantage of having a DAS instead
of a SAS is that if the primary cable fails, the secondary cable can still be
used for data communications.

FDDI LANs support speeds of 100 Mbps. They can support up to 1,000
workstations, and although they require a repeater for each two kilometers
of cable, the total cable length can be 200 kilometers (approximately 120
miles). This makes FDDI LANs capable of covering a greater geographic
distance than any other type of LAN. They are thus suitable for creating
backbone networks (such as those used to interconnect workstations in
multiple buildings on a university campus) and metropolitan area networks (MANs).

Because token-passing LANs are very common, there are multiple vendors for MAUs, NICs, cabling, repeaters, transceivers, and all the other components needed to implement them. This promotes competition among
vendors and helps keep the costs of implementing and maintaining these
LANs relatively low.

As a general rule, Arcnet LANs are among the lowest-cost LANs available (based on average cost per workstation). CSMA/CD (Ethernet) LANs
often have a lower average cost per workstation than token-ring LANs.
FDDI LANs are usually the most expensive to implement, and the average
cost per workstation of a wireless LAN is usually somewhere between that
for token-ring and FDDI LANs.

◆ LAN STANDARDS AND STANDARD SETTING ORGANIZATIONS

As you have undoubtedly noticed, it is almost impossible to discuss LAN protocols without discussing LAN standards. Thus far in this chapter, you have already come across references to IEEE and ANSI standards as well as to the ISO. As many experts have said, if you like standards, you'll love LANs because there are so many to choose from. In this section, we will provide a brief overview of the major LAN standards and identify the organizations that developed them.

THE NEED FOR LAN STANDARDS

Soon after LANs began to become popular in business organizations, the computer and communication industries recognized the importance of developing standards. LAN standards were needed both to define LAN implementations and to direct the development of new information technologies capable of interfacing and communicating with one another. The concept of open architectures is endorsed by the major standard setting organizations. Since the interface specifications for systems with open architectures are published, multiple vendors can develop and market LAN components. Systems with closed architectures do not publicize interface standards, in order to limit the number of network component vendors.

THE OSI REFERENCE MODEL

Probably the most widely known and widely endorsed open architecture model is the International Standard Organization's OSI (Open Systems Interconnection) reference model. Like all open systems models, the OSI model makes technical recommendations for data communication interfaces. If hardware and software vendors follow these recommendations, the job of developing products that are able to operate and communicate effectively with one another is made easier.

The OSI reference model is a seven-layer framework around which standard protocols (such as CSMA/CD and token-passing) can be defined. The communications tasks carried out at each of the seven layers are summarized in Table 5-4. Essentially, this model specifies how software should handle the transmission of a message from one workstation (or application program) to another. End users only have to be concerned with the application layer; the messages they have are appropriately formatted (packaged) for transmission as they descend the layers, and are actually transmitted to other devices over communication media at the physical layer. At the receiver's end, the message ascends the layers and in the process is successively reformatted (unpacked) until it reaches the application layer in an understandable form. Figure 5-7 illustrates this process.

Layer	Description
1. Physical	Handles voltages, electrical pulses, connectors, and switches so data can be transmitted over network media.
2. Data link	Controls grouping data into blocks (message packets) and transferring blocks from one point in the network to another.
3. Network	Controls data and message routing through network channels.
4. Transport	Controls data transfer for the complete transmission path, from sending point to receiving point.
5. Session	Establishes and terminates communications links between computers.
6. Presentation	Formats data for transfer between different systems. Provides user interfaces with the systems.
7. Application	Provides network services to users and user applications, including file transfer.

TABLE 5-4   The Open Systems Interconnection (OSI) model

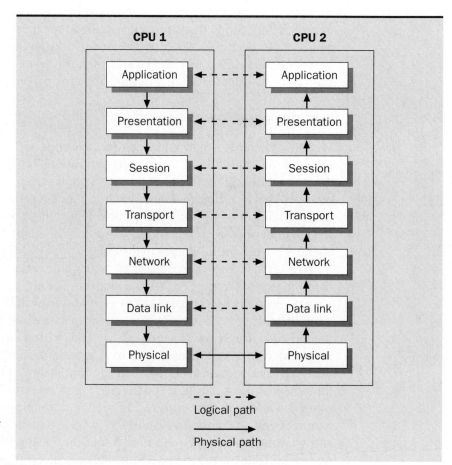

FIGURE 5-7

Message transmission and reception recommended by the OSI reference model

Figure 5-7 indicates that the only real connection between senders and receivers occurs at the physical level. The other levels (layers 2 through 7) establish virtual or logical links between senders and receivers.

The purpose of the seven layers is to segment the functions that must be carried out for two machines to communicate with one another. The model is primarily used as a guide for hardware and software designers and programmers to use for developing new equipment, communications protocols, and software used in LAN and other network environments. The OSI reference model should never be considered a program, network operating system, or communications protocol. It should only be viewed as a recommended set of guidelines for developing these important networking technologies as well as the computing and communications equipment that they run on.

IEEE STANDARDS FOR LANs

The Institute of Electrical and Electronics Engineers (IEEE) was one of the first standard setting organizations to endorse open architectures and the OSI reference model. In 1980, this organization started a special project, *IEEE Project 802*, to develop standards for LAN implementations and applications; the project number itself refers to the year (80) and month (2—February) in which the committee was formed. A number of IEEE 802 subcommittees were formed to facilitate the development of standards. The responsibilities of these subcommittees are summarized in Table 5-5.

As you can see in Table 5-5, IEEE sets standards for all the major types of media access control methods found in LANs. It also has developed standards (or is in the process of developing standards) for the types of LANs that are likely to become increasingly common in the years ahead (such as LANs using fiber optic cabling and LANs that integrate voice, data, and video communications). By developing standards for emerging LAN technologies and applications, IEEE is helping to ensure that hardware and software components will be designed to enable effective communication between workstations and other network devices.

OTHER IMPORTANT STANDARD SETTING ORGANIZATIONS

While IEEE has developed the most widely known and comprehensive set of standards for LANs, other standard setting organizations have also made important contributions. Some of these organizations and their most significant contributions to LANs are summarized below.

ANSI

The American National Standards Institute (ANSI) serves as a coordinating organization for a variety of information systems standards in the U.S. Its

IEEE Subcommittee	Focus/Objectives
802.1	**High-Level Interface.** This subcommittee focuses on issues related to the higher OSI layers such as network architecture, network interconnections, and network management.
802.2	**Logical Link Control.** The IEEE subdivides the data link layer of the OSI model into two sublayers: the logical link control (LLC) and the media access control (MAC) levels. It is at the MAC level that media access protocols such as CSMA/CD and token passing are implemented. The objective of the LLC is to provide a consistent interface to the MAC level so that protocols above the data link level are able to function no matter which MAC level protocol is used.
802.3	**CSMA/CD.** This subcommittee sets standards for the various types of CSMA/CD implementations found in LANs. Table 5-2 in this chapter lists several of the major types of implementation standards that have been developed by this subcommittee.
802.4	**Token Bus.** This subcommittee develops standards for token-bus networks including standards for message frames, establishing priorities in the address table, adding new stations, and recovering from lost tokens.
802.5	**Token Ring.** This subcommittee develops standards for token-ring networks including standards for token and message frame formats, adding new stations, and recovering from lost tokens.
802.6	**Metropolitan Area Networks (MANs).** This subcommittee sets standards for LANs that cover the same types of geographic distance and media speeds as those specified by ANSI's FDDI standard. IEEE standards for such LANs cover voice and video transmission in addition to data transmission.
802.7	**Broadband Technical Advisory Group.** This subcommittee provides guidance and technical advice to other IEEE subcommittees that are charged with developing broadband LAN standards (such as the IEEE 802.3 subcommittee for 10Broad36 LANs).
802.8	**Fiber Optic Technical Advisory Group.** This subcommittee provides guidance and technical support to other IEEE subcommittees charged with developing standards for LANs using fiber optic cabling (such as the 802.6 subcommittee).
802.9	**Integrated Data and Voice Networks.** This subcommittee sets standards for networks and LANs that carry both voice and data. For example, it has set interface standards for Integrated Services Digital Network (ISDN) networks and ISDN-capable PBXs.

TABLE 5-5 Major IEEE 802 subcommittees

role is to coordinate the development of national standards and to interact with the International Standards Organization (ISO) so that U.S. standards will be in compliance with ISO recommendations. ANSI has developed standards for many popular programming languages such as BASIC, COBOL, and SQL. In LAN environments, ANSI's FDDI standard is widely recognized and followed. FDDI was discussed previously in this chapter.

Electronics Industries Association (EIA)

The EIA is an ANSI-accredited organization that has developed a variety of standards for equipment and components found in LAN environments. The standards for the RS232 and RS449 connector plugs are among its best known standards.

Arcnet Trade Association (ATA)

The ATA was designated by ANSI to define and set standards for the Arcnet (Attached Resource Computing Network) protocol. As discussed, Arcnet LANs use a baseband token-passing protocol. The ISO is expected to recognize the Arcnet protocol and endorse it as an international standard.

AN IMPORTANT NETWORK-LAYER PROTOCOL: TCP/IP

You will soon see that there are a number of other protocols that network managers need to be aware of, such as IPX, SPX, SNMP, and CMIP. Knowing about these is especially important for LAN managers in organizations that have interconnected networks, especially if the networks being interconnected employ different network-layer protocols. At present, TCP/IP is arguably the most important network-layer protocol that network managers should be aware of.

The *TCP/IP* (Transmission Control Protocol/Internet Protocol) is one of the oldest network standards; it has been used for more than 20 years. It was developed for the U.S. Department of Defense (DOD) as part of its Advanced Research Project Agency Network (ARPANET). ARPANET connected computers at the DOD to those at universities and corporations involved in defense contracts and related research. TCP/IP was designed to allow these interconnected computers to exchange large files over available long-distance telephone communications lines. Since the various locations had a large range of computers and since the available communications media could be unreliable, a robust protocol was needed that could handle relatively error-free transfer of large files between different systems. TCP/IP proved to be quite good at handling such exchanges, and as a result it has emerged as a major protocol for handling message exchanges between different networks.

ARPANET has grown and evolved over the years into what is now called the Internet, and TCP/IP is the main protocol used on the Internet. In 1994, the Internet consisted of more than 21,000 networks connecting some 2 million computers in more than 60 countries. The user base expands at the rate of 7 percent to 10 percent per month and will be one of the key components in the "information superhighway" that is planned for the U.S. TCP/IP has been modified several times to keep up with the growth and ever-higher transmission speeds available on the Internet.

Business interest in the Internet is increasing as more and more businesses are using this network of networks for some of their business activities. Many colleges, universities, secondary schools, and government agencies, and libraries are also connected or considering connection to the Internet. Hence, LAN managers in many types of organizations have to be aware of TCP/IP and the NOSs for LANs that support it.

◆ LAN OPERATING SYSTEMS

As noted, a network operating system (NOS) for a LAN is essentially an extension of the workstation's operating system. Like other operating systems, a NOS used in a LAN carries out a number of input/output (I/O) tasks for users, such as retrieving or saving files stored on secondary storage devices (for example, disk units, optical disk storage units) or directing data to a printer or other output device. However, instead of carrying out such functions just for devices that are attached to the motherboard of the user's workstation (as the workstation's OS would do), a LAN NOS retrieves and saves from disks controlled by network servers and directs output to print servers, or to shared printers and output devices. In addition, all such tasks are carried out for users in a manner that insulates them and their applications from the details needed to perform I/O operations and other housekeeping chores such as primary memory management. All the users need to know how to do is to make requests for such services, and the NOS carries them out.

In a LAN environment, the system software needed to carry out network functions is found in both workstations and servers. While the bulk of the NOS files are installed in servers, very important pieces are found in the workstations. The latter work with the workstation's usual OS and help it determine when network services are being requested by the user.

IMPORTANT NOS MODULES WITHIN LAN WORKSTATIONS

Three major NOS modules are installed in each workstation attached to a LAN. These are the network redirector module, the application program interface (API), and the network interface module. These are illustrated in Figure 5-8.

Network Redirector Module

A workstation's OS (such as DOS, OS/2, Windows, Macintosh System 7) controls only interactions with peripheral devices (such as printers, disk drives, input devices) that are physically attached to the workstation. For example, it may be able to retrieve files stored on floppy disk drives A and B, hard disk drive C, and CD-ROM drive D. However, if the user requests a

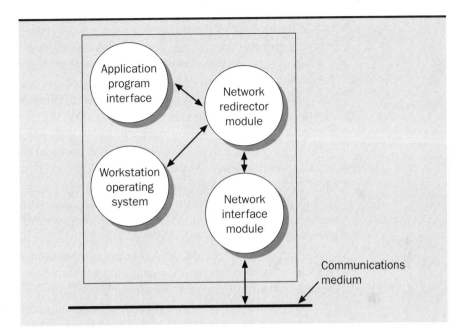

FIGURE 5-8
Some important
components of a
LAN OS

file from a shared disk drive located on the network (from drives F, G, H, and so on), a software module must be available at the workstation that detects this command for network services and directs it to the appropriate network server that can fulfill the user's request. This software module is called different things by different LAN OSs, but it is generally referred to as the ***network redirector module.***

The network redirector module's primary job is to intercept all I/O requests before they are received by the workstation's OS. If the request involves devices directly attached to the workstation, the network redirector module passes the request to the workstation's OS. If the request is for services from shared devices in the network, the network redirector module sends the request through the workstation's NIC and over the network to the appropriate server or device.

As you can see, the network redirector module is an additional system software program stored at the workstation that has a key job in network operations. If this module did not intercept user requests for network services before they got to the workstation's OS, the workstation's OS would return an error message to the user stating "device not found."

Application Program Interface

An application program communicates with the network through a set of programming and formatting rules known as an ***application program interface (API).*** The primary functions of an API are to accept a request for network services from the application, put the request into a standard API format being used in the network, and send the request over the network to the appropriate device needed to fulfill the user's request.

Part of this task may involve attaching sender and receiver addresses to the request. When the request has been carried out and a response is sent back to the workstation, it is the job of the API to reformat the response into a form that is understandable by the application that requested it.

Some of the most common APIs found in LAN environments are listed in Table 5-6. Among these, Novell's IPX/SPX protocol is the most common.

Most of the general-purpose application software packages (such as word processing, spreadsheet, and database packages) as well as many special-purpose application programs (such as accounting and inventory management) that are on today's market can run in LAN environments. However, each of these may operate correctly in some types of LANs but not others. This is often due to API compatibility among workstations and servers; to work correctly, they must have a common API through which they can communicate and transfer appropriately formatted data to one another. The potential for API incompatibility difficulties should be taken into account when selecting LANs, LAN NOSs, and the actual application packages that will be used.

Network Interface Module

The medium interface module is responsible for placing data onto the network and receiving data from the network. It is responsible for putting the message into the format needed for transmission over the network and for putting the data in a form expected by servers and other network devices. This software module interfaces directly with the network interface card (NIC).

Like an API, the **network interface module** is responsible for message format. However, while the API is concerned with format translation between an application and the network, the network interface module is concerned with putting the data into the form needed for the media access control method in use (such as the message frames needed for token passing or CSMA/CD).

Name	Developer
Advanced Program to Program Communications (APPC)	IBM
Internetwork Packet Exchange/Sequenced Packet Exchange (IPX/SPX)	Novell
NetBIOS	Microsoft
Pipes (UNIX)	AT&T
Xerox Network Systems (XNS)	Xerox

TABLE 5-6
Some LAN APIs

NIC Drivers

As noted in Chapter 4, another type of software used in networked workstations is network adapter card drivers, or *NIC drivers*. These essentially enable workstations to send messages through the NIC and over the communications network's communications media. Because of this, the driver used with the NIC must be consistent with the workstation's OS, the NOSs used in the network, the network's media access control methods, and other protocols. The NIC drivers that must be used in interconnected networks are usually more complex than those used in stand-alone LANs because they likely will need to be compatible with multiple NOSs and protocols.

You should note that the actual names for the system software components that we just described vary among the LAN NOSs currently available. The foregoing discussion generally provides the generic names for these programs and a general description of the functions that they perform.

IMPORTANT NOS MODULES FOUND AT SERVERS

As we have previously noted, the bulk of the system software found in LANs is located within the servers. This software is often more sophisticated and complex than that found within workstations because in many LANs, servers must be able to accommodate numerous concurrent users. In addition, the system software modules installed in LAN servers are responsible for the general operations of the entire LAN.

Two main categories of LAN operating system software are used within servers. The first category includes LAN operating systems that combine NOS functions and server system software in one integrated package. Novell's NetWare is the most widely used example of this. The second category includes NOSs that run under operating systems such as Windows or OS/2; for example, Banyan's Vines runs under UNIX, and IBM's LAN Server runs under OS/2. We will be discussing some of these more popular LAN operating systems later in the chapter.

Integrated packages have captured the lion's share of the LAN operating systems market. Novell's NetWare is used in approximately 60 percent of all LANs. LAN Server, LAN Manager, and AppleTalk each account for about 10 percent of the market and Banyan's Vines is used in about 5 percent of all LANs. In the years ahead, UNIX is expected to gain an increasing share of the LAN operating system market, and Windows NT is also expected to post significant gains. Some experts predict that NetWare, UNIX, and Windows NT will be used in more than 90 percent of all LANs by the year 2000.

The NOS modules found within servers are often responsible for carrying out a number of important network functions, including access optimization, concurrent access control, fault tolerance, workstation-to-workstation communication, file transfer, access to other networks, network security, and network management. Each of these functions is briefly described below.

Access Optimization

One of the main responsibilities of a file server is to provide users with relatively transparent access to shared files and data. To most efficiently handle such requests in a timely manner, this task should be optimized. Since these types of requests demand input/output operations, access optimization is often called *I/O optimization.*

Two of the major approaches used to optimize user access to shared files and data are seek time enhancement (or disk seek enhancement) and disk caching. Seek time enhancement involves arranging concurrent user requests in an order that minimizes the distance a disk drive's read/write heads have to be moved when the NOS finishes one user's request and starts on the next. Because the mechanical movement of read/write heads is one of the most time-consuming parts of reading from or writing to a disk, minimizing movement increases server performance.

Disk caching works similarly to cache memory in individual workstations. It is based on the fact that gaining access to data in primary memory is faster than gaining access to data on disks. A disk cache keeps the data that users are most likely to request in a portion of the server's primary memory called a cache. When a user request is received, the disk cache is searched first; if the data is not found in the cache, the server goes to the disk to retrieve it.

Many LAN NOSs and LAN management software packages make it possible to use both I/O optimization and disk caching within the same server. The combination of these techniques can help ensure that user requests for shared files, software, and data can be handled very efficiently.

Concurrent Access Control

Concurrent access control allows the server to handle two or more simultaneous user requests for the same file, application software, or data. The NOS should have an effective mechanism for handling such requests.

Many general-purpose application packages (such as word processing or spreadsheet packages) were initially designed for use on stand-alone microcomputer systems. If such programs are accessible to LAN users, problems can occur when two or more users try to access the package at the same time.

Many LAN NOSs and LAN management software packages deal with such problems through a routine that tracks whether or not the software package is already in use. If the package is in use when a second user attempts to access it, the second user receives a "file already in use," "file busy," or "file locked" message. Similar tracking mechanisms are often used to ensure that site license agreements are not violated. Site license agreements stipulate how many users can concurrently use the application package. These written agreements are signed when LAN managers purchase versions of general-purpose application packages that have been designed for use in network environments; allowing more than the specified number of users to simultaneously use the software is a prosecutable violation.

If LAN users are sharing data files or databases, file locking or record locking mechanisms may be used. Each of these can help ensure that two or more users cannot attempt to simultaneously update an individual record or data element. They can also be used to ensure that only authorized users can access or update particular files or records. Besides being classified as access controls, file and record locking mechanisms are often considered security mechanisms.

Fault Tolerance

Fault tolerance is concerned with the reliability of the LAN and its ability to survive and recover from systems failures that could cause it to crash. The key word in most fault tolerance mechanisms found in LAN environments is *backup;* in LANs, fault tolerance is usually achieved through some combination of backup software and hardware. Because backups are important to recovery from network failures, they are discussed more fully in Chapter 14.

Two frequently used backup mechanisms are making backup copies of important software and having a backup server available should the primary server crash or malfunction. The major approaches used with each of these are outlined below.

Backup copies of important software and data can be made in several ways:

1. *Writing backup copies of important software and data on alternative disks.* Copying critical LAN NOS files, application software packages, and user data files to alternative disks can be done periodically. This ensures that if the primary copies are lost or become inaccessible due to server failure, secondary copies are still available.

2. *Use of disk mirroring.* With disk mirroring, the same data, files, or software is simultaneously written to two or more disks. If one of the disks or disk drives fails, the data can still be accessed from the other disk. Such systems can also be used for I/O optimization as both disks can handle user requests for files or data.

    A slightly more sophisticated approach than disk mirroring is disk duplexing. Whereas in disk mirroring a single bus channel is used between the file server and the disks, with disk duplexing, there are separate bus channels to each disk. This provides additional redundancy and fault tolerance by avoiding the possible failure of the single bus channel (or bus channel controller) found in disk mirroring systems.

3. *Use of RAID systems.* Redundant arrays of independent disks (RAID) are high-volume secondary storage systems with disk mirroring and duplexing capabilities. Data can be written to or read from two, three, or more disks in RAID systems. RAID systems that have been specially designed for LANs are becoming more common.

4. *Tape backup systems.* Tape backup systems are available for virtually all sizes of computers. They can be used in LAN environments to back up some or all of the shared data and files on a periodic or continuous basis. In many larger networks, LAN managers perform a full tape backup every night.

5. *Virtual disk systems.* Virtual disk systems are a special class of LAN utilities that provide real-time continuous backup services. Such systems perform backups continuously, automatically, and transparently, eliminating the need for LAN managers to perform a daily backup. When a crash occurs, the LAN manager only needs to turn back the clock to a few seconds before the crash and all data (even that in open files) is recreated as it existed at that moment. Vortex Systems, Inc. is a leading vendor of virtual disk systems.

In addition to disk mirroring and RAID systems, *duplexed servers* are another very good type of hardware/software backup combination. With duplexed servers, if the primary server fails, the second can take over LAN operations and keep the LAN functional. A copy of all the software needed to keep the LAN operational is located at the second server. In addition, copies of all shared application software and data files can also be stored on the backup server.

Ideally, the backup server is comparable in power and processing capacity to the primary server. However, LAN managers should always remember that while this is a desirable characteristic, it may not be necessary. Often a less sophisticated machine can be used as a backup server.

Workstation-to-Workstation Communications

Software that is comparable to the API and network interface modules located at the workstations must also be located at LAN servers. Such modules accept user requests for shared resources transmitted over the communications medium, pass them to the appropriate application for processing, and then send the results back to users.

To most efficiently handle concurrent user requests, *multitasking* capabilities are desirable. With multitasking, the server is able to work on multiple requests at the same time. For example, it may transmit results to one user while the read/write heads of the disk unit are being moved to the data needed to satisfy another user request.

File Transfer

As we have noted on numerous occasions in this textbook, the ability to transfer data or files from one workstation to another is one of the most fundamental LAN functions. All true LANs enable users to do this. File transfer software exists in some form at all workstations and servers in a LAN. At a minimum, software capable of handling file transfers must be located at the sending and receiving workstations, as well as at any workstations between them that receive the file or data while it is in transit. Since file servers are involved in almost all the major file transfers that occur in client/server LANs, it should not be surprising to learn that they contain file transfer software.

Access to Other Networks

Increasingly, LANs provide users access to other computer networks. As we have already noted, interconnectivity is a major trend of the 1990s. For example, connecting a LAN to a WAN or to the Internet is often desirable.

The software that makes it possible for one network to interface with another is usually located at servers. Such software handles all necessary transmission speed and protocol conversions. Software that allows users to access a large host computer is also usually located at a server in the LAN.

Network Security

Many of the security features found in LANs are found in software located at servers. Most LAN security software aims to ensure that users can access the shared resources they need, and restricts unauthorized users from accessing resources they don't need. For example, only one user (or a very limited number of users) should be able to access a private file. For other files, some users may have "read only" privileges, others may have update privileges, and some may even have delete privileges. Most of these types of security issues are software controlled and most of the security-oriented software modules are located in the servers that control access to shared data, files, and equipment. LAN security and network security software will be discussed in more detail in Chapter 14.

Network Management

LAN managers need a variety of utility and LAN management software. Such software is usually classified as system software and is usually located in servers. This software helps the LAN manager add and delete users, create and modify user profiles, set user security levels, monitor LAN performance, and detect, diagnose, and correct problems. Network management is a theme that runs throughout this text and network management software will be discussed more fully in Chapters 13 and 15.

◆ LEADING LAN OPERATING SYSTEM PRODUCTS

There are many NOSs available on the market for LANs. It would be nearly impossible to discuss them all in this textbook. However, a summary of some of the most popular LAN operating systems is found in Table 5-7. Some of the NOSs listed in Table 5-7 are categorized as client/server NOSs, and the others are classified as peer-to-peer NOSs. We will provide a very brief description of the most popular LAN operating systems falling into these two categories.

CLIENT/SERVER NOSs FOR LANs

Client/server NOSs dominate the market for LAN system software. The client/server NOSs are found in nearly 90 percent of all LANs that have been implemented; Novell NetWare alone is used in nearly 60 percent of all LANs. As we noted previously, some of these are integrated packages while

CLIENTSERVER NOSs

Product	Vendor	Topology	Media Access Methods
LAN Manager	Microsoft, Inc.	Bus, Ring	CSMA/CD, token passing
LAN Server	IBM Corp.	Ring	Token passing
NetWare	Novell, Inc.	Ring, Bus	Token pasing, CSMA/CD
StarLAN	AT&T Corp.	Star	CSMA/CD
TOPS	Sun Microsystems	Bus, Star	CSMA/CD
Vines	Banyan Systems	Bus, Ring	CSMA/CD, token passing
Windows NT	Microsoft, Inc.	Bus, Ring	CSMA/CD, token passing

PEER-TO-PEER NOSs

Product	Vendor	Topology	Media Access Methods
AppleTalk	Apple Computers	Bus	CSMA/CA
LANtastic	Artisoft	Bus	CSMA/CD
NetWare Lite	Novell, Inc.	Bus	CSMA/CD
Windows for Workgroups	Microsoft, Inc.	Bus, Ring	CSMA/CD, token passing

TABLE 5-7   Popular LAN OSs

others run under existing OSs. A brief description of the most popular client/server LAN operating systems follows.

Novell NetWare

This is the most widely used LAN operating system. It is usually implemented as a token-passing ring or an Ethernet (CSMA/CD) bus.

Like all major software packages, several versions of NetWare have been released in response to advances in microcomputer processor technologies. For example, NetWare 2.xx versions were designed for Intel 80286 processors and NetWare 3.xx versions were designed for Intel 80386 processors. NetWare 4.xx is the latest release; this is the first release designed for enterprise-wide LANs and for interconnecting two or more NetWare LANs.

As NetWare has progressed through its various versions, the number of users that can be supported in a single LAN has increased. For example, NetWare 3.xx versions can support from five to several hundred users in a single LAN while NetWare 4.0 can support up to 1,000 users per LAN.

Fault tolerance focused on mirrored disks has been supported in all released versions. Also, disk caching and disk directory support services are available to optimize server performance. NetWare also offers a variety of interfaces to other computers (IBM minicomputers and SQL servers) and networks (X.25 packet-switching networks). NetWare also supports the TCP/IP protocol.

Windows NT

Windows NT (new technology) is Microsoft's most recently developed NOS for client/server LANs. It is also commonly referred to as Windows NT Server or Windows Advanced Server. LANs currently running other NOSs can often switch to Windows NT quite inexpensively. This is largely because making such a change does not usually require changing the NICs that have already been installed in the machines—often only new NIC drivers are required.

Because Windows and Windows-based applications have very large and well-established user bases, Windows NT is expected to continue to pick up larger and larger segments of the LAN operating system market.

LAN Server

LAN Server is a product of IBM Corporation. As we noted previously, LAN Server is not an integrated package like Novell NetWare. It runs under IBM's OS/2 operating system. OS/2, like many of the microcomputer OSs being used in the 1990s (including Windows) incorporates several data communication capabilities, including terminal emulation and LAN interfaces.

LAN Server only supports token-ring (IEEE 802.5) LANs. However, it also supports a variety of asynchronous terminal emulation protocols, interfaces to IBM host computers, and access to X.25 networks.

LAN Manager

Microsoft's LAN Manager was designed to compete head-to-head with LAN Server. Like LAN Server, LAN Manager runs under the OS/2 microcomputer operating system. It provides the same functionality as LAN Server but supports both IEEE 802.5 (token-ring) and IEEE 802.4 (token-bus) LANs.

TOPS

Sun Microsystem's TOPS is a LAN operating system that runs under UNIX. UNIX is a family of operating systems that run on all sizes of computers. UNIX incorporates a fairly wide range of data communication functions that allow different types and brands of computers to communicate with one another. Because it facilitates interconnectivity, UNIX is predicted by many experts to become one of the dominant OSs in network environments (including LANs) in the years ahead.

Today, both UNIX and TOPS are used almost exclusively in LANs built around RISC (reduced instruction set computing) servers and workstations. Such hardware is often found in engineering subunits where graphical applications (such as CAD—computer-aided design) are used and where graphical files are exchanged.

Vines

Banyan Systems' Vines is used in LANs that are built around UNIX-based servers. It can be used in large networks and supports network interconnections quite well. Vines includes a database (StreetTalk) in which all network resources, both local and those in interconnected networks, are identified. This facilitates e-mail, program-to-program communications, and general network optimization and control.

PEER-TO-PEER LAN NOSs

As may be observed in Table 5-7, there are several popular NOSs for peer-to-peer LANs. Among these, AppleTalk is the most widely used. Significant numbers of NetWare Lite and LANtastic LANs also exist. Windows for Workgroups is the newest product listed in Table 5-7. It is expected to grow in popularity because of the large established base of Windows and Windows NT users. Each of these products has already been described or mentioned in this or previous chapters (especially Chapter 3). A brief summary of each is provided below.

AppleTalk

Almost 10 percent of the microcomputers found in U.S. businesses are Apple computers; the percentage is much larger in educational (especially secondary school) settings. Most of the Apple microcomputers in businesses are in the Macintosh family and when a fairly large number are concentrated in a relatively small area, the trend has been to network them.

Macintosh microcomputers have a built-in network feature (AppleTalk) that enables them to be connected to other Macintoshes and compatible peripheral equipment. The cabling system used is called LocalTalk. Local-Talk uses CSMA/CA as the cable access method. The maximum cable length is 1,000 feet, the maximum number of devices in the LAN is 32, and the transmission speed is a very slow (for a LAN) 230 Kbps.

LANtastic

Of the peer-to-peer LAN OSs listed in Table 5-7, LANtastic is capable of supporting the largest number of workstations and shared peripheral devices (several hundred) in a single LAN. For performance and user satisfaction, LANtastic is one of the top-rated peer-to-peer LAN OSs.

NetWare Lite

NetWare Lite (recently renamed NetWare Personal) is Novell's alternative for organizations or subunits with a relatively small number of users and where a dedicated server is really not needed or feasible. It is designed to enable users at one workstation to access resources at other workstations (both software and peripheral hardware). In addition, NetWare Lite provides an interface to client/server LANs that run NetWare and incorporates a variety of network management utilities such as logon security, the ability to restrict user access to certain resources, disk and directory mapping, message distribution, and print spooling.

Windows for Workgroups

As noted above, Windows for Workgroups is the newest peer-to-peer LAN operating system product listed in Table 5-7. Although it is too early to know for sure, experts expect it to grow in popularity largely because of the extremely large and growing number of Windows and Windows NT users. This product provides the familiar Windows graphical user interface and a relatively easily mastered set of communications modules. Microsoft has made it easy for LAN managers to switch to Windows for Workgroups by making it possible to continue using existing NICs. Microsoft has also made it possible for organizations to start off with the peer-to-peer version of Windows for Workgroups and then switch to a client/server version built around Windows NT servers.

◆ PRINT SPOOLERS

Another important type of system software found in LANs is print spooling software. This is the system software that enables users to share printers. Only one user's file can be printed out at a time, but it is quite common for users to simultaneously request services from the same printer. The function of print spooler software is to accept users' print files from the network medium, put them in a print queue, store them on disk until the print job can be performed, and remove them from the queue after they have been printed.

Spoolers for printers include some type of priority scheme to determine the order in which user print files are printed. For many, the print order is the same order in which the print files were received. For others, when print files are received within the same time frame (say within the same 30 seconds), the smallest job is printed first, then the second smallest, and so on. Still others follow a predefined user-assigned priority list where some classifications or users have their print files placed at the beginning of the queue no matter how many other print jobs are already lined up.

Once a user file has been printed, it is often removed from the disk in order to make room for other print jobs. However, in some systems, the job may be held on the disk for printing again at a later time, for printing on a different printer, or so users can look over the document at their workstations before actually printing (for example, using the "view" or "preview" functions found in most word processing packages).

There are many ways to configure shared printers in a LAN. In some instances, one or more printers are physically attached to a file server and the spooler software maintains the print queue for each. In other arrangements, one or more printers are attached to either dedicated or nondedicated print servers. In this case, the spooler software is located at the print server itself. In peer-to-peer networks, spoolers are located at the individual workstations whose attached printers are shared by other LAN users.

When the spooler is responsible for more than one print queue, it may also be responsible for routing jobs to more than one type of printer. For example, there may be an inexpensive "draft" printer, and one or more higher quality printers (such as laser printers or color printers) available for polished documents. This arrangement allows users to use a less expensive resource (such as a relatively inexpensive dot matrix printer) to make sure that the print job is in the appropriate format before routing it to a more expensive resource to produce the final document. Often the less expensive printer is the default printer and users must specifically request the higher quality printer.

To assist the LAN manager, modules may be included in spooler software to help him or her monitor printer usage, review print queues, and produce statistical reports. These can help the network manager identify problems and observe trends in user printer usage.

◆ OTHER IMPORTANT SYSTEM SOFTWARE IN LANs

Printer spoolers are probably the most common of the special-purpose system software found in LANs. However, as organizations downsize and LANs assume more and more of the information processing tasks that have traditionally been performed on larger centralized systems, LAN managers must learn about an increasing variety of system software associated with special-purpose servers and applications.

For example, online database applications and database servers are becoming more common in LANs. When this occurs, LAN managers must learn about the interfaces between the NOS and the database management system (DBMS). In order to minimize the amount of time it takes to respond to user queries and data requests, the NOS and DBMS must work efficiently with one another.

If an image processing system is implemented in a LAN, there must be proper interfaces between the scanner server (the computer that accepts scanned-in document images from the scanner and puts them in the format needed for transmission over the network medium) and the NOS. Also, because document images are often stored on optical disks, there must also be efficient input and output interfaces between the LAN operating system and the optical disk storage system that is used.

In brief, during the 1990s, mastering LAN operating system concepts (and specific NOSs) is necessary but often not sufficient. More and more LAN managers are being challenged to learn special-purpose system software that operates the particular types of servers that migrate to their LANs. These new devices and applications will run most efficiently when there are appropriate interfaces between their special system software modules and the NOS.

In some cases, the addition of new servers and devices doesn't require a great deal of new learning because they function similarly to existing devices. For example, the spoolers used on plotters and other special-purpose output devices carry out essentially the same functions as those used for printers. But in other cases, the inclusion of new servers and applications that bear little resemblance in operation and function to those already in use can present LAN managers with a steep learning curve and many challenges.

◆ TRENDS IN LAN OPERATING SYSTEMS AND PROTOCOLS

While we have already covered quite a number of issues concerning LAN protocols and operating systems, this chapter would not be complete without discussing some of the important trends that may be observed for these. A summary of these trends is provided in Table 5-8.

Trends in LAN NOSs	Trends in LAN Protocols
Ability to support a larger number of users and workstations	Ability to interface with a variety of interconnection protocols
Graphical user interfaces (GUIs)	Ability to run on a variety of network media
Multitasking or parallel processing capabilities	Ability to efficiently handle large and graphically based files
Ability to interface with a variety of workstation OSs	Ability to simultaneously transmit multiple types of data
Support for a variety of interconnection and media access protocols	
Enhanced backup and security mechanisms	
Enterprise-wide networks	

TABLE 5-8 Major trends in LAN NOSs and protocols

TRENDS IN LAN NOSs

One of the major trends in LAN operating systems is support for a larger number of workstations and shared devices. For example, we have already noted that as Novell's NetWare has progressed from the 2.xx versions through the 3.xx versions to NetWare 4.xx, the number of workstations that could be supported in a single LAN increased from less than 100 to 1,000. Versions of the major peer-to-peer NOSs have followed a similar trend. For example, LANtastic can now support several hundred workstations and shared devices in a single LAN.

Other trends in LAN operating systems parallel the evolution that has been occurring in microcomputer OSs. During the 1990s, graphical user interfaces (GUIs) have become the dominant user interface for microcomputer OSs. Apple Computer's Macintosh System Software established this trend during the 1980s, and when Microsoft (with Windows) and IBM (with OS/2 2.0) followed suit, GUIs soon became standard features in microcomputer OSs.

With GUIs, users can use "point and click" devices to point to icons (onscreen symbols) of the operations that they want to perform or packages that they want to use. This is usually considered more user-friendly than typing in commands for the same operations, which is necessary when using most versions of MS-DOS. Most versions of the application packages that are designed for use with operating systems with GUIs (for example, the Windows version of WordPerfect) also make it possible to point to icons to carry out many operations.

GUIs are gradually becoming more common among NOSs. While some still require typing in commands in a manner very similar to the DOS versions that were available in the 1980s, the interfaces for most are becoming more graphically oriented. One of the most graphically oriented NOSs on the market today is Windows for Workgroups, and LAN managers should expect to see most LAN operating systems begin to provide similar interfaces.

Multitasking (the ability to concurrently run two or more programs) has also become more common among microcomputer OSs. For example, an OS with multitasking capabilities enables a user to transmit a large file while working on a word processing document or spreadsheet. Multitasking is a standard feature in the latest versions of most popular microcomputer OSs including Windows, OS/2, and Macintosh System Software, and the latest releases of most of the popular LAN operating systems have been designed to take advantage of these capabilities. This means that most of the current releases of LAN NOSs also have multitasking capabilities. For example, many can accept a user request for services at the same time they are retrieving the data needed to satisfy a previous user request.

Another trend in NOSs is the ability to interface with a variety of workstation OSs. This feature recognizes that users are often most comfortable using one OS and may be reluctant to learn another just to be connected to a LAN. Most of the popular LAN operating systems make it relatively easy to specify the preferred OS when creating user profiles. (We will discuss user profiles more fully in Chapter 14.) Hence, in many LANs you may have some users running MS-DOS, others Windows, others OS/2, and so on. This means that more than one version of general-purpose application software packages may have to be included on file servers (for example, both DOS and Windows versions of WordPerfect), but that may be a small price to pay to make users comfortable working in a LAN environment.

Many NOSs are also moving toward the ability to support multiple media access methods (such as CSMA/CD and token passing) and protocols (such as TCP/IP and X.25). We have already mentioned the role that NIC drivers can play in such support and why such support may be advantageous. Users typically don't care what media access method or protocol is used as long as the network does what they need it to do. Hence, NOSs often try to make such support as transparent as possible.

A related trend is for NOSs to support multiple media access control methods and protocols for a variety of media. For example, many wireless networks support either CSMA/CD or token passing. Also, a version of FDDI for LANs using coaxial cable or twisted pair has been created. So not only are LAN operating systems likely to continue to support multiple protocols and media access methods, they are likely to do so for multiple media.

The number, variety, and sophistication of backup and security features included in NOSs for LANs have increased as new versions have been released. This has been true for both client/server and peer-to-peer LAN operating systems. Most experts agree that fault tolerance has been enhanced for most NOSs over time and that LAN security has also been tightened. Experts also agree that these improvements are due in part to increased concern about these issues on the part of LAN managers.

One other important trend in NOSs is toward coherent enterprise-wide support of LANs. As we have already noted several times in this book, one of the important trends of the 1990s is the interconnection of LANs to other LANs and to WANs. When organizations start to do this, they often discover that it is difficult to establish transparent user interconnections. To address such challenges, Novell released NetWare 4.0 in 1993. This was intended to make it possible to interconnect all LANs running NetWare in an organization and to enable users physically connected to one LAN to easily access needed resources in other LANs. Hence, NetWare 4.0 has been marketed as an enterprise-wide networking solution. Banyan's Vines is also advertised as being able to support all of an organization's UNIX-based LANs and WANs.

All of these trends in LAN operating systems indicate that these technologies are maturing and that data communication functions are being handled more efficiently and effectively in LAN environments. While there are many other trends, especially technical trends, that could be discussed in this section, we hope that we have provided you with a general sense of where NOSs for LANs seem to be headed. LAN managers are in a better position to make decisions with long-range implications when they have some idea of current trends.

TRENDS IN LAN PROTOCOLS

Two of the trends for LAN protocols, summarized in Table 5-8, have already been covered in the previous section on NOS trends. These are the ability to support multiple interconnection protocols (TCP/IP, X.25 protocol, and those used in IBM's Systems Network Architecture and Systems Application Architecture networks), and the ability to be used over a variety of media (copper-based FDDI and wireless versions of CSMA/CD and token passing). However, there are some other important trends in LAN protocols that LAN managers should be aware of.

First, many LAN protocols are being modified to handle large graphics files in an efficient manner. Files containing computer generated graphics, document images, and the like tend to be very large; a single graphic or image may be several megabytes in size. In order to transmit them efficiently between workstations, high-speed media are needed. In addition, the protocols used must be able to handle such files. The protocols most widely used in LANs were initially designed to handle text and data; in order to allow users to exchange graphics, images, and faxes, many have had to be modified. Even interconnection protocols such as TCP/IP are being modified to enable the efficient exchange of such files. As graphical and multimedia applications become increasingly common, LAN managers should expect to see further protocol modifications.

Paralleling this trend is the ability to simultaneously handle multiple types of data. Several vendors are developing and marketing products that enable network managers to provide both voice and data communications over LAN media; this essentially makes a telephone handset another workstation peripheral device. Some vendors are developing and marketing desktop-to-desktop videoconferencing products; this means that real-time video images are transmitted between workstations and displayed on user monitors. Other new products provide LANs with ISDN (Integrated Services Digital Network) interfaces; these enable users to simultaneously exchange audio, data, fax, and video signals. Like the movement toward the exchange of multimedia files, the increasing popularity of these products creates pressure to modify existing LAN protocols.

◆ SUMMARY

The computer programs that are found in LANs can be broken down into two main categories: system software and application software. Application software includes both the general- and special-purpose user-oriented programs that enable users to carry out their work activities. System software consists of the background programs that enable application software to run the hardware; this software provides an interface between application software and computer hardware and carries out the "housekeeping" tasks (such as copying files or retrieving and storing files on disk) that users need on a regular basis. Some examples of system software found in LAN environments are workstation operating systems, LAN operating systems, backup and utility programs, and print spoolers.

Protocols enable networked workstations to communicate with one another, with servers, and with shared peripheral devices. A protocol governs the flow of data between sending and receiving workstations; it consists of the set of rules that determine how and when workstations are allowed to exchange data, data format, message addressing, and error checking procedures.

Data link protocols are most important in LANs. Two of the most important media access control methods found in LANs are contention and token passing. Polling, another type of protocol, is found infrequently in LANs.

When a contention protocol is used, each workstation monitors the medium to determine if it is in use; this is called carrier sense (CS). Because any workstation can transmit data when it detects that the medium is not being used, the LAN is said to have multiple access (MA). When two or more workstations begin transmitting at the same time, the message signals interfere with one another and become garbled; the result is called a collision.

Because collisions are possible, the workstations must be able to detect and recover from them; this ability is called collision detection (CD). In light of these characteristics, it should not be surprising to learn that the most common contention protocol found in LANs is called carrier sense multiple access with collision detection (CSMA/CD); this is frequently used in LANs with bus topologies. CSMA/CA is another type of contention protocol; the CA stands for collision avoidance.

Token passing is another media access control method widely used in LANs, especially LANs with bus and ring topologies. In this approach, a small packet (a token) is passed around the ring from one device to the next. When a workstation has a message to send, it captures the token and sends its message to the next device in the LAN. FDDI (Fiber Distributed Data Interface) is a token-passing protocol that is used in true ring topologies with fiber optic cabling. Arcnet LANs are low-cost LANs that use a token-bus protocol. IBM's Token Ring LANs are really star topologies in which workstations are connected to multistation access units (MAUs).

When polling is used, a central computer (the supervisor) asks (polls) each of the workstations attached to it whether it has data to send; this is done in a round-robin fashion. If a workstation has a message to transmit, the message is forwarded to the supervisor, which then sends it to the workstation to which the message is addressed.

There are numerous LAN standards and standard setting organizations. Most of the standard setting organizations endorse the concept of open architectures in order to ensure connectivity between data communication hardware and software. Among the most widely known and endorsed open architecture models is the International Standards Organization's OSI (Open Systems Interconnection) reference model. Like all open system models, the OSI model makes technical recommendations for data communication interfaces. When hardware and software vendors follow these recommendations, their products typically can communicate with one another.

The OSI reference model is a seven-layer framework around which standard protocols can be defined. This model specifies how software should handle message transmission from one end-user workstation or application program to another. The purpose of the seven layers is to segment the functions that must be carried out for two machines to communicate with one another. The model is primarily used as a guide for hardware and software designers and programmers in developing new equipment, communications protocols, and software used in LAN and other network environments.

The Institute of Electrical and Electronics Engineers (IEEE) was one of the first standard setting organizations to endorse open architectures and the OSI reference model. IEEE Project 802 was started in 1980 to develop standards for LAN implementations and applications. A number of IEEE

802 subcommittees were formed to facilitate the development of standards. Standards for all the major media access control protocols found in LANs have been developed by the IEEE. This organization has also developed standards for the types of LANs that are likely to become increasingly common during the 1990s, such as fiber optic LANs and LANs that integrate voice, data, and video communications.

ANSI (the American National Standards Institute) serves as a coordinating organization for a variety of information systems standards in the U.S. Its role is to coordinate national standards development and interact with the International Standards Organization (ISO) so that U.S. standards will be in compliance with ISO recommendations. ANSI's FDDI standard is widely recognized and followed. The Electronics Industries Association (EIA) is an ANSI-accredited organization that has developed a variety of standards for equipment and components found in LAN environments; it developed the standards for the RS232 and RS449 connector plugs. Arcnet Trade Association (ATA) was designated by ANSI to define and set standards for the Arcnet (Attached Resource Computing Network) protocol; the ISO is expected to recognize the Arcnet protocol and endorse it as an international standard.

TCP/IP (Transmission Control Protocol/Internet Protocol) is a network-layer protocol that has been used for more than 20 years. TCP/IP was initially designed to allow computers at the Department of Defense, defense contractors, and universities to exchange large files over standard communication lines (telephone lines). It is now the most important protocol used on the Internet. The Internet will be one of the key components in the information superhighway that is being built in the U.S. More and more businesses are using the Internet for some of their business activities, and many colleges, universities, secondary schools, libraries, and government agencies are also connected to this network.

A LAN operating system (NOS) is essentially an extension of the workstation's operating system. Like all operating systems, a NOS carries out a number of input/output (I/O) tasks for users, such as retrieving files from disks controlled by network servers and directing output to shared printers.

In a LAN environment, the system software needed to carry out network functions is found in both workstations and servers. Most NOS files are installed in servers but some very important pieces are found in workstations; these work with the workstation's OS and help determine when the user is requesting network services.

Three major NOS modules are installed in each workstation. These are the network redirector module, the application program interface (API), and the network interface module. The network redirector module's primary job is to intercept all I/O requests before they are received by the workstation's

OS. If the request involves devices directly attached to the workstation, the network redirector module passes the request to the workstation's OS. If the request is for services from shared devices in the network, the network redirector module sends the request through the workstation's NIC (network interface card) and over the network to the appropriate server or device. The job of an API is to appropriately format a user's request for network services so that it can be transmitted over the network and routed to the appropriate device; part of this task often involves attaching sender and receiver addresses to the request. The network interface module is responsible for placing data onto the network and for receiving data from the network; this software module interfaces directly with the NIC. NIC drivers are another important network interface software.

Two main categories of NOS software are found within servers. The first category includes NOSs that combine LAN operating system functions and server system software in one integrated package (such as Novell's NetWare). The second category includes LAN operating systems that run under operating systems such as Windows, UNIX, or OS/2 (Windows for Workgroups, Banyan's Vines, and IBM's LAN Server). The NOS modules found within servers are responsible for carrying out a number of important network functions including access optimization, concurrent access control, fault tolerance, workstation-to-workstation communications, file transfer, access to other networks, network security, and network management.

Client/server NOSs dominate the market for LAN system software and are found in nearly 90 percent of all LANs that have been implemented; Novell NetWare alone is used in nearly 60 percent of all LANs. Other important client/server LAN operating systems include IBM's LAN Server, Microsoft's LAN Manager, Banyan's Vines, Sun Microsystem's TOPS, and Windows NT Server. Among the most popular peer-to-peer NOSs are Apple Computer's AppleTalk, Artisoft's LANtastic, Novell's NetWare Lite, and Microsoft's Windows for Workgroups.

Another important type of system software found in LANs is print spooling software. This is the system software that enables users to share printers. As organizations downsize to LANs, LAN managers must learn about a variety of other system software found in servers including database management systems (DBMS) software found in database servers and document management system software used in image processing systems.

Several trends may be observed in LAN protocols and operating systems. For example, NOSs are able to handle an increasing number of workstations. LAN operating systems are moving toward graphical user interfaces (GUIs), multitasking, and simultaneous support for a variety of workstation OSs. In addition, NOSs are moving toward support for multiple media access methods and internetworking protocols as well as the ability to handle the exchange of multimedia files. Over time, the backup, security, and LAN management features associated with NOSs are becoming more sophisticated. A goal of many NOSs is coherent, enterprise-wide support for interconnected LANs.

Two important LAN protocol trends are the ability to support multiple interconnection protocols and the ability to be used over a variety of media (including wireless media). In addition, protocols are being modified so that they can better handle large graphical files and multimedia files. LAN protocols are also moving toward support for simultaneous transmission of data, fax, voice, graphics, and video signals.

✸ KEY TERMS

Application program interface (API)
Application software
Contention
CSMA/CA
CSMA/CD
FDDI
IEEE Project 802
Network interface module

Network redirector module
OSI reference model
Polling
Protocol
System software
TCP/IP
Token passing

✸ REVIEW QUESTIONS

1. What is the difference between application software and system software? What is the difference between general-purpose and special-purpose application software? Identify examples of software that falls into each category.
2. What is a protocol? Why are protocols important?
3. What is polling? In what types of LANs is it likely to be used?
4. What are the differences between CSMA/CD and CSMA/CA? Identify examples of the major LAN products using these protocols.
5. What are the characteristics of token-passing LANs?
6. How do token-passing rings function?
7. What is a MAU? What is the MAU's purpose in a token-passing ring?
8. How do token-passing bus networks operate?
9. What are the differences between Arcnet LANs and MAP (or TOP) LANs?
10. What are the characteristics of FDDI LANs?
11. What are the differences between open and closed architectures?
12. What is the OSI reference model and why is it important?
13. How is logical link control different from media access control?
14. Identify the IEEE standards that cover the token-passing and contention protocols that are most commonly found in LANs.

15. What is TCP/IP and why is it important?
16. How does a LAN operating system differ from a workstation OS? How do NOSs and workstation OSs work together to carry out user requests?
17. Briefly describe the major functions carried out by the NOS software found in servers.
18. What functions are performed by the network redirector module?
19. What functions are carried out by application program interfaces? What are some of the APIs found in LANs?
20. What functions are performed by the network interface module?
21. What are the characteristics of NIC drivers?
22. What are the characteristics of integrated NOSs? How do these differ from other types of NOSs?
23. Identify the network functions carried out by the LAN system software located within servers. Briefly describe how each of these functions can be implemented.
24. Identify and briefly describe each of the major client/server NOSs.
25. Identify and briefly describe each of the major peer-to-peer NOSs.
26. What are print spoolers? Briefly describe the different ways in which print spooling can be configured in a LAN.
27. Briefly describe why LAN managers are having to learn about DBMS and document management system software.
28. Briefly describe each of the major trends for NOSs.
29. Briefly describe each of the major trends for LAN protocols.

✳ DISCUSSION QUESTIONS

1. What are the major issues that LAN managers are likely to consider when choosing between a contention-based LAN and a token-passing LAN? Discuss why each should be considered a major issue or selection criterion.
2. What issues and criteria are LAN managers likely to consider when choosing between token-ring, token-bus, and FDDI LANs? Discuss why each should be considered a major issue or selection criterion.
3. Why are internationally recognized LAN standards important? What are the managerial benefits of implementing a LAN that complies with internationally recognized LAN standards?
4. Discuss the advantages associated with using an integrated NOS in place of a NOS that runs under workstation OSs. Also discuss the possible advantages of using a NOS that runs under a workstation's OS instead of an integrated package.
5. Describe what you believe is the ideal set of mechanisms for ensuring the fault tolerance of a LAN.

6. Discuss how the job of a LAN manager is becoming more complicated as a result of changes taking place in NOSs and protocols.

✳ PROBLEMS AND EXERCISES

1. Visit a LAN and find out what protocols, topology, and NOS are being used. Ask why each of these was selected. Share your findings in a report or presentation.
2. Visit or obtain information about an actual FDDI LAN. Identify the applications that are run and the reasons why this type of LAN was selected. Share your findings in a report or presentation.
3. Obtain the network interface cards and media connector hardware for both a token-passing ring and and an Ethernet LAN. Identify their similarities and differences and share these in a report or presentation.
4. Bring one or more multiple access units used in token-passing ring LANs to class. Demonstrate and explain how workstations are connected to the MAU and how MAUs can be "daisy chained" to increase the size of the LAN.
5. Using trade periodicals and catalogs, identify the major vendors of NICs, connector hardware, cabling, and MAUs found in token-passing ring and Ethernet LANs. Develop price ranges and tables listing the specifications for the major products and share these findings in a report or presentation.
6. As an in-class or lab exercise, install a client/server LAN NOS (such as NetWare, LAN Manager, or LAN Server) on a server and one workstation. Also install an application (such as a word processing software package) on the server and demonstrate how the workstation would request and receive the application package from the server.
7. As an in-class or lab exercise, install a peer-to-peer LAN operating system (NetWare Lite, LANtastic, or Windows for Workgroups) on two workstations. Also install an application (such as a word processing software package) on one of the workstations and demonstrate how the other workstation would request and receive the application package from that workstation.
8. From trade publications, periodicals, or vendor documentation, identify the actual names that are used for the network redirector, application program interface, and network media interface modules for one or more NOS products. Briefly explain how the functions performed by these modules are described in these sources and share your results in a report or presentation.
9. From trade publications, periodicals, or vendor documentation, identify how access optimization, concurrent access control, fault tolerance,

workstation-to-workstation communication, file transfer, access to other networks, network security, and network management are handled by one or more client/server NOS products. Share your results in a report or presentation.

10. From trade publications, periodicals, or vendor documentation, identify how access optimization, concurrent access control, fault tolerance, workstation-to-workstation communication, file transfer, access to other networks, network security, and network management are handled by 1) a client/server NOS product and 2) a peer-to-peer LAN OS product. Identify the similarities and differences in how the products carry out these functions and share your results in a report or presentation.

11. Using a NOS product, demonstrate how print spooling operations can be implemented and modified. Identify the types of printer spooler configurations that the NOS product can support as well as the types of menus or other interfaces that are created to enable the user to select from two or more shared printers.

12. Using a NOS product, demonstrate the types of user interfaces that are supported.

✳ REFERENCES

Barrett, E. "The Critical Steps to Hub Selection." *LAN Technology* (October 15, 1992): p. 39.

Bates, R. J. "Bud." *Disaster Recovery for LANs*. New York: McGraw-Hill, 1994.

Berg, L. "The Scoop on Client/Server Costs." *Computerworld* (November 16, 1992): p. 169.

Burns, N. and K. Maxwell. "Peering at NOS Costs." *LAN Magazine* (May 1992): p. 69.

Chacon, M. E. "More Horsepower." *LAN Magazine* (November 1994): p. 53.

Chang, A. and J. Wall. "Making Important Bus Architecture Decisions." *LAN TIMES* (July 8, 1991): p. 33.

Derfler, F. "DOS Based LANs Grow Up." *PC Magazine* (June 25, 1991): p. 167.

———. "Ethernet Adapters: Fast and Efficient." *PC Magazine* (February 14, 1993): p. 191.

———. "Network Printing: Sharing the Wealth." *PC Magazine* (January 26, 1993): p. 249.

———. "Peer-to-Peer LANs: Teamwork Without Trauma." *PC Magazine* (May 11, 1993): p. 203.

Dern, D. P. "TCP/IP for Ma and Pa: A Simple Solution at Last." *LAN TIMES* (August 9, 1993): p. 35.

Fitzgerald, J. *Business Data Communications: Basic Concepts, Security, and Design*. 4th ed. New York: John Wiley & Sons, 1993.

Frank, A. "Networking Without Wires." *LAN Technology* (March 1992): p. 51.

Frenzel, C. W. *Management of Information Technology*. Boston: Boyd & Fraser, 1992.

Gantz, J. "A Practical Guide to Cutting Network Cost." *Networking Management* (May 1991): p. 24.

Goldman, J. E. *Applied Data Communications: A Business-Oriented Approach*. New York: John Wiley & Sons, 1995.

Henderson, T. "The Multimedia Menace." *LAN Magazine* (November 1994): p. 138.

Jain, R. "FDDI: Current Issues and Future Plans." *IEEE Communications* (September 1993): p. 98.

Janusatis, B. "LAN Management." *Computerworld* (January 27, 1992): p. 91.

Keen, P. G. W. and J. M. Cummins. *Networks in Action: Business Choices and Telecommunications Decisions*. Belmont, CA: Wadsworth, 1994.

Kindal, S. "Networks That Work." *Financial World* (April 28, 1992): p. 65.

Kine, B. "Understanding the Requirements of 10BaseT." *LAN TIMES* (May 11, 1992): p. 27.

Kozel, E. R. "Commercializing the Internet: Impact on Corporate Users." *Telecommunications* (January 1992): p. 11.

Lamb, J. "The New NetWare: SMP Workhorse." *LAN Magazine* (November 1994): p. 42.

Moad, J. "One True Operating System Out, Diversity In." *Datamation* (May 15, 1994): p. 44.

Molta, D. "Network Standards: Setting and Enforcing the Rules of the Road." *Network Computing* (February 1992): p. 62.

Panchak, P. L. "Network Alternatives: Making the Right Choice." *Modern Office Technology* (March 1992): p. 22.

Parker, C. and T. Case. *Management Information Systems: Strategy and Action*. 2nd ed. New York: Mitchell/McGraw-Hill, 1993.

Rains, A. L. and M. J. Palmer. *Local Area Networking with NOVELL Software*. 2nd ed. Danvers, MA: Boyd & Fraser, 1994.

Rigney, S. "Wiring Hubs: The Low Cost Alternative." *PC Magazine* (November 10, 1992): p. 335.

Salamone, S. "Test Shows Which NOS Performs Best for You." *Network World* (January 20, 1992): p. 1ff.

Schatt, S. *Data Communications for Business*. Englewood Cliffs, NJ: Prentice Hall, 1994.

Schnaidt, P. and D. Brambert. "NetWare Express." *LAN Magazine* (June 1992): p. 36.

Snell, N. "Where's UNIX Headed?" *Datamation* (April 1, 1994): p. 24.

Stamper, D. A. *Business Data Communications*. 4th ed. Redwood City, CA: Benjamin/Cummings, 1994.

———. *Local Area Networks*. Redwood City, CA: Benjamin/Cummings, 1994.

Strole, N. C. "The IBM Token Ring Network—A Functional Overview." *IEEE Network* (January 1987): p. 1.

Tolly, K. "Token Ring Adapters Evaluated for the Enterprise." *Data Communications* (February 1993): p. 73.

Volvino, J. "LAN Operating Systems." *Computerworld* (December 2, 1991): p. 75.

Case Study:
Southeastern State University

Largely because microcomputers and networking equipment acquisitions have been the responsibility of the colleges and departments within colleges, quite a variety of network operating systems is found within the LANs at SSU. The majority of the networked microcomputers are found in LANs running Novell NetWare (typically version 3.xx). However, a number of other NOSs have a foothold at SSU including Banyan Vines, AppleTalk, LAN Manager, Windows NT, and Windows for Workgroups. Of these, Banyan Vines, LAN Manager, and AppleTalk have the largest presence. Both Windows NT and Windows for Workgroups are relative newcomers at SSU and each has been implemented only once in relatively small administrative areas.

Banyan Vines is used in a UNIX LAN developed by the Math and Computer Science Department as well as in the small network of SPARC workstations found in the College of Engineering. As you might expect, the College of Education's affinity with Apple Computer products has led to the creation of two AppleTalk LANs (using LocalTalk) in the education building. LAN Manager is being used to network the point-of-sale (POS) systems found in university-owned stores and shops in the Student Union building. In addition, the Department of Geography received a federal grant to develop an FDDI LAN in order to investigate the pedagogical value of student access to geographic information systems (GISs)—mapping systems with graphically-oriented output.

Ethernet LANs are the most common. This is primarily due to the price differentials between the network interface technologies for Ethernet and those for token-ring LANs; the academic units at SSU that have implemented Ethernet LANs found that for the same amount of money, they could network 50 to nearly 80 percent more workstations than they could with token ring. There are more 10Base2 Ethernet LANs at SSU than any other type. However, there are several 10BaseT LANs and one 10Base5 Ethernet LAN. The LAN that is now using Windows NT was once a 1Base5 Ethernet LAN.

The operating systems with the greatest potential for growth at SSU are Novell NetWare, UNIX, and Windows NT. Novell NetWare, as the most common LAN NOS on campus, is well positioned for further growth. Networking personnel (the LAN managers) at SSU collectively have more experience and knowledge of NetWare than any other NOS. Hence, when new LANs are proposed, NetWare is at the top of everyone's list, especially when the proposed LANs are in areas that have no previous experience with LANs. The widespread knowledge and use of NetWare make it appear to be the least risky alternative, especially for the uninitiated.

UNIX-based systems are also likely to become more common. Several forces drive this, but the biggest one is external. As noted in previous segments of this case, the state's university system made the decision to have the university computer network become part of the Internet. Because TCP/IP is the primary protocol on the Internet, the state has adopted TCP/IP as the protocol for exchanging data communications among the university system's colleges and universities. Since UNIX has provided some of the best TCP/IP support among NOSs, the Computer Services Department of SSU has not discouraged the creation of UNIX-based LANs, and has, in fact, promised limited support for implementers. In addition to university system and the Computer Services Department's support for UNIX-based systems, the College of Business Administration wants its MIS students to have experience developing applications in UNIX environments. To this end, it persuaded the university to purchase an RS-6000 server running AIX (IBM's proprietary version of UNIX) that students taking COBOL and database applications courses will have access to.

Windows NT is likely to become more popular largely because of the increasing use of Windows and Windows-based applications by administrative and clerical workers and some faculty. Since this NOS provides graphical user interfaces with which users have grown quite comfortable, it is easy to secure user support for using Windows NT. Using Windows-based applications does not change very much with Windows NT, and most experienced Windows users feel that learning and using the communications modules is not difficult.

CASE STUDY QUESTIONS AND PROBLEMS

1. Using the information provided in this segment of the case study and the concepts covered in the chapter, develop an inventory of the types of media and media access control methods found in the LANs at SSU. Use three columns for this inventory: LAN Type, Media Type, and Media Access Control Method.
2. If you wished to interconnect LANs in the College of Education with a Novell NetWare LAN in another building, what networking obstacles would have to be addressed?
3. SSU's Computer Services Department is encouraging the managers of all stand-alone LANs that want access to the VAX computers on campus and the Internet to adopt a NOS that supports TCP/IP; ideally they

would like all LAN-to-VAX communications to utilize TCP/IP. Identify and briefly discuss the advantages to LAN managers of adhering to this request. Also identify and discuss any disadvantages.

4. In light of the information presented in this and the previous segments of the case study, to what extent are the trends in NOSs and communication protocols at SSU consistent with the general trends discussed in this chapter?

Middleware

CHAPTER OBJECTIVES

After completing this chapter, you will be able to:

◆ Describe what middleware is and the role it plays in LAN environments.

◆ Briefly describe the functions performed by middleware found in workstations, servers, and intermediate nodes.

◆ Describe the differences among the major types of middleware found in computer networks.

◆ Identify some of the major middleware vendors and products available on the market today.

◆ Briefly discuss how middleware products compare on criteria important to LAN managers.

◆ Provide several examples of how organizations are using middleware.

In this chapter, we will focus on middleware. *Middleware* is software found in client/server networks that helps establish a connection between distributed clients and servers. Middleware can be thought of as an intermediary between clients and servers; it often enables users at client workstations to use the operating system that they are most comfortable with (DOS, OS/2, Windows, or UNIX) and still be able to access resources on servers that use other operating systems. Middleware is also used by application programmers to develop applications that can be run on networked computers using a variety of operating systems. Many experts see middleware as playing a crucial interconnectivity function. Several have referred to middleware as a "glue" that can bind together pieces of applications distributed among computers in diverse networks. Because of the emergence of client/server networks as the dominant network computing platforms of the 1990s, and because of the special role that middleware is expected to play in such networks, we felt that it was important to devote a chapter of the textbook to this software.

◆ WHY IS MIDDLEWARE IMPORTANT?

Middleware is essentially a layer of software running between applications and the network. It shields users and application developers from the complexity of underlying communications protocols, operating systems, and hardware configurations that may exist. As discussed in Chapter 5, there is a variety of internetworking (for example, TCP/IP) and media access protocols (such as CSMA/CD and token passing) in LAN environments. In addition, we have noted throughout this text that organizations have grown increasingly interested in interconnecting their LANs (as well as in establishing LAN to WAN connections) so that users in one network can access the computing resources available in other networks. Some industry watchers report that by 1995, interconnected LANs will outnumber stand-alone LANs by more than two to one. Many experts expect stand-alone LANs to be a rarity by the turn of the century.

Ideally, the interconnections among LANs and other networks should be transparent to users. That is, users should be responsible only for requesting the resources that they need; they should not have to worry about how the request gets routed through the network to the appropriate server(s). When users have to go through a complicated series of steps to obtain needed computing resources, they are more likely to perceive the network as being unfriendly or unusable. Even if the capabilities are in place to enable users to access needed resources, the more complicated it is to do so, the less likely they are to try. This can defeat the whole purpose behind giving them access; the cost savings and potential productivity increases that could accrue from shared resources may not be realized.

Similar arguments apply to the programmers who develop applications in client/server environments. This is especially true for the application developers who write programs that enable users to share resources or that make it possible for the same application to be run on a variety of user workstations and workstation operating systems. Programmer productivity is likely to be slowed if they have to write the code needed to convert from one protocol to another (and from one operating system to another) in addition to writing the code for the actual applications. If all that programmers have to worry about developing is the actual application, they should be able to get applications to users more quickly. This, in turn, should help organizations reap the benefits of client/server computing that so many industry experts are raving about.

Software vendors who see client/server networks as the dominant computing platforms of the 1990s (and beyond) have turned their attention to middleware products. These products are essentially geared toward application developers, but users are the ultimate beneficiaries. The products provided by middleware vendors shield application developers from the complexities of underlying networks, network operating systems, and communications protocols, so that they can concentrate on building and enhancing applications. Developers should be able to write better programs, especially programs that allow users to integrate and share resources spread throughout the organization's LANs and other networks. Also, programmers should be able to develop applications more quickly. Some vendors claim that distributed applications can be written between 4 and 10 times faster with the assistance of middleware.

The potential for such improvements in client/server application programming is made possible because middleware makes it unnecessary for application programmers to become network experts. Middleware products provide programmers with a common set of application programming interfaces (APIs) that enable different applications to communicate across networks, regardless of how diverse their hardware platforms and networking protocols may be. Many middleware products also provide services (including directory, name, and authentication services) needed to support and manage distributed computing environments.

Many experts have predicted that middleware is destined to be a focal point in helping organizations realize the benefits of client/server computing. Middleware is expected to make it easier for organizations to integrate heterogeneous computing environments. It is also expected to make it easier for programmers to write robust enterprise applications.

Many organizations have already found middleware tools invaluable when downsizing applications traditionally run on mainframes to LAN environments. Other organizations have used middleware to support reengineering (of business processes) efforts, for deploying client/server applications, and for integrating customers and mobile workers into corporate computing environments.

◆ WHERE IS MIDDLEWARE FOUND IN NETWORKS?

The exact location of middleware can vary with the network configuration. However, middleware modules are found at clients (user workstations), the servers at which shared resources are located, and at all of the intermediate nodes between clients and servers through which user requests for network resources may be routed. This is illustrated in Figure 6-1.

FUNCTIONS PERFORMED
BY MIDDLEWARE AT WORKSTATIONS

The middleware modules installed in workstations carry out a number of important functions. These are summarized in Table 6-1. The primary job of middleware is to ensure that the user request is routed to the correct server so that the request can be executed and results sent back to the user. At the workstation, middleware modules facilitate this process by creating the message, identifying the server that can provide the services being requested, establishing the connection between the workstation and the server (or intermediate node), and converting the message into the format (demanded by the media access protocol) for transmission over the communications medium.

Once the message is on its way, the middleware at user workstations often performs message tracking and message arrival functions (this is

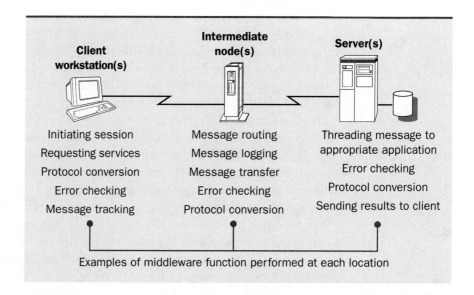

FIGURE 6-1
Middleware locations in interconnected client/server networks

Client workstation(s)	Intermediate node(s)	Server(s)
Initiating session	Message routing	Threading message to appropriate application
Requesting services	Message logging	Error checking
Protocol conversion	Message transfer	Protocol conversion
Error checking	Error checking	Sending results to client
Message tracking	Protocol conversion	

Examples of middleware function performed at each location

TABLE 6-1

Functions
performed by
middleware at
client
workstations

◆ Establishment/closing of the communications session between the workstation and the server and/or intermediate node(s)

◆ Identification of the server(s) that can satisfy the user's request for services

◆ Initiation of the message exchange between the client and the server(s)

◆ Protocol conversion

◆ Error detection

◆ Message arrival tracking

◆ Failure notification

especially important when messages have to be routed through two or more intermediate nodes). In addition, middleware will inform users of communications or message transfer failures.

After a user request has been processed, middleware facilitates the acceptance of the results shipped back from the server and checks them for communications errors. On the return route, middleware also converts the results from the format demanded by the media access control protocol into the format required by the application program interface (API). In this latter role, middleware carries out many of the same functions that would be performed by APIs and network interface modules in a stand-alone LAN (these were described in Chapter 5).

FUNCTIONS PERFORMED BY MIDDLEWARE AT INTERMEDIATE NODES

When LANs are interconnected (or when LANs are connected to WANs and other networks), it may be necessary to route a user's request for services from a particular server through one or more intermediate nodes. In such situations, middleware modules located at the intermediate nodes are likely to facilitate a number of important communications chores; some of these are summarized in Table 6-2.

TABLE 6-2

Functions
performed by
middleware at
intermediate
nodes

◆ Accept (and respond to) messages received from user workstations that are addressed to particular servers

◆ Identify the best route (and alternate routes) to the appropriate server for the message

◆ Establish a connection (session) with the appropriate server or the next intermediate node

◆ Protocol conversion

◆ Message logging and buffering

◆ Send the message to the server (or the next intermediate node)

◆ Error checking

Once a request is received by an intermediate node, middleware often assists in checking for communications errors; if no errors are found, an acknowledgment may be sent to the sender indicating that the message has arrived at the intermediate node. Middleware may also facilitate message logging at intermediate nodes. In message logging, the message is copied and placed in a special file called a **message log**. Message logging makes it possible to store the message; it can be sent later if the intermediate node is unable to immediately pass the message to the server. Incidently, when results are being passed from servers to client workstations, logging also typically occurs at intermediate nodes. Message logging capabilities are also called **store-and-forward capabilities**. Such capabilities allow both clients and servers to do other work until the message can be transmitted or the request can be satisfied. Message logging can also be important in network recovery after a failure by ensuring that copies of active messages are available even if a server, node, or communications line fails.

Middleware found at intermediate nodes may also determine the best route to the addressed server as well as alternate routes in case the best route cannot be used. When intermediate nodes are responsible for identifying and selecting from among alternate message routes, a **dynamic routing** scheme is said to be used. Dynamic routing can help ensure that messages are delivered rapidly, efficiently, and cost-effectively.

Once the best route and alternate routes from the node to the server have been identified, middleware establishes a "session" (connection) with the server and passes the message along to the middleware at the server. Middleware at intermediate nodes is often responsible for putting the user's message in the format needed for transmission over the media connecting the nodes and the servers. This may mean converting the message format from the protocol used to transmit the message to the immediate node (such as TCP/IP) to the protocol needed to send the message from the node to the server (such as X.25).

MIDDLEWARE FUNCTIONS PERFORMED AT SERVERS

Once the message arrives at the server, it is checked for errors, converted to the API needed for the application requested, and passed to the application for processing. Once processing has been completed and results have been generated, middleware begins transmitting the server's response back to the user, including the routing of the results through intermediate nodes. The activities carried out by middleware modules stored at servers are summarized in Table 6-3.

As noted in Chapter 1, peer-to-peer client/server networks are considered to be the most sophisticated of the client/server computing alternatives. In these types of client/server networks, the individual computers

◆   Receives user's request for services

◆   Passes user's request to the appropriate application

◆   Formats results of user's request from application for transmission over network

◆   Sends results to user

◆   Error checking

◆   Protocol conversion

TABLE 6-3
Functions carried out by middleware at servers

serve as both clients and servers, and they cooperate (and interact) when processing an application. To many, this arrangement represents the true meaning of the phrase "the network is the computer."

In peer-to-peer client/server networks, the application is not located in a single server. Instead, parts of it are distributed to two or more servers. Needless to say, middleware typically must play an even greater role in such arrangements, especially when the servers with different pieces of the application are located in networks that use very different protocols and operating systems. For these types of distributed applications, middleware can truly be considered the glue that binds the different parts of the applications together. It is beyond the scope of this book to discuss in depth client/server computing at this level of sophistication. However, Figure 6-2 has been included to provide a graphical summary of such arrangements.

◆ TYPES OF MIDDLEWARE

Because middleware is an emerging set of technologies, universal agreement on what is (and what is not) middleware is difficult to find. However, the general consensus has three fundamental types of middleware: remote procedure calls (RPCs), message-queuing software, and object request brokers (ORBs). Each of these technologies has its own strengths and weaknesses in supporting distributed applications in interconnected, but heterogeneous, networks. In addition, the origins of each approach are somewhat different, and each seems best suited to addressing certain types of network integration problems. However, because the majority of the middleware products available from vendors are built on one (or more) of these three core technologies, we will discuss each in this section.

REMOTE PROCEDURE CALLS

Remote procedure calls (RPCs) are probably the best-known type of middleware. RPC technologies are the oldest and most mature type of middleware for distributed applications. Products built around RPCs typically provide programmers an application program interface (API) able to work

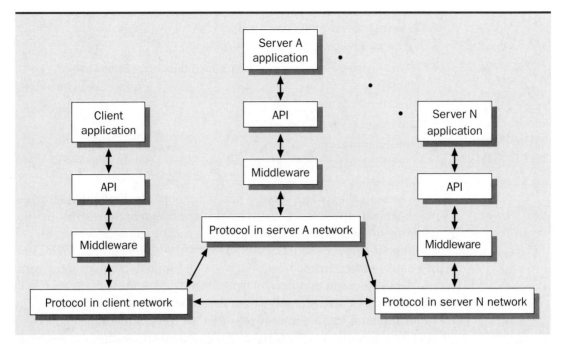

FIGURE 6-2 Middleware in a client/server network with peers using a single application distributed among servers

with a variety of communications protocols. RPCs are essentially an extension of the procedure-based programming concepts with which the majority of programmers are familiar. When RPCs are used, the program executes a procedure call in the same way that it calls a local subroutine. However, the middleware directs the call to the remote server, which executes the procedure and returns the results. In brief, RPCs essentially extend conventional application programming models to network environments. The basic RPC process is illustrated in Figure 6-3.

Products built around RPCs make it possible for programmers to develop applications for client/server networks in a familiar way. Programmers are able to code applications without having to devote a lot of time to writing the code needed to establish the communications links to various network services. The middleware makes sure that the RPCs embedded in the code for the application are routed to the correct server for execution, and that needed results are returned.

When writing RPCs, programmers work with the middleware's RPC development language, generically known as Interface Definition Language. This is used to write the program modules linking the application to middleware so that communication tasks can be performed. These modules specify how the client "calls" the application at the server and essentially inform the server what data it will be sent and in what format the

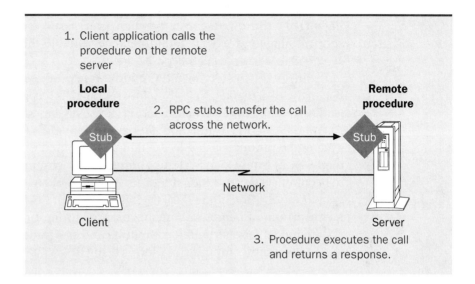

1. Client application calls the procedure on the remote server

Local procedure

2. RPC stubs transfer the call across the network.

Remote procedure

Stub

Stub

Network

Client

Server

3. Procedure executes the call and returns a response.

FIGURE 6-3
A remote procedure call

data will be. The resulting program modules are embedded in the application source code and, once the program is compiled, middleware becomes part of the resulting application.

A special program compiler (RPC compiler) is used by application developers. Once the application is developed, the program code is run through the RPC compiler. The compiler generates a segment of code (that is, a set of program statements) called a *stub*. The stub is installed in both the program doing the calling (the client program) and the application program at the server (the server program). The stubs generated by the RPC compiler are what constitute middleware; they (logically) are found between the client and server applications and establish the connections necessary for them to exchange data.

Because different platforms (operating systems) handle data differently, the program stubs generated by the RPC compiler must make sure that data is translated into—and out of—a common format. This is called *external data representation (XDR)*. XDR ensures that data from the client is presented in the correct format needed by the server, and that results transmitted by the server are in the format needed by the client.

Because RPCs use a call/response method of communication between clients and servers, response time, network performance, and network and server availability are crucial factors when this type of middleware is used. Once an RPC has been issued, the client workstation is unable to do any processing until it receives a response from the server. For this reason, RPCs are best suited to high-speed, fault-tolerant LANs that have sufficient bandwidth to ensure adequate response time. RPCs are less suited to WANs and interconnected LANs and WANs, as response times are likely to be slower and the potential for unavailable communications links is higher. In addition, because the client can't issue the request until the server is able to process it, fault tolerance (often in the form of duplexed servers) and server availability are critical.

Some of the best-known RPC middleware products are listed in Table 6-4. In the RPC market, NobelNet, Incorporated, is recognized as one of the leading vendors because of its offerings for UNIX-based systems (EZ-RPC), NetWare LANs (RPCWare), and Windows-oriented LANs (WinRPC). However, of the products listed in the table, Distributed Computing Environment (DCE) from Open Systems Foundation, Incorporated, is the closest to becoming a *de facto* RPC standard. DCE is resold by a number of major system vendors including IBM, Digital Equipment Corporation, and NCR. A consortium of nearly 350 organizations is promoting DCE as an open system model for distributed computing and as a framework for integrating multivendor computing platforms using RPCs.

MESSAGE QUEUING

While RPCs are best used for linking tightly-coupled applications (such as those found in cooperative processing applications, where the key parts of a single application program are spread across multiple processors), message-queuing middleware excels at connecting loosely coupled (and relatively autonomous) applications dispersed throughout the organization. These may be applications developed to support activities in particular departments but which need to share data (such as inventory and order processing).

Messaging-oriented middleware (MOM) is the second-most widely used technology for building distributed applications. MOM allows client and server segments of an application to communicate with one another via messages. The messages passed between the client and server consist of requests from clients and responses from servers. These are typically handled

Product	Vendor
Distributed Computing Environment (DCE)	Open Systems Foundation, Incorporated
EZ-RPC	NobelNet, Incorporated
Network Computing System	Hewlett-Packard Company
Open Systems Computing	Sun Microsystems, Incorporated
RPC Tool	Netwise, Incorporated
RPCWare	NobelNet, Incorporated
WinRPC	NobelNet, Incorporated

TABLE 6-4
RPC-based middleware products

by an interprogram messaging system. Similar to e-mail, interprogram messaging uses a store-and-forward mechanism to send messages between clients and server. However, unlike e-mail (which consists of relatively unstructured data such as text, audio, and video), interprogram messages consist of structured data transmitted in packets that include (header) information telling the receiver what to do with the data.

The interprogram messaging systems used with MOM utilize message queues, buffers, or logs capable of storing messages at intermediate nodes while en route between clients and servers. Such queuing and store-and-forward mechanisms virtually guarantee message delivery, even if there is a network or node failure. Messages can be stored in a queue until the receiver (or network) is available and the message can be delivered. Cost-optimization can therefore be built into the messaging system; for example, the most cost-efficient routes to the receiver can be selected and low-priority messages can be held for non-peak times. In addition, queuing, logging, and store-and-forward mechanisms usually provide fault tolerance.

One of the main differences between MOM and RPCs is that messaging products support high-level application program interfaces (APIs) that shield programmers from the complexities of underlying network protocols and operating systems. These high-level APIs consist of an average of four major commands and a dozen error conditions. The role of these high-level APIs in the message-queuing process is illustrated in Figure 6-4.

Relative to RPCs, MOM products provide application developers with greater flexibility in controlling how client and server applications interact. For example, messages can be passed among clients, servers, and intermediate nodes in an asynchronous or synchronous fashion (asynchronous being the most common). In fact, both transmission modes may be used for the same transaction. For example, a client may transmit an asynchronous message to an intermediate node that, in turn, passes the message

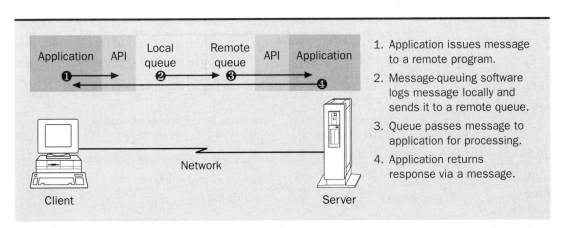

FIGURE 6-4   A message-queuing application

along in a synchronous fashion to the server. Some MOM products also allow users to broadcast messages to all devices on a network and to route a single message to nodes on a predefined mailing list.

Some of the more sophisticated MOM products provide distributed networking services and RPCs in addition to messaging. Distributed networking services may include naming services that allow any application located in the network to be easily found.

As noted, interprogram messaging is best used to connect loosely coupled applications. This makes MOM products desirable to organizations that wish to establish electronic links with their customers and suppliers. MOM is also well-suited to building work flow applications (such as just-in-time manufacturing or delivery systems) and for integrating departmental applications (such as purchasing, inventory, and shipping).

MOM is not as widely used as RPCs primarily because it is a relatively new set of technologies. However, there are quite a few MOM vendors (see Table 6-5), and a vendor alliance, the Message-Oriented Middleware Association (MOMA), has been formed. These vendors are very interested in developing a standard API that would make it easy for applications developed using one MOM product to communicate with applications developed using other MOM products.

OBJECT REQUEST BROKERS

The third and currently least-widely used type of middleware is a set of products collectively referred to as *object request brokers (ORBs)*. ORBs are an offshoot of object-oriented programming systems (OOPS). As middleware technologies, ORBs are less mature than both RPCs and message-oriented middleware. However, many experts view ORBs as promising, just as they see OOPS becoming the dominant programming systems by the end of the century.

Product	Vendor
Communications Integrator	Covia Technologies
DECmessageQ	Digital Equipment Corporation
Message Express	Horizon Strategies, Incorporated
Datatrade	IBM
X-IPC	Momentum Software Corporation
Pipes	Peerlogic, Incorporated
Suitedom	Suite Software
ezBridge Transact	Systems Strategies, Incorporated

TABLE 6-5 Message-oriented middleware (MOM) products

Object-oriented programming enables application developers to develop modular components (objects) that can be reused, which substantially reduces the time it takes to develop applications. Objects consist of data as well as the basic operations used to manipulate the data. Objects also hide (encapsulate) their contents behind an interface that is understandable to all objects. These properties make it possible for application developers to change the contents of an object without affecting other objects or the way in which objects communicate with one another.

ORBs bear some similarity to both RPCs and MOM products but are more similar to message-queuing software. ORBs are used in distributed object computing environments in which resources, applications, and program components are defined as objects. They relieve object-oriented programmers of having to know the location or properties of remote objects for the objects to exchange information with one another. In fact, object-oriented programmers are shielded from messaging functions by this type of middleware; they are therefore able to develop distributed applications in the same way that they would in a stand-alone environment.

Essentially, ORBs enable objects to interact through a high-level API that permits objects to exchange information across a network. An ORB accepts requests from client objects, finds the remote objects that can fulfill those requests, invokes the requests (that is, causes remote objects to fulfill the requests), and passes the results back to the client objects that made the request. This process is illustrated in Figure 6-5. The similarity of this process to the RPC process is no accident. Many of the ORBs implemented use RPC-like transport mechanisms for controlling the interactions between networked objects.

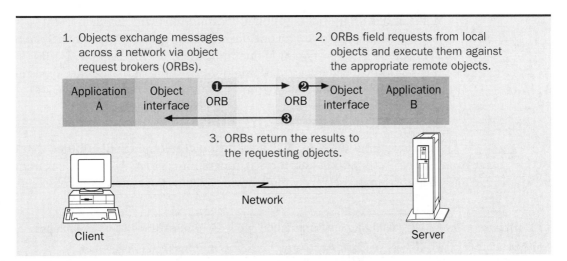

FIGURE 6-5 The role of ORBs in the exchange of messages between objects

ORBs are the heart of a set of products called ***distributed object management systems (DOMS)***. In general, DOMS are used to support message exchanges between large objects such as databases and applications. ORBs are responsible for the routing, integrity, and security of messages passed between distributed objects.

The de facto standard for ORBs is ***CORBA (Common Object Request Broker Architecture)***. This has been defined by the Object Management Group (OMG), a consortium of nearly 400 vendors of OOP products. CORBA is essentially a standard API through which clients can access ORBs.

Several vendors have released ORB products (see Table 6-6). However, the only products that fully adhere to the CORBA standard are from HyperDesk Corporation. These include Domain Object Management System (HD-DOMS) and NetWare for DOMS.

Industry experts expect both DOMS and ORBs to become more common as OOPS become more popular and as object-oriented operating systems (such as Microsoft's Cairo or Taligent Corporation's Pink) gain acceptance. Most experts expect DOMS services to eventually be embedded in object-oriented operating systems.

◆ OTHER TYPES OF MIDDLEWARE

As with other new information technologies, there is no universal agreement as to what middleware is and is not. While most industry experts concur that RPCs, MOM products, and ORBs should be classified as middleware, others argue that additional technologies should be classified as middleware. These include tools used to develop SQL (Structured Query Language) applications in distributed database environments, gateways between diverse databases, mail-embedded (or mail-enabled) applications, and even products (dubbed *local middleware*) that allow software packages designed to run on one operating system (say, Macintosh System 7) to run on other workstation operating systems (for example, UNIX). In this section we will take a very brief look at some of these other types of software with middleware-like functionality.

SQL TOOLS FOR DISTRIBUTED DATABASES

Middleware products for SQL application developers in distributed database environments enable the SQL programmers to make an SQL call from clients to remote servers without having to know the underlying protocol(s).

TABLE 6-6 Some vendors who sell ORB products	◆ Digital Equipment Corporation	◆ Internation Business Machines
	◆ Helwett-Packard Company	◆ Ions Technologies, Ltd.
	◆ HyperDesk Corporation	◆ SunSoft, Incorporated

Such middleware figures out the location of the database server that can satisfy the client request, sends the request to the server, and ensures that the client receives the response from the server. This is similar in function to an RPC.

Like other types of middleware, middleware products for distributed SQL applications shield programmers from underlying protocols and provide them with the ability to access remote database servers across a network, regardless of its location and operating system. Some of the products that provide SQL programmers with such middleware-like functionality include Borland's SQLConnect, Information Builders' Enterprise Data Access/SQL (EDA/SQL), Trifox's Vortex, and Oracle Corporation's SQL*Net.

DATABASE GATEWAYS

These include point-to-point gateways between LAN workstations and specific database servers as well as SQL gateways. Point-to-point gateways provide users in a LAN that uses a LAN NOS (say, Novell NetWare) with access to data stored on a specific database server (say, an IBM 3090 whose operating system is MVS and whose database management system is DB2). Examples of such products include Database Gateway from Micro Decisionware, Inc. and Showcase Gateway/Server from The Rochester Software Connection, Inc.

SQL gateways provide access to two or more different relational database products that each run their own version of SQL. These products provide clients with access to these different database servers and databases by translating client requests into the appropriate SQL syntax needed by the different database servers. SequelLink from TechGnosis, Inc. is an example of an SQL server; it can provide clients with access to DB2, Ingres, Informix, Oracle, Sybase, and other relational databases.

MAIL-ENABLED APPLICATIONS

Several vendors have developed products that allow application developers to use e-mail systems to exchange data between clients and servers. These usually include an API and middleware-like software that can reformat data from one e-mail system to another. An example is a product from Soft-switch, Inc.: Directory Synchronization/cc:Mail, which automatically links the user directory of Lotus Development Corporation's cc:Mail with the user directories of other major e-mail packages including IBM's Professional Office System (Profs) and DEC's All-in-one.

Several vendors are pushing to develop a standard API for message-enabled applications. Apple Computers, Inc., Lotus Development Corp., and Novell, Inc. are supporting a standard called Vendor Independent

Messaging (VIM). Microsoft has proposed its Microsoft API (MAPI) as a standard API for message-enabled applications.

LOCAL MIDDLEWARE

Local middleware products essentially insulate an application (usually a commercially available software package) from the workstation operating system. This makes it possible for an application package designed to run on one operating system to be used on a workstation running a different operating system. For example, Insignia SoftPC allows DOS and Windows applications to run on UNIX. Liken, a local middleware product from Xcelerated Systems, allows UNIX users to run Macintosh programs.

◆ COMPARING MIDDLEWARE OPTIONS

Many criteria could be used to weigh the trade-offs associated with each of the major types of middleware. Among these are support for both synchronous and asynchronous transmission modes, the presence of standards, the number of vendors, the learning curve for traditional application developers, the ability to recover from failure, and the ability to support reengineering efforts. The manner in which RPCs, MOM products, and ORBs stack up against these criteria is summarized in Table 6-7 and briefly discussed in the following sections.

SYNCHRONOUS AND ASYNCHRONOUS TRANSMISSION

As discussed in Chapter 2, two very common transmission modes used in computer networks are synchronous and asynchronous transmission. Synchronous transmission is used to send multiple data bytes as a packet from sender to receiver; it is desirable when the files to be exchanged are large. Asynchronous transmission is used to send one byte at a time from sender to receiver; it can be used effectively when messages are relatively short.

TABLE 6-7
Comparison of the major middleware technologies

Attribute	RPCs	MOM	ORBs
Asynchronous transmission	No	Yes	Yes
Synchronous transmission	Yes	Yes	Yes
Standards	Many	Few	Very few
Vendors	Many	Few	Very Few
Learning curve	Easy	Steep	Very steep
Failure recovery	Poor	Great	Unknown
Reengineering support	Yes	Some	Maybe

New, very-high-speed approaches for asynchronous communications (including ATM—asynchronous transfer mode) are likely to cause a resurgence of interest in this transmission mode.

As may be observed in Table 6-7, both MOM products and ORBs support both synchronous and asynchronous transmissions. However, asynchronous communications are most common for MOM products.

Because of their makeup, RPCs are used almost exclusively with synchronous transmission. This is because once the client has issued a call for a routine to be performed on a remote server, it suspends work on the client until the requested results are received from the server. By sending both the call and results over communications media using a high-speed synchronous protocol, client wait time is minimized. This is why RPCs are almost invariably based on synchronous transmission modes. However, the availability of high-speed asynchronous transmission protocols such as ATM may change this in the years ahead.

STANDARDS

In our discussion of RPCs, MOM products, and ORBs, we pointed out that the standards for middleware technologies are either *de facto* or proprietary in nature. Still, when supported by vendors, the presence of standards helps ensure that the products they develop are interoperable (able to communicate/interface with one another). There is evidence of progress toward universally accepted standards for each of the major categories of middleware, but that for RPCs is many steps ahead of that for MOM products and ORBs.

It is safe to call Open Systems Foundation's DCE (Distributed Computing Environment) a *de facto* standard for RPC technologies. DCE-based products have been developed by a number of major vendors including DEC, Hewlett-Packard, IBM, and NCR.

MOM vendors have recently formed an alliance, the Message-Oriented Middleware Association, and are developing a standard API for message-oriented middleware products, as well as standards for systems management, message prioritization, user account management, and systems security. However, to date, true standards have not been developed. As this new MOM vendor alliance solidifies, the pace of progress toward a standard API and other needed standards for these products should increase.

For ORBs, the CORBA standard developed by the Object Management Group (OMG) seems to be emerging as the *de facto* standard. However, as there are few products on the market developed to comply with CORBA guidelines, it is still a little early to know whether it will stand the test of time.

AVAILABILITY OF VENDORS

LAN managers do well to lean toward network technologies for which there are a large number of vendors. When there are numerous suppliers, competitive pricing is usually the rule rather than the exception; adequate customer service is also more likely to be found. In general, there are more likely to be numerous vendors when there are well-entrenched standards.

In terms of vendor availability, RPCs currently seem to have a decided edge over MOM products and ORBs. RPCs have been around the longest and have the most well-established vendor base; the number of both MOM and ORB vendors is dwarfed by that for RPCs.

LEARNING CURVE

The ability of traditional application developers to learn how to incorporate middleware (or other new information technologies) in their programming is a consideration for IS managers in many organizations. Many experienced programmers were trained to develop applications that would run on centralized, host-based systems using third-generation programming languages (such as COBOL and FORTRAN). These individuals are still adjusting to distributed client/server computing environments and the use of fourth-generation languages (4GLs) and high-level application program interfaces.

Traditional programmers often find it easier to adapt to RPCs than to MOM products or ORBs, because they see the similarity between an RPC and local procedure calls that traditionally have been included in programs written in third-generation languages (3GLs). RPC-based middleware essentially allows them to write programs as they always have with only minor modifications.

For MOM, the learning curve for traditional programmers is often steeper. With MOM, programmers have to learn how to work with the API associated with the message-oriented middleware product that they are using. This is often easier for programmers who have used 4GLs than for those who have programmed only in 3GLs.

Extensive retraining is often needed to get traditional programmers comfortable with ORBs and OOPS. Object-oriented programming is quite different from programming in procedurally-oriented 3GLs, and many traditional programmers have had considerable difficulty learning these new approaches. Many experts have stated that a completely different mindset and programming approach are needed with OOPS.

In sum, traditional application programmers are more likely to quickly learn how to use RPCs than how to use either MOM products or ORBs; the implementation of RPC middleware is likely to be easier than that for MOM or ORBs.

FAILURE RECOVERY

As you can see in Table 6-7, MOM products are typically best at recovering from node, communications media, and network failures. This is because of the message (transaction) logging and store-and-forward mechanisms commonly associated with these products.

RPCs are typically most susceptible to system failures. If a client request is unable to be passed along by an intermediate node, if the communications channel fails, or if the server is unavailable, the client cannot complete its work on the application. The same holds true for results being sent back from the server. Without a response from the server, the client cannot continue its work on the application; in the absence of a response, work must be suspended.

The ability of ORBs to recover from server or network failures is still uncertain. However, because of the similarity in how RPCs and ORBs function, some experts feel that ORBs may be as vulnerable to system failures as RPCs.

SUPPORT FOR REENGINEERING

As it is often easier for traditional application developers to learn how to use RPCs than MOM or ORB technologies, they are usually able to make a significant fast contribution in efforts to reengineer applications. Because RPCs are easier to implement, the organization is often able to make significant, fast progress toward downsizing (moving applications from large, centralized systems to smaller, distributed systems). Modifications to existing applications (sometimes called *legacy systems*) may only necessitate the substitution of RPCs for (some of the) local procedure calls.

The reengineering support provided by MOM products and ORBs is less certain, but both possess the potential to significantly assist reengineering processes, especially database processing. Because MOM can help shift thinking toward the ability of applications to interact via messaging, it may cause the IS staff to rethink the need to have applications or data stored in specific locations. The results of such rethinking can streamline existing processes and restructure work flow. In addition, OOPS and ORBs are often described as having the potential to provide tremendous assistance in reengineering business information processing. The reusability of OOP code and the self-defining nature of objects provide compelling support for this view. However, as a programming system, OOPS is still new, and the jury is still out as to whether it can live up to its billing.

◆ THE NEED FOR MIDDLEWARE MANAGEMENT

Middleware products are still relatively immature (but emerging) technologies. Many products include diagnostic utilities that allow LAN and other network managers to control message queues, change network configuration tables, and monitor application/transaction performance. However, few middleware products incorporate the ability to monitor and control middleware components from a single administrative workstation or through commercially available LAN/network management software packages.

Some LAN/network management software vendors have developed products that will manage middleware in addition to carrying out traditional network management functions. Traditional network management software packages, however, are designed to manage devices (such as printers and servers) and events; they are not designed to manage software and applications. The addition of capabilities to manage middleware and applications is a significant change in the functionality of these packages.

Creative Systems Interface (CSI) has developed a package that allows network managers to manage middleware either centrally or locally. This package makes it possible for users to develop programs that automatically manipulate middleware services including configuration tables and diagnostics. Such automatic manipulation can occur at predefined times or in response to specific events, such as an intermediate node or communications link failure. For example, if a communication link fails, CSI's product enables network managers to automatically reroute traffic around the failed link. Also, CSI's product will automatically generate reports from diagnostic information. Similarly, Covia Technologies, Inc. is developing modules that will gather diagnostic information about its middleware product's performance (such as the status of message queues, transaction performance, and errors).

Other vendors, including Horizon Strategies, Inc., are working on middleware debugging tools that could be used for troubleshooting distributed client/server applications. This company is also working on middleware that would allow application developers to partition the distributed computing environment into production (*live*) and test networks; this would enable the programmers to test out new or modified applications in the test network without disrupting or jeopardizing the production network.

These and other middleware vendors agree that the key to improving middleware management probably will be the development of standards for middleware management. One suggestion has been to define a standard Management Information Base (MIB) for middleware. Once an MIB is defined, it could be used to develop the modules needed for interfaces to commercially available LAN/network management products.

It is quite clear that network managers should expect further enhancements in middleware products, which are likely to include a wider variety of network and application management functions. In addition, middleware management also seems destined to become part of many of the commercially available network management software packages.

◆ HOW ORGANIZATIONS ARE APPLYING MIDDLEWARE

Thus far we have provided an overview of middleware and how it works, but we really haven't provided examples of how it is being put to work by organizations. In this section, we will summarize several companies' uses of middleware.

USING MIDDLEWARE TO DISTRIBUTE APPLICATIONS

Unum Life Insurance is best known as a major provider of disability insurance. It has developed a message-based communications infrastructure that will enable workers to access programs and data from almost anywhere in its global network. A key part of this communications system is a MOM product from PeerLogic, Inc. The middleware is used to link applications across the organization's diverse set of protocols, network operating systems, and hardware platforms. A MOM product was selected instead of an RPC product because network professionals at Unum felt that MOM offers greater flexibility for enterprise-wide applications.

One application enables agents at any location to instantly write and print new policy contracts and booklets for any Unum insurance product. In the past, it could take days or months to do this because the programs for doing so were scattered among many computers that could be thousand of miles apart. Agents agree that because of middleware and the new system that middleware made possible, they now have better information, more quickly, and in a format that is easier to interpret.

USING MIDDLEWARE TO IMPLEMENT JUST-IN-TIME MANUFACTURING

Chaparral Steel (headquartered in Midlothian, Texas) is using middleware to support an integrated just-in-time (JIT) manufacturing operation. Applications for order entry, inventory, shipping, and other business functions are directly linked to the melt shop where steel is made. It is management's desire to utilize JIT from start to finish in its steel manufacturing (including having a truck ready to pick up the order as it rolls off the mill).

To support JIT, a DECnet LAN and Digital's Pathworks Software are used to tie together the different systems developed for the various functional areas of the firm. For accounting and inventory applications, a VAX minicomputer running the VMS operating system is used. IBM-compatible microcomputers are used in the sales department, while Macintosh computers are used to handle quality assurance. RISC-based workstations are used for the applications run in the melt shop.

Prior to implementing middleware, Chaparral Steel had trouble getting the applications for the different functional areas to communicate with one another even though they were all available via the DECnet LAN. Programmers employed by Chaparral initially wrote a number of programs to handle message delivery between functional areas (which is essential for JIT). However, management felt that middleware could provide a superior solution.

Chaparral selected SuiteTalk from Suite Software (Anaheim, California)—a MOM product. With the middleware, programmers created real-time, interactive links among the key applications needed to implement JIT. The speed and accuracy of information exchanges between applications became better than ever before. To complete the JIT system, Chaparral is now using middleware to help establish electronic data interchange (EDI) links between the steel mills and suppliers and customers.

USING MIDDLEWARE TO IMPROVE CUSTOMER SERVICE

Bankers Trust Company, headquartered in New York City, is using middleware to link Lotus Notes to databases. Lotus Notes is used as the main form of message exchange and e-mail among users at Bankers Trust. However, prior to implementing middleware, it was awkward to use Lotus Notes to retrieve data needed to address customer requests from the bank's various databases. In addition, the bank found it too slow and expensive to write the custom-built Notes applications needed to speed up its ability to quickly answer customer requests.

In order to overcome these communication difficulties, Bankers Trust selected a middleware product called InfoPump from Channel Computing, Inc. InfoPump is server-based software that moves data among all the bank's major databases and applications (no matter where they are located) and automatically performs all necessary reformatting of data. Now all customer requests that can't be addressed on the spot (by customer service representatives) are entered into the InfoPump system. Messages are automatically routed to appropriate individuals, and answers to customer questions are obtained much more quickly.

USING MIDDLEWARE TO INTEGRATE DATA

The Center for Disease Control (CDC) in Atlanta, Georgia, implemented a middleware product called Entire Net-Work from Software AG of North America, Inc. It allows users on LANs running Novell NetWare to use a standard graphical user interface to access data from the CDC's, Adabase (Software AG), and SAS (SAS Institute, Inc.) databases as well as flat files stored on an IBM mainframe. User requests flow transparently across the different operating platforms without user awareness of the underlying steps. As one manager notes, "All they know is that they send it off into the void and answers come back." IS managers have also noted that the middleware solution was far less expensive than the programming effort that would have been needed to develop custom-based programs with the same level of functionality.

These are just a few examples of how organizations are using middleware to overcome interconnectivity difficulties that are holding back more efficient work operations. The variety of potential uses reflected in these examples helps one understand why the market for middleware products is likely to have increased from $50 million in 1993 to $1.2 billion by the year 2000.

◆ SUMMARY

Middleware is a layer of software located between applications and the network. It shields users and application developers from the complexity of the network's underlying communications protocols, operating systems, and hardware configurations.

Middleware makes interconnections among LANs and other networks transparent to users. It also enables application developers (programmers) in client/server environments to focus on writing code for distributed applications without having to worry about writing the communication modules needed for remote clients and servers to interact. This enables programmers to quickly write programs that allow users to integrate and share resources spread throughout the organization's LANs and other networks; in some cases, distributed applications can be developed four to ten times faster with the assistance of middleware.

Many organizations have also found middleware tools invaluable when downsizing applications traditionally run on mainframes to LAN environments. Organizations are using middleware to support reengineering (of business processes) efforts, for deploying client/server applications, and for integrating customers and mobile workers into corporate computing environments.

Middleware modules are found in clients (workstations), servers, and at all the intermediate nodes between clients and servers through which user requests for network resources may be routed. At the workstation, middleware modules help create the message, identify the server that can provide the services being requested, establish the connection between the workstation and the server, and convert the message into the format needed for transmission over the communications medium.

Middleware at user workstations often performs message tracking and message arrival functions; in addition, middleware facilitates the acceptance of the results shipped back from the server and checks them for communications errors.

The middleware found at intermediate nodes performs many important functions, including the routing of messages from clients to servers. Middleware at intermediate nodes usually checks all messages received from clients and servers for errors and transforms them into a form that can be understood by the receiver. Often, message logging functions are performed at intermediate nodes to ensure that messages aren't lost if a network error occurs. In addition, middleware at intermediate nodes may store a message for later transmission if the receiver or communications medium is unavailable. Dynamic routing may be provided at intermediate nodes to ensure that messages are delivered rapidly, efficiently, and cost-effectively.

At servers, middleware checks client requests for errors and converts the requests to the application program interface (API) needed by the application being requested. Once processing has been completed and results have been generated, middleware initiates the transmission of the server's response back to the client, including the routing of the results through intermediate nodes.

Because middleware is considered to be an emerging set of technologies, there is no universal agreement on what middleware is and is not. However, the general consensus is that there are three fundamental types of middleware: remote procedure calls (RPCs), message-queuing software, and object request brokers (ORBs).

Remote procedure calls (RPCs) are the oldest, best-known, and most mature type of middleware for distributed applications; these are essentially an extension of traditional procedure-based programming concepts. When RPCs are used, the program executes a procedure call in the same way that it calls a local subroutine; however, the middleware directs the call to the remote server, which executes the procedure and then returns the results.

When writing RPCs, programmers work with the middleware's RPC development language. This is used to write the program modules that link the application to middleware so that communications tasks can be performed. A special program compiler (an RPC compiler) is also used by application developers. Once the application is developed, the program code is run through the RPC compiler. The compiler generates a segment of code (that is, a set of program statements) called a *stub*. The stub is installed in both the program doing the calling (the client program) and in the application program at the server (the server program).

Because RPCs use a call/response method of communications between clients and servers, response time, network performance, and both network and server availability are crucial factors. Once a remote procedure call has been issued, the client workstation is unable to do any processing until it receives a response from the server. For this reason, RPCs are best suited to high-speed, fault-tolerant LANs that have sufficient bandwidth to ensure adequate response time.

Messaging-oriented middleware (MOM) is the second-most widely used type of middleware used to build distributed applications. MOMs allow client and server segments of an application to communicate via messages—requests from clients and responses from servers. Such inter-program messaging systems utilize message queues, buffers, or logs able to store messages at intermediate nodes while en route between clients and servers; these store-and-forward mechanisms help to ensure message delivery. Since MOM is best used to connect loosely coupled applications, it is best-suited for organizations that wish to establish electronic links with customers and suppliers, build work flow applications, or integrate departmental applications.

Object request brokers (ORBs) are an offshoot of object-oriented programming systems (OOPS). As middleware technologies, ORBs are less mature than both RPCs and message-oriented middleware. ORBs are used in distributed object computing environments that have implemented OOPS and that have defined resources, applications, and program components as objects. They relieve object-oriented programmers of having to know the location or properties of remote objects for the objects to exchange information with one another. The *de facto* standard for ORBs is CORBA (Common Object Request Broker Architecture). This has been defined by the Object Management Group (OMG).

While RPCs, MOM products, and ORBs are the best-known classifications for middleware, some experts feel that other technologies should be classified as middleware. These include tools used to develop SQL (Structured Query Language) applications in distributed database environments, gateways between diverse databases, mail-embedded applications, and products called local middleware that allow software packages designed to run on one operating system to run on other workstation operating systems.

There are many criteria that could be used to help LAN managers determine the trade-offs associated with using the different types of middleware. Among these are support for both synchronous and asynchronous transmission modes, the presence of standards, the number of vendors, the learning curve for traditional application developers, the ability to recover from failure, and the ability to support business reengineering efforts.

Some LAN/network management software vendors have developed products to help network managers manage middleware in addition to traditional network management. The addition of capabilities to manage middleware and applications is a significant change in the functionality of network management software.

Organizations are using middleware in a variety ways to improve their operations and ability to develop distributed applications. These include using middleware to improve access to data stored in dispersed databases, to implement just-in-time manufacturing, and to improve customer service.

✳ KEY TERMS

CORBA

Distributed object management
  systems (DOMS)

Dynamic routing

Local middleware

Message log

Messaging-oriented middleware
  (MOM)

Middleware

Object request brokers (ORBs)

Remote procedure call (RPC)

Store-and-forward capabilities

✳ REVIEW QUESTIONS

1. What is middleware? Why is it important?
2. What functions are carried out by middleware at client workstations?
3. What functions are carried out by middleware at intermediate network nodes?
4. What is message logging? What are the benefits of message logging?
5. What is dynamic routing? What are the advantages of dynamic routing schemes?
6. What functions are carried out by middleware at servers?
7. What are remote procedure calls? How does a client interact with a server by using an RPC?
8. For what types of applications and networks are RPCs best suited?
9. Identify several examples of RPC-based middleware products.
10. What is DCE? Why is DCE important?
11. How do clients and servers communicate when message-oriented middleware is used?
12. In what ways is MOM more flexible than RPCs?
13. How is fault tolerance built into message-oriented middleware?
14. What types of applications is MOM best suited for?
15. Identify several examples of MOM products.
16. What standards are currently available for MOM products? What standards are likely to be available in the future?
17. What are ORBs? For what types of programming environments are ORBs most appropriate?
18. How do ORBs work—what is their role in exchanges between distributed clients and servers?
19. What are DOMS? What are the similarities and differences between DOMS and ORBs?
20. What is CORBA? Why is it important?
21. Identify several examples of ORB products.
22. Describe the differences among SQL tools for distributed environments, database gateways, mail-embedded applications, and local middleware.

23. Identify the criteria that may be used by LAN managers to compare middleware alternatives. Briefly discuss why each criterion is important. Also, identify the type(s) of middleware that get the best rating(s) for each criterion.
24. How does middleware management software differ from traditional LAN/network management software? Identify some of the functions performed by middleware management software.
25. Identify some middleware management software products.
26. Briefly describe some of the ways in which middleware is being used by organizations. Provide examples.

✳ DISCUSSION QUESTIONS

1. Why is it desirable to shield application developers from the complexities of underlying networks, network operating systems, communication protocols, and database management systems? Discuss how this can help them do their jobs more efficiently.
2. Why is it desirable to have standards for each of the different types of middleware products? What are the benefits of standards for vendors and users? How has standard setting been approached by vendors for each of the different types of middleware products?
3. In the immediate future, why are RPC middleware products more likely to be implemented than either MOM products or ORBs?

✳ PROBLEMS AND EXERCISES

1. From vendors, obtain literature on two or more RPC middleware products. Identify the workstation OSs, LAN OSs, and communications protocols that each supports. Identify the advantages of each of these packages. Summarize the information you obtain in a presentation or report.
2. From vendors, obtain literature on two or more MOM products. Identify the workstation OSs, LAN OSs, and communications protocols that each supports. Identify the advantages of each of these packages. Summarize the information you obtain in a presentation or report.
3. From vendors, obtain literature on two or more ORB products. Identify the workstation OSs, LAN OSs, and communications protocols that each supports. Identify the advantages of each of these packages. Summarize the information you obtain in a presentation or report.
4. From vendors, obtain literature on one or more of the middleware management software programs identified in the chapter. Summarize the functions performed by the package(s) and share the information you obtain in a presentation or report.

5. Identify a local company that has implemented middleware. Interview an IS manager at the company (or have the manager make a presentation to your class) to determine why middleware was selected, what type of middleware is being used, what benefits have been realized, and how middleware is likely to be used in the future. Share the results of your interview with your class in a presentation or report.

6. From the trade literature, find three or more articles on organizations that have implemented middleware (these must be different from those mentioned in the chapter). For each organization, summarize what type of middleware is being used, why middleware was selected, how middleware is being used, the benefits that have been realized from using middleware, and how middleware is likely to be used in the future. Share your summaries in a presentation or report.

✳ REFERENCES

Baum, D. "Managing LAN Application Development." *Infoworld* (July 19, 1993): p. 53.
———. "Middleware: Unearthing a Software Treasure Trove." *Infoworld* (November 30, 1992): p. 46.
Butler, J. "Middleware Needed to Plug C/S Holes." *Software Magazine* (July 1993): p. 55.
Chisholm, J. "Clawing Your Way to the Middle." *Unix Review* (January 1993): p. 15.
Conniff, M. "Middleware: Networking's Postal Service." *Network World* (September 7, 1992): p. 59.
"Middleware Helps Net Managers Build Distributed Applications." *Data Communications* (October 15, 1992): p. 185.
DeBoever, L. R. and M. Dolgicer. "Middleware's Next Step: Enterprise-Wide Applications." *Data Communications* (September 1992): p. 157.
Dolgicer, M. "The Middleware Muddle: Getting the Message." *Data Communications* (October 1993): p. 33.
Dolgicer, M. and L. R. DeBoever. "Taking Middleware One Step Beyond." *Data Communications* (April 1993): p. 35.
Eckerson, W. "Wanted: Middleware Management Capabilities." *Network World* (March 1, 1993): p. 19.
———. "Smack-Dab in the Middle." *Network World* (June 21, 1993): p. 43.
Gibbs, M. "Making Sense of Collaborative Computing." *Network World* (January 10, 1994): p. 23.
Hackathorn, R. D. and M. Schlack. "How to Pick Client/Server Middleware." *Datamation* (July 15, 1994): p. 53.
Horwitt, E. "Middleware Gaps Closing." *Computerworld* (July 11, 1994): pp. 1, 14.
King, S. "Middleware!" *Data Communications* (March 1992): p. 58.
LaPlante, A. "Workflow Tool Improves Bank's Customer Service." *Infoworld* (April 5, 1993): p. 62.
"Trifox Takes the 'Middle' Road to Link Database Applications." *Network World* (October 12, 1992): p. 33.
Radding, A. "Five Types of Tools to Get Data from Legacy Systems." *Software Magazine: Client/Server Computing Special Edition* (January 1993): p. 31.

Case Study:
Southeastern State University

A proposed system is causing a great deal of excitement at SSU. A task force composed of representatives from across campus is developing a set of networked multimedia kiosks that can be used by students, university personnel, and visitors to access a wide variety of information. Each kiosk will include touch-screen interfaces, both video and voice output, and braille keyboards (to ensure compliance with the Americans with Disabilities Act —ADA). One kiosk will be placed at the main entrance of the campus, just outside the university's Welcome Center. Others will be placed in high foot traffic areas of campus, including one near the student union building and another outside one of the major student restaurants.

Ideally, an individual will be able to walk up to a kiosk and be able to access a wealth of information about SSU. Through a touch-screen menu system similar to ones found in kiosk systems at Disney attractions and shopping malls, users will be able to access athletic and cultural event schedules, the campus addresses of administrative offices and university staff members, the campus phone directory, class schedules and locations, upcoming continuing education courses, on-campus movie schedules, meeting schedules for student groups, information about special events for students in the student union, and a variety of other information about SSU.

First-time visitors to SSU will be able to utilize the kiosks to run a multimedia demo about the university, complete with video clips, still photos, and audio output; this would be similar to the multimedia demonstration used by recruiters from SSU's Admissions Department. Visitors will also be able to use the kiosk's multimedia mapping system to find their way to particular buildings. The mapping system will include color photos of the campus buildings, audio narrations on the history and functions found within each building, visitor parking options, the locations of ticket offices for campus events, and even the locations of the other kiosks on campus in the event that the visitor should get lost on the way.

The committee determining the functions and information to be included in the kiosks wants them to be operational 24 hours a day, 7 days a week. It also wants students and university staff members to be able to access the kiosk information system from the workstations on campus that are attached to LANs. University personnel and authorized students will be able to electronically transmit information to kiosk servers for updating or correcting campus information (such as location changes or cancellation of scheduled meetings) and for posting information about newly arranged meetings or events. Remote (off-campus) access to the kiosk information system has also been suggested.

It should be apparent from this brief description why the proposed system is generating a great deal of interest. Many members on the design committee see the kiosk system as a way to effectively use information technology to spread the word about activities on campus among students as well as to assist visitors in navigating the campus. If SSU student reactions and use of the system are similar to those observed at other universities that have implemented kiosk systems, utilization rates should be very high.

Personnel of the Computer Services Department have been given the responsibility to implement this system. They realize that in order to implement the proposed system, some major obstacles will have to be addressed. One of these is making it possible for LAN users to access information stored on the kiosk information system. A second is how to make it possible for users in various LANs to create and electronically transmit information for posting to the kiosk servers. Another challenge is presented by the fact that currently, the information that the design committee wants to see incorporated in the kiosk system is spread across a variety of VAX and microcomputer databases that are "owned" and maintained by different SSU subunits.

CASE STUDY QUESTIONS AND PROBLEMS

1. Gathering all you know about the microcomputer technologies found in the LANs of SSU's student computer labs, how realistic is the goal of making all the information that will be available on the kiosk system available to LAN users? Provide several specific reasons to justify your answer. Also, describe what would have to be done to make all of the kiosk applications available to the workstations in student LANs.
2. What types of servers do you think should be included in the kiosk network? Briefly justify each of your selections. Also, describe where the servers should be located.
3. What type(s) of media should be used to network the kiosks with the kiosk servers? Briefly justify your selection(s).
4. Based on the discussion in the chapter, identify and briefly discuss at least three ways in which middleware could be used to implement the proposed kiosk system at SSU.

LAN
Application
Software

CHAPTER OBJECTIVES

After completing this chapter, you will be able to:

◆ Describe the differences between general-purpose and special-purpose application software.

◆ Identify the most popular types of application software found in LANs.

◆ Describe the major types of workgroup software and groupware.

◆ Explain the implications of the interaction of different kinds of application software on the network.

◆ Identify and briefly discuss the management issues associated with installing and maintaining application software on a LAN.

◆ Identify and briefly discuss the management issues related to software licensing in LAN environments.

◆ Discuss the implications of remote access to software.

◆ Identify approaches used to ensure adequate levels of software performance in networks.

◆ Identify the trends in LAN application software.

Every network administrator should strive to provide compatible application software in a LAN environment. In addition to providing compatible applications, priority should be given to customizing as much as possible to meet the needs of individual users and departments. This involves knowing the capabilities of each application software package and its ability to work with other applications and the network operating system. Also, these tasks must often be performed invisibly; most users are not interested in the software configuration problems that the LAN administrator must handle. Users are usually only interested in what the application can do to help them perform their jobs.

◆ GENERAL-PURPOSE VS. SPECIAL-PURPOSE APPLICATION SOFTWARE

A wide variety of application software is available for networks. Typewriters have all but disappeared from the office environment and have been replaced by personal computers sporting the latest word processing software. Word processors are used by the secretary, the accounting professional, the marketing professional, and the engineer. Each of these different types of employee has special requirements, but all are likely to take advantage of word processing software in their respective jobs. Because of this, word processing software is often considered to be ***general-purpose application software***; such software is useful to workers throughout the organization, not just to those in particular departments or functional areas. Chapter 1 summarizes some of the most widely used general-purpose application packages found in LANs.

Word processors now offer adjunct packages, such as spelling checkers for the engineering discipline or for legal professionals. Among the other available adjunct packages are equation editors for mathematicians and statisticians.

Another popular general software item is the electronic spreadsheet. This application is less than 20 years old, making it relatively new to the computing scene. Spreadsheets provide a means for arranging numeric values in meaningful form. Stored formulas automatically calculate the figures, thus reducing the inaccuracies that often trouble human calculation. Graphing and other visual representations of data are becoming very popular, and spreadsheets now commonly provide very good graphing capabilities.

Database management software, while less widely used than either word processing or spreadsheet software, is also usually categorized as general-purpose application software. Database packages facilitate the creation and maintenance of databases and typically include report generation modules that enable stored data to be retrieved in different formats. Many include graphing capabilities.

Database management software has become a mainstay of network administration because administrators can easily store and retrieve network data. Most network management systems incorporate database management modules so that a network database can be created and maintained. Many other specialized software packages have the database as the foundation on which the rest of the package is built. For example, many accounting systems are now relational databases that have been customized to fit the needs of accounting professionals.

Communications software is becoming more commonplace in corporate environments. In the past, almost all data communication functions were handled by computing specialists. Now, communications software for end users enables them to engage in a variety of communications-related activities such as the following.

◆ Sending images in fax form over telephone lines.

◆ Accessing electronic bulletin board systems (BBSs), information services such as CompuServe, Prodigy, and America Online, and external databases such as Lexis and Nexis.

◆ Exchanging e-mail with coworkers and colleagues in other organizations.

◆ Accessing centralized hosts or LANs from remote locations.

There are several well-known general-purpose communication packages, including ProComm and Crosstalk. While many workers still consider communications software to be special-purpose application software, its increasing use by workers in all the functional areas of organizations also makes it possible to classify these packages as general-purpose application software. Because of its growing importance in organizations, communications software is covered more fully in Chapter 15.

Software to manage personal contact information (such as names, addresses, telephone numbers, e-mail addresses, meeting schedules, and written notes) is becoming more widely used by managers and office workers. This software is desktop organizer or ***personal information management (PIM) software***. The packages that fall into this category, such as Sidekick and Lotus Organizer, manage the personal contact information entered by users and carry out other tasks as well. For example, if the computer and telephone are connected, the software can dial telephone calls for the user and automatically redial the number at preset time intervals if the line is busy. Because this type of software can benefit such a wide range of business workers and professionals, it is often considered to be general-purpose application software.

In contrast to general-purpose application software, ***special-purpose application software*** consists of software packages useful to a more limited range of workers, such as those in a single functional area of an organization.

Accounting packages such as DACeasy, Peachtree Software, and Accpac Plus are examples of special-purpose application software. Just about any package used exclusively (or almost exclusively) by workers in a specific functional area of an organization is likely to be classified as special-purpose application software. Figure 7-1 illustrates the use of several accounting applications in a LAN. Table 7-1 shows some other tasks for which special-purpose application software has been developed.

◆ WORKGROUP COMPUTING SOFTWARE AND GROUPWARE

Many tasks performed within today's organizations lend themselves to a group of people rather than a single worker. In many instances, information technology and data communications have made the physical distance between coworkers insignificant; even continents apart, coworkers can still communicate and work collaboratively on the same project. As a result, it has become more common to ignore physical location when creating work teams. For example, in a publishing firm (or a newspaper), a group of geographically dispersed people can edit the same manuscript—the electronic links enable the individuals to see and edit words and graphics. They can jointly work toward the creation of computer-generated, camera-ready copy. While computer-generated, camera-ready copy has been available for some time, only recently have geographically dispersed workers been able to review and edit text and graphics in real time as a group.

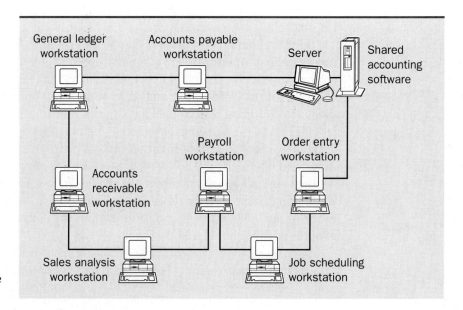

FIGURE 7-1
Tasks handled by a special-purpose application

Business Area	Special-Purpose Software
Finance/accounting	Accounts payable
	Accounts receivable
	Budgeting
	Financial statement compilation
	Fixed assets control
	Payroll
	Portfolio analysis
Marketing	Market analysis
	Order entry
	Sales tracking
Production	Computer aided manufacturing (CAM)
	Inventory control
	Manufacturing scheduling
	Materials resources planning (MRP II)
Research and development	Computer aided design (CAD)
	Patent searches
	Specialized engineering functions
Personnel/human resources management	Employee evaluation
	Employee tracking
	Government report generation
	Recruiting

TABLE 7-1
Examples of special-purpose software

WORKGROUP COMPUTING SOFTWARE

Workgroup computing software enables members of a workgroup (such as a project team) to work collaboratively or cooperatively with one another. Such software is becoming more commonly used, and a wider variety of applications (both general- and special-purpose) is being supported. Software such as Windows for Workgroups, which facilitates collaborative and cooperative computing among the members of a workgroup, is also sometimes classified as workgroup computing software.

Windows for Workgroups is a peer-to-peer network operating system that enables collaboration among workers using the same word processing documents, spreadsheets, and databases. This software allows workers to connect to printers and disk directories that reside on other computers. Being able to access the disks and applications used by other individuals gives common access to word processing documents, spreadsheet files, and databases, thereby facilitating the completion of workgroup tasks.

When using Windows for Workgroups, after logging onto the network, you can connect with other specified networked device(s). For example, you can connect to a shared network drive on another computer attached to the

LAN; you do so by selecting the drive letter for the shared network drive from a menu. Once connected to the drive on the other computer, you can access the applications and files stored there; for example, you might select a word processing file that you are jointly working on with the individual whose workstation contains the shared disk drive. Often, the application options available on accessible shared drives are represented by icons (in familiar-looking Windows formats). When you wish to disconnect from the shared drive, you can do so by using a Windows for Workgroups' Disconnect Drive dialog box.

With Windows for Workgroups, **shared directories** can be created on a workgroup member's disk drive; the directories are accessible to other members of the workgroup. Different levels of access can be assigned to files in the directory from read-only access to full access (which allows workgroup members to modify, create, and delete files). Password access is also supported.

An important feature of most workgroup computing software systems is the mail function. It enables the members of the workgroup to exchange messages electronically. In Windows for Workgroups, the mail function that group members connect to is called the **post office**. The post office is in a shared directory and manages the messages exchanged between members of the workgroup. The post office modules enable group members to create, retrieve, and archive messages to one another. Note that the mail function in workgroup computing systems is somewhat different from other e-mail systems in that it focuses on message exchanges between group members.

Besides peer-to-peer NOSs such as Windows for Workgroups, there are office automation software suites such as WordPerfect Office and New-Wave Office (from Hewlett-Packard) that enable networked workers to share files, exchange mail, and generally work in sync with one another on various office chores. These packages run under NOSs, but are often perceived by users *as* NOSs because the interaction of the application with the NOS is transparent.

Enhanced e-mail is available with many workgroup computing and groupware packages. Such enhancements may include the ability to develop personal mailing lists, mail forwarding, delivery confirmation, mass mailing, and other customization features. Some enhanced e-mail systems also provide for integrated voice mail and e-mail services; for example, through the use of add-in voice boards, group members can leave voice mail messages with their fellow team members.

Meetingware or *electronic meeting/conferencing* software allows two or more workgroup members to simultaneously engage in electronic communication via their personal computers. These packages are sometimes known as *interactive work applications*. Representative products include Aspects (from Group Technologies, Inc.), Instant Update (from On Technologies, Inc.), and VisionQuest (from Collaborative Technologies Corporation).

Group scheduling software, which is sometimes called *calendaring* software, is another type of workgroup computing software. This software is used to maintain both individual and group meeting/appointment calendars. Group leaders can use this to schedule group meetings. The software will automatically scan each group member's calendar for conflicts with the specified meeting time. If no conflicts are found, the software will automatically add the meeting to each group member's calendar. Network Scheduler (from Powersource) and Action Plus (from Action Plus Software) are examples of group scheduling software packages.

Project management software, when used in a network environment, may also be used to support collaborative or cooperative work among group members. Such packages enable project team managers to use automated forms of the program evaluation and review technique (PERT), the critical path method (CPM), and Gantt charts. These tools help project managers depict the interrelationships of project tasks and estimate project completion times. They can also be used to determine project costs and work schedules for group members. Harvard Project Manager and Microsoft Project are examples of project management software packages.

Workflow automation software allows geographically dispersed co-workers to work together on project teams. These packages have taken into account that team membership, team member locations, and individual work assignments may change over the course of completing a project. When such events occur, work flows over the network to the new member, member location, or to the member(s) with new assignments. The workflow automation software used by professional and consulting organizations also can track the time spent by individuals on particular projects; this information is used for billing or charge-back purposes.

Document coauthoring/management software allows two or more group members to work concurrently on the same document. Such software makes it possible for individual group members to check out a document (as from a library) for review or modification. This software manages the check out process, monitors the changes made, and records the identity of the individual making the changes. It also allows other group members to review the changes and records other suggested modifications made by group members.

Document review/sharing software is a specialized type of workgroup computing software that keeps track of the individuals who must review a document as well as the order in which they must review it. These systems typically include an "electronic routing slip." The documents reviewed using such systems include technical reports, proposals, bids, and contracts and other legal documents.

Group decision support system (GDSS) applications, now referred to as *group support system (GSS) applications*, are also examples of workgroup computing systems. In most such systems, each group member has a workstation, which is used to provide input and can be given to the total project. GSS packages vary in sophistication; many support group voting, ranking, and rating systems. Some support brainstorming, the nominal group technique (NGT), and Delphi decision making approaches. GSSs can be implemented in both LANs and WANs, and teleconferencing may be used to visually link decision makers in dispersed decision rooms.

Workgroup computing support is being incorporated into more general-purpose application software packages. For example, Borland's Quattro Pro spreadsheet package enables spreadsheets to be transmitted to shared directories in a network. Borland calls this procedure "publishing." Publishing facilitates the collaborative construction of a spreadsheet by making it accessible to all the members of the workgroup.

GROUPWARE

Like workgroup support systems, *groupware* consists of software packages designed to support the collaborative efforts of a group of coworkers. These packages seek to take advantage of networks to enhance group productivity. They are designed to facilitate group decision making and collaboration on documents, spreadsheets, and other files. Many of the features mentioned in the previous section on workgroup computing systems are also supported by groupware packages including enhanced e-mail, group calendaring, project management and workflow automation services, and desktop organizers. They may also include shared "to-do" lists, voice mail services, computer conferencing services, fax services, bulletin board services, and access to online databases and telecommunication services such as CompuServe.

From this description, you should be able to see that there is no clear distinction between groupware and workgroup computing systems. In spite of the lack of a clear distinction between these concepts, many experts have noted that groupware packages are likely to integrate the functions available in two or more different types of workgroup support applications (which are often available in stand-alone form).

At present, Lotus Notes is the best-known groupware product and commands the lion's share of the groupware market. Figure 7-2 shows a Lotus Notes screen. An increasing range of third-party add-on programs is available, and these are only serving to enhance Notes' stance as the market leader. Most experts recognize Microsoft's Windows for Workgroups, Apple's Open Collaboration Environment (AOCE), and the workgroup support being incorporated into Borland International's application suite as the emerging contenders to Notes in the groupware market.

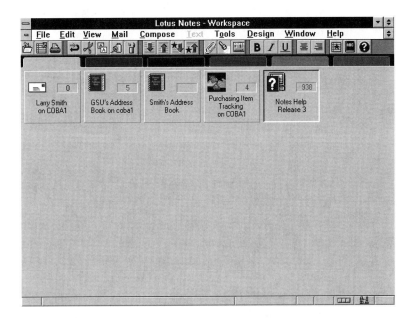

FIGURE 7-2
A Lotus Notes
screen

◆ APPLICATION SOFTWARE FOR CLIENT/SERVER SYSTEMS

Client/server networks have the potential to provide exceptional support for work teams. Most of the types of workgroup support systems discussed are used in LANs, often in stand-alone LANs. In contrast, client/server networks are likely to consist of interconnected LANs, if not interconnected LANs and WANs. Further, a diverse set of NOSs, computing platforms, and workstation operating systems is likely to be found in client/server networks. Also, because client/server networks are often more complex and sophisticated than are LANs (especially stand-alone LANs), so are client/server workgroup support applications.

Many of the general-purpose, special-purpose, and workgroup support applications found in client/server networks have been mentioned in this chapter. However, others are particularly well-suited to client/server environments, especially shared database applications. As noted in the discussion of middleware in Chapter 6, software is now available to retrieve data from several databases in different subnetworks to generate a report needed by an individual or workgroup.

Software is also available that allows programmers connected to different subnetworks within a client/server network to work collaboratively on application development. The development of an increasing percentage of commercial software is being done in this fashion. The ability to electronically share the expertise of geographically dispersed programmers is providing new ways to ensure the development of high-quality software.

◆ MANAGING SOFTWARE IN NETWORKS

In addition to ensuring that network users have the software needed to perform their jobs, network managers must contend with a number of other software management issues. These include ensuring that available applications do not interfere with one another while they are running, enforcing site license agreements, and providing remote access to application software. These and related software management issues are discussed next.

MANAGING THE INTERACTION OF APPLICATION SOFTWARE ON THE NETWORK

Most LANs are composed of workstations generally used for a limited number of applications. A particular worker may use word processing, spreadsheet, scheduling, and personal information management software. However, another user in a nearby office may spend the majority of her time using computer aided design (CAD) packages, while the user next door to her may spend most of the day on database applications. In addition, users can access and share an application in different ways, ranging from shared configuration files to individual configurations for each user workstation (see Figure 7-3). In general, managing individualized configurations is a greater chore than managing shared configurations.

Different applications place different demands on the LAN. For example, the word processing, spreadsheet, scheduling, and personal information management software execute almost entirely within the workstation. The workstation's interaction with the file server for these applications primarily consists of transmitting the software to the workstation and saving user files on network disks. These operations create minimal traffic on communications media and therefore make minimal demands on the LAN.

Running CAD on the network, however, may place greater demands on the network depending on whether the workstation running CAD has a local hard disk of sufficient capacity to store some (or all) of the voluminous files created by this graphically-oriented application; if it does not, the traffic on the network could be greatly increased. Most CAD workstations today have local disks that store both user files and some of the most frequently used CAD modules; this reduces the volume of data that must flow over the network between the workstations and the server. However, CAD users lacking sufficient hard disk space can slow network traffic to a crawl. Even networks with transfer rates of 10 Mb or greater take some time to move the large graphics files produced by CAD users. Such problems occur when users save or retrieve large CAD files from network drives. Supplying sufficient hard drive capacity to user workstations can be an effective way of addressing such problems.

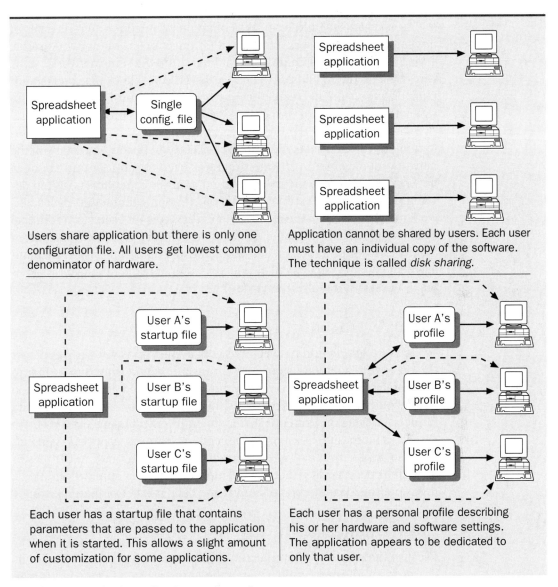

FIGURE 7-3 Multiple application configurations

With many database applications, users frequently retrieve and save data; this can generate considerable network traffic. In a multiuser environment in which several users access data from the same set of files, the NOS must work with the database management system to respond to all of the user requests. It also must cooperate with the database manager in locking certain records so that concurrent update problems can be avoided. This type of processing requires near-constant network traffic and the potential for response time degradation is high unless the network has been carefully planned.

As you can see, different types of applications put different types of demands on the network and have different impacts on network traffic. Trying to support all of these applications within the same network can be very challenging. To address the network traffic problems that such simultaneous support may create, the network administrator must attempt to estimate the amount of traffic from each source and project the cumulative effect. Capacity planning, network simulation, and workload generator tools are often used to develop such estimates. In addition, many of the network management software programs used to monitor and enhance network performance incorporate trend analysis capabilities used to develop network traffic estimates. The information generated enables network managers to provide resources such as dedicated servers, additional local disk capacity, additional workstation memory, and additional server memory to handle anticipated levels of network data traffic. Network management systems, capacity planning tools, and network tuning approaches are discussed more fully in Chapter 13.

FILE CONVERSIONS

To facilitate interaction between network applications, the LAN manager should ensure adequate support for translating data from one format to another. Users are likely to want to move data created in one application to a second application and, unless the applications have identical data formats, conversion from one format to another will be necessary. Fortunately, an increasing number of application packages, especially general-purpose application packages, can translate user files into ASCII files or delimited character files that can be imported by other applications. While this is not the ideal method of converting data, it is better than rekeying.

LICENSE AGREEMENTS

Because the majority of application software found within LANs is purchased from software vendors, enforcing software license agreements is an important software management issue in LAN environments. Essentially, when software is bought from a software vendor, the purchaser actually obtains only the right to *use* the software; the actual ownership of the copyright or patent stays with the vendor. Many software vendors require that an individual license be purchased for each personal computer in a network; however, it is often possible to acquire a license for a limited number of concurrent users rather than for each workstation. Other levels of licensing are discussed in Chapter 9. Some of these are:

◆ single user, single computer
◆ single user, multiple computers

◆ multiple users, multiple computers restricted to a quantity of concurrent users

◆ a license permitting an unlimited number of users on single or multiple servers

Most of the software sold commercially contains only ***compiled code*** in machine-readable form. Compiled code is ready to execute on the computer. Software typically does not include the ***source code*** for the software—the English-like statements that the programmer actually keys into the computer in a programming language. Source code is converted to compiled code by special software called *compilers*. If the software vendor were to sell the source code, the transaction could uncover the trade secrets and confidential methods the vendor uses to distinguish the software from competitors' software. The only way to determine these secrets from compiled code is through a process called ***reverse compilation***; it is typically used only by vendors producing competing products.

For mission-critical applications, such as accounting packages or specialized packages used in particular industries, the source code may be sold. Some vendors require the purchaser to protect the source code by placing it in a safety deposit box controlled by a mutually agreed-upon custodian. In addition, most vendors will not allow the purchaser to sell the source code to other companies. Some companies want to purchase source code because they fear that the software vendor will go out of business, leaving them with no support for their mission-critical applications. Others purchase source code to customize the software to better fit their computing needs.

METERING SOFTWARE

An increasingly important type of software used by network managers to enforce license agreements is ***metering software***. This software monitors and regulates the number of concurrent users of the same application.

Many LAN managers study application utilization statistics to determine how many concurrent requests for an application package there are likely to be. Such information can help minimize software licensing costs by ensuring that you are not paying to use more copies of the software than are needed. For example, if your LAN has 48 workstations and users employ a variety of applications, it is not likely that you will need a license to use 48 copies of any of the application packages supported on the LAN. Instead, you are likely to license no more than the peak volume of concurrent users expected.

Once license levels for individual products are determined, metering software can be used to monitor concurrent users of the different products.

For example, if the software license is for 36 concurrent users, the metering software will not allow a thirty-seventh user to access the application. Metering software can also be used to assemble statistics needed to set concurrent usage levels so that appropriate maximum levels can be determined. It may also provide information needed to modify existing software licenses.

PROVIDING REMOTE ACCESS TO LAN APPLICATION SOFTWARE

The organizational computing landscape now extends beyond the office. Many national firms have started programs whereby employees who work extensively outside the office are no longer assigned private offices. Rather, large cubicle working areas that can be reserved in advance are available when office time is needed. For many mobile workers, the primary piece of office equipment is the portable computer. This computer is used for customer presentations, managing personal contact information, submission of orders through a modem, and even the printing of letter-quality documents on a portable printer. Other organizations are encouraging certain workers to do at least some of their work at home. This concept, called *telecommuting*, requires that the employee have access to the computing equipment and software to effectively accomplish the tasks expected. Laptop computers make the raw processing power available, and remote access to LANs and other computers in the home office can provide the connectivity necessary to make the information accessible to these remote employees.

Along with convenience, remote access also brings the potential for security breaches. Enabling telecommuters and mobile workers to have dial-in access to LANs and host computers in the home office can also provide access to persons who want to violate the security of the system. Even amateur hackers have broken security measures. In addition, the advent of wireless LANs and wireless remote access to cable-based LANs has escalated the challenge of securing LANs.

To prevent unauthorized users from accessing an organization's computers and data, many LANs now have stringent security mechanisms in place that meet Department of Defense standards or those standards promulgated by the largest financial institutions. However, these measures must be enforced. Some of the measures used to prevent unauthorized access are:

◆ required passwords longer than five characters

◆ required password changes at regular intervals

◆ multiple password levels for highly sensitive data

◆ encryption of passwords and all messages that pass over the network's communications media

◆ terminal profiles and the required terminal identification for some applications

◆ time frame restrictions on the use of certain applications or systems

◆ audits of users' files by independent security specialists

While no network should ever be considered secure, the implementation of these security mechanisms may help to prevent security breaches by unauthorized users. Network security is addressed more fully in Chapter 14.

ENSURING SOFTWARE PERFORMANCE LEVELS

For mission-critical functions, there can be no compromise in standards, and network managers are playing an increasing role in ensuring compliance with standards. For example, the approval process for transactions must not be short-circuited because an employee is absent from headquarters on business-related travel. If the vice president's signature is required for a check over a certain amount, there should be a way for this officer to send approval electronically. Of course, the procedures must include safeguards to assure that such remote approvals are, in fact, genuine.

Most organizations have found it necessary to standardize on application software because it is unreasonable to expect LAN managers and other IS professionals to support whatever applications users want. Such standardization helps ensure some degree of consistency across the organization and facilitates the network administrator's job of managing the interaction of applications on the network. Without standardization, even a simple task (such as taking a document prepared on one word processor and loading it into another word processor) can be a time-consuming and costly procedure. While users may have their personal preferences for application packages, individual taste should not be accommodated at the expense of network efficiency and control.

In addition to standardizing on application software, it is often desirable to standardize on hardware. Once again, personal preferences interfere with network performance and the ability of other LAN users to perform their jobs. In many organizations, battles persist between the IBM-compatible fans and the devotees of the Macintosh. While Novell NetWare enables Macintosh computers to be connected to a LAN consisting primarily of IBM-compatible workstations, NetWare really facilitates little compatibility beyond copying files back and forth between the hardware platforms.

Some application software vendors, such as Microsoft, are going beyond the ability to transfer files between IBM-compatible PCs and the Apple Macintosh. Microsoft Word files, for example, can be converted from Macintosh to IBM format prior to transfer within the application. This is also true for Microsoft Excel files. In fact, both platforms share the same user manual. This convenient approach may lead other software vendors to do the same thing.

IBM and Apple are working together to overcome some of their incompatibility problems in a partnership company named Taligent. Taligent's agenda is to manufacture computers capable of running software developed for either the Apple Macintosh or IBM-compatible platforms. If this strategy is implemented, another avenue to compatibility will be opened.

Even within the IBM-compatible world, software incompatibilities may exist. For examples, users of Windows applications often experience annoying incompatibilities when attempting to convert data from corresponding DOS applications (for example, from a DOS version of Lotus to the Windows version). Also, once the files have been loaded into the Windows version of an application and modified, the user may not be able transfer them back to the DOS version with all of the modifications intact.

Network managers need to be aware of the incompatibilities among application software programs if they are to ensure adequate levels of network efficiency and performance. The ability of LAN users to transfer files may be compromised if they do not stay on top of potential incompatibilities and the ways in which they can be addressed.

◆ TRENDS IN LAN APPLICATION SOFTWARE

As LAN applications evolve, they are becoming increasingly network aware. Network awareness may be observed in several features that make it easier to use the software in network environments. Some of these features are:

- the ability to identify the user
- the ability to detect the use of print queues as a part of a network
- the ability to detect the simultaneous use of data files by more than one user
- providing a count of concurrent users of the application
- online documentation for the software, giving detailed instructions to the user on how to use the network features of the software

In addition to these trends, most LAN application software packages are incorporating graphical user interfaces (GUIs), which means that they are larger and increase network traffic when program files are transmitted from the server to the workstation. Also, more multimedia applications are finding their way into LAN environments, which is putting pressure on network managers to study transmission speeds as well as server and workstation storage capacities.

◆ SUMMARY

From the LAN user's perspective, application software is the most important software supported on the network because it helps them perform the job. The variety of application software found in LANs includes general-purpose application software, special-purpose application software, workgroup support software, and groupware.

General-purpose application software is used by workers in most functional areas of organizations. Word processing packages, spreadsheets, and database packages are typically classified as general-purpose applications. In addition, due to the increasingly widespread use of communications software and personal information management software, these packages may also be classified as general-purpose applications. Special-purpose application software includes programs used by workers in particular functional areas of an organization, but not in others. Accounting programs and CAD (computer aided design) programs are examples of special-purpose applications.

Workgroup support programs and groupware include software that enables members of a workgroup to use the network to work collaboratively or cooperatively. Peer-to-peer NOSs, such as Windows for Workgroups, specifically designed to support collaborative work activities, are often considered to be workgroup support or groupware products. Office automation suites, enhanced e-mail systems, meetingware, group scheduling software, project management software, workflow automation software, document coauthoring/management software, document review/sharing software, and group support system (GSS) applications are other examples of workgroup support software or groupware. Lotus Notes is currently the best-known groupware product.

In addition to ensuring that network users have access to needed application software, a number of other software management issues face network managers. These include managing the interaction of network applications, enforcing software license agreements, providing remote access to application software, and quality assurance.

Different types of application software place different demands on the network. General-purpose applications often place the least pressure on the network because most of the processing is done at the workstation. Graphically-oriented, special-purpose software such as CAD programs have more significant effects, because files transmitted over the network are large, and because the storage requirements for such applications can be large. Database applications can affect network performance and traffic when high numbers of retrievals and queries are needed. In LANs that support a variety of applications, the manner in which they interact, the need for file conversion between applications, and their cumulative effect on network traffic are important factors that network managers must consider.

Enforcing license agreements may also be the responsibility of network managers. In many LANs, the number of copies of an application package that can be used concurrently is less than the number of workstations in the LAN. This occurs because application utilization statistics indicate that all LAN users rarely request the same application at the same time and because software licensing costs can be reduced by purchasing fewer licenses. Metering software can monitor concurrent usage of the same application and ensure that license agreements are enforced.

Providing remote access to LAN resources supports a mobile workforce and telecommuting. However, such access can also give hackers and unauthorized users an avenue for breaking into an organization's computing resources. Hence, if providing remote access by employees is desirable, adequate security mechanisms should also be in place.

Ensuring the integrity of mission-critical applications is another important software management issue for network managers. Standardizing on software and hardware and minimizing application incompatibilities are often desirable options for network managers to pursue.

LAN application software is generally becoming more network aware. In addition, more programs are incorporating graphical user interfaces (GUIs), and an increasing number of multimedia applications are being supported by LANs; both of these changes are creating the need for increased secondary storage capacities and higher transmission speeds.

✳ KEY TERMS

Compiled code
Document coauthoring/management software
Document review/sharing software
Enhanced e-mail
General-purpose application software
Group scheduling software
Group support system (GSS) application
Groupware

Metering software
Meetingware
Personal information management (PIM) software
Post office
Project management software
Reverse compilation
Shared directories
Source code
Special-purpose application software
Workflow automation software

✳ REVIEW QUESTIONS

1. What are the differences between general-purpose and special-purpose application software? Identify several examples of each.
2. What types of work-related activities may involve the use of communications software?
3. What are the characteristics of personal information management software?
4. What are the characteristics of workgroup computing support and groupware packages?
5. How may shared directories facilitate collaboration among group members?
6. What are the characteristics of enhanced e-mail services?
7. What is meetingware? What types of worker collaboration are possible with meetingware?
8. What are the major features of group scheduling software programs?
9. What are the characteristics of project management software?
10. What are the characteristics of workflow automation software?
11. What are the characteristics of document coathoring/management systems and document review/sharing software?
12. What are the characteristics of group support systems?
13. What types of features may be found in groupware packages?
14. What issues are associated with managing the interaction of application software in networks?
15. Why must network managers be concerned with file conversions?
16. What are license agreements? Why are these important?
17. For what is metering software used?
18. Why is compiled code for application packages more common than source code?
19. How is network security related to providing remote access to network resources?
20. Why are hardware and software standards important in network environments?
21. Identify and briefly describe the trends in application software for LANs.

✳ DISCUSSION QUESTIONS

1. Explain what effect remote access software could have on the career of a programmer, a salesperson, or an author. Explain some of the effects of remote access to software platforms.

2. The editor of the *Kiplinger Letter* has just called and ask you to write an article of 500 words or less on what changes will take place regarding the places most people will work in the next ten years. Cover the effects you foresee with communications software and remote access software.

3. Why are standards of software usage necessary in a large company? What would happen if there were no policy about which word processor and spreadsheet were adopted by the company as a whole? Is the adoption of such policies indicative of a return to centralized data processing? Why or why not?

✳ PROBLEMS AND EXERCISES

1. Find copies of software license agreements of at least three software vendors. List a summary of the basic provisions of each. Contrast the provisions in each agreement.

2. Use recent periodicals to evaluate the three leading Windows-based products in each of the following categories: word processing, spreadsheets, database, and personal information management. Summarize the strengths and weaknesses of each product.

3. Find an extensive current write-up of an e-mail system in a computing or networking periodical or visit a company that uses an e-mail system. Write an article appropriate for your local newspaper explaining to the average worker the operations of an e-mail system and how best to use it.

4. Find several articles in computing or networking periodicals on add-on packages to Lotus Notes. Summarize the contents of the articles in a paper or presentation.

5. Find several articles in computing or networking periodicals on how Lotus Notes is being used by organizations. Summarize the contents of the articles in a paper or presentation.

✳ REFERENCES

Brandel, W. "Licensing Stymies Users." *Computerworld* (April 18, 1994): p. 1.

Cafasso, R. "Legent Tools Close in on Integration." *Computerworld* (April 25, 1994): p. 4.

Canter, S. "The Dawn of Network Applications." *PC Magazine* (March 15, 1994): p. NE.1.

Coffe, P. "Software Shouldn't Feel Like Playing a Chess Game." *PC Week* (September 6, 1993): p. 36.

Comaford, C. "User-Responsiveness Software Must Anticipate Our Needs." *PC Week* (May 24, 1993): p. 61.

Crichton, M. "Installer Hell." *Byte* (September 1993): p. 294.

Daly, J. and E. Scannell. "OS/2 Developers to Get First Whiff of Taligent." *Computerworld* (November 8, 1993): p. 37.

Dell, T. "How Long Will We Say 'Cross Platform'?" *LAN TIMES* (June 28, 1993): p. 54.4.

Dussault, R. "The Dilemma of Software Upgrades." *The Business Journal Serving Greater Sacramento* (November 29, 1993): p. 13.

Editor. "Bits and Bytes." *Business Week* (April 19, 1993): p. 110B.

Editorial. "Cross-Platform Technology: No Simple Cure-All." *PC Week* (July 20, 1992): p. 55.

Frenkel, G. "Software Distributors: The Electronic Way." *PC Magazine* (September 28, 1993): p. NE11.

From, E. "Industrial Control Software Evolves From Proprietary to Open Systems." *PC Week* (March 29, 1993): p. 73.

Graziano, C. "Companies Sign Flat-Fee Licensing Agreement." *LAN TIMES* (June 26, 1993): p. 73.

———. "Alternative Software Licensing Available: OURS Delivers New Licensing Guidelines to Let Vendors Know They Have a Choice." *LAN TIMES* (May 24, 1993): pp. 84, 86.

Gunnerson, R. "The Search for Software Solutions." *Global Trade & Transportation* (January 1993): p. 41.

Hamilton, K. "Users Consider Multiple Factors in Software Purchases." *Service News* (May 1994): p. 2.

Horwitt, E. "Client/Server Complicates Licensing." *Computerworld* (February 14, 1994): p. 57.

Khan, S. "Schedulers vs. Networked PIMs: What's the Difference, Anyway?" *PC Week* (November 29, 1993): p. 101.

Koegler, S. D. "Case Study: Payroll Management Turns From Wasteful to Efficient." *LAN TIMES* (October 18, 1993): pp. 51, 54.

Machrone, B. "Getting Organized." *PC Magazine* (June 15, 1993): p. 87.

Miller, M. J. "Applications Integration: Making Your Programs Work Together." *PC Magazine* (March 30, 1993): p. 108.

Musich, P. "Attachmate Tightens Its PC-Host Ties." *PC Week* (August 30, 1993): p. 37.

Raskin, R. "Usability Wars." *PC Magazine* (June 29, 1993): p. 30.

Scannell, E. "Taligent Goes Public With Operating System." *Computerworld* (March 28, 1994): p. 4.

Seymour, J. "Style and Focus in Software Suites." *PC Magazine* (November 9, 1993): p. 101.

Sostrom, J. D. "A Call for Vendors to Deliver Cross-Platform Apps." *PC Week* (March 29, 1993): p. 81.

Sullivan, K. B. "Group Schedulers Run Gamut." *PC Week* (November 29, 1993): p. 101.

Venditto, G., C. Cline, et al. "9 Multiuser Databases: Robust and Ready to Share." *PC Magazine* (March 31, 1992): p. 289.

Vizard, M. "Software Upgrade Costs Rising." *Computerworld* (June 21, 1993): p. 1.

Waltz, M. "Counting on Mobile Ware." *LAN TIMES* (August 8, 1994): p. 68.

Whitmire, D. E. "Selecting Software for Word Processing: Today's Brands Have the Flexibility to Produce Professional-Looking Documents." *The Office* (November 1992): p. 15.

Case Study:
Southeastern State University

The managers of most of the LANs in student computer labs at SSU are interested in implementing Windows-based applications as soon as they can. They are aware that most business organizations have moved toward Windows applications and desire to make it possible for SSU students to have hands-on experience with a variety of Windows-based general-purpose application software packages. However, these LAN managers also recognize that it is likely to be a number of years before Windows-based applications will be the norm, rather than the exception, in student computer labs at SSU.

In many of the LANs in student computer labs, few workstations are capable of running Windows applications. Although Windows may be used on machines with 80286 processors, most LAN managers bemoan the slow speeds at which Windows executes on these machines. They consider it feasible to make Windows applications available on machines with only 386, 486, and Pentium processors. While the number of microcomputers with these processors is increasing in the student labs at SSU, in most instances, they are a minority of the machines found in a typical computer lab. Hence, in most labs, providing quality support for DOS applications does the greatest good for the greatest number of users.

Windows-based applications have been installed in a 10BaseT LAN with 48 stations in a student computer lab in the College of Business Administration. However, the installation has been far from smooth. Although all the workstations in the lab are 80386 machines, several steps have had to be taken to provide adequate support for Windows applications. For example, the RAM in each of the machines has been increased from two to four megabytes so that Windows and Windows applications will run more efficiently. Also, the hard drive capacity of each workstation has been increased so that Windows can be locally stored and retrieved, rather than transferred from the file server. In addition, Windows applications have been transferred to their own server; this has increased network response time for both Windows applications and the DOS-based applications supported on the network. These and several other changes (such as modifications to both the number and parameters of print servers) occurred in response to slow network performance and too-frequent server and network crashes after installing Windows and Windows-based applications on the existing file server.

In spite of these modifications, network performance degrades when the LAN nears its 48-user capacity and the majority of the users are employing Windows-based applications. The LAN manager and several instructors who intend to teach or use Windows applications in their courses are concerned that the network may crash if they use the LAN for software proficiency tests. In the past, proficiency tests for DOS-based applications have slowed network response to a crawl and have occasionally caused network crashes; some faculty are extremely wary of what might happen when giving a proficiency test on a Windows-based application.

A proficiency test is usually a 30- to 45-minute timed, hands-on test whereby a student at a workstation is asked to use the software to perform a number of fundamental operations; it is essentially a test to demonstrate that the individual student knows how to use the software to perform specific functions. For example, a proficiency test on WordPerfect often includes inserting, deleting, indenting, centering, rearranging, and spell checking the text in an existing document. The number of operations that must be performed in the given amount of time typically makes it impossible for test takers to take advantage of the application's on-line help functions and still get a passing grade on the test.

Because the proficiency tests are timed, all test takers try to access the application at the same moment. If the network is going to crash during the test, this is when it is most likely to happen; it is also at this time that network performance slows to a crawl—even for some DOS-based applications, some students do not get the application to come up on their machine until five minutes (or more) after the test has started. Needless to say, this has increased student stress and frustration levels and has caused them to question the fairness of proficiency testing in this environment.

Apart from their concern about the stability of the network during proficiency tests on Windows applications, some faculty members are questioning the fairness of giving assignments that require the use of Windows-based applications. This is primarily due to the limited number of Windows-capable workstations that students can access. The lab in the College of Business Administration in which proficiency tests are given is available to students on a limited basis, that is, when it is not being used to instruct students in the use of the software or to administer proficiency tests. Often, it is inaccessible to students during every class hour of the day (because faculty members have scheduled classes or proficiency tests in the lab) and is available only a few hours each evening. In addition, the computer labs on campus that are open 24 hours a day have very few machines supporting Windows. Hence, if a student is given an assignment to complete using a Windows-based application or wishes to prepare for a proficiency test on a Windows-based application, it may be extremely difficult to find an appropriate workstation to use.

This situation is causing a great deal of consternation among both students and faculty. Things will get worse before they get better as every college on campus (with the exception of the College of Education) is moving toward increasing use of Windows-based applications by its majors.

CASE STUDY QUESTIONS AND PROBLEMS

1. Explain why network performance degrades and the probability of network crashes increases at the start of proficiency tests in the student lab (hint: review Chapter 5's description of the media access control method used in 10BaseT LANs).
2. Based on the information provided in this segment of the case study, do you feel that students should be trained and tested in Windows-based applications at SSU? Explain.
3. In addition to being used as a server for Windows-based applications, how else could the second server in the College of Business Administration's student lab be used?
4. In addition to the modifications made, what else would you recommend to the LAN manager of the student lab in the College of Business Administration to enhance network performance for Windows-based applications?
5. What type of application configuration do you think is used in the College of Business Administration's lab (see Figure 7-3)? What type of application configurations are likely to be found at other computer lab LANs at SSU? Justify your selections.

PART THREE

Network Planning and Installation

Network Planning

CHAPTER OBJECTIVES

After completing this chapter, you will be able to:

◆ Explain why careful planning of network implementation is required.

◆ Discuss why outside consultation is sometimes necessary to enhance network plans.

◆ Describe the difference between a network simulator and a workload generator.

◆ Identify the key components of site preparation plans.

◆ Discuss the factors that must be taken into account when installing LAN cabling.

◆ Identify the approaches used to test newly installed or expanded LANs.

◆ Discuss the experience and training that LAN installers should have.

◆ Identify the elements of maintenance plans for workstations, servers, and other network components.

◆ Discuss the managerial implications of interconnecting LANs with other networks.

◆ Identify staffing requirements for new or expanded LANs.

One of the most significant LAN characteristics is its ability to expand or contract in incremental steps. Once installed, a LAN is likely to change. Many of these changes in LANs occur in response to other changes. For example, the processing requirements of organizations typically change over time, and these changes may demand new computing platforms. Information technologies continue to evolve, and new computing and software products continue to be developed. In addition, new communications protocols and transmission speeds may be needed. These factors (and others) may result in modifications to a LAN.

Valid business reasons for abandoning an existing system also exist. Some of these include inefficiency beyond repair, lack of capacity to handle increasing user work loads, incompatibilities with new information technologies needed by users, inadequate security controls, and requirements for new forms of electronic communications that the current system cannot handle. However, no matter what level of system change is needed (minor, major, or abandonment), action should not be taken without appropriate analysis and planning.

◆ WHY IS NETWORK PLANNING NEEDED?

Some of the major responsibilities of a LAN manager include the planning, implementation, monitoring, and maintenance of the LAN. Network administrators should be able to identify and implement needed changes. They should also be able to develop plans for making changes to the LAN. Such plans may involve equipment and software upgrades and interconnection schedules. They are also likely to involve training programs for users.

Many of the demands initially made on a proposed LAN can be predicted, as can the nature of the general changes that will have to be made over time. By being able to predict major changes and when they will occur, modifications can be made with minimal inconvenience to system users.

As noted, the major hardware components of a LAN consist of servers, workstations, media, connectors, and network interface cards (NICs). During network planning, the limits of connectivity for these components must be taken into account as must compatibility issues. For example, there are finite distances beyond which a given Ethernet system cannot send a signal without signal regeneration; trying to expand the network beyond this limit without repeaters could cause network failures within both the expanded and original sections.

Without appropriate planning, a network difficult to use and manage could be created. By taking a long-range view of the types of computing resources the network should support in the future, and of how users' computing needs are changing, it becomes possible to design and implement an appropriate network and to avoid potential problems.

◆ EVALUATING THE FEASIBILITY OF A SYSTEM

The flowchart in Figure 8-1 summarizes some of the activities associated with network planning. One of the first steps in network planning is an assessment of the feasibility of a new network or network change. This may involve conducting a *feasibility study* to determine whether a proposed network (or network change) can be implemented, operated, and maintained

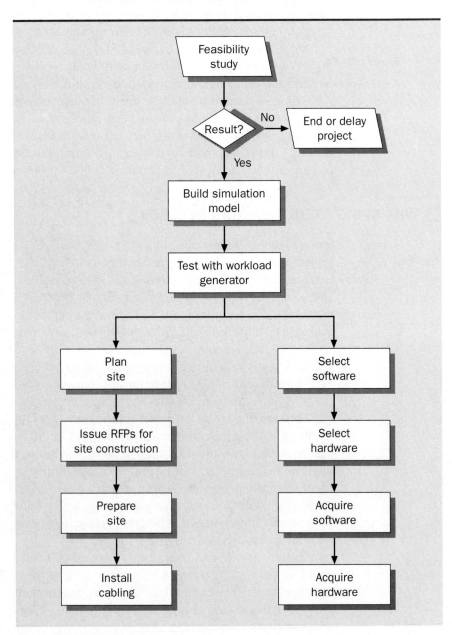

FIGURE 8-1
A flowchart for
network planning

in a cost-effective manner. Often, a feasible system is one that is both afford-able and able to meet the computing needs of users—that is, it is a system whose anticipated benefits are greater than its anticipated costs. For this rea-son, cost-benefit analysis is often an important aspect of a feasibility study.

In many instances, it is helpful to hire an outside consulting organiza-tion to perform the feasibility study. This may be essential when the pro-posed network or network change involves specialized technologies with which the organization has no prior experience. In addition, there may be some political value in securing outside assistance from a respected firm; its unbiased opinion about the feasibility of the proposal may have a strong impact on users, managers, and information systems personnel. This outside help may also be used to define budget requirements, appro-priate software, and required hardware for the proposed system.

If an outside organization is called upon to provide input, it should be assigned the responsibility for constructing the technical definition for the network and for identifying needed hardware and software. The outside organization may also be called upon to generate an implementation plan and schedule and for developing new policies and procedures for the new system. It may also be called upon to develop and implement training pro-grams for both network staff members and users.

◆ CAPACITY PLANNING APPROACHES

Several types of software tools have been developed to help network plan-ners. Some of the most important to LAN managers are called *capacity planning tools*. The two major categories of capacity planning tools are network simulation tools and workload generators.

NETWORK SIMULATION TOOLS

Simply connecting the wires, loading the application and system software, and successfully transferring an application file from a server to a work-station is rarely sufficient to determine whether the network as a whole can be successfully installed, operated, and maintained under normal work loads. Such a test is like trying to predict the maximum load a truck can carry without actually testing the truck's performance with the load onboard. More systematic and rigorous assessments are needed to predict long-term success.

Network simulation models enable the network administrator to ana-lyze how the network will perform under a set of prespecified conditions. These can be used to predict response times, processor utilization, line contention, and potential communications bottlenecks.

Simulations have been used in computing for some time. Simulation makes it possible to observe (using a combination of preprogrammed conditions and random number generation schemes) what the magnitude of the network workloads will be and whether peak work loads will reach or exceed the capacity of the network. The results of such a simulation can be used to help make decisions about adding new applications, adding new workstations or peripherals, installing higher-capacity cabling, and other potential changes to the existing system.

These simulation models can vary in complexity and in the manner in which the user defines the work load of the system. Some of the quantitative inputs used in network simulations include data transmission speed, the protocol(s) used, the number of messages (or packets) per user, and the number of current workstations in the network.

In comprehensive network simulation models, both a network configuration file and a data file may be included as input. The configuration file describes the hardware configuration in terms of disk drive access times, the speed of the disk drives, the speed of the line, the data link protocols, workstation characteristics, and locations of the files. In addition to these hardware-oriented variables, a comprehensive model may include software performance measurements, NOS overhead, and command execution times. From these inputs, the model can offer expected response times, average percentage of line utilization, average percentage of full server utilization, disk response time, and other quantitative measurements.

Most network simulation models also indicate such things as the number of collisions, network response times, queue sizes, server memory or disk utilization, and packet transfer rates. One of the key measures of hardware performance used in network simulation is *throughput*, the total amount of data transmitted over the network in a given period of time. Network throughput is often measured in kilobytes per second. In LANs, the two most important hardware determinants of network throughput are the NIC and the server. The NIC/server combination selected and installed plays a large role in determining the maximum throughput for the LAN. For example, if a particular NIC/server combination can deliver 10MB/second throughput on the network, a single user theoretically could achieve that full data transfer rate. However, if there are multiple concurrent users on the network, the throughput is likely to be less because they share the communications media. For example, the number of collisions is typically higher in Ethernet LANs as the number of concurrent users increases—as the network has to recover from these, the effective data transfer rate for the network decreases.

Network simulation can be likened to using a computer program to simulate a wind tunnel for an aircraft model—the simulation allows the user to assess how the aircraft (system) is likely to perform without ever physically constructing it. In addition, system simulation allows the user to plug in numbers reflecting expected growth rates for the LAN to determine when anticipated work loads will exceed current capacity. For example, if the results of a simulation indicate that server utilization rates will exceed current capacity, the network manager could plan to add resources to the existing server or add servers to share the burden of the program execution.

As noted in Chapter 13, simulation tools make it possible for network managers to build and execute simulation models. While these vary considerably in complexity, they can be very useful in network planning.

WORKLOAD GENERATORS

Many software vendors have learned that products may perform extremely well in a stand-alone network having little traffic, but when used in an operating LAN that is interconnected to other networks, the performance of the same software may be inadequate.

Workload generators can be used by network planners to determine what software and network performance is likely to be in a proposed network configuration. This is done by building an imitation of the proposed configuration so that the current system does not have to be affected. Once the imitation of the proposed network is created, a *workload generator* program is used to duplicate the pattern of user work executed on an actual system. This enables the network manager to foresee whether a proposed configuration is likely to be superior to the current network configuration.

Workload generators can also be programmed to test for stress points in a proposed configuration. By incrementally increasing the user work load placed on the network, the network manager can determine at what level the network will crash. In Chapter 13, you will read that workload generators can also be used in the network tuning process.

◆ PLANNING NETWORK CONFIGURATION

By using network simulation and workload generators, the LAN administrator can get a clearer picture of what types of changes are likely to be needed to satisfy projected user computing and communications requirements. However, even when needed modifications are identified, their implementation must be carefully planned and scheduled. Often the changes are implemented incrementally to see if problems result from each addition or deletion. If a problem results when changes are made in combination, it can be very difficult to determine the problem's cause.

Most effective LAN administrators write implementation descriptions for the system to be built or expanded. This written description often becomes one of the primary planning documents used to develop or change the network. The administrator is also likely to justify in writing the system change or proposed network. In general, network managers should solicit user input and feedback on these documents; incorporating user input typically enhances the probability that an appropriate system (or system change) is implemented.

PLANNING FOR HARDWARE AND SOFTWARE ACQUISITION

Regardless of how the network plan is developed, it often is used to generate a request for proposal (RFP) that is sent to vendors able to provide components needed in the proposed configuration. RFPs and other factors associated with the vendor selection process are discussed in Chapter 9.

The contents of RFPs and other documents sent to vendors of network components vary considerably because of the wide range of LAN alternatives (and plans) that exist. However, for many of the LANs currently being developed, a number of general network planning guidelines are being followed by network managers. These are briefly summarized in the following paragraphs.

◆ Be sure that the workstations have adequate RAM to execute the applications you are planning. Remember that most (or much) of the application processing is done at the workstation (client) in client/server networks.

◆ Select a server that has a high-performance design, including high-performance disk drives. SCSI disk drives are being increasingly recommended for servers.

◆ Use high-performance NICs designed for modern workstation and server performance.

◆ Plan ahead in your network design. Currently, 10BaseT seems to be the means of running 10MB/second Ethernet with the potential of running 100MB/second Ethernet (Fast Ethernet) in the future. Other paths to higher transmission speeds will certainly develop.

◆ Observe the limitations of your planned configuration. For example, Thin Ethernet currently permits only segments not exceeding 607 feet. There can be up to 5 trunk segments connected through 4 repeaters. Only 3 of the segments can contain network nodes, and the entire network cannot exceed 3,035 feet. There is a minimum distance between stations of 1.5 feet, and a trunk segment must be terminated at both ends with a terminating resistor matching the impedance of the cable. One end of the cable must be grounded. If you are planning a Thin Ethernet LAN, these specifications must not be ignored. Of course, there are similar specifications for other types of LANs.

◆ Generally speaking, if the number of workstations exceeds 30, consider using more than one segment with the estimated traffic spread equally over the number of segments you choose to use.

Network managers are likely to encounter other network planning guidelines derived from the experience of other network administrators. These almost always contain some pearls of wisdom that should not be ignored.

SITE PLANNING

Preparing the physical facilities where the network will be installed is another important network planning task. Many networks experience intermittent problems that could have been avoided if proper site planning and preparation had been done. In general, the network site should be as hospitable for computing as possible; a hostile environment can be a nightmare for the LAN administrator.

In site planning and preparation, the two most potent enemies of computing equipment should be eliminated: dirt and inconsistent power. If network equipment must be installed where dust and dirt abound, the server at least should be protected by isolating it in a clean environment. In dusty and dirty environments, diskless workstations are quite common; this is because disks and disk drives are especially vulnerable to dust and dirt. Likewise, the disk drive system(s) in a file server will not survive long in a dusty or dirty environment.

In site planning and preparation, the network manager should include mechanisms for avoiding inconsistent power levels. Inconsistent power can be the cause of (repeated) server failure. Like the RAM in microcomputer workstations, server RAM is volatile; its contents are lost if the server loses power. The server and network are also likely to experience problems if power levels are inconsistent. *Uninterruptible power supplies (UPS)* are used to sustain electricity in the server and network in the event of a power failure. In many situations, a UPS with adequate power to sustain the server for at least 30 minutes after a power failure is desirable; with a UPS, the server can be brought down in an orderly manner. A *surge protector* or power filter can prevent inconsistent variations in power levels; the surge protector can avoid the damage to electronic chips that power inconsistencies can cause. In addition, the electrical circuits for the server and LAN must be properly grounded, using specially designed electrical receptacles made for this purpose. Site preparation plans should include all of these considerations.

Another important part of site planning and preparation is the development of cabling diagrams and cabling installation plans. *Cabling diagrams* serve as blueprints that illustrate how the cabling will connect network components. Such diagrams are essentially detailed maps of the network topology that include the locations of all hardware. When developing cabling diagrams, the network manager should be mindful of how the network can be expanded in the future; cabling plans that accommodate the ability to expand and to interconnect with other networks in the future are usually more effective than those that do not.

Local building and wiring codes must also be considered in the development of site and cabling plans. Indeed, most municipalities have codes governing the kinds of electrical cabling permitted in buildings. For example, most municipal codes preclude, in areas near air ducts and ceiling areas that serve as pathways for forced air systems, the use of cabling materials that emit toxic gases when burned. In addition, many municipalities require the use of plenum cable, or running the cabling through metallic conduits.

Some types of cable installation are not allowed. For example, local regulations may not allow holes to be drilled in existing walls and ceilings. In addition, environmental regulations may prohibit disturbing existing structures if they contain hazardous materials such as asbestos. Also, for buildings that are part of the Historic Register, there are a host of other regulations that must be taken into account during the site planning and preparation process.

Because of the local, environmental, and historic regulations that may have to be considered when developing site and cabling plans, it is often wise to find a local computer cabling specialist. Note that contractors who specialize in computer cabling are not electrical contractors; while electrical contractors know how to install electrical circuitry, most are not familiar with the special requirements for computer network installation.

Just as for any other type of contractor, references for all network cabling contractors should be obtained and checked. The goal is to identify the network cabling specialist likely to do a good job at a reasonable price. A reasonable price is not necessarily the lowest price; almost any contractor can find materials that will meet minimum specifications and undercut the price of a competitor. However, if another contractor is willing to do the work at a slightly higher price, has better references, and has a reputation for better work, it may be wise to avoid the cheapest contractor. The key is to minimize the cost over the life of the system, not just to minimize the initial cost.

A final factor that should be considered in site plans is the likelihood that work will have to be done on network equipment in the future. For this reason, it is important to make the components as accessible as possible. Planning for accessibility can facilitate the future installation of new NICs, connectors, cabling, and so on, and facilitate troubleshooting. These activities can be performed faster and with less impact on users when network personnel can easily get to and work on network hardware.

INSTALLATION PLANNING

Once site and cabling plans have been developed, it is possible to proceed with installation planning. Installation planning involves developing a schedule for installing the cabling, installing network interface technologies, connecting workstations and servers, and installing software. These activities are summarized in Figure 8-2. In many ways, installation planning involves scheduling the implementation of the network.

In most instances, network interface cards (NICs) will have to be installed in the expansion slots of each of the computers to be attached to the network. However, in some cases, the workstation vendor may install the NIC cards for you at the factory. The NICs will have to be configured by jumpers, dip switches, or EPROMs to match the configuration of the personal computer. NIC configuration is likely to go much more smoothly by analyzing workstation configuration in advance; there are electronic tools available to help you do so. For example, the MSD.EXE program that accompanies Windows 3.1 and DOS 6.0 and 6.2 is an elementary tool that can usually provide a reliable outline of the machine's configuration.

During installation planning, it should be remembered that running cable, equipping existing computers with NIC cards, and installing new equipment cannot occur without some interruption of work if the network site is already occupied. If installing in an existing office, it generally is better to begin, install, and finish a small group of workstations, then move on to another small group. By doing this, only one part of the total office environment is disrupted at any point in time. Incremental changes such as these are typically more successful than trying to make all the changes at once.

The installation of the server(s) is the most critical installation operation for a new network or for the expansion of an existing network. Good planning for this is especially important. When expanding an existing network, it is frequently best to add a second server, leaving the first server available for existing operations.

When upgrading a major online system having numerous current users, it may be necessary to install the new server(s) offline until you are ready to convert to the new system. When upgrading online systems, most system administrators will refuse to bring a new system up for live use until the new system has been run parallel with the old system and is fit to take over for the old system. Updating a currently operating multiuser system requires very careful planning, especially if data files have to be converted to new formats and fully up-to-date when the crossover from the old to the new system occurs.

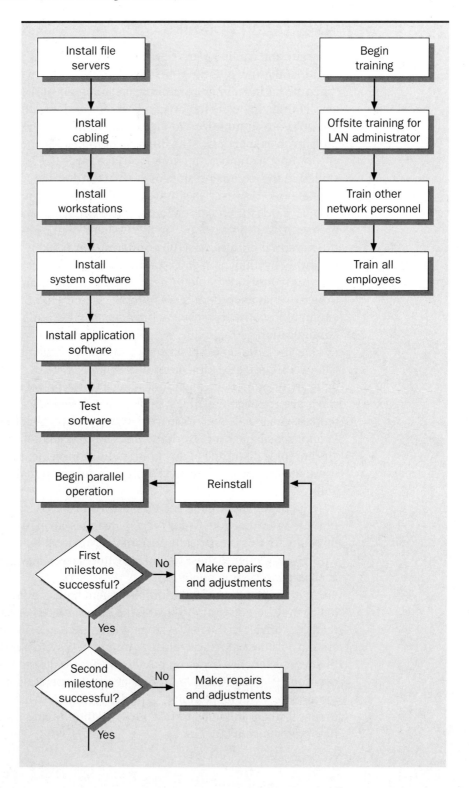

FIGURE 8-2
Installation
activities

TEST PLANS

Good testing procedures can help a LAN manager know if a new portion of a LAN, or a significant upgrade to the LAN, is ready for day-to-day operation. An important aspect of effective testing procedures is user input and feedback. Users can help network managers design good tests for new networks or for network upgrades, especially if they are familiar with the applications that will be used in a new network. Users can also help in the development of network simulation models and representative workload generators. In addition, once a network (either live or offline) has been implemented, users frequently identify problems missed by network designers or installers. In general, it is wise to involve users throughout the network testing process, from developing testing plans to testing the system after it has been installed.

For some types of network applications, it is very important to perform a thorough evaluation of the outputs produced by the old and new systems. By running the old and new systems in parallel prior to converting to the new system, the outputs can be compared. If the outputs that should be equal are not, the problem(s) causing the inequality must be solved. By testing the outputs of all key applications in an incremental and systematic manner, network managers will be able to assess when to convert to the new system.

Server testing is an important part of any good network test plan. This is especially important when adding new applications to an existing server or when adding new servers to the LAN. Among the server tests needed are tests of the alert and alarm systems for indicating that server capacity levels are being reached (alert and alarm systems are discussed more fully in Chapter 13).

Sometimes it is difficult to make the transition to using more than one server. If possible, a multiple-server system should be tested offline in a trial network. The purpose of the trial network is to isolate any conflicts that may occur from the interaction of the multiple servers. While the creation of a trial network may demand purchasing more servers than will be needed in the implemented system, server prices are going down (making this a more feasible option), and the additional servers can be used as network backups.

◆ TRAINING PLANS

Training should begin long before the implementation of a new LAN or LAN upgrade. If applications execute differently on the new or improved network, learning curves may be long and steep for both users and network personnel.

TRAINING FOR NEW NETWORKS

When introducing a network to a work environment, significant training is likely to be needed for both network personnel and for users. Training should start with the person selected as the LAN administrator.

Ideally, the person selected as the LAN administrator will have at least a modest networking background. Such experience can provide a foundation on which concepts learned in training can be built. However, LANs have become sophisticated enough to require up to a year's training for LAN managers, even if they already possess some experience with networking. It is also important for LAN administrators to acquire both solid technical training for the type of networks they will manage and training in how to actually manage the LAN on both day-to-day and long-term bases. Acquiring technical knowledge about the LAN is rarely sufficient; LAN managers should also be trained in network planning, staffing, and budgeting, as well as in a number of other managerial functions.

Many organizations take advantage of training and certification programs offered by vendors of networking technologies. For example, Novell offers a broad array of authorized courses through affiliated organizations nationwide. These courses are consistent from place to place; Novell monitors the quality of the presentations that are made and certifies the instructors. Through passing certification exams that the courses are designed to prepare students for, Novell awards the Certified NetWare Administrator (CNA) designation, the Certified NetWare Engineer (CNE) designation, the Enterprise Certified NetWare Engineer (ECNE) designation, and the Certified NetWare Instructor (CNI) designation. Each of these levels of certification is more difficult to attain than the previous.

These and other networking certification programs (described more fully in Chapter 17) have value recognized in the marketplace. Many employers offer bonuses for attaining these certifications and offer to pay for (at least part of) the cost of the training. However, some employers have paid for training only to have newly certified network managers leave the organization for higher-paying positions. As a result, some organizations now refuse to pay for training that is part of a certification program. Organizations must weigh the risks associated with losing newly trained personnel against those of having an insufficiently trained network staff.

Once trained, the LAN administrator can begin training for network staff members and network users. However, training for both network personnel and users can also be provided by vendors or network consultants. For large or relatively complex LANs, other trained workers are typically needed to assist the LAN manager; in Chapter 17, a number of different networking job titles are identified. The likelihood of employing many network staff members is even reflected in software packages' literature. For example, Novell software version 3.11 (and later versions) recognizes that roles such as Supervisor, Supervisory Equivalent, User Account Manager, Console Operator, Print Server Operator, Print Queue Operator, and Workgroup Manager may each be played by a different staff member. All of these roles are intended for persons working below the LAN Administrator position but needing additional access to the system software to perform the tasks associated with that function.

In addition to network staff training, network users may have to be trained (and retrained). They need to be trained in how to logon to the network and how to use the network to help them perform their jobs. Because training can help users be more productive, the content and delivery methods for user training programs should be carefully planned. This is especially important when users are first being trained to use a network.

TRAINING FOR NETWORK EXPANSIONS

When a network is being expanded, training plans are likely to be less extensive. The LAN administrator will be familiar with the tasks to be performed in this position, and the amount of new training needed may be minimal. However, if a major hardware or software change will take place, LAN managers and other network personnel may need a significant amount of new training. A major hardware change may include the addition of new servers and peripherals, or interconnecting the network with other networks. A major software change may involve choosing a new NOS or upgrading to a new version of the existing NOS. For example, because of the differences between the 3.xx and 4.xx versions of Novell NetWare, many organizations have found that significant levels of training are often needed for network personnel.

User training programs at network expansion tend to follow the same pattern as that for networking personnel. If the change is simply an expansion of an existing system, there may be very little, if any, change in the way that users utilize the network to perform their jobs. However, if the change involves major hardware or software modifications, users are likely to need extensive training.

◆ NETWORK MAINTENANCE PLANS

Network maintenance plans are outlines of the steps that will be taken to keep the LAN operational over the long term. They typically include elements of adaptive maintenance (for example, how changes will be made in response to changes in the organization's computing requirements), corrective maintenance (for example, plans for how "bugs" and glitches will be eliminated from the network), perfective maintenance (for example, plans for tuning the network and enhancing its performance levels), and preventive maintenance (for example, plans for replacing key network components before they wear out and cause serious network problems).

In network environments, maintenance services can often be obtained from vendors or third-party firms. However, the nature of the maintenance available from outside sources has changed, especially that available from vendors. See Figure 8-3 for some typical maintenance options.

Many organizations first developed their microcomputer maintenance plans when software vendors provided virtually unlimited free telephone support. Now, most software vendors have ended free support or will do so soon. If telephone support is desired from a vendor, it is now likely to be available only for a fee. However, there are still some sources of free maintenance assistance. For example, memberships in the national bulletin board systems or the Internet can put you in touch with individuals who offer free advice on maintaining your network. By describing the problem you have encountered, managers of similar networks may be able to tell you how they solved the problem.

The type of maintenance support that can be subscribed to typically has several levels. For example, there may be one fee for up to three calls per month (with more than three at an additional charge), a second fee for up to ten calls per month, and the highest fee for unlimited calls per month. There may be one rate for a designated caller and a much higher fee if all network personnel are able to call for assistance. Such fees are usually paid by annual subscription, and some are high enough to add considerable expense to your budget.

In today's network environment it is possible to have a maintenance plan combining third-party and vendor support. An increasing number of consulting organizations can help manage a network from their offices via a telephone. Such organizations can provide "help desk" functions at a fraction of the cost associated with setting up a help desk in-house.

FIGURE 8-3
Maintenance
options

Hardware and software:
◆ Flat rate
◆ Time and material
◆ Non-contract

Or

Hardware:	**Software:**
◆ Flat rate	◆ Flat rate on site
◆ Time and material	◆ Time and material on site
◆ Non-contract	◆ Time and material by telephone

In most instances, more time and effort will be devoted to software maintenance than to hardware maintenance. However, hardware maintenance cannot be overlooked. There will be hardware failures, and these will range from ones that are easily isolated and repaired on-site to those that can be found only with sophisticated testing equipment in a vendor's maintenance shop. Because of the inevitability of hardware failures, the LAN administrator must weigh the cost of repairing older equipment against that of replacing the malfunctioning component with new equipment (which may carry one to three years of manufacturer's warranties to protect against further maintenance costs).

A key issue in a network maintenance plan deals with the maintenance and repair of servers. When server maintenance or repair is needed, the server usually has to be temporarily detached from the network. Prior to doing so, its contents should be backed up; tape backup systems or backup servers are common ways of doing this. Backup servers are often the preferred option because these can be used to keep the network operational while the primary server is being repaired. Disk storage components are the server parts most likely to malfunction, and if SCSI disk drives are used (and they commonly are), maintenance expenses can be significant.

Some of the most difficult challenges to maintenance planning and maintenance programs are situations in which it is unclear whether the malfunction is software related, hardware related, or a combination of both. This is particularly challenging if the failed components have been purchased from multiple vendors who are under contract to provide maintenance services. It can be very difficult to coordinate maintenance services from multiple vendors in the wake of a network failure, and you should expect them to be pointing toward each other's products as the cause of the failure.

Good network maintenance plans often list who is responsible for repairing malfunctioning network components. The list should include those persons' phone numbers. If preventive and perfective maintenance activities are included, the dates and times when these will occur should be listed; it is also wise to share these types of maintenance schedules with users so that they can plan their work activities around them.

PLANNING INTERCONNECTIONS

Increasingly, LAN managers are responsible for managing networks that are part of other networks. For example, the LAN could be part of a backbone network, MAN, or WAN. Figure 8-4 illustrates some typical interconnections. Stand-alone LANs are now being interconnected with other networks and the percentage of stand-alone LANs is decreasing.

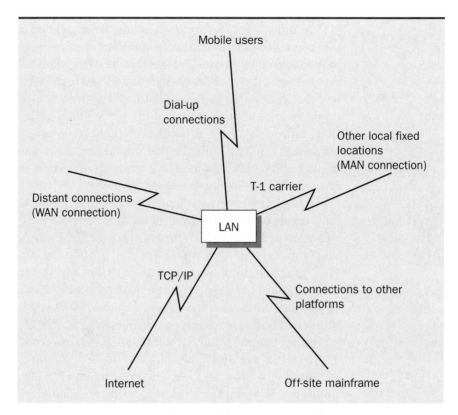

FIGURE 8-4
Interconnections

Fortunately, most NOSs are designed to allow servers in one LAN to be interconnected to servers in other LANs. For example, Novell's NetWare 4.01 enables some servers to manage other servers in other networks. Such interconnections enable users in one network to access computing resources available only in other networks.

If interconnections are planned, it is important to select NOSs and protocols (such as TCP/IP) that can be used in interconnected network environments. Potential conflicts between servers in the different networks to be interconnected should be anticipated and tested.

Cabling interfaces must also be planned, especially if the different networks targeted for interconnection utilize different types of cables. Some of the other interconnection issues that must be considered follow.

♦ What microcomputer operating systems are in use in the networks to be interconnected?

♦ What applications will be shared across networks? Which will continue to be available only in specific networks?

♦ What interfaces or conversion programs now in use will be used in the expanded system?

◆ How will data traffic within and across networks be affected by the planned interconnections?

◆ How will network response times be affected by the interconnections?

◆ Where are potential conflicts between servers and network configurations most likely? How will these be addressed?

◆ What test procedures will be used to assess the compatibility of interconnected networks?

◆ Who is responsible for addressing network failures? Who is responsible for failure of individual workstations?

In Chapters 15 and 16, you will see that there are a number of other important interconnection issues. However, for now, is should be quite apparent why interconnection planning has become such an important part of network planning.

◆ STAFFING PLANS

Probably the most significant factor in the construction or expansion of a LAN is the quality of the staff that operates and maintains the network day to day. The dedication and commitment of the networking staff is critical in the quick resolution of network problems. This is one of the reasons why organizations have found it necessary to develop plans to attract and keep qualified networking personnel.

In general, staffing plans cannot be finalized until the other aspects of network planning have been done. Network managers must be aware of network configurations, what applications will be supported, how the network will be maintained, and so on, before they can determine how many and what types of staff members are needed. Once they have some idea of the size of the network staff and the skills needed, they have to identify acceptable salary levels. This is often done by determining what other organizations are paying for comparable positions. Next, recruiting, training and development, and job performance measurement plans for needed staff members can be formulated. However, it may be easier to develop such plans than to implement them.

As client/server networks emerge as the dominant computing platform within organizations, the availability of qualified staff for such networks has not kept pace. Individuals with client/server networking experience are in high demand and organizations are paying high salaries to attract and keep them on staff.

Turnover rates among the networking staff are likely to be high, especially in organizations reluctant to pay market rates for qualified staff members. In addition, there is such a strong demand for middle managers of networks that a qualified person can quickly move up through many organizations and garner a significant raise at each step. Needless to say, such high turnover can have a dramatic effect on the ability of network managers to maintain consistent network performance levels.

Chapter 17 takes a closer look at networking careers. From this chapter, you should have gotten a better idea of why the demand for networking personnel is so high, as well as some of the steps that can be taken to ensure that the people who are hired can keep their skills and knowledge up to date.

◈ SUMMARY

Network planning is needed for organizations that intend to implement one or more LANs for the first time. It is also important when the organization intends to expand or significantly upgrade the hardware or software in existing networks. As both of these activities have become commonplace, network planning has become an important aspect of network management.

One of the first activities found in many network planning processes is the assessment of the feasibility of the proposed network. To this end, a feasibility study may be conducted.

If the proposed network is determined to be feasible, a more detailed look at the suggested system is usually undertaken. Through the use of capacity planning tools such as network simulation models and workload generators, network managers are able to get a clearer picture of the ability of the proposed network (or network change) to handle anticipated network traffic and user work loads. One of the key measures used in these assessments is network throughput.

Network planning also includes determining the network's configuration and plans for implementing the configurations. Plans must be made for acquiring and installing network hardware and software. There must also be plans for preparing the site at which the network will be installed; this often involves the development of cabling and cabling installation plans that are consistent with municipal, environmental, and historic preservation regulations. Network installation plans also include schedules for installing network interface technologies, software, servers, and for testing the network prior to conversion.

Training plans for network managers, network staff members, and users are other important parts of network planning. When LANs are implemented for the first time, extensive training may be needed. Significant training may also be needed for network expansions or for major hardware or software upgrades; however, training is likely to be more specific and focused than that needed for first-time network implementations. Certification programs for network managers and network staff often include substantial training components.

Network maintenance plans are developed by network managers to outline the steps that will be taken to keep the LAN operational over the long term. These may include adaptive, corrective, preventive, and perfective maintenance activities. An important part of maintenance planning involves determining the amount of maintenance support that will be provided by vendors and third-party organizations.

Planning for the interconnection of networks is an important network planning activity in more and more organizations. The compatibility of the networks to be interconnected is one of the most important issues that must be addressed in these network plans.

Staffing plans are also important in network environments. Due to the general lack of qualified network staff, skilled network personnel are in high demand and turnover rates for these employees are high. This has put pressure on organizations to pay competitive salaries and to develop programs to recruit, train, and retain qualified networking personnel.

✳ KEY TERMS

Feasibility study
Network maintenance plans
Network simulation

Throughput
Workload generator

✳ REVIEW QUESTIONS

1. Why are LANs likely to change or be replaced over time?
2. Why is network planning important?
3. Explain the purpose of a feasibility study.
4. What is the difference between network simulation tools and workload generators? What types of decisions do these help network managers make?
5. What is network throughput?
6. What types of written implementation descriptions are likely to be developed by network managers?
7. Identify some of the planning guidelines that are currently being followed by network managers.
8. What is site planning and why is it important? What activities are associated with site planning?
9. What types of configurations are best for dusty or dirty computing environments?
10. What problems can inconsistent electric power levels create? How can such problems be avoided?

11. What are cabling diagrams and cabling installation plans? Why is each important?
12. How may building codes, wiring codes, and historic preservation codes affect site plans?
13. What activities are likely to be included in installation plans?
14. Why is installation often done in incremental steps rather than all at once?
15. What are test plans and why are they important? Why is user input important in the development and implementation of test plans?
16. Why are training programs for new networks likely to be more extensive than those for network expansions or upgrades?
17. What is the role of certification programs in network training?
18. What are maintenance plans and why are they important?
19. What are the differences among adaptive, corrective, preventive, and perfective maintenance activities?
20. How may the level of maintenance support provided by vendors vary?
21. What is involved in interconnection planning? Why has this become more important?
22. Why is it important for organizations to have staffing plans for network personnel?

✳ DISCUSSION QUESTIONS

1. How and why has network planning become more complex? What planning activities and issues are becoming increasingly important? Which do you expect to continue to increase in the years ahead? Why?
2. Why are training programs likely to be more extensive for new LANs than for LANs that are being expanded or upgraded? Discuss how the content of the training programs (the concepts that students are exposed to) is likely to differ for these two situations.
3. Discuss the advantages and disadvantages of running an old and new online network in parallel prior to converting to the new network. Be sure to include the managerial, financial, and staffing issues associated with this.

✳ PROBLEMS AND EXERCISES

1. Ask a local network manager to come to your class to describe a good feasibility study. Use the manager's comments to develop a list of items that should be addressed in a feasibility study.
2. Draw the topology of a LAN or other network at your school or in a nearby organization. Include every major component in the network.

3. Draw a cabling diagram for an existing LAN or network. Critique its readiness for upgrading or expansion.
4. Interview a local LAN administrator about the type of training that he or she has received and that scheduled for the future. Also, ask the LAN administrator about the training available for network staff and users.
5. Invite a local network manager to discuss the network maintenance plans and activities in his or her organization. Be sure to ask about the types of maintenance services from vendors or third-party firms to which the organization subscribes.

✳ REFERENCES

Bates, R. J. "Bud." *Disaster Recovery for LANS*. New York: McGraw-Hill, 1994.

Editor. "Network Planners Loyal to Home-Office Brand Vendors." *Computer Reseller News* (April 25, 1994): p. 90.

Farris, J. "Network Administration's Growing Role." *LAN TIMES* (August 9, 1993): pp. 28, 29.

Graziano, C. "NetWare 4.0 and Corporate Politics: Large Firms Slow to Upgrade—Could Politics Be the Problem?" *LAN TIMES* (August 23, 1993): p. 7.

Horwitt, E. "Enterprise Push to Cut Server Costs." *Computerworld* (May 16, 1994): p. 14.

Klett, S. P. "Vendors Plan LAN/WAN Links." *Computerworld* (April 18, 1994): p. 72.

Koegler, S. "In-House Staff vs. Vendor Support: The Best Solution Utilizes Both, But It's a Constant Tug of War." *LAN TIMES* (March 28, 1994): p. 43.

Smith, B. "How to Negotiate a Service Contract." *LAN TIMES* (December 6, 1993): p. 90.

Stamper, D. A. *Local Area Networks*. Redwood City, CA: Benjamin/Cummings, 1994.

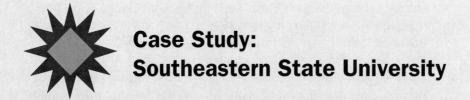

Case Study:
Southeastern State University

Two types of network planning have been done at Southeastern State University: that performed by the Computer Services Department and that done within the different colleges and departments. As has been noted, the Computer Services Department is charged with planning, operating, and maintaining the large systems on campus, especially those interconnected to systems on other state college and university campuses. As you might expect, the plans developed by the Computer Services Department tend to be generated in a systematic manner and result in a number of formal planning documents and diagrams.

The networking plans developed within the different colleges, and in some cases within departments within colleges, vary considerably in completeness and complexity. Many have been assembled hastily in response to the sudden availability of funding, and some consist of little more than wiring diagrams developed by the local supplier of a particular vendor's equipment.

The diversity among plans for LANs has, of course, contributed to the diversity found among LANs at SSU. This is somewhat ironic in that the general planning and budgeting process used on campus follows the same general pattern from college to college. For example, department chairs assemble input from faculty to develop their budget requests to present to deans, and deans, in turn, aggregate the budget requests of department chairs in developing their budget requests, submitted to the vice presidents and to the president. In many instances, for whatever cause, it seems that the systematic nature of planning and budgeting that may be observed within the general administration is left by the wayside once requests for new LANs or for LAN expansions or upgrades are approved. After the money is allocated, all decisions about how it is spent (and on what) are left to the individuals charged with implementing the LAN. Those individuals are often more interested in implementing something as quickly as possible than in planning what they should be implementing.

The need for more standardization in the LAN planning process has been formally identified as a priority by the Computer Services Department. This has resulted from an increasing number of requests from managers of stand-alone LANs to establish connections to SSU's VAX computers, and the

Internet. As one Computer Services Department member stated, "This is ridiculous, every configuration is different. There is no consistency in cabling, so each of the interfaces that we have to develop is unique. Half the LANs have primitive backup systems, and some put the entire network in jeopardy because they lack UPS systems. And half the so-called 'network managers' have no idea what they are doing. They don't have a clue about site preparation or network incompatibilities."

Many of these complaints stem from the fact that budget requests for LAN manager positions usually follow budget allocations for new LANs or LAN changes by at least one full calendar year. The deans of the colleges generally won't ask for these important staff positions until their request to establish the LAN has been approved. The contingency plan within most colleges is to utilize a faculty member with some technical knowledge about microcomputers and an interest in networking as an unofficial LAN administrator for the first year or two. It becomes the responsibility of this individual to implement the LAN. In more than one situation, the temporary LAN manager has been later permanently appointed to serve in this capacity.

While the Computer Services Department has had to confront the problems that this pattern of LAN acquisition and staffing has caused, there is no indication that things will change in the near future. The university has no intention of modifying the general planning and budgeting process used in the colleges and academic departments.

CASE STUDY QUESTIONS AND PROBLEMS

1. How is the use of network simulation, work load generators, and other capacity planning processes likely to differ from the Computer Services Department to the colleges at SSU? How are maintenance plans and processes likely to differ?
2. How could training for temporary LAN managers alleviate the network planning inconsistencies that SSU is experiencing? What types of training would you recommend? Why? What types of training do you think members of the Computer Services Department would recommend?
3. Based on the information provided in this and previous segments of the case study, what types of interconnection issues should LAN managers at SSU be exposed to in training sessions?

Selecting LAN Components

CHAPTER OBJECTIVES

After completing this chapter, you will be able to:

◆ Discuss why the evaluation and selection of LAN components and vendors is important.

◆ Describe the vendor selection process.

◆ Identify the sources of information relevant to LAN and vendor selection.

◆ Describe the differences among requests for proposals, requests for quotations, and requests for bids.

◆ Describe the approaches used to evaluate vendor proposals, quotations, and bids.

◆ Discuss the importance of contracts and agreements with vendors.

◆ Describe the differences among purchase contracts, leases, and maintenance and support agreements.

After network plans have been developed, an organization is ready to take the first steps toward network installation. Among these steps is the selection of vendors to supply the components needed for the network. In many instances, the organization must decide among both hardware and software vendors. Network managers must decide who they want to do business with, as well as what types of formal business agreements (contracts) are needed between the organization and the vendor(s).

In this chapter, we will focus on vendor selection and contracting. Included in the discussion are general guidelines for these important processes.

◆ WHY IS VENDOR SELECTION IMPORTANT?

Organizations and vendors develop formal agreements or contracts. These agreements and contracts essentially commit the organization to using particular hardware, software, and services for a specified time period. Usually, such agreements are legally binding, and failure to live up to them can be grounds for a lawsuit. Hence, the vendor selection and contracting process should not be taken lightly; organizations and their network managers should choose vendors wisely and should be aware of what is (and is not) included in any formal business agreement signed.

When narrowing the list of potential vendors for network components, it is important to consider the fact that there may be a high volume of communication between the organization and the vendor. LANs and most other data communication networks are not trouble-free computing platforms. Network managers will likely encounter problems and glitches with which they have no experience, and when they do, they will likely turn to vendors for assistance in solving the problem. Problems are most likely to be rectified quickly when it is easy for network managers and vendors to communicate with one another. Hence, during the vendor selection process, it is important for network managers to consider how easy it is likely to be to communicate with each of the potential vendors. After all, once a contract is signed, the organization may be committed to dealing with that vendor for a long time.

◆ VENDOR SELECTION ISSUES

Many of the criteria and issues commonly used in the vendor selection process parallel those used to evaluate LAN alternatives. A number of criteria used to choose from among LAN alternatives are discussed in Chapter 3; these are summarized in Table 9-1. From these, important questions that network managers need answered by vendors can be derived.

◆   How many users in a single LAN can your products support? Can your products handle the current number of users and the number that we expect to have in the future?

◆ Number of users

◆ Application software availability

◆ Types of workstations supported

◆ Printer support

◆ Expandability

◆ Communication speed

◆ Media options

◆ Geographic limitations

◆ Interconnectivity with other networks

◆ Adherence to widely accepted standards

◆ Hardware availability

◆ Vendor support

◆ LAN management software availability

◆ LAN security options

◆ Overall cost

TABLE 9-1 LAN selection criteria

◆ Do your products support the necessary application software? Can needed application software be purchased or leased? Will needed application software have to be developed in-house?

◆ Will your products accommodate the range of workstations in our organization? Are special provisions or configurations needed to handle a diversity of workstation types?

◆ Do your products easily support a wide range of printer types and print server configurations? Are special provisions or configurations needed for laser or color printers?

◆ Do your products easily handle the addition of new input, processing, storage, and output technologies to the LAN? What is the upper limit to the incorporation of new computing technologies? How much time and effort is involved in adding new technologies?

◆ What communication speeds are possible with your products? Can speeds be increased without extensive modifications?

◆ Can your products be used with a variety of communications media? Can your products be used in high-speed LANs such as FDDI, CDDI, ATM, and Fast Ethernet?

◆ Can your products be used across diverse, interconnected computing platforms and client/server architectures? Do your products interface with the variety of communications protocols used in our interconnected networks?

◆ Are your products designed in accordance with widely accepted standards? If so, which ones?

◆ Are the LAN hardware components needed for your products available from a variety of sources? If so, which sources are the most reliable and best priced?

◆ What types of support do you provide for your products? Do you have a helpline? Where are your closest service and customer support offices located? What is your typical response time to customer problems? How much do you charge for such support?

◆ What types of LAN management functions are found in your products? Do your products interface with popular network management systems? If so, which ones?

◆ What backup and security features are included in your products? How can fault tolerance be achieved with your products?

◆ How much would it cost to implement your products in our LAN(s)? How are upgrades priced?

Needless to say, this is not a comprehensive or exhaustive set of questions for vendors. Appropriate questions will, of course, vary from company to company and from network to network. However, these are some of the major questions for which you might want clear answers before signing a contract with a vendor.

◆ OTHER IMPORTANT VENDOR SELECTION ISSUES

Besides these questions, experienced network managers are often interested in learning more about the compatibility of the vendor's products with existing systems, the overall financial strength of the company, the extent to which the vendor will assist with implementation and user training, and the experience that other companies have had in using the vendor's products. Each of these issues is briefly discussed in the following sections.

COMPATIBILITY WITH EXISTING SYSTEMS

Most organizations developing LANs already have computer systems in place; in many instances, a wide variety of computer equipment may be in use. To make the best use of investments in computing technologies, many organizations want to use as many of these existing technologies as possible in any LANs developed. In other words, many firms are more interested in incorporating existing technologies into LANs than they are in building new LANs from scratch. This is true for firms upsizing to a LAN by networking the microcomputers already in place. It is also true for many firms downsizing (moving applications traditionally run on mainframes or minicomputers to LANs); often, such firms wish to continue to use the mainframe or minicomputer in some capacity in the new network to get as much as they can from the money invested in these machines.

When compatibility with existing systems is important, potential buyers should critically evaluate claims that the vendor's products will work with the systems already in place. Buyers should ask for proof that inoperability will not be a problem. This may involve asking vendors to identify companies that incorporated similar technologies in the LANs they developed; it may even involve contacting or visiting these companies to find out how well their older systems interface with the vendor's products in LAN environments. Potential buyers should be especially attentive to how much additional cost, time, effort, and maintenance is needed to incorporate existing technologies into the LAN(s) they wish to create. They should also determine whether the interfaces between preexisting technologies and the LAN can be a recurring source of operational problems.

When obtaining answers to compatibility questions, LAN managers should not be surprised to learn that it is often more cost effective in the long run to build a LAN with new equipment than it is to incorporate existing technologies into LANs. While this revelation may not sit well with general managers who wish to extend the useful life of existing systems as long as possible, the fact remains that a well-designed LAN built with quality components is likely to have fewer operational and maintenance problems than are LANs that attempt to incorporate older technologies.

VENDOR'S OVERALL FINANCIAL STRENGTH

The vendor's overall financial strength often indicates the vendor's ability to stay in business and to provide you with the service and support needed to keep the network up and running. Network managers should recognize that they may be taking a risk if they decide to sign a contract with a new, relatively unknown vendor. Regardless of whether they have excellent products, new companies are not entrenched in the market. This can make purchasing products from new companies somewhat risky and, as a result, many network managers are more reluctant to do business with a new vendor than with a vendor who has been selling LAN products for a long time.

ASSISTANCE WITH LAN IMPLEMENTATION AND TRAINING

Organizations that have never implemented a LAN are often interested in knowing how much help they are likely to receive from the vendor when installing the LAN. Some vendors have authorized sales and maintenance outlets that will help buyers with virtually all phases of the implementation process, including the installation of cabling, network interface cards (NICs), connections, LAN operating system software, application software, network management systems, and backup hardware and software. Some will even help convert data to the format needed in the network and will help configure printer servers, fax servers, and other shared, special-purpose computing resources. Vendor assistance with these various tasks is not necessarily free, but costs can often be included in the total purchase price. In any event, the availability of help during the LAN installation process may be an important issue for organizations inexperienced in LAN installation. It may also be an issue for organizations that have already implemented LANs, especially if the LAN to be installed is different from those currently in place.

Besides the availability of help with LAN installation, the location of the reseller's business may be important. If the reseller is located in another city, it may be more difficult and more costly to take advantage of the help that is available. Many potential purchasers are likely to consider it an advantage to do business with a vendor that has a local reseller/maintenance outlet.

The extent to which potential vendors can train users and network managers may also be a key factor in the vendor selection process. If an organization requires such training, it is important for it to determine the nature (content, type of instruction, instructor's background, duration, and so on) of available training as well as related costs. As overall cost and quality of training can vary widely, an objective investigation of available training options and their effectiveness may be warranted.

Organizations often want both users and network personnel to be trained. Of these, user training is often the easiest and least costly to implement. When training for network personnel is desired, vendors that provide formal training or certification programs (such as Novell) have a competitive edge. For example, as noted, Novell offers a number of training courses for network managers; most of these have been developed to prepare participants to take the exams needed to become Certified NetWare Administrators (CNAs) or Certified NetWare Engineers (CNEs). These and other network-related certification programs will be discussed more fully in Chapter 17.

EXPERIENCE OF OTHER COMPANIES WITH VENDOR'S PRODUCTS

When deciding which vendor(s) to do business with, it is almost always desirable to find out how other organizations feel about the vendor(s). Checking with current and former users of a vendor's products can be a very effective way to find out whether the vendor's products are good and whether the vendor lives up to promises made to potential buyers. However, it is generally wise to check beyond the references provided by the vendor; such references are sure to be positive and convince you that this is a good vendor with which to do business. A different story may be told by other organizations currently using the system; it may take some time to discover these organizations, but the effort should be worth it. Organizations that once used the vendor's products (but are now using someone else's products) can also provide some valuable insights about the vendor's products and services.

◆ THE VENDOR SELECTION PROCESS

The major steps in the vendor selection process are illustrated in Figure 9-1. Table 9-2 summarizes some of the major activities involved with vendor selection and contracting. As may be seen in Figure 9-1, like the LAN selection process, the vendor selection process begins with deciding what kind of LAN is needed. Once the LAN type has been chosen, attention can be turned to identifying potential vendors of the necessary hardware and software components. After potential vendors have been identified, initial screening can weed out the unsuitables. Vendors that make it through the initial screening may be asked to respond to a request for proposal (RFP), a request for quotation (RFQ), or to submit a bid. Vendor proposals, quotations, or bids are evaluated by the organization and, as a result, one or more vendors are selected. The vendor(s) and the organization then make contracts and formal agreements; once these are formulated, the work of actually installing the products begins (the installation process is discussed in Chapter 10).

IDENTIFYING POTENTIAL VENDORS

Network managers may use several approaches to identify potential vendors for LAN products. These include doing one's own research, contacting other

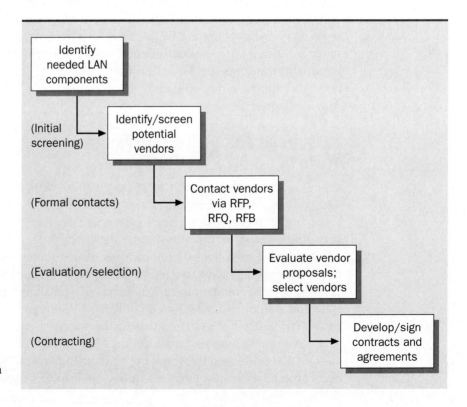

FIGURE 9-1
Vendor selection process

1. Determine what type of LAN is needed.

2. Develop list(s) of potential vendors for all major LAN components.

3. Perform initial screening of the potential vendors identified in Step 2; narrow the possibilities to a manageable number.

4. Contact the vendors that make it through the initial screening and ask them to respond to an RFP, RFQ, or RFB (request for bid).

TABLE 9-2
The vendor
selection process

5. Evaluate vendor proposals, quotations, or bids and make final selection(s).

6. Develop formal agreement or contract with the vendor(s).

companies in the same line of business to find out what products they are using, contacting "known" vendors of the types of products needed and asking them to identify their major competitors, reviewing trade publications, and hiring a LAN/network consultant. Before discussing each of these approaches, you should realize that using a combination of vendor identification processes is often superior to using a single approach. By combining two or more of the approaches discussed below, the marketplace consensus about which vendors to contact and which to avoid is likely to emerge.

Doing Your Own Research

While doing your own research may be time-consuming, it can be one of the best ways for a LAN manager to learn what the options are. Research may involve contacting other companies, contacting known vendors, and reviewing trade publications. However, it may also include looking through the Yellow Pages for the major cities in your area, browsing through *Computer Shopper*, reviewing data communication product catalogs (such as those from *Black Box* and *Inmac*), talking to the owners or managers of local microcomputer stores, and contacting the faculty or IS staff at nearby technical schools, colleges, or universities.

Expect to spend a substantial amount of time on the phone. Be sure to develop a good set of notes. As is true for any of the approaches discussed in this section, be alert for objective, unbiased information, and be wary of the advice of people who have a vested interest in trying to convince you to see things their way. In addition, look for patterns that emerge in the information you select. For example, if the same vendor name is recommended by several independent sources, there is usually a good reason behind it; similarly, if several sources warn you to stay away from certain vendors, you probably should.

Contacting Firms in the Same Industry

Part of the vendor identification process may involve contacting other organizations in the same line of business or industry. Establishing such contacts can provide the potential purchaser with a sense of how far along similar businesses are with networking, which types of networks are most common, what types of benefits have been achieved by establishing LANs, and which vendors have been most responsive to problems and the easiest to work with. Such contacts can also alert the potential buyer to the existence of vendors whose products are especially well-suited to his or her particular line of business or industry.

If possible, visit similar businesses that have already implemented LANs. This can provide valuable insights about facility (site planning), cabling, and personnel and staffing requirements that will have to be addressed.

Reviewing Trade Publications

Objective comparisons of LAN products can often be found in trade periodicals such as *Communications Week, Computerworld, Data Communications, Infoworld, LAN Magazine, LAN TIMES,* and *Network World.* Many of these include a section comparing vendor products for a particular type of LAN technology (for example, servers, bridges/routers, RAID storage systems, groupware, e-mail, LAN operating systems, network management systems, and so on). Occasionally, objective LAN product comparisons can also be found in more general computing publications such as *Byte, PC Magazine, PC Week,* and *PC World.*

The value in reviewing articles comparing particular types of LAN products is that relatively objective and unbiased information can be obtained. The products are typically evaluated on a variety of criteria (such as speed, throughput, ease of use) and the published ratings are usually based on data supplied by actual users. In many instances, vendor service issues such as responsiveness and helpfulness are included in the comparisons; these can be particularly important issues for organizations considering implementing LANs for the first time.

Many public and college-level libraries subscribe to these publications, so it is usually not difficult to obtain copies of such product comparisons. In fact, LAN managers may qualify for free subscriptions to some of these publications (including *Data Communications, Communications Week,* and *Network World*).

Other trade publications also contain objective information on networking technologies. For example, detailed product specifications and market share information can be found in publications such as McGraw-Hill's *Datapro Reports* or *Faulkner Information Services.* Large organizations' IS areas often have subscriptions to one (or both) of these IT product directories. Smaller organizations are likely to find the subscription costs for these too high, but still may be able to find these publications in local libraries.

Hiring a LAN or Network Consultant

A LAN or network consultant may also help to identify potential vendors of LAN products. Actually, individuals or firms who specialize in LAN consulting may be able to provide assistance throughout the vendor selection process (from identifying potential vendors to contracting). They may also be used to help determine what type of LAN is needed by the organization, oversee LAN implementation, and develop and deliver training programs. Of course, the organization will be charged for the services performed, but a good consultant is well worth the cost.

Before hiring a LAN or network consultant, the organization should ask for a list of previous clients, some of whom should be contacted. Contacts with previous clients can help the organization verify the consultant's objectivity, ability to provide quality service(s), and the fairness of charges to the organization. A formal contract specifying the services to be performed by the consultant should be developed and signed.

No matter which of these approaches (or combination of approaches) is used, organizations should give themselves sufficient time to identify the vendors of products needed for the network(s). After doing a good investigation, organizations should have some assurance that viable potential vendors have not been overlooked and that the vendors included on the final list will be able to satisfy the organization's needs.

INITIAL SCREENING

The goal of this phase of the vendor selection process is to pare down the list(s) of vendors of needed LAN components to a manageable level. The completion of this stage is signalled by the development of a "short list" of vendors for each major LAN component. The vendors included on the short list are those who will be contacted by the organization and invited to submit a proposal, quotation, or bid.

In practice, the initial screening process often coincides with the vendor identification process. Typically, as the list of potential vendors is developed, evaluative information is also collected. For example, when contacting organizations in the same business or industry about potential vendors, you not only learn who potential vendors are, you often also hear opinions on how good or bad they are. Also, if you review trade periodical articles comparing vendor products, you will obtain more than a list of potential vendors; the published ratings imply that some products are better than others. In brief, it may be impossible to cleanly separate the vendor identification and initial screening phases of the vendor selection process.

Some organizations attempt to use a systematic approach to the initial screening process by developing a list of screening criteria such as that in Table 9-3. The development of such a list essentially forces the organization

◆ Vendor has a local service office.

◆ Vendor's products score well in independent tests.

◆ Vendor has developed a reputation for helpful service.

◆ Vendor's products are recommended by current users.

◆ Local service is praised by current users.

◆ Vendor's products are used by similar organizations.

◆ Vendor's products meet all required specifications for the LAN.

◆ Vendor sells a wide range of networking products.

TABLE 9-3
Potential screening criteria

◆ Prices for vendor's products are acceptable.

◆ Vendor's products are compatible with other needed LAN components.

to come to grips with the major qualities they seek in a vendor or in the products needed for the network. The set of initial screening criteria that is formulated can be put in the form of a checklist. Once this is done, vendors or vendor products can be evaluated against each of the items on the checklist. Those vendors and products that satisfy all (or most) of the criteria on the initial screening checklist are most likely to be included on the organization's final short list.

Note that the initial screening checklist is only an aid in determining the vendors and products included on the short list. At best, it separates more suitable from less suitable candidates for the LAN components in question. There may be valid reasons to keep a vendor on the short list even if screened out by using the checklist; similarly, there may be valid reasons to exclude a vendor from the short list even if it satisfies all the items on the initial screening checklist. In short, using an initial screening checklist does not guarantee success. However, it is a systematic approach for developing a short list of potential vendors, and that is precisely what this phase of the vendor selection process is about.

CONTACTING VENDORS: RFPs, RFQs, AND RFBs

Firms may contact the vendors of LAN products in a variety of ways. Some of these initial contacts may be quite informal. For example, once a short list of potential vendors has been identified, it is possible to establish contact simply by calling local representatives (or the company's 800 number) to arrange a time to meet and discuss what the organization seeks. Other initial contacts are more formal and systematic. These usually include developing an outline of what is needed by the organization and sharing it with potential vendors; vendors then have the opportunity to develop a formal response to the organization's needs. In larger organizations, a purchasing department may establish the formal contacts between the organization and potential vendors. However, no matter what contact approach is used, vendors should be encouraged to phone or visit the organization to improve the quality of their proposals.

Three of the more common systematic vendor contact approaches are named after the main documents sent to vendors. These are the request for proposal (RFP), request for quotation (RFQ), and request for bid (RFB) approaches. Each of these is briefly discussed in the following sections.

Request for Proposal (RFP)

One of the more common and best known methods for informing potential vendors of the organization's desire to implement a new computing system or network is to develop and distribute a document called a *request for proposal (RFP)*. For example, when an organization is interested in implementing a LAN, the RFP will identify the firm's needs; it will also request all interested vendors to submit a formal proposal showing in detail how they will satisfy these needs. An RFP may range in length from a few pages to hundreds, depending on the size and complexity of the project.

If an organization is considering LANs for the first time, it is likely to distribute a flexible RFP to potential vendors which gives them considerable leeway. Generally, this involves sending vendors possible LAN alternatives and a list of constraints and objectives that must be met. This list addresses performance objectives rather than specific types of LAN hardware, software, and services. An example of such a list is provided in Table 9-4. Interested vendors then prepare and submit a proposal that satisfies the terms in the list.

◆ An overview of the organization, its current computing systems, and its interest in LANs

◆ Price, budget, and time constraints

◆ LAN alternatives potentially acceptable to the organization

◆ General LAN requirements including size, expected growth, reliability requirements, implementation timetable, interconnectivity requirements, service requirements, and backup and security requirements

◆ Processing requirements including input, output, processing, and storage volumes, the mix of applications that must be available to LAN users, response time requirements, and print output speed and quality requirements

◆ A list of issues and considerations that must be addressed before the proposal will be considered

◆ A request for a blank copy of the vendor's standard contract

◆ Explicit instructions regarding to whom and by when the proposal should be submitted

◆ Criteria that will be used when selecting among vendor proposals

---

TABLE 9-4   Information supplied to vendors in a flexible RFP

RFQs and RFBs

When the organization knows exactly what type of LAN it wants (and what types of components are to be included), it is more likely to send vendors a document called a *request for quotation (RFQ)* or a *request for bid (RFB)*. RFQs and RFBs tell vendors exactly what types of LAN components (hardware, software, and services) are required. Virtually everything is spelled out for the vendor, and all that is required is a quotation and a commitment to meet the buyer's terms. An RFQ for the purchase of 100 LAN workstations, for instance, would likely include items such as those in Table 9-5. This RFQ might be sent to the workstation manufacturer, distributors, wholesalers, mail-order suppliers, or retail stores.

WHICH APPROACH SHOULD BE USED?

The advantage of the RFQ or RFB approach is that vendors can respond much more quickly, as most of the guesswork is taken out of the request. Relative to an RFQ or RFB, an advantage of the flexible RFP approach is

---

Hardware Requirements

◆ 100 Compaq Deskpro XL units; each unit should contain a 66 MHz Pentium processor with 16 megabytes of RAM (expandable to 64 MB), one 3½" diskette drive, one 240 MB hard drive, a PCI local bus, and integrated 32-bit Fast SCSI-2 and Ethernet capabilities

◆ 100 Compaq QVision monitors with 1280 x 1024 SVGA graphics with videocards and required cabling

◆ ODI drivers and NICs to support both 10BaseT Ethernet and 16 million bps token ring

◆ Connectors and media interfaces needed for the NICs

Software Requirements

◆ Site licensing and preloaded copies of the latest versions of Windows, OS/2, MS-DOS, Lotus Notes, and Lotus SmartSuite

Other Details to Be Included in Bid

◆ Purchase price for each item (both single unit and in specified quantity)

◆ Warranty period and terms

◆ Types of service support offered; cost of each type of service

◆ Shipping/installation charges

◆ Delivery terms

TABLE 9-5
Example
RFQ/RFB

All bids must be postmarked by January 31.

---

that vendors have more leeway to compete and it may uncover an attractive alternative that the buyer has not considered. For example, a hardware distributor might suggest to an organization considering the acquisition of 100 Compaq Deskpros with Pentium processors that equal functionality might be achieved with 100 Gateway computers with 486DX2 processors.

One of the disadvantages of using a flexible RFP is the task of evaluating several, often diverse, proposals. This can be much like comparing apples and oranges. However, when the organization develops an RFP that is clear and specific about proposal format, this potential problem can be avoided.

No matter which of these approaches is used, vendors should know that you are shopping around. This puts vendors on a competitive footing, and generally the buyer is more likely to find good solutions at good prices when vendors know that they are competing for the business. As a general rule, as the money to be spent on the system increases, the larger the number of vendors contacted (the larger the short list) should be.

Turnkey System Vendors

Organizations can sometimes acquire a complete ready-to-use LAN (including hardware, software, and a service contract) from a single vendor. Such LANs are often called *turnkey systems*, reflecting the fact that once installed, the user can presumably "turn the key" and have the system ready to go. Vendors of turnkey systems will install the LAN, load all software, train users, and essentially take care of all implementation and conversion chores; most will also provide system maintenance services once the LAN is installed.

Turnkey systems are available for an increasing variety of organizations and applications and are viable options for LAN solutions. Turnkey system vendors should not be overlooked when soliciting proposals or quotations.

◆ EVALUATING VENDOR PROPOSALS, QUOTATIONS, AND BIDS

Soon after the deadline has passed for vendors to submit their proposals or bids, the organization begins to evaluate them. The major outcome of this phase of the vendor selection process is determination of which vendor(s) the organization will do business with.

When an RFQ or RFB is used, the evaluation process is often quite straightforward and is driven by the prices that have been submitted. In fact, many governmental and municipal organizations must adhere to *low bid policies*—the organization is obligated to do business with the vendor that has met all the conditions of the RFQ or RFB at the lowest price. Other organizations not locked in to low bid policies can consider more than price when evaluating vendor quotations or bids.

As noted, when a flexible RFP is used, the proposals submitted by vendors may be quite diverse; this can make the proposal evaluation process more difficult. To ensure that a relatively objective approach is used, organizations often establish criteria for evaluating proposals in advance. Two useful procedures that many organizations use to aid in this process are vendor rating systems and benchmark tests.

VENDOR RATING SYSTEMS

A widely used tool for evaluating LAN proposals submitted by vendors is the *vendor rating system*. When a vendor rating system is used, vendor proposals are scored with respect to how well their proposed networks stack up against a specific set of criteria.

An example of a vendor rating system for LAN proposals is shown in Table 9-6. The criteria and weighing scheme used to rate vendor products should be developed in advance. In practice, many organizations determine the criteria while they are developing the RFP; such criteria may be included in RFPs distributed to vendors. Ideally, the criteria used and their weights represent the consensus of the organization's managers, users, network managers, and IS professionals—that is, the input of each of these sources should be considered.

In Table 9-6, each criterion deemed important in the vendor selection process is given a relative weight. In this example, the weights range from 1–10. The criteria assigned higher weights are more important. Vendor proposals are rated on each criterion.

To obtain a total score for a vendor proposal, the proposal's ratings on the criteria are multiplied by the set of criterion weights. The products

Selection Criterion	Weight	Vendor 1 Score	...	Vendor N Score
Application software	10	7		8
Speed/throughput	10	5		10
Expandability	9	8		6
Printer support	9	10		7
Interconnectivity	8	8		5
Vendor support	8	7		7
Security	8	8		6
Component availability	9	6		9
LAN management software	7	6		9
Overall cost	5	8		4
Total score		602	...	605

TABLE 9-6
A vendor rating system

(results of multiplication) are then summed to obtain a total score for each vendor proposal. The total scores can then be used as a starting point for determining which proposal is most desirable.

When a vendor rating system is used, there is no guarantee that the vendor proposal with the highest total score will be the one selected. Sometimes, the group responsible for developing the RFP and evaluating the proposals received may have their recommendation vetoed or ignored by higher-level managers. For example, a company president who has always had good experiences with UNIX networks may only consider a UNIX-based LAN, regardless of what the vendor rating system suggests. Also, if two or more vendors are close in the final ratings (as are Vendors 1 and N in Table 9-6), the group evaluating proposals and making a recommendation might consider the close scores a tie; they are then likely to look beyond the criteria used in the rating system for additional justification to select one vendor over the other.

In the private sector, the objective of the rating system is not necessarily to be fair—rather, it is to help the organization select the best vendor. In the public sector, where equity is a prime consideration and decisions often must be objectively justified, the results of a vendor rating system may, in fact, largely determine the vendor.

Rating systems such as that illustrated in Table 9-6 are relatively simple to use, and they bring some degree of objectivity to the vendor selection process. For example, if someone on the vendor selection committee has a bias toward a particular vendor, it is more difficult to reflect that bias through a rating system that forces the vendor to be evaluated across a variety of different criteria. However, vendor rating systems are, at best, crude mathematical models, and it is very possible that the best vendor proposal may not wind up with the highest score.

BENCHMARK TESTS

Suppose that, based on the results of a vendor rating system, one or two particular vendors emerge as the best options. What then? Many companies, before committing themselves fully, will make a final decision based on the results of benchmark tests.

A ***benchmark test*** is a test in which one or more application programs (or data sets) supplied by the potential buyer are processed on the LAN hardware and software included in the vendor proposal. The collection of programs and data submitted by the potential buyer is called a *benchmark*. The benchmark should reflect the type of work for which the vendor's proposed LAN will actually be used, thereby providing an indicator of how well the LAN would actually perform for the buyer. For example, in the purchase of a LAN-based transaction processing system, the benchmark may consist of a representative sample of the types of transactions that would actually be processed. By performing a benchmark test, the buyer should get an indication of how fast transactions can be processed on the LAN, how the LAN handles errors, and how easy the system is to use. If, however, the LAN is being implemented to support engineering applications, the benchmark might consist of a number of tests assessing how rapidly large CAD files can be transferred between workstations, how the LAN responds when several engineers simultaneously run simulated tests for different designs, and the like.

Benchmark testing can be expensive and the results cannot always be trusted. It can be expensive because the buyer must take the time to prepare a suitable benchmark, because it may require travel to a site with an existing LAN that can be used for the benchmark test, and because it may be necessary to pay the vendor to perform the benchmark test on its equipment. Benchmark tests are not always accurate; there are many real-life situations the benchmark cannot check. For example, what happens when all LAN users request access to the same LAN resources at the same time? Also, the results of the benchmark test are highly dependent on the quality of the benchmark itself. In short, rather than looking at benchmark tests as tools that can eliminate the risk of a bad decision, network managers should recognize that benchmark testing only reduces such risk.

◆ VENDOR CONTRACTS

No matter what approach is used, once a vendor is selected, it is time to sign a contract. If you have decided to go with a turnkey vendor or to have a single vendor supply all of the LAN components that you need, only a single contract is needed. If you have chosen several vendors (for example, different vendors for hardware and software), it is usually wise to have a contract with each. Of course, an attorney is the best source of information about whether a contract is needed, as well as what should be included in the contract.

A *contract* is a document, enforceable in court, that spells out such things as what the basic agreement is; who the parties to the agreement are; what goods, services, and monies are being exchanged; what types of continuing expectations are to be met; and what action will be taken if either of the parties signing the contract fails to live up to expectations. A good contract should make the terms of the agreement clear to everyone.

Although most people sign contracts with good intentions, a number of things can go wrong and end up in litigation. For instance, the vendor may claim a system can deliver a level of performance that, in real life, it does not come close to achieving. Or, the vendor may deliver a system with a critical bug that causes the LAN to malfunction and the buyer to suffer a severe financial loss. Or, a LAN that is promised to be in place and fully operational by mid-November does not get installed until late January. There are thousands of possible causes for problems. A good contract should put all important intentions in writing and specify a suitable recourse when the vendor fails to meet such intentions.

Many firms retain lawyers with special expertise in contract law and technology matters to help them prepare or review contracts. When substantial sums of money can be lost, having a "computer lawyer" at one's disposal is really the only solution. The layperson's interpretation of right and wrong is often something altogether different from the legal interpretation, which bases its judgments on complicated federal, state, and local statutes and on matters of precedent.

During the creation of a contract, the buyer should visualize as many scenarios as possible concerning how the future might evolve with a vendor and make sure that each of these scenarios is accounted for in concrete, quantifiable terms in the contract. That means putting such vague phrases as "excellent throughput," "fast response time," or "little downtime" into objective, measurable terms, as well as making a list of everything that could possibly go wrong and determining how the most important of these situations should be resolved.

Organizations may enter into numerous types of contracts with vendors of network components. Some of the most important are purchase contracts, standard vendor contracts, support/maintenance agreements, and software licences. These are briefly described in the following sections.

PURCHASE CONTRACTS

When purchasing major LAN components from a vendor, it is important to have a clear understanding of what will be supplied, when it will be supplied, and what results are expected. The well-defined purchase contract states the responsibilities and expectations of both parties and helps eliminate ambiguities and disagreements; it also helps protect buyers from frustration with what vendors deliver.

A *purchase contract* should describe in detail what, where, when, and how items purchased from the vendor are to be delivered. When purchasing hardware, the contract should specify the vendor, quantity, model (or part) numbers, and unit cost for each component. It should also include the total cost for all items to be purchased. It should specify for what (if any) items the vendor can provide substitutes (or equivalents) and what models or part numbers the buyer is willing to accept as an equivalent. When purchasing software, the purchase contract should, at a minimum, specify the version of the software to be provided, the functions that the software is expected to perform, and the hardware that the software will be run on.

Purchase contracts should also include a payment schedule if the components are to be delivered and installed in phases. It should state that the final payment (often 15–30 percent of the total purchase price) will not be made until the vendor has satisfied all conditions of the contract.

Purchase contracts may also include *penalty* (protection) *clauses* that reduce the total purchase price, include monetary damages, or specify actions that must be taken if the vendor fails to meet the conditions of the contract. Some of the problems addressed by penalty clauses are summarized in Table 9-7. In spite of the additional protection that penalty clauses can provide, network managers should be aware that vendors may refuse to sign contracts that include such clauses.

STANDARD VENDOR CONTRACTS

Established vendors of LAN products often have standard contracts that potential buyers can sign. However, actually accepting such a contract should not be a snap decision. While standard vendor contracts often contain language that protects the buyer from potential problems, keep in mind that these contracts have been developed by the vendor's attorneys; their primary purpose is to protect the vendor's interests. Before signing a contract supplied by a vendor, it is wise to have it reviewed and approved by your organization's attorney. After reviewing the vendor's contract, your attorney may suggest changes or the development of a substitute contract that adequately protects both you and the vendor.

SUPPORT/MAINTENANCE AGREEMENTS

If your organization is installing a LAN for the first time, you are likely to need a lot of support during the first year or so. Technical support can often be provided by vendors, consultants, or third-party firms. Software support is likely to be particularly important, especially if it is software new

---

TABLE 9-7
Potential problems addressed by penalty clauses

- Vendor fails to meet delivery schedules.
- Component (system) does not meet performance requirements specified in contract.
- Vendor fails to deliver all components specified in contract.
- Software fails to meet performance requirements specified in contract.
- Software does not provide the functionality specified in contract.
- Vendor does not deliver the version of the software specified in the contract.
- Software is incompatible with (or interferes with the normal operation of) existing applications.

to the organization (for example, a LAN OS, a LAN management software package, groupware, and so on). By signing a *support/maintenance agreement*, you are more likely to be able to solve software problems quickly and to learn how to use the software most efficiently. As noted in Chapter 8, such agreements can be an important aspect of network maintenance plans and programs.

Support/maintenance agreements are not free. Often, vendors offer varying levels of assistance and maintenance. The more you want, the more you will have to pay. For example, the lowest level of support may consist of telephone support, say, one free telephone consultation per week and additional charges for more than one call per week. A second level may consist of unlimited telephone support for a substantially higher fee. A third level might include unlimited telephone support, plus a one-hour on-site maintenance visit per month during regular business hours. The highest level (and the highest cost) might include unlimited telephone support as well as unlimited on-site maintenance calls as needed, day or night.

Network managers must carefully weigh the costs and benefits of support/maintenance agreements before signing on the dotted line. It is usually unwise to agree to (and pay for) more or less support than you are likely to need.

SOFTWARE LICENSE AGREEMENTS

Organizations typically sign *software license agreements* for both the LAN system software and application software they purchase for use in LANs. The wording of the license agreement itself is often found within the shrink-wrapped packaging for the application software diskettes or LAN system software. The wording usually states that by installing and using the software, you are agreeing to abide by the terms and conditions of the printed license agreement.

While the content of software license agreements varies from product to product, many will preclude you from making copies of the software to use in other LANs. Some impose limits on how many LAN users may simultaneously use the product, and many state that you cannot modify the source code without violating vendor support agreements, that you cannot offer modified source code for resale, and so on. The bottom line, however, is that when you purchase the software, you do not get unconditional use of the product; in fact, all you own is the diskette, not its contents.

Carefully review and understand all the conditions of the license agreements during the software selection process. Even if a software product can meet your organization's information needs, a restrictive software license agreement may be enough to steer you toward a competing product with a less restrictive license agreement. In addition, substantial differences in the licenses and pricing of competing products often exist, so cost may also become a major factor. For example, if a license for five simultaneous users of a LAN database application package from one vendor is the same price as a license for an unlimited number of users from a competing LAN database package, which product are you going to select?

Because no standards exist for software license agreements, you are likely to encounter different license agreements for competing products (as well as different agreements for different products sold by the same vendor). To avoid potential lawsuits, peruse the provisions of each agreement prior to purchase. After installing the software, monitor its use to ensure that the license agreements are not being violated. Numerous organizations, including colleges and universities, have been forced to pay hefty fines for violating the terms of software agreements (usually for illegally copying or failing to control the copying of software). Such violations have sometimes been uncovered by unannounced "raids" of the operating sites of alleged violators by the Software Publishers Association in conjunction with the Federal Bureau of Investigation. During such raids, the contents of all employee diskettes and hard drives are examined; fines can be imposed for each illegal copy that is found. To help organizations abide by software license agreements, products are now available that enable companies to perform their own internal audits of license violations.

As mentioned in Chapter 7, there can be great diversity among software license agreements; several general categories may be observed. These are briefly discussed below.

Single Workstation Licenses

Single workstation licenses are among the most restrictive; the software can be used only on one workstation by one user. If an organization needs the software to be installed on multiple workstations, multiple copies of the software package must be purchased. Vendors may enforce single workstation agreements by including a counter in the installation modules that makes it impossible to install the program on more than one machine.

Multiple Workstation Licenses

Multiple workstation licenses allow the purchaser to install the software on a fixed number of workstations. Like single workstation agreements, vendors may enforce this by including a counter in the installation modules for the software.

Restricted Number of Concurrent Users Licenses

This type of agreement is often undertaken when application software is installed in a LAN. In most instances, a single copy of the software is installed on a server; the license agreement dictates how many users may simultaneously use the software. The price the company pays for the software often correlates to the number of concurrent users specified in the license agreement; the higher the number of authorized concurrent users, the higher the price of the software.

This type of software agreement is often enforced by metering software (Chapter 7) or a NOS program module that monitors the number of users simultaneously using the software. This module contains a counter that is increased by one each time the user requests a copy of the application; the counter is decreased by one each time a user exits the application. A user requesting the application will be able to access it if the counter is at fewer than the maximum number of concurrent users specified by the agreement. When the counter is at the maximum number of authorized users, a user requesting the application will receive an error message that the application is not available.

Server Licenses

A *server license* makes it possible for an application to be installed on a single server. Once installed, all workstations directly attached to that server may use the application (concurrently, if need be). Note that this type of software license is restricted to a single server. If there are multiple servers and the organization desires to install the application on each one, multiple server licenses are necessary.

Site Licenses

A *site license* essentially gives the organization the right to install the software on all of its workstations and servers at a given site. For multiple LANs, concurrent access by all users attached to all LANs is possible. For a combination of stand-alone workstations and LANs at the site, this type of agreement allows the software to be installed on both the stand-alone machines and LAN servers. Copies of the software could also be installed on individual machines within a LAN in addition to being available on the server.

Corporate Licenses

A *corporate license* essentially gives all users at all of an organization's operating locations access to the software (concurrently if necessary). Such agreements are structured similarly to site licenses but extend access privileges to all sites. Unlike many of the previously discussed license agreements, corporate agreements may also allow the organization to reproduce unlimited numbers of the software's documentation.

As you might expect, the price of the software license can increase dramatically from one type of agreement to another. Single workstation agreements are usually included in the purchase price of the software package. You will typically pay more (often a fixed fee per workstation) to install the software on multiple workstations. With restricted number of concurrent user agreements, pricing increases as the number of concurrent users increases (many vendors base such prices on ranges: for example, 5 or fewer, 10–25, 25–50, 50–100, 100–250, and so on). Server licenses can be more expensive than restricted number of concurrent user licenses, and site licenses are often more expensive than server licenses. Corporate licenses, while providing the greatest installation flexibility, are typically the most expensive of all these options.

Purchasers should keep in mind that software license agreements are primarily designed to protect the interests of the vendor. However, many will allow the purchaser to request a refund if the software is defective. Some agreements enable the purchaser to transfer the license to another user or company (such as in the case of a merger or acquisition by another firm); however, it is often necessary to get vendor approval before making the transfer. Most purchasers can terminate the agreement by destroying the software and its documentation. In addition, some states and countries allow software purchasers to sue for damages if the software causes business losses.

◆ INSTALLATION

Vendor contracting is the final phase of the vendor selection process. Once all the important contracts and agreements are signed, it is time to concentrate on site planning and preparation, installation schedules, and all the tasks associated with installing the LAN and making it operational. Some of the planning issues associated with these activities are discussed in Chapter 8; installation activities themselves are discussed in more detail in the next chapter.

◆ SUMMARY

One of the most important activities that occurs after network plans have been developed and prior to network installation is the selection of vendors and the specific components needed to implement the network. During the vendor selection process, network managers determine who will supply network components, as well as what types of formal business agreements (contracts) are needed between the organization and the vendor(s).

The vendor selection process results in legally binding agreements and contracts between the organization and vendors that essentially lock in the organization to using particular networking products for a specified time period. Failure to live up to the conditions of these agreements and contracts can be grounds for a lawsuit.

The criteria and issues commonly used in the vendor selection process parallel those used to evaluate LAN alternatives. When selecting vendor products, consider factors such as: how many users can be supported; the compatibility of the products with other existing hardware; communications speeds; media flexibility; protocol support; adherence to standards; security and backup; and cost. In addition, when selecting vendors, consider factors such as: the types (and costs) of pre- and post-implementation support they provide; the overall financial strength of the vendor company; and the experience that similar organizations have had with vendor products and support.

For organizations upsizing or downsizing to LAN computing platforms, compatibility of vendor products with existing systems and equipment can be very important. Such compatibility should be critically evaluated. This may involve contacting similar organizations using the vendor's products, as well as contacting organizations that once used the vendor's products but have switched to other products.

Once an organization knows what type of network it wants to implement, vendor selection consists of four steps: identification and initial screening; development and distribution of a request for proposal (RFP), request for quotation (RFQ), or request for bid (RFB); evaluation of vendor proposals, quotations, or bids and selection of vendors; and contracting with selected vendors.

Several approaches may be used to identify potential vendors for networking products. These include doing your own research, contacting other companies in the same line of business or industry, identifying the major competitors of "known" vendors, reviewing trade publications, and hiring LAN/network consultants. In practice, a combination of these approaches is more common than is the use of a single approach.

The goal of the initial screening phase of the vendor selection process is to pare down the list of vendors to a manageable short list. In practice, this phase often coincides with the vendor identification phase. The development of a list of screening criteria can facilitate this process.

Once an organization has developed a short list of potential vendors, it may contact the vendors either informally or formally. Three of the more formal types of contacts include requests for proposals (RFPs), requests for quotations (RFQs), and requests for bids (RFBs).

An RFP is developed by a buyer and distributed to vendors. The RFP identifies the firm's needs and asks interested vendors to submit a formal proposal showing how they would satisfy such needs.

When the organization knows exactly what type of LAN and LAN components it wants, it is more likely to send vendors a request for quotation (RFQ) or a request for bid (RFB). RFQs and RFBs spell out for vendors exactly what specific types of LAN components (hardware, software, and services) are required; all that is required of the vendor is a quotation and a commitment to meet the buyer's terms.

After vendors have submitted their proposals or bids, the buyer begins the process of evaluating them. The major outcome of the evaluation phase is determination of which vendor(s) the organization will do business with. Vendor rating systems are often used during this phase; with these, each vendor proposal is scored with respect to how well their proposed network stacks up against a specific set of criteria. Benchmark tests may also be used to select vendors. These involve running one or more applications on the hardware/software specified in the vendor proposal.

Once vendors and vendor products have been selected, network managers must concern themselves with a number of contracting issues. A contract is a legally enforceable document that spells out such things as what the basic agreement is; who the parties to the agreement are; what goods, services, and monies are being exchanged; what types of continuing expectations are to be met; and what action will be taken if either of the parties signing the contract fails to live up to expectations. An attorney acting on behalf of the buyer should be involved in the contracting phase of the vendor selection process.

Organizations may enter into several types of contracts and agreements with vendors of LAN products. These include purchase contracts, standard vendor contracts, support/maintenance agreements, and software license agreements.

Once contracts and agreements are signed, the vendor selection process is concluded. The organization should then concern itself with site preparation and network installation. These activities are the focus of the next chapter.

✳ KEY TERMS

Benchmark test	Request for quotation (RFQ)
Contract	Server license
Corporate license	Site license
Low bid policy	Software license agreement
Penalty clause	Support/maintenance agreement
Purchase contract	Turnkey systems
Request for bid (RFB)	Vendor rating system
Request for proposal (RFP)	

✳ REVIEW QUESTIONS

1. Why is the vendor selection process important?
2. Identify and briefly discuss the issues that network managers should consider when selecting vendors and vendor products.
3. What steps can an organization take to ensure that vendor claims of compatibility and interoperability are valid?
4. Briefly describe each of the steps in the vendor selection process. Identify the major activities carried out during each step.
5. Describe each of the approaches that organizations use to identify potential vendors for network components. Briefly discuss the benefits of each.
6. What are the characteristics of screening checklists? Why might they be used during the initial screening phase?
7. What are the differences among RFPs, RFQs, and RFBs? Briefly describe the circumstances when each is likely to be used.
8. What are the characteristics of turnkey systems? Why may these be attractive options to potential buyers?
9. What is meant by a low bid policy? How does a low bid policy impact the evaluation of vendor proposals, quotations, or bids?
10. What is a contract? Why is it important to have an attorney review any contracts or written agreements between the buyer and the vendor?
11. What are the characteristics of purchase contracts? What are the differences between payment schedules and penalty clauses?
12. What are the characteristics of standard vendor contracts?
13. What are the characteristics of support/maintenance agreements?
14. What are the characteristics of software license agreements?
15. Describe the differences among the various types of software license agreements.
16. Describe the rights that buyers may have when they enter a software license agreement.

✳ DISCUSSION QUESTIONS

1. Many issues and criteria are used when selecting vendors and vendor products. Identify the five issues or criteria that you feel should be part of any organization's vendor selection process, and discuss why you feel that each should always be considered.
2. Vendor screening and rating systems often play an important role in the vendor selection process. Discuss why this is so and describe who should be involved in determining what vendor screening and rating criteria should be used.

3. Discuss how vendor selection criteria may differ between an organization upsizing to a LAN and a second organization downsizing to LAN computing platforms.

4. Many government and municipal organizations (including public colleges and universities) operate under low bid policies. What impacts can these have on the vendor selection process? What are the long-term advantages and disadvantages of low bid policies?

✳ PROBLEMS AND EXERCISES

1. Invite a local network manager to come to your class to discuss the vendor selection process in his or her organization. Be sure to ask which criteria are generally most important in the selections made.

2. In small groups, develop a set of eight to ten criteria and criteria weights used to screen network operating system products for use in a LAN in a computer lab at your school. Each group should justify the inclusion of each criterion as well as its weight. The instructor (or a designated student) should record the criteria and weights so that differences can be observed and discussed by the members of the class.

3. Obtain and review copies of actual contracts between organizations and vendors of networking components. Briefly summarize the formats and types of provisions they contain.

4. In a networking or computing periodical, find a current article comparing the support provided by vendors of networking components. Identify the criteria used to compare vendors and how the criteria were derived. Summarize the contents of the article in a paper or group presentation.

5. Obtain copies of actual software licenses for products used in a LAN. Briefly summarize the formats and types of provisions they contain.

6. Obtain copies of actual support/maintenance agreements between organizations and vendors of networking products. Briefly summarize the formats and provisions that these contain.

✳ REFERENCES

Appleton, E. L. "Put Usability to the Test." *Datamation* (July 15, 1993): p. 61.

Bartholomew, D. "Killer Consultants." *Informationweek* (July 25, 1994): p. 42.

Campbell, R. J. "Is There a Consultant in the House?" *LAN Magazine* (September 1994): p. 53.

Didio, L. "A Textbook Example of Network Expansion." *LAN TIMES* (August 9, 1993): p. 35.

Fitzgerald, J. *Business Data Communications: Basic Concepts, Security, and Design.* 4th ed. New York: John Wiley & Sons, 1993.

Frank, A. "How Much Technical Support Do You Get?" *LAN Magazine* (September 1994): p. 42.

Frenzel, C. W. *Management of Information Technology.* Boston, MA: Boyd & Fraser, 1992.

Goldman, J. E. *Applied Data Communications: A Business-Oriented Approach*. New York: John Wiley & Sons, 1995.

Moad, J. "Successful Benchmarking With the Team Approach." *Datamation* (February 1, 1994): p. 53.

Rains, A. L. and M. J. Palmer. *Local Area Networking with NOVELL Software*. 2nd ed. Danvers, MA: Boyd & Fraser, 1994.

Schatt, S. *Data Communications for Business*. Englewood Cliffs, NJ: Prentice Hall, 1994.

Stamper, D. A. *Business Data Communications*. 4th ed. Redwood City, CA: Benjamin/Cummings, 1994.

———. *Local Area Networks*. Redwood City, CA: Benjamin/Cummings, 1994.

Woek, P. and J. LeBlanc. "Taking the Guess-Work Out of Network Design." *Data Communications* (December 1992): p. 95.

Case Study:
Southeastern State University

Like the network planning process, there is considerable variation in how vendors are selected and screened at SSU. The process used for campus-wide applications and the larger networking systems for which the Computer Services Department is responsible is quite formal. It typically involves a sizable committee, consisting of Computer Services staff members as well as administrators and users from the departments of the university affected by the new network or network change. The screening and selection process typically used involves the development of RFPs, vendor rating schemes, and benchmark testing. SSU's Legal Office is involved in all contracts and license agreements developed after vendors are selected.

Some of the RFPs developed through this process have been lengthy. For example, one developed in 1990 for the new student information system was almost 200 pages in length, and some of the proposals submitted in response to this RFP were nearly three times that long. For this particular system, the committee collected and reviewed RFPs developed for similar systems by other colleges and universities. Needless to say, it took a considerable amount of time for the committee to develop and agree on the RFP that was sent out and to develop and agree upon the vendor rating system and benchmark tests that were utilized to select the system.

The vendor selection processes for the LANs purchased and implemented by the various colleges on campus are typically much less time-consuming and formal. Often, the entire process is handled by the individual who has been designated to be the LAN administrator. If a committee is formed to do this, it is usually small. In fact, a committee is only likely to be formed when fairly sizable new LANs (typically with 40 or more workstations) are to be created; for LAN expansions or upgrades, the entire process is typically handled by the LAN manager.

Vendor screening is often limited to vendors who have distribution and maintenance outlets in the immediate area, especially for critical LAN components such as servers. Usually, a request for quotations is developed by network managers and is distributed to vendors by the university's Purchasing Department after it has been approved.

LAN administrators interested in a particular model of a product from a particular vendor often have experienced difficulty getting their RFB approved by the Purchasing Department. Because SSU is a state-supported university, the Purchasing Department is required to purchase networking and other computer equipment from the state's list of approved contractors (even if the prices on the approved list exceed those available for the same products from local sources not included on the list). If the vendor or product the LAN manager desires is not on the approved list, the RFB is returned unapproved to the LAN manager for revision. The RFB may also be returned for revision if it is structured to exclude minority contractors from being able to submit bids.

In general, the Purchasing Department prefers that RFBs be as generic as possible. For example, it is acceptable to specify processor types, bus architectures, and backup requirements, but it is generally unacceptable to request specific brands (vendors) or models. The more specific the RFB is, the more likely it is to be returned for revision.

Having RFBs returned for revisions has angered many LAN managers at SSU and has generally led to hostility between the Purchasing Department and networking personnel. LAN managers have said that the Purchasing Department has too much control over the acquisition of networking equipment and forces them to buy equipment with dubious reliability from suppliers with poor service records only because they appear on the approved state contract list. Purchasing agents counter that they are complying with state regulations, and that local suppliers should be encouraged to get on the approved list of state contractors.

LAN managers also regularly get miffed at the Purchasing Department after bids are returned. Even when the set of bids includes equipment that the LAN manager likes, the Purchasing Department will not approve the items on a specific bid unless it is the lowest bid returned that meets all the conditions specified in the RFB. This is because the state university system operates under low bid policies that are very difficult, if not impossible, to waive. Once again, purchasing agents (who have the ultimate authority to approve the purchase) feel that they are doing their jobs correctly by enforcing the rules, while LAN managers often feel that equipment they don't want is being crammed down their throats.

The Purchasing Department is also involved in reviewing and approving all contracts, licenses, and agreements between the university and networking component vendors. However, it is usually LAN managers who are responsible for ensuring that vendors live up to the provisions specified in the contracts signed as part of the purchase. And, history has shown that many vendors are much more interested in making the sale than in abiding by the agreements they have signed.

CASE STUDY QUESTIONS AND PROBLEMS

1. Discuss the advantages and disadvantages of the vendor selection process used for large systems and networks at SSU.
2. How have the purchasing policies for LAN equipment at SSU contributed to the diversity of workstations and computing equipment found in LANs in student computer labs? What networking challenges does SSU face as a result?
3. What approaches would you recommend to administrators at SSU to reduce the animosity between LAN managers and the Purchasing Department?

Network
Installation

CHAPTER OBJECTIVES

After completing this chapter, you will be able to:

- ◆ Identify and discuss the administrative tasks related to a network installation.

- ◆ Describe the characteristics of effective installation contracts.

- ◆ Describe the sequential steps of system installation and the challenges associated with each.

- ◆ Discuss the issues related to accepting a system and the potential consequences of a premature acceptance.

- ◆ Identify and discuss the tasks associated with site preparation.

- ◆ Identify the activities associated with cable installation.

- ◆ Describe the installation and configuration of major hardware and software components.

- ◆ Discuss network testing and the elements of successful implementation.

- ◆ Identify and discuss the different kinds of user training that must take place during installation.

Once vendors and networking products have been selected, a number of tasks must be accomplished before the network can be implemented and made operational. This chapter concerns the installation activities that begin after you have selected networking software and hardware. The installation phase continues until your LAN is turned over to users.

The moment at which the system becomes operational is not easily pinpointed. Most network managers regard the system as operational when all hardware, system software, application software, user setups, and security setups have been completed and the users are able to use the network to execute their daily tasks. However, in practice, the installation of software, the configuration of user setups, and security implementation are ongoing tasks. Therefore, in this chapter the primary focus is on the following topics:

◆ installation administration

◆ site planning and hardware installation

◆ testing and acceptance procedures

◆ initial system training for users and managers

◆ initial training for users and managers

◆ ADMINISTRATIVE TASKS

Because a large volume of tasks must be accomplished in the installation of a LAN, maintaining accurate documentation can be a critical administrative task. Documentation is vital during each phase of LAN planning and implementation, and, as you will read in Chapter 13, documentation plays an important role in ongoing network management and maintenance. In addition, when selecting the components of the LAN, the purpose and rationale behind the selection of each component should be recorded.

As noted in Chapter 9, after network hardware and software components have been selected, the next step is to negotiate a contract with the vendors that will be supplying the components. If only one vendor has been selected to provide all the components, you have one agreement on which to focus. However, in most instances, network components will be supplied by multiple vendors and multiple contracts will be necessary. In any contracting situation, the advice of an attorney who has experience with the installation of computing and networking equipment can be invaluable.

In general, installation projects are easier to manage with appropriate project management tools. Project management software can be particularly valuable in keeping up with project tasks and in determining which

tasks can be done concurrently. Most project management software includes modules for Gantt and PERT/CPM charts such as those shown in
Figures 10-1 and 10-2.

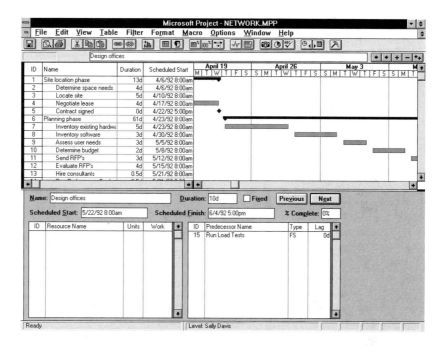

FIGURE 10-1
A Gantt chart

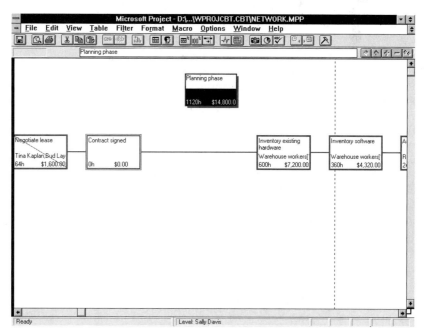

FIGURE 10-2
A PERT/CPM
chart

◆ INSTALLATION CONTRACTS

An **installation contract** for a LAN or network should clearly define the activities that will be performed by the vendor or the third-party firm that will be installing the system. Penalty clauses may be included in the contract to ensure that vendors keep installation schedules.

Installation contracts may also include clauses to ensure that software operates in an acceptable manner; this often calls for thorough testing of the software in the environment in which it will be used. Ideally, the tests will include data that is exactly like that for the live system. And, if the software being tested depends on other software to complete its tasks, all of the software involved should be tested at one time. You should test the software under conditions of peak work volumes when the number of multiple concurrent users is at its maximum; such tests let you know if the software functions acceptably in extreme situations. Failure to thoroughly test software prior to accepting it from the vendor can lead to one or more of the following problems:

◆ the software does not provide all the functions you requested

◆ the functions provided are much more difficult to use than you anticipated

◆ the software could be an older version than you thought you were purchasing

◆ the software may not be compatible with other software that will be used

◆ if the software is new, it may not be sufficiently tested and may contain irregularities (bugs)

Because of the importance of the software components of networks, installation contracts should call for thorough testing in the network environment prior to user acceptance.

An installation contract should specify the exact conditions under which the vendor has completed the installation obligations. Once again, there is no substitute for a skillfully worded contract negotiated by an experienced attorney who is familiar with your business and your requirements.

An installation contract should also precisely describe product warranties and hardware you expect the vendor to install. Terms like "IBM-compatible" are rarely sufficient. In the information systems industry, hundreds of brands and models of computers claim to be "IBM-compatible." However, in reality, there are varying degrees of compatibility. In fact, the term "IBM-compatibility" is itself ambiguous—which IBM computer will you use as the standard? The point is that the installation contract should specify exactly the brand and component parts (including model numbers) that you expect to be delivered and installed. However, if you are comfortable with the substitution of an "equivalent" component, say so in the contract. Otherwise you should expect to install the exact component for which you have contracted.

As noted in Chapter 9, payment schedules are often included in purchase contracts. These are typically set up to ensure that installation is complete prior to the final payment. Some network managers compare the installation of a LAN system to the installation of a swimming pool in your back yard. If you agree to pay 75 percent of the cost of the pool when the hole is dug, the vendor may not have sufficient incentive to finish the pool in the stated time frame. And once a hole of the magnitude required for a swimming pool is dug in your back yard, you have little latitude to make changes or substitute suppliers in the middle of the project, and it is unlikely that you will be able to find someone to finish the job for 25 percent of the price you agreed to pay.

LANs, especially complex LANs, are installed in incremental stages in the following sequence:

◆ preparation of the site where the LAN will be installed

◆ installation of electrical power and lighting

◆ installation of network cabling

◆ installation of network hardware

◆ installation of software

◆ software and network testing

◆ system conversion

◆ network acceptance and operation

Weeks or months may pass in the interim between the initial phases of installation and the final acceptance and operation of the network by users. Vendors have costs that must be paid during the installation period; therefore it is reasonable and customary for vendors to receive some compensation for work that has been satisfactorily accomplished along the way. A payment schedule should be arranged in advance for making interim payments to vendors during network installation. It should specifically describe the tasks that must be completed in order for interim payments to be made. Many payment schedules involve withholding a substantial portion of the payment (25 percent to 35 percent) until the system is accepted by the customer. If you do not withhold a sufficient incentive at the end, some vendors may move on to their next projects and let some of the final details slide. A substantial final payment can represent a significant chunk of the vendor's gross profit for the project, which most cannot afford to ignore.

An installation contract may also include penalties for failing to meet the project's schedule—a common event in large projects. Other penalties may be included for failing to install all the network components, for installing the wrong components, or for failing to fully test the network before turning it over to users. These penalties should be stiff and should give the vendor an incentive to avoid the conditions that would invoke the penalties. However, it is often easier to include penalty clauses in installation contracts than it is to enforce or collect on them. Very few penalties are ever enforced without protracted litigation. The legal fees and loss of time associated with litigation may be expenses that you never fully recover. Rarely does a company contribute anything to profit by going to court, and most of the time, litigation simply recovers a small portion of the actual costs of a poorly installed system.

The installation contract should clearly describe the conditions necessary to declare the project complete. Bland and ambiguous language simply referring to the "installation of hardware and software" is woefully inadequate. As much as possible, precise and measurable descriptions of the completed network should be included.

Starting to use the system before it has been officially accepted can cause problems because use can be legally interpreted as informal acceptance. Companies who jump ahead eagerly can legally forfeit their right to enforce the vendor's responsibility to complete the installation of the system. The contract should clearly state that the system is the vendor's responsibility until final acceptance.

PREPARING THE INSTALLATION SITE

As noted in Chapter 8, planning the layout of the LAN includes stipulation of the changes to the location where the LAN will be installed. This step essentially lays out the floor plans showing every detail of the LAN. An example of a physical site drawing is shown in Figure 10-3. On these drawings the following items should be identified:

♦ location and identification of specially grounded electrical power outlets

♦ location of servers

♦ location of workstations

♦ location of printers

♦ location of cable runs

♦ environmental modifications such as temperature, humidity, and lighting controls

♦ identification of factors that could be affected by local building codes

♦ applicable occupational safety code items

♦ location of telephone line terminations

A key item in this large group of physical items is the location of the servers. Unlike workstations and other user equipment, with servers there is less concern about ease of general user access. Of more concern is ease of

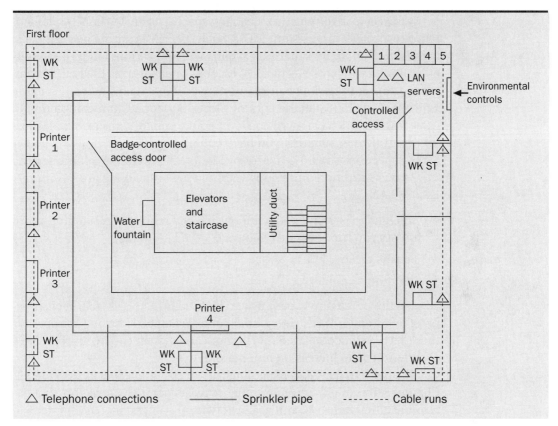

FIGURE 10-3   A physical site drawing

access by authorized network personnel to perform maintenance activities and other network supervision tasks that demand access to the server. Servers must be located in secure areas where physical access can be controlled. Server installation should include mechanisms that ensure that electrical power cannot easily or accidently be turned off.

In general, security needs must be balanced against the need for ease of access to key network components. For example, if external connections are a part of the LAN, access to communications lines and ease of movement with the terminal equipment is necessary. Lack of easy access to key network components by authorized personnel can make it more difficult to troubleshoot problems as well as to expand or upgrade the LAN.

CABLE INSTALLATION

Installing network cabling is one of the major tasks performed during LAN implementation. In the future, wireless connections may save considerable time and effort. Currently, however, barriers such as concrete and steel can be difficult obstacles to overcome.

It is wise to involve network cabling specialists during LAN implementation, especially for first-time LANs. Such specialists are the best source of information about how to install cabling to avoid various kinds of interference. Even if you have contracted with one firm to oversee the entire installation project, the prime contractor may decide to use a network cabling subcontractor to install the communication media. Upon completion of cable installation, most municipalities require you to go through an inspection routine to ensure that you have complied with local building codes. Such inspections should be performed before access to the cable is sealed.

As has been noted previously, three types of cabling are commonly used in LANs:

♦ shielded twisted pair (shielding is usually required by building codes unless you run separate conduits for the LAN cable)

♦ coaxial cable

♦ fiber optic cable

Some types of telephone cabling can be used for LANs as well, such as cabling required for ISDN (Integrated Services Digital Network) services. As these services become available in more areas, such cabling used is likely to be found in an increasing number of LANs.

The three major types of cabling vary in cost and capacity. As noted in Chapter 3, the shielding and number of twists per inch are factors that determine different levels (and costs) of twisted pair cabling. In Chapter 5, it is noted that both thin and thick coaxial cable can be found in Ethernet LANs and that Fast Ethernet cabling capable of transmission speeds of 100 million bits per second is available. Different grades of fiber optic cabling are also available.

When selecting the type of cabling to be installed, it is important to consider the types of speeds and applications that will be run on the LAN in the future. For example, if you plan to run applications within three years that demand 100 Mbps transmission speeds, it may be best to pay the extra cost and install the cabling needed to support such speeds now. It is usually cheaper in the long run to install more than you need now than to replace existing cabling later.

Laying (often called running) the cabling is tedious work. Sometimes it is difficult to install a piece of cable without bending, crimping, or kinking it to the point of failure. Hence, once it is laid, it is necessary to test the cable to assure that the installation process itself has not caused a failure; specific test equipment is designed for this purpose.

Because a large number of system failures are ultimately traced back either to faulty cabling or faulty terminations, testing the cable after installation should be the rule rather than the exception. Testing the cable can warn you about the following maladies:

◆   actual cable breaks

◆   kinked cables

◆   interference from sources such as magnetic fields, motion, and temperature changes

◆   faulty terminations

If problems are found, it is usually best to correct them before installing network hardware and software and, if the installation facility is new, before furniture is moved in. It is typically much easier to check and repair cabling problems at this point.

Because fiber optic cabling is usually more difficult to work with than either twisted pair or coaxial cable, you should utilize installers who have had experience implementing fiber optic computer networks. This helps ensure that this more expensive cabling is installed correctly the first time, saving the organization both time and money in the long run.

◆ INSTALLING HARDWARE

Some hardware installation procedures can begin while the site is being wired. However, as noted above, equipment should not be put in places that will hinder the people pulling cable. Doing so will only slow the progress of the entire project, because you cannot completely install the hardware until the cabling is run and tested.

INSTALLING NETWORK INTERFACE CARDS

One of the time-consuming and tedious steps in hardware installation is the placement of the network interface cards (NICs). This involves opening the box that protects the system unit, installing the NIC, installing the connectors needed to interface with the cabling, installing the NIC drivers, connecting the workstation to the cable, and testing the operation of workstation.

When installing the NIC cards, it is important to ensure that you have no IRQ, port, or memory settings that interfere with the other operations of the computer. If you have ordered coordinated hardware and NIC cards, setting the first one may take some time, but the rest of the units are less time-consuming. If there is a wide variety of hardware and NIC cards, this installation can be quite demanding; some cards have jumper settings, while more recent cards can be configured by software that accompanies the card.

More and more computers are being shipped with NIC cards permanently attached to the motherboards and the software (NIC drivers) needed to run them preconfigured on the hard disk. There are positive and negative aspects to this type of configuration. For example, if the NIC card is an integral part of the motherboard (along with the video board, serial, and parallel ports), any failure of one of these components could cause you to replace an entire motherboard just because one component has failed. However, vendors of these products claim that longer mean times between failure (a measure of reliability) and the efficiency of having these components preinstalled outweighs the potential repair expenses. If the product you are using has a three- or four-year warranty, vendor claims of lower expenses in the long run may be justified.

INSTALLING WORKSTATIONS, SERVERS, AND OTHER HARDWARE

Networks that involve the use of GUIs, multimedia, and e-mail can have some tremendously complex configuration problems. One of the first solutions to this problem (recommended by most vendors) is to ensure that you have plenty of RAM in each workstation. Particularly troublesome are some MS-DOS or PC-DOS environments, in which there is a hard limitation of 640KB of RAM even though many more megabytes of extended/expanded memory are available for certain components to operate successfully. Managing these challenges can be troublesome, and if you are not experienced in addressing such problems, you had better hire an installer who is. Otherwise, installing the workstations and servers may take you a long time.

It is also important to ensure that adequate memory is available in servers. As network utilization increases, more and more demands are placed on server memory operations, and network managers should monitor server memory utilization to avoid problems. Figure 10-4 illustrates how server memory utilization can be monitored by Novell NetWare.

Each of the network vendors publishes a minimum RAM under which their product will execute. In all cases, these minimum memory figures provide little realistic information, however. Here are some of the published minimums for RAM memory:

- NetWare 3.11 or 3.12   4MB
- NetWare 4.00 or 4.01   4MB
- Banyan Vines   4MB
- LAN Manager   5MB
- LAN Server   6MB
- Ungermann-Bass Net/One LAN   5MB

There is ample evidence to suggest that server RAM should be well above such published minimums. For example, according to industry statistics, over 50 percent of the maintenance calls to server vendors are related to insufficient server memory. Also, many LAN managers have found from practical experience that many seemingly unsolvable server problems are

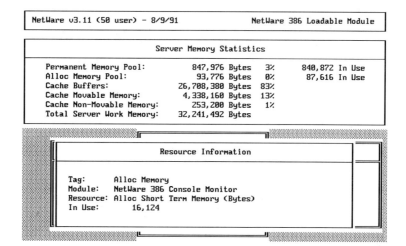

FIGURE 10-4
Monitoring
server memory
utilization

immediately corrected when RAM is increased; even vendors do not always understand why. So if you anticipate that your LAN applications will continue to grow, you should seriously consider installing double or triple the minimum RAM required to run the server. By doing so, you are providing some room for server utilization to grow and may be able to avoid some maintenance problems. Many experienced LAN managers refuse to install a server without at least 16MB of RAM.

In addition to installing servers with sufficient RAM, you should make sure that the servers you install have adequate disk space and disk management controls. There are many reasons why it is desirable to install servers with high-capacity disk systems. For example, most LAN operating systems require at least 40 to 60MB of disk space for the system files alone. Some LAN operating systems, including NetWare, also enable the files from multiple microcomputer operating systems to be stored on server disks. The NameSpace feature of NetWare can, for example, allow Macintosh, DOS, OS/2, Windows, and UNIX files to be stored, and all of these are quite large. In addition, printer spooler files and other utility files as well as the executable files of the application software, e-mail, and groupware must also reside on the server's disks. Applications with graphical user interfaces can consume many megabytes of disk space, and a maze of temporary files may be created by applications for redundancy, printer files, and configuration files.

If you fail to take measures to control the disk space available to users, you may suddenly find everything from a user backing up his or her local hard disk to your server to a proliferation of game software being duplicated from one user subdirectory to another. Fortunately, most NOSs include mechanisms to control the maximum amount of server disk space that each individual user can access outside of shared applications.

Even if the server has adequate disk space, you may choose to install workstations with hard disk storage units. One reason local hard disks may be desirable is because the new GUI interfaces, particularly Windows, Windows for Workgroups, and so on, load large graphics files that can create high volumes of network traffic if they can only be accessed from the server. Network performance is likely to be enhanced if these files are placed on workstation hard disks.

Some of the additional types of hardware that will have to be positioned and installed include printers, backup tape drives, facsimile servers, communication servers, printer servers, and even CD-ROM servers. Sometimes the physical positioning of this equipment is logistically important. For example, when installing large networks, many network managers will attempt to place as many servers in a single location as possible. One reason is that the backup task for all servers may be best accomplished from one workstation or one server. Also, by locating all servers in close proximity to one another, multiple backup units may be available in case one fails.

When you have wired the premises, placed the hardware into position, and installed the NICs and other network hardware, you can begin to connect the server and workstation units to the network cabling. And once the hardware is in place, the software can be installed.

◆ INSTALLING SOFTWARE

Usually the software used in a LAN is installed in the following order:

1. LAN operating system software
2. Workstation operating system software
3. Application software
4. Utility software

In general, LAN operating system software is becoming easier to install. However, as may be observed in Figure 10-5, the AUTOEXEC files that must be constructed can be intimidating if your previous experience has been limited to constructing and editing the AUTOEXEC files for microcomputer operating systems.

If you are using existing equipment to which network interface technologies are being added, the installer will have to configure each server and workstation. If you are installing a new LAN with new servers and workstations, software installation procedures may be much easier. Many vendors of LAN workstations and servers now provide the option of preinstalling the LAN software of your choice. When this is coupled with preinstalled NICs and NIC drivers, the time savings can be considerable. Many hardware vendors are now installing ***open system drivers*** that are

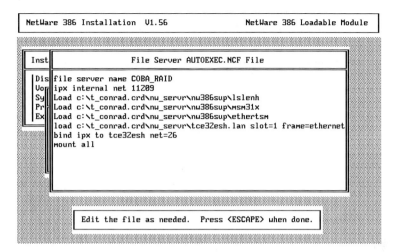

FIGURE 10-5

A server
AUTOEXEC file

well-suited to client/server networks because they are capable of supporting multiple NOSs; these are also easier to reconfigure if network changes are made (which can happen frequently in large client/server networks).

The easy installation of microcomputer operating systems depends on the types of workstations that are to be used. If the workstation is only capable of supporting DOS applications, precision is needed in the construction of the automatic boot up configuration files. Two files, AUTOEXEC.BAT and CONFIG.SYS, must be carefully constructed in order to properly set up the workstation. This can be complicated because of the current DOS limitation of 640KB of RAM. Workstations that are not bound by the 640KB limitation, such as those running UNIX, OS/2, and Windows NT, have their own configuration problems. So, unless you are able to keep your knowledge and skills up to date, it may be advisable to use an expert in this installation activity, because if the workstation configuration is done incorrectly, it will not interface successfully with the network.

The installation of multiple operating systems in the same workstation is becoming more commonplace and, as you might expect, this can make the installation much more complex. Once again, involving an installer who has had experience with such configurations may be wise.

Installing the network's system software includes construction of the printing environment. Most LAN configurations manage the printing function centrally; however, individual printer configurations for particular work areas can usually be accommodated. Novell NetWare provides for the management of a large number of print queues and the ability for a particular document to be printed on any available printer.

As has been noted in previous chapters, a print queue is a temporary file created on the server to store the file to be printed until it can be sent to the printer and printed out. The printing configuration can also provide for the printing of special forms (such as invoices and checks) on printers that have been designated for high volume jobs. The software can even notify a specific workstation that the paper in the printer needs to be changed.

From the user's point of view, the fact that a new LAN may have been installed should not significantly change the interfaces they see in application packages. Ideally, users should scarcely be aware that the system has been changed or upgraded. To accomplish this, it may be necessary to construct the boot-up procedures so that users are automatically guided to the files they need to access. For example, when a user signs onto the system, the interface should move to a selection screen where the user can select from the list of applications he or she can access. Simple selection procedures such as inputting an "A", "B", and so on, or clicking on the option with a mouse, should be all that a user has to do to access an application. Users should not have to know the disk location of files, the directory structure, or which servers store the applications needed by the user. In addition, network applications such as e-mail should be constructed as simply as possible; for example, users should not have to type in long address sequences to send messages to other network users with whom they communicate regularly—selecting the recipient of a message should be as easy as selecting an application. While user-friendly network interfaces may have to be done for each individual user, the effort is likely to pay off in positive user reactions to the new system.

INSTALLING UTILITY SOFTWARE

Every manufacturer of LAN operating system software provides utilities to add additional functionality to the software. For example, you may decide to purchase and install additional utility software to provide menus or metering software to measure the number of concurrent users in each type of application software.

In addition, there may be third-party packages designed to work with the software. For example, network management systems (discussed in Chapter 13) are often purchased from third-party companies.

Utilities that could reasonably be expected to be useful on a network include archiving, archive restoration, LAN management, LAN administration, file transfer, statistical, and diagnostic programs. Figure 10-6, for example, shows a utility for NetWare that helps monitor and maintain products currently supported on the network.

The exact configuration of utilities that you may acquire is contingent upon the scope of your network and the type of reporting you will need. For example, if your network is a stand-alone combination of servers, workstations, other hardware, and software, your choice may include the whole spectrum of tools available. If your network is interconnected to

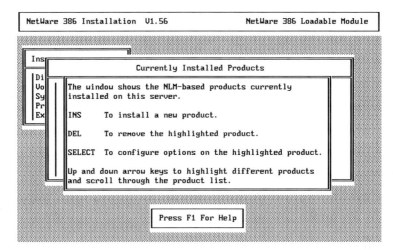

```
┌─────────────────────────────────────────────────────────────────────┐
│ NetWare 386 Installation  V1.56              NetWare 386 Loadable Module │
└─────────────────────────────────────────────────────────────────────┘
```

```
┌ Ins ┐
│ Di │    ┌─────────── Currently Installed Products ───────────┐
│ Vo │    │ The window shows the NLM-based products currently  │
│ Sy │    │ installed on this server.                          │
│ Pr │    │                                                    │
│ Ex │    │ INS      To install a new product.                 │
│    │    │                                                    │
│    │    │ DEL      To remove the highlighted product.        │
│    │    │                                                    │
│    │    │ SELECT   To configure options on the highlighted product. │
│    │    │                                                    │
│    │    │ Up and down arrow keys to highlight different products │
│    │    │ and scroll through the product list.               │
│    │    └────────────────────────────────────────────────────┘
```

FIGURE 10-6
NetWare product
monitoring
utility

```
        ┌─────────────────────┐
        │  Press F1 For Help  │
        └─────────────────────┘
```

other networks, you may choose to limit your selection because tools are available elsewhere on the network.

◆ CONVERTING DATA FOR THE NETWORK ENVIRONMENT

After the installation of application, system, and utility software has been completed, arrangements must be made for the entry of data. If the network is new, it may only be necessary to enter data for the applications that have been installed. However, installing a new system without having to convert existing data is rare. Most of the time, data from previous applications must be converted to the format required by the system being installed.

An aspect of data conversion and storage is the determination of which users will have access to which data. If the data is intended to be available to any user, it can be stored in the network's public directory. However, if the data is required to have limited availability, then proper security mechanisms must be constructed (such as trustee assignments and password protection). In addition, for data that can be concurrently accessed by several users, appropriate file and record locking mechanisms (such as the one illustrated in Figure 10-7) must be incorporated so that deadlock situations (such as that illustrated in Figure 10-8) can be avoided.

Decisions concerning general and selective access to data can be complex. Many organizations require the identification of a ***data owner***. The owner of the data is responsible for controlling access to it. In a LAN environment, data owners are given more access privileges than are granted to users; they also may control password systems for the applications that they control.

When data will be available to select users, the security provisions that are installed must be tested to ensure that only authorized users are given

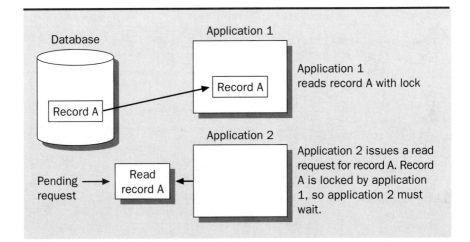

FIGURE 10-7
Waiting for lock release

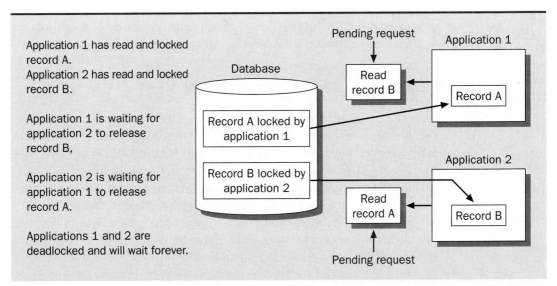

FIGURE 10-8 Deadlock in a shared database

access to the data. The concurrent access controls (such as file and record locking controls) that have been installed should also be thoroughly tested. These are just some of the tests that are performed subsequent to the installation of hardware and software.

◆ NETWORK TESTING AND ACCEPTANCE

After the initial installation of hardware and software, testing procedures must be performed in order to confirm the proper installation and the interaction of different network components. As mentioned earlier, purchase and installation contracts should specify the nature of the testing procedures and the conditions under which the network will ultimately be judged acceptable. The testing should include at least two major components—functional testing and testing of the performance of the network.

FUNCTIONAL TESTING

Functional testing includes testing the operation of all individual components as well as that of the entire network (with all components participating). Just because an application worked on a previous system is no guarantee that the same application (or an upgraded version of the application) will work successfully in the network that has been installed.

It is important to assure that the applications have been correctly installed at both the system level and the user level. On a network, unlike a stand-alone computer, simply placing the program files on the hard disk does not ensure that the user can access the application. User profiles and logon scripts should be tested to ensure that users can access the applications and data that they need, because during user logon procedures, the system checks to see if the user is authorized to access the application. Some networks even check the workstation's identification number to confirm that the user is using the authorized workstation.

Logging on as the supervisor of a system and being able to successfully access the application does not guarantee that the user logon will progress successfully. Hence, tests of security mechanisms and application access must be performed at the user level. This is especially important when the user utilizes two or more applications that interact with each other; the user must have access to each of the applications without one interfering with another or causing security conflicts.

Another important set of tests are those for printing configurations. Some applications may require that the user output to a particular printer. In most graphical interfaces, the printer used for output must be capable of producing the fonts required by the individual application.

Networks that include applications that require concurrent access to the same application or data must be tested with concurrent users. This concurrent testing should be done for every concurrent user with each user using each component that will be required for daily work. There can be many combinations of usage in these cases and it is important to test each one. As experienced network managers have noted, the one combination that you fail to test will be the one that causes severe problems later on.

PERFORMANCE TESTING

It is one thing to test a racing automobile individually on a street within legal speed constraints, and quite another to race the automobile on a track with other racing cars in competition. While each of these are performance tests, the latter is a much better assessment of how the racing vehicle is likely to perform in a race.

When testing a LAN, the installation contract should specify the measurements that will confirm that the entire system is operating properly and will be accepted. Many network users have been dismayed to find out that although each application works individually on the network, the applications would not operate concurrently when the network was loaded with all users. Testing individual applications to ensure that they will run on the network is rarely sufficient; this is like testing the race car on a street within legal speed limits. It is usually necessary to test all the applications at once to ensure that they interact appropriately and do not interfere with one another. Testing all the applications at once with all workstations in use is a form of *stress testing*. This is used to confirm that the system will function when fully loaded with users who are making maximum demands on the network.

Performance testing often requires expertise and expense. If you choose to test the system using the actual employees who will be using the network, you will be taking them away from the work that they ordinarily perform. However, by using this type of test, you will be able to closely match the duties that will be performed once the network becomes operational. Alternatively, you can purchase software (workload generators) that will simulate the performance of users using network applications to test the network. While such simulated tests are not perfect (real users can be very creative in their mistakes), they are better than not testing the network under loaded conditions at all.

Performance tests cannot be an unorganized event. The sequence of tests should be well planned and well executed. Just as they are useful throughout the installation process, PERT diagrams and Gantt charts can be used to organize and coordinate performance tests. They should be orchestrated to ensure as much as possible that the network will operate in an acceptable manner.

In practice, **performance tests** are used to determine if the network satisfies the conditions of an acceptable system specified in the installation contract. Hence, they should be conducted prior to allowing users to use the network to perform their jobs (to avoid the premature acceptance problems mentioned earlier in the chapter) and they should be sufficiently extensive for you to feel confident that it will operate properly when loaded with users.

CUTTING OVER TO THE NEW NETWORK

Once the network has been accepted, it can be put into operation. Users can be allowed to officially logon to the network and users of the old system can be moved to the new one. To facilitate a smooth user transition to the network, it may be desirable to run both the old and new systems in parallel until all users have been adequately trained and feel comfortable with the new network.

◆ TRAINING ISSUES

Generally speaking, there are three groups of LAN users who must be trained. First, LAN managers must be trained in the operation and use of the system. Second, group managers must be trained in user group operations as well as in some of the managerial tasks associated with the network. Third, the users themselves must be trained. As noted in Chapter 8, more extensive training is likely to be required if networks are being installed for the first time. However, if you are expanding or upgrading an existing network, there is probably a significant amount of "network literacy" among networking staff members and users, and their training is likely to be more focused and less time-consuming.

Training can be done either in-house (using employees as trainers) or through the use of trainers from outside the company. Whether you train in-house or externally, it is typically necessary to isolate the employees being trained from their normal work routines. This enables them to focus on the training and learning how to use the network. If the workers cannot concentrate on the training, the instruction may be of little value. While some organizations are large enough to have their own full-time professional trainers on staff, most organizations turn to outsiders for part or all of their training.

It is difficult to create an in-house training program in organizations that are in transition from one system to another. The networking staff is likely to have its hands full with the transition and insufficient time or resources to develop a good training program. However, professional trainers from outside the organization are equipped to organize the information in a form that can be easily learned and mastered by students.

The focus of the training will vary for the three groups mentioned above. Training programs for network managers should have a strong emphasis on operational procedures and security. Group managers need to be trained in how to manage the users in their group as well as how to interface with network managers. User training should be focused on using the network to perform daily job duties.

NETWORK STAFF TRAINING

Generally speaking, managing a dynamic LAN is easier than managing a dynamic mainframe or minicomputer installation. This is because the LAN can grow or shrink in relatively inexpensive incremental steps; similar changes in large centralized systems are typically very costly and demand much greater expenditures of time and effort. However, even for small LANs, at least two people should be trained in the fundamental operations and maintenance of the networks. If only one person knows the procedures to reinitialize the LAN or make repairs, the organization can find itself in big trouble if this person is unavailable when the network crashes.

LAN managers and network staff can keep their skills and knowledge of networking technologies up to date in several ways. This can be partially accomplished through reading literature and on-the-job training. However, the most valuable training is usually done in the training facilities of reputable LAN consultants. Hundreds of good seminars on LANs and networking are held each year. While somewhat expensive, they provide network staff members with a brief respite from day-to-day network operations and enable them to concentrate on learning about new technologies that may be valuable to their organizations. In addition, as noted in Chapter 8 (and discussed more fully in Chapter 17), LAN managers and network staff personnel can take advantage of vendor-sponsored certification programs for training and expanding their knowledge of networking technologies. These too can be expensive, but the depth of knowledge that can be gained about specific networking products may not be available from any other source.

The administration of medium to large size LANs can be a full-time job for one or more people. The amount and kind of training required depends on the level of complexity of the system. Many organizations fail to see the degree of expertise required and the complexity of the systems they have installed; this may cause them to downplay the importance of ongoing training and development for their networking staff. This often results in high turnover rates among networking personnel and continued frustration in attempting to stabilize a system about which they have insufficient knowledge. LAN managers must know the technical and managerial information needed to function as a network administrator. They also must know the functions of group managers, user interfaces, and interactions with the network. Training programs for network managers should be comprehensive enough to cover all of these areas.

In addition, LAN managers need to learn how to develop a vision for the future of the network. The network will have to evolve and change as the organization evolves and changes. New networking technologies will be announced, some of which are useful to the organization and some of which are not. LAN managers should be trained in how to evaluate these systems, test prototypes, consult with vendors, and determine when, what, and how to recommend network changes. Many times such changes are substantial and involve both technical and economic evaluations. Through appropriate training, bad decisions and costly errors are less likely to be made by LAN managers.

USER TRAINING

User training may vary somewhat depending on the nature of the software used. For some older DOS applications, users may be required to provide their initials, and some may require users to select their choice of printers, and so on. These nuances are usually not difficult for users to master. Applications with graphical user interfaces often store the user's profile of preferences in initialization files that are kept in the user's own directory. However, some of the following issues may require extra attention during user training:

◆ changing the destination printer characteristics

◆ changing passwords

◆ setting file or directory security

◆ cleaning existing applications when logging out

◆ setting search path/drive mappings

◆ finding resources on other servers/networks

◆ running e-mail

◆ using groupware

To avoid confusion and a host of start-up problems, information about how these tasks are accomplished should be presented to users in advance of the system being placed into service.

Training, especially user training, is an ongoing concern. There will inevitably be turnover and training will be needed for new users and network personnel. Even if outside trainers are used to initially provide network training, several employees should receive sufficient training so that they can serve as in-house trainers in the future. Also, part of any contract with outside trainers should include the right to use the training materials in ongoing training after the initial training is completed. It may also be worthwhile to consider contracting with someone to write self-paced training materials for users and network personnel. These may either be paper-based or available on magnetic or optical disk so that interactive computer-based training can be used.

◆ SUMMARY

After components for the network have been selected, installation activities can begin. The main purpose of installation is to prepare the network for implementation and operation. The administrative tasks performed by network managers during installation include installation administration, site planning and preparation, testing and acceptance procedures, and training network staff and users.

Because vendors or third-party firms are likely to be involved in network installation, it is important to have an appropriately worded installation contract. This should clearly define all installation activities that will be performed and the conditions under which the network will be accepted by the organization. Such contracts may also include payment schedules and penalty clauses to ensure that all installation work is performed in a timely manner.

When installation activities begin, they are usually performed in incremental stages. Often the sequence consists of site preparation, installation of electrical circuits, network cabling, hardware, and software and testing, conversion, and acceptance.

Site preparation consists of planning the physical layout of the location where the network will be installed. This includes determination of the locations of all cable runs and network hardware. It also involves ensuring that all building and electrical codes are followed.

Network cabling should be installed by contractors who have experience implementing computer networks. Care must be taken to avoid damaging the cabling during installation. It is advisable, and more cost-effective in the long run, to install cabling with higher capacities than are currently needed.

Installing hardware includes the installation of workstations, network interface technologies, servers, printers, and other peripherals that will be included in the network. It is important to install servers and workstations with sufficient RAM and disk storage to both satisfy current requirements and provide some room for growth.

The installation of software is usually performed in the following sequence: network operating system installation, microcomputer operating system installation, application software installation, and installation of utility software. The NOS and microcomputers' OSs must be configured to interact with one another. In addition, application software should be installed and configured in a manner that facilitates user access and use. Data that is used by applications often must be converted into the format needed by the network at this time.

After network hardware and software is installed, testing begins. This includes both functional testing (ensuring that each component operates correctly) and performance testing (ensuring that all the components interact correctly under real or simulated working conditions). If such tests are successful, the organization typically accepts the network and puts it into operation.

Training often coincides with network installation and typically starts before the organization converts over to the new system. Training for network managers, network personnel, and users should be provided. Training can be conducted by either in-house trainers or professional trainers. Training for network personnel focuses on the technical and administrative functions associated with running the network on a day-to-day basis. User training concentrates on using the network to perform daily job duties.

✳ KEY TERMS

Data owner
Functional testing
Installation contract

Open System drivers
Performance tests

✳ REVIEW QUESTIONS

1. Identify all of the conditions needed for a network to be considered operational.
2. Identify and briefly discuss the major activities associated with network installation.
3. Identify and discuss the characteristics of well-constructed installation contracts.
4. Explain the purpose of payment schedules and penalty clauses in installation contracts.
5. Explain the dangers of a customer company accepting a network prematurely.
6. Identify the steps followed in most network installations.
7. Identify the activities associated with site preparation.
8. Why are experienced network cabling installers commonly used to implement networks?
9. Why is it advisable to install higher capacity cabling than you currently need?
10. Why is cable testing needed?
11. What activities may be involved in the installation of network interface cards?
12. What are the advantages and disadvantages of installing workstations and servers with preinstalled network interface cards?
13. Why are servers with RAM and hard disk capacities in excess of minimum requirements installed in many networks?
14. In what sequence is software installed in a network?
15. What types of interfaces can be used to ensure users easy access to the network applications?
16. What types of utility software may be installed in LANs?
17. What activities may be associated with the conversion of data for the network?
18. What are the differences between functional testing and performance testing?

19. What role may workload generators play in network testing?
20. How is the focus of training different for network managers, group managers, and users?
21. Who may provide network training?
22. What topics should be addressed in training programs for network managers?
23. What topics should be addressed in user training programs?

✳ DISCUSSION QUESTIONS

1. If you were considering the installation of a network that would be used to sell tickets to athletic events at your school, describe the characteristics of an acceptable system.
2. Using a parallel conversion approach, describe the process that you feel should be used to convert data from the old system to the new LAN. Justify your description.
3. If you are installing a LAN for the first time, why should you be concerned about the financial stability of the vendor, the length of time the vendor has been in business, the nearest location of a service center, and the specific skills of installation personnel?

✳ PROBLEMS AND EXERCISES

1. Ask a local computer vendor to come to your class to demonstrate a general accounting system. Note the main points made by the vendor. Be sure to ask how the system would operate differently in a network environment.
2. Obtain information from businesses that offer computer training programs. Identify the content of the programs and the instructional methods used.
3. Ask a local network manager to visit your class to discuss the major activities associated with network installation in his or her company.
4. Ask a local network manager to come to your class to discuss the manner in which training is handled for users and network staff. Be sure to ask how the content of training differs for these two groups, and ask about training for LAN administrators.
5. Research the training available from LAN vendors that offer certification programs. Identify the activities required for certification. How many courses are required and what are their titles?
6. Conduct a role play using the following characters: MIS manager, LAN administrator, printer operator, user department manager, and controller of the company. The purpose of the meeting is to determine

what types of training are to be made available to your company with 20 employees. Some of the employees are users, some department managers, and some are Information Systems employees. Determine in advance what the agenda of the meeting should be, but allow for some spontaneity.

✳ **REFERENCES**

Bates, R. J. "Bud." *Disaster Recovery for LANs*. New York: McGraw-Hill, 1994.

Berg, A. "How to Revise Server RAM Needs: New Guidelines Established for NetWare Servers With More Than 2GB of Storage." *LAN TIMES* (June 27, 1994): p. 93.

Bozman, J. S. "Is the OSF's DCE Ready for the Big Time?" *Computerworld* (May 2, 1994): p. 73.

Cummings, S. "The Changing of the Help Desk: With Enhanced Features and Added Complexity, These Systems Are Not Just for Customer Support Anymore." *LAN TIMES* (April 5, 1993): p. 65.

Didio, L. "A Textbook Example of Network Expansion: Medical Center Stays a Step Ahead of Its Growing Infrastructure." *LAN TIMES* (August 9, 1993): p. 35.

Durr, M. "Switching: What Do You Need?" *LAN TIMES* (February 28, 1994): p. 74.

Hoffman, T. "Brokerage Replaces DASD Storage With Disk Arrays." *Computerworld* (May 2, 1994): p. 73.

Koegler, S. D. "Case Study: Payroll Management Turns From Wasteful to Efficient." *LAN TIMES* (October 18, 1993): p. 51.

Levine, R. "Guide for Creating a Network Help Desk." *LAN TIMES* (December 6, 1993): p. 78.

Loudermilk, S. "Solectek Plans Extension of Wireless System." *LAN TIMES* (Ausust 8, 1994): p. 14.

Stamper, D. A. *Local Area Networks*. Redwood City, CA: Benjamin/Cummings, 1994.

Case Study:
Southeastern State University

Recently a failure in a LAN in a student computer laboratory at SSU caused frustration from students preparing semester assignments. The student newspaper interviewed the campus Director of Computer Services, who was quoted as saying that the failure was caused by a poorly installed network interface card in one of the LAN workstations. The LAN manager countered that the NIC was thoroughly tested when first installed and that it and all the workstation's connectors passed all troubleshooting tests with flying colors; he was convinced that the NIC was not the problem and voiced his suspicion that the problem was caused by a computer virus that had entered the workstation from the main campus network (this was one of the labs that had established a connection to the Internet through the VAX computers operated and maintained by the Computer Services Department).

In response to having his opinion questioned in the student newspaper by a lowly LAN manager, the Director of Computer Services refused to investigate the problem in the LAN and also refused to spend money on software that could be used to detect and eradicate viruses in LANs. It was only after other LANs connected to the VAX computers experienced similar problems and after the LAN manager had paid for virus eradication software out of his own pocket—which, once installed, confirmed the presence of a virus that was affecting NIC drivers—that the Director of Computer Services agreed to expend the funds for a university-wide site license for virus detection and eradication software. However, no retraction or correction was ever made in the student newspaper and any damage that had been done to the LAN manager's credibility was not publicly undone.

Faulty installation, especially inappropriately selected NICs, has been the cause of problems within LANs at SSU. In one instance, an entire LAN had to be recabled because the work had been contracted out to an electrical contractor who had never installed computer cabling before and whose cable testing consisted solely of ensuring that the main cable could carry an electric current from one end to the other (before any of the connectors going to the workstations had been attached). During the installation of the connection hardware, the network cable was so badly crimped that more than half of the workstations were unable to access the server. The university had to pay the electrical contractor for the faulty work as well as bear the costs of recabling because the installation contract contained very ambiguous wording about the qualifications of installers and the testing procedures that would be used.

Over time, the LAN managers and technicians at SSU have learned which NICs work best in each of the different types of microcomputers used as workstations as well as which vendor products to avoid. Unfortunately, this information is shared informally among network administrators because there are no mechanisms on campus to promote regular formal interaction among networking personnel. As a result, NICs with known configuration, performance, or reliability problems are still included in purchase orders approved by the Purchasing Department. In addition, the potential still exists for inexperienced installers to be used to implement LANs on campus because the university has yet to develop a standard LAN installation contract.

CASE STUDY QUESTIONS AND PROBLEMS

1. Do you feel that the Director of Computer Services reacted appropriately to the LAN manager's suspicion that a virus was causing problems experienced in the LAN? Why or why not?
2. What steps should SSU take to ensure that faulty cabling installation does not occur in the future? Justify your recommendations.
3. What types of formal communication mechanisms do you feel should be established at SSU to make it possible for LAN managers to share their knowledge and experience with one another? What value is there in creating formal communication mechanisms among SSU's LAN managers?
4. Based on the information provided in this segment of the SSU case, what types of installation and management issues should be covered in network manager training programs?

Downsizing
to LANs

CHAPTER OBJECTIVES

After completing this chapter, you will be able to:

- ◆ Describe the differences among downsizing, upsizing, and rightsizing.

- ◆ Discuss the major reasons why organizations are interested in downsizing.

- ◆ Describe the aspects of downsizing that managers should be concerned about.

- ◆ Identify the potential advantages and disadvantages of distributed processing systems.

- ◆ Discuss why downsizing plans are important.

- ◆ Identify examples of applications that are both suitable and unsuitable for downsizing.

- ◆ Identify some of the factors that can derail a downsizing program.

- ◆ Describe the impacts of downsizing on the jobs of network managers.

A major issue that has become a driving force in network planning and decision making is downsizing. However, because downsizing is a term that has multiple meanings in organizational settings, it is important to understand what we will focus on in this chapter.

◆ WHAT IS DOWNSIZING?

Among general managers, human resource management specialists (personnel administrators), and workers "downsizing" is a term for reducing the number of managers, professional staff members, or other employees in an organization. It is another word for "layoff" or "reduction in force." Downsizing often results in a winnowing of middle management ranks or the elimination of middle management positions and levels, or what hardline managers refer to as "trimming the deadwood."

Often, the rationale given for these types of downsizing includes the desire to reduce the distance between top-level managers and rank-and-file workers (which should improve communication between these levels), to further empower lower-level workers by asking them to assume tasks that their superiors have traditionally performed, and, of course, to reduce labor and managerial overhead costs. Such flattening of the organizational structure has, in many instances, resulted in leaner and more cost-effective operations. An organization may achieve a competitive advantage from downsizing.

Among IS professionals and network managers, ***downsizing*** possesses a second meaning (please note, however, that they are fully aware of what downsizing means to general managers). This second meaning refers to the movement of applications which have traditionally been run on mainframes and large, centralized, host-based systems to midrange systems (minicomputers), LANs, and high-end workstations. Many organizations have realized that they can equal or surpass the computing power of their existing mainframes by moving to clustered or networked minicomputers and other distributed computing platforms. In addition, some organizations have learned that they can save money by migrating to smaller computing platforms.

For MIS professionals, downsizing is often considered the opposite of upsizing. As we have already noted, ***upsizing*** usually refers to the creation of LANs by networking stand-alone workstations or to the migration from stand-alone LANs to interconnected client/server networks. ***Rightsizing*** is another term that is often heard in conjunction with downsizing; this term is used by MIS professionals to refer to making sure that a particular application is matched to an appropriate computing platform (be it a mainframe, midrange system, LAN, or workstation). The philosophy behind rightsizing is to ensure that there is not too much or too little computing power available for the applications. This implies that rightsizing may involve downsizing some applications. For example, if an application that has previously been run on a mainframe could be run as efficiently on a minicomputer or a server in a LAN, rightsizing (or "rehosting") it may involve downsizing. Figure 11-1 illustrates some of the different ways in which organizations may downsize or upsize.

The increasing availability of networkable minicomputers and workstations is causing most organizations to reassess their need for mainframes and other "big iron" systems. Some experts have already declared mainframes "dinosaurs" and maintain that only the world's largest companies still need mainframes. High-end workstations have started to rival low-end supercomputers in FLOPs, MIPS, and other measures of throughput. And several experts have predicted that 80 percent of the new applications which will be developed between now and the turn of the century will be for client/server networks. Computing platform changes that will place less reliance on centralized mainframe systems and more reliance on distributed processing, client/server computing, and interconnected LANs and WANs appear to be inevitable, and these will undoubtedly have an impact on MIS plans, operations, and staffing patterns.

As applications are downsized from mainframes to minicomputers, LANs, and high-end workstations, there are likely to be changes in the MIS staffing mix. For example, fewer computer operators and mainframe-based application programmers may be needed, and the need for network managers and client/server application developers is expected to increase. The migration of applications from mainframes to other platforms is likely to result in more department level and end-user application development, which means MIS professionals who can facilitate such development (especially LAN-based applications) may be needed. In sum, downsizing may result in smaller MIS departments and a refocusing of MIS tasks and activities.

◆ WHY ARE ORGANIZATIONS DOWNSIZING?

Organizations are interested in downsizing for a variety of reasons. Before we discuss these reasons, we must note that we are basing our discussion on patterns that may be observed across large numbers of organizations. A particular organization's decision to downsize may include a combination of the reasons that we will discuss and other reasons beyond the scope of our discussion.

COST SAVINGS AND COST AVOIDANCE

As a general rule, mainframe computing platforms are more expensive than either minicomputer or LAN-based computing platforms. In addition, minicomputer platforms are typically more expensive than LAN-based systems.

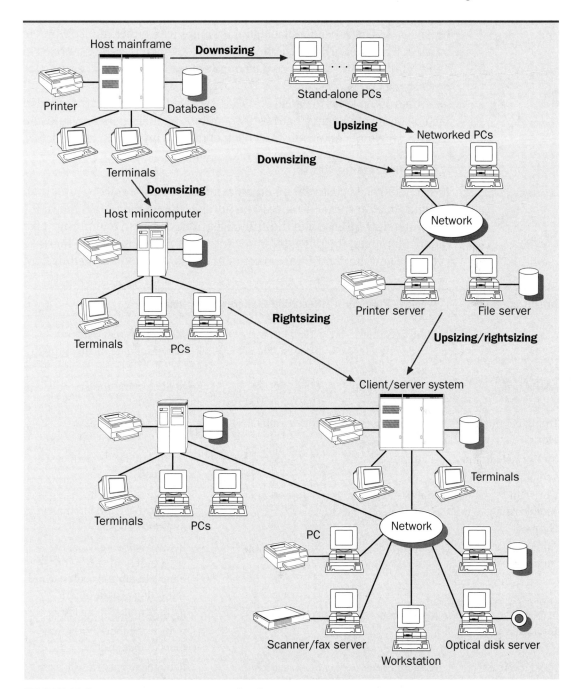

FIGURE 11-1 Downsizing, upsizing, and rightsizing

These rules generally hold true in spite of dramatic decreases in the costs of mainframes and mainframe storage systems over the last few years; a PC-based network can offer organizations equivalent processing power of mainframes at lower costs. Because of these potential cost savings, many organizations have been replacing mainframes with either minicomputers or LANs. Examples of organizations that have downsized are summarized in Table 11-1; as you can see, cost savings are frequently cited reasons for downsizing.

Some of the areas in which costs can be saved and future costs avoided include hardware leasing, hardware operations and maintenance, software leasing, and application software. Each of these is briefly discussed below.

Hardware Leasing Costs

It is generally less costly for an organization to lease a minicomputer than a mainframe. In some instances, the purchase price for a high-end mini-computer may be less than the annual leasing costs for a mainframe. Like microcomputers, the power of minicomputers has continued to increase over time and the computing power of many minicomputers is equivalent

Organization	Previous Platform	Resulting Platform	Reasons
Motel 6	Minicomputer	Client/server; interconnected LANs	Minicomputer could not accommodate further growth
AMR Corp.	Mainframes and minicomputers	Client/server; UNIX-based servers	Reduce computing costs
Orange County, Florida	Mainframe	Token-ring LAN	Reduce computing costs
McDonnel Aircraft Company	Mainframe	UNIX-based servers	Accommodate growth cost-effectively
Christian Broadcasting Network	Mainframe	PC-based LAN	Reduce software licensing fees and hardware maintenance/operations costs
Indian Head Industries, Inc.	Mainframe	Minicomputer	Reduce software licensing costs and hardware mainte-nance/operations costs; avoid retraining programmers
Napco International	Mainframe	PC-based LAN	Reduce maintenance costs; reduce number of program-mers and operators
FirstLine Trust	Minicomputer	PC-based LAN	Reduce cost per transaction; increase user productivity, improve customer service
WorldCorp	Mainframe	Client/server LANs	Reduce software maintenance fees and hardware leasing costs

TABLE 11-1   Organizations that have downsized

to that for a great number of mainframes, especially older models. By replacing a mainframe with a minicomputer, hardware leasing costs can often be substantially reduced. However, by replacing a mainframe with a PC-based LAN, hardware leasing costs may be eliminated altogether.

Hardware Operations/Maintenance Costs

Large, centralized, mainframe-based systems often have high overhead costs for operations and maintenance. These overhead costs include both utility costs and costs associated with the operations staff. The operations staff may consist of *computer operators* (the people who keep the mainframe going, mount tapes or disk units, monitor computer processing, bring the system back up after a crash, and so on) and computer center managers who supervise computer operators. The bulk of hardware maintenance costs for mainframe-based systems is often absorbed by annual maintenance contracts with the mainframe vendors.

Some organizations have found that they can save money by consolidating multiple data processing centers, moving toward highly automated, unattended data centers, and outsourcing noncritical data processing tasks. However, the fact remains that hardware operations and maintenance costs are almost always lower for minicomputer and LAN-based computing platforms. The operations staff needed for such smaller systems is typically less numerous; this can mean significant reductions in labor costs. In addition, annual maintenance contracts for minicomputer and LAN-based networks are typically less costly.

Hardware Upgrade Cost Avoidance

Some organizations have opted to downsize when the computing capacity of their larger systems has been reached. Rather than upgrading the larger systems, such organizations have chosen to replace them with PC-based LANs or minicomputers that can be expanded more easily (and incrementally) as their computer processing needs increase. These smaller systems are often modular and this facilitates adding computing capacity a little at a time. For example, many minicomputer systems are designed so that "clusters" can be created by attaching one or more new minis to those already in place.

When this downsizing strategy is pursued, organizations may not only avoid costs associated with hardware upgrades (including higher hardware leasing, maintenance/operations, and software leasing costs) but also experience immediate reductions in overall hardware and software costs.

Software Leasing Costs

Many organizations lease much of the software that is run on their larger systems. Both system software (including the operating system) and application software (accounting systems, database systems, and so on) may be leased. As you might expect, the cost to lease such software for a mainframe-based system is often substantially more than the cost to lease software for minicomputer-based systems. Hence, by downsizing from a mainframe-based system, an organization may be able to realize substantial savings in its annual software leasing expenses. Some organizations have reported annual savings in software leasing costs of more than $1 million by "pulling the plug" on their mainframes.

Application Software Costs

As a general rule, costs to purchase or develop application software for mainframe and minicomputer-based systems are higher than for microcomputer-based LANs. For example, database packages, spreadsheets, and e-mail systems for mainframes and minicomputers are typically more expensive than the equivalent software for microcomputers. In addition, software maintenance costs are typically higher for mainframes and minicomputers than for PC-based LANs. While an organization may be able to realize some application software savings by downsizing from a mainframe to a minicomputer, even larger savings may result if the organization were to downsize to a microcomputer-based LAN.

Labor Cost Reductions/Avoidance

As we have already noted, by moving away from centralized, mainframe-based systems, an organization may be able to shrink the size of its computer operations staff. This can result in substantial labor cost reductions for the organization. However, this is not the only staffing area that may be affected by such downsizing. Another area that may be affected is the programming staff. By downsizing, fewer application programmers (including maintenance programmers) may be needed. This is especially likely if the organization downsizes to one or more PC-based LANs, because the application software used in such platforms is much more likely to be purchased from vendors than developed in-house by application programmers.

It should come as no surprise that as the number of mainframes used in the U.S. continues to drop (at an annual rate of close to 10 percent), so does the number of application programmers employed by organizations. In fact, as the result of downsizing, some organizations have been able to reduce their programming staff size by more than 50 percent. In addition, downsizing is contributing to changes in the jobs performed by programmers; this is largely due to the fact that downsizing makes application packages from vendors more attractive. Many experts predict that in the future, the jobs of most application programmers employed by organizations will consist of evaluating and customizing vendor packages rather than writing new lines of code.

DISTRIBUTED PROCESSING

The term ***distributed processing*** implies that an organization's computer resources and information processing activities are performed at more than one location. In the 1990s, it also means that an organization's distributed computing platforms are interconnected. The creation of distributed processing systems has been under way since the 1960s when minicomputers were first introduced and departmental computing systems were initially created. Over the years, increasingly powerful microcomputers and workstations have made distributed processing more and more attractive to organizations. Such organizational interest in developing distributed processing systems has contributed to the prevalence of downsizing.

There are many potential advantages to moving away from centralized, mainframe-based systems toward distributed processing systems consisting of minicomputer- and LAN-based systems; some of these are summarized in Table 11-2. Many organizations see greater flexibility and better satisfaction of local information processing needs through tailor-made computing platforms to be among the most important strengths of distributed processing systems.

However, as may be observed in Table 11-2, some significant disadvantages may be associated with the creation of distributed processing systems. For example, some loss of control over computing expenses and resources may occur. Also, the sophistication of the application packages available on larger centralized systems is often greater than that available for distributed database systems and LANs.

In spite of these potential disadvantages, most organizations are moving away from relying on large, centralized systems to satisfy their information

Potential Advantages	Potential Disadvantages
◆ Better satisfaction of local needs	◆ Some loss of control in managing computing expenses, buying discounts, and in enforcing standards (for example, standards for data security and integrity)
◆ Faster response time	
◆ More flexibility in reacting to changes	
◆ Better tailoring of facilities to specific applications	◆ Sophisticated applications may not run on smaller computer systems
◆ Lower communications-related costs	◆ Duplication of resources and staff
◆ Fewer problems with intersite data flow (for example, between foreign affiliates)	◆ Integration of applications may be easier on centralized systems
◆ Better control over data needed for local decision making	
◆ Reduced size of computer operations staff	

TABLE 11-2 Potential advantages and disadvantages of distributed processing systems

processing and communications needs. The creation of client/server networks and the implementation (and interconnection) of LANs contributes to the development and enhancement of distributed processing systems. Downsizing is also consistent with the trend toward distributed processing.

SIMPLIFICATION OF COMPLEX NETWORKS

The computing systems and networks that have evolved in organizations with multiple operating locations often consist of a hodgepodge of mainframes and minicomputers as well as stand-alone LANs and workstations. When looking at a detailed graph of the system that is in place, one is likely to observe the presence of a variety of large computers from numerous vendors, that are interconnected (if they are interconnected at all) via a number of different communications media (for example, regular dial-up phone lines, T-1 lines, satellite links, and so on). Among the first impressions created by such a graph is that the architects of the systems in place seem to have lacked any kind of grand plan and concern to keep things simple. Yet, upon asking, valid reasons can be given for the acquisition of each system and communications link.

In order to move away from such messy configurations, some organizations have been downsizing to simpler and more elegant computing architectures. One company that is doing this is Spartan Mills, Inc., a South Carolina-based textile firm. Part of its downsizing efforts consist of phasing out its diverse mixture of mainframes and minicomputers across its operating locations and phasing in a client/server network built around LANs that are centered NCR UNIX-based servers. The planned network is simpler, more understandable, and much more pleasing to the eye when depicted graphically. Further, the company will be able to get out of an array of hardware and software leasing and maintenance contracts and will save millions of dollars annually by doing so. The resulting system will be more consistent (and compatible) across locations, less expensive to operate, and easier to manage. Spartan Mills is just one of many organizations that have found that downsizing can facilitate the creation of computing networks that are both straightforward in design and cost-effective.

GREATER USER CONTROL OF DATA AND RESOURCES

Downsizing often provides greater local control over data and computing resources. Many business managers prefer such local control, especially control over their most important application systems. Localized control enables business managers to have greater responsibility over their information processing costs. In addition, it provides them with greater flexibility and the ability to respond quickly to changing market conditions.

Such increased control and flexibility can be accomplished by moving applications off mainframes and implementing them on departmental minicomputers, LANs, or individual workstations (via downsizing). In some cases, such changes may be made to improve customer service, user productivity, and decision making even if overall organizational cost savings are not anticipated (such was the case with Canada's FirstLine Trust, summarized in Table 11-1).

OTHER REASONS FOR DOWNSIZING

Providing an exhaustive list of all the reasons why organizations downsize is beyond the scope of this book. However, a few more reasons that have not been adequately addressed in the previous sections should also be mentioned.

Some companies downsize in order to increase the variety of software that they can make available to users. Relative to software available for microcomputers, application software for minicomputers and mainframes is limited. Further, because the software available for larger systems is written to run on specific brands and models, software that will run on one computer may not be available for another. This if often true for data files and database management systems as well. Many of these software and data incompatibility problems can be avoided by downsizing to microcomputer-based LANs; this can give users access to a much wider range of application software.

Along these same lines, downsizing to PC-based LANs can promote application packages from vendors as computing solutions in place of custom applications written in-house by application programmers. Because in-house development of custom programs can take a long time, users may often have solutions to their computing problems much more rapidly by purchasing application programs from vendors. In the long run, promoting the purchase of application packages over in-house development can save money and reduce software maintenance expenses.

When downsizing results in moving data, applications, and computing resources closer to end users, there is a greater likelihood that end users will begin to develop their own applications and computing solutions (including innovative ones). Such end-user development can help reduce the lengthy application backlogs that are present in many organizations. In addition, it can give the IS staff more time to concentrate on critical or enterprise-wide applications. Finally, it helps to promote higher levels of computer literacy among users in all parts of the organization.

◆ AREAS OF MANAGEMENT CONCERN ABOUT DOWNSIZING

In spite of its potential advantages, downsizing presents organizations and network managers with a number of challenges. We will discuss some of these in this section. It is important to note that the manner in which these are addressed by organizations and IS managers (including network managers) will have a significant impact on the overall success of their downsizing efforts.

POTENTIAL DISADVANTAGES OF DOWNSIZING

Downsizing has a number of potential disadvantages. Some of the major ones are summarized in Table 11-3; each of these is discussed below.

Diminished Central Control

The movement toward distributed processing systems and client/server computing has often diminished the power and clout of IS managers. By giving users greater control over data and computing resources, the ability of the IS area to enforce adherence to standards for application development, hardware/software acquisition, and security/backup/recovery is often reduced. As the result of this diminished central control, there is greater risk of problems among the systems that user departments implement.

Unexpected Costs

While cost savings and cost avoidance are among the major reasons organizations downsize, the movement toward smaller computing platforms often includes costs that organizations do not anticipate at first. For example, overhead costs associated with multiple, distributed systems may exceed those for a centralized system. In addition, the cost associated with ensuring adequate backup and security mechanisms may be higher in decentralized systems than in centralized ones.

Table 11-4 summarizes some of the costs associated with developing client/server systems. As may be observed in Table 11-4, downsizing (especially to PC-based client/server LANs) is not cost-free. Although organizations may realize substantial savings over the long haul by downsizing, there can be substantial start-up costs for companies that decide to migrate to client/server computing platforms. These costs may equal or exceed those associated with the systems they are replacing.

◆ Diminished central control	◆ Enhanced parochialism
◆ Unexpected costs	◆ New software maintenance challenges
◆ Higher user workloads	◆ Resistance by key managers
◆ Increased need for effective data management	◆ Retraining of IS staff
◆ Unnecessary duplication	

TABLE 11-3 Potential disadvantages of downsizing

- ◆ Server hardware
- ◆ Internetworking hardware and software
- ◆ Application development tools for client/server systems
- ◆ Training for programmers, users, managers, and so on
- ◆ New PCs, workstations, printers, and so on
- ◆ Upgrades to underpowered PCs/workstations to be used in network
- ◆ Off-the-shelf application software packages
- ◆ Consulting services and systems integrators
- ◆ Outsourcing

TABLE 11-4
Some major costs in creating client/server networks

Higher User Workloads and Skill Levels

When distributed systems are created and computing resources are moved to user departments, user workloads and skill levels may increase. When such systems are implemented, users must continue to perform their daily work tasks; in addition, they may be called upon to back up and maintain the systems in their areas, develop some of their applications, and develop other computer-related talents that previously were not part of their jobs. In order to develop these additional skills, user training programs usually have to be implemented.

The enhanced user control over data and computing resources that often results from downsizing and the development of distributed systems can also place more demands on user managers. This is especially true when user managers assume responsibility for managing the application portfolio or data center in their work area. Discharging these new responsibilities may divert managerial attention from the main tasks associated with managing the work area, and has the potential to detract from the primary function of the department.

In short, downsizing and the creation of distributed client/server systems can add new tasks, stresses, and strains to the jobs of users and user managers. If these changes are not anticipated and adequately addressed they can contribute to a decline in productivity and morale, and other problems within the subunits in which the new systems are implemented.

Increased Need for Effective Data Management

Downsizing can exacerbate data management problems. If an organization possesses a large centralized database, some of the major challenges that may be faced during downsizing concern how the database can be divided into distributable sections. The creation of distributed databases is often associated with downsizing and distributed databases are typically more difficult to manage and maintain than centralized/integrated databases. Furthermore, downsizing often results in the distribution of data management responsibilities among user departments.

Downsizing and the creation of distributed databases has the potential to lead to data redundancy (the storage of the same data elements in multiple datastores), increased data storage costs, and asynchronous database states (which occur when updates made in one part of the database are not reflected in the other parts). In addition, database integrity and security may be compromised if update, backup, and recovery procedures are not consistently followed across user departments.

Since an organization's databases can be critical elements in its competitive advantage and success, it is important for organizations to effectively address the potential database problems that can result from downsizing. Overlooking such potential problems can be disastrous.

Unnecessary Duplication

Downsizing and the creation of distributed systems can lead to unnecessary duplication, especially in the area of application development. Users in two or more areas could independently develop essentially the same application. To avoid "reinventing the wheel," good communication across user areas is needed. If a user in one area needs an application that he or she knows has already been developed in another part of the organization, all the user may have to do is request a copy of the application and modify it slightly to meet his or her needs. Needless to say, this will lead to faster satisfaction of the user's computing needs than developing the application from scratch.

Enhanced Parochialism

The movement of systems to smaller computing platforms in user work areas may lead to situations where subunit goals are seen as more important than corporate goals (parochialism within organizations is often observed when subunit performance is favored at the expense of corporate performance). Giving users and user managers more control over data and computer resources can contribute to the development of parochial attitudes. It may also lead to the creation of "islands of technology" that satisfy the computing needs of a subunit quite well but are difficult to integrate with the rest of the organization's computing systems.

New Software Maintenance Challenges

With downsizing, software maintenance may become more challenging. In a centralized, mainframe-based system, software malfunctions may be easier to detect and correct because the software is located in one place and is the responsibility of one group of people (the IS staff). As a result of downsizing and the creation of distributed systems, malfunctioning software may have to be corrected in multiple locations and different groups of people may be responsible for making the corrections.

Software distribution and upgrades may also be more difficult to contend with in distributed systems than in centralized systems. In a centralized system, the upgrade can be done once; when the upgrade is complete, all connected users have access to it. With decentralized systems, the upgrade may have to be performed on numerous servers, and in some cases, hundreds (or thousands) of user workstations.

Resistance by Key Managers

Downsizing initiatives within an organization may be met by resistance from both functional area and IS managers. Downsizing, after all, represents a significant change in how information is processed within an organization, and like any other change, downsizing is likely to be resisted.

User (functional area) managers may resist downsizing because of the added responsibilities that they and the workers they supervise may have to absorb. As we noted above, downsizing can add new computer-related responsibilities to the duties that users and user managers are already expected to perform. Resistance may also result from user managers' reluctance to play a greater role in data management activities. They may also be reluctant to devote a significant portion of their budgets to the equipment, software, training, and services needed to move applications and application control to their subunits.

IS managers may resist downsizing because they perceive it as a threat to their job security and that of their subordinates. In addition, IS managers may resist downsizing because they do not feel that reduced IS control over application development, data management, and computing resources is desirable.

Since managerial resistance to downsizing can result from a wide variety of factors, it is impossible for us to adequately address it in this text. However, it is important to recognize that resistance to change spawned by downsizing is likely and can present some major challenges.

Retraining the IS Staff

Downsizing and the creation of distributed client/server networks not only entails changes in the jobs of users and user managers, it also affects the jobs of IS staff members. As we noted previously, some organizations justify downsizing by the reductions in total size of the IS staff that could result. The jobs of those IS staff members who are left are also likely to be changed. These remaining staff members often have to be educated about client/server networks and data communication systems. They may have to be trained in how to use middleware, how to develop applications for client/server systems, and how to use network management and client/server application development (CASE) tools. Such retooling of the IS staff may be costly and may entail the development of new thought patterns that represent significant departures from those needed for centralized, mainframe-based systems.

As you might expect, other potential problems may be associated with downsizing and the creation of distributed client/server networks. For example, as we noted in Chapter 9, organizations that desire to interconnect diverse subunit networks often have to consider middleware if they wish to enable users in one area to have access to computing resources in other areas. In addition, when working through the vendor selection process, organizations may find that the hardware and software choices for the applications that they want to downsize are extremely limited and that the functionality of downsized applications is likely to be reduced.

◆ THE NEED FOR DOWNSIZING PLANS

Our previous discussion of the potential problems with initiating downsizing underscores the need for sound downsizing plans. Without solid migration plans, downsizing programs can be more problematic and unsettling than they need to be.

While the elements of a downsizing plan are likely to vary from organization to organization, many include the following:

◆ A "blueprint" for how hardware and software will be deployed in the resulting computing network.

◆ Identification of the applications to be downsized and the sequence in which they will be moved.

◆ Identification of hardware to be removed and the timetable in which this will be done.

◆ Reduction in force or retraining programs for IS staff members who support the large systems currently in place and their applications.

◆ Training plans for users and user managers whose work will be directly affected by the new system.

◆ Identification of new application software, middleware, application development tools, network management software, and so on, that will be needed, and the order in which these will be acquired.

◆ Identification of additional computing hardware, hardware upgrades, networking hardware, communication media, and so on, that will be needed, and the order in which these will be acquired and installed.

◆ Identification of backup and security systems that will be needed for the new system and the order in which these will be acquired, tested, and implemented.

Merely developing a downsizing plan is not usually sufficient. It should also be communicated to all parts of the organization that will be affected by it and followed (to the extent that it can be) after it is developed. Like any other organizational plan, its success will be contingent on how well it is executed in addition to the extent to which it can easily accommodate midcourse corrections. The need to make midcourse corrections is likely to be the norm, rather than the exception, with any plan that involves rapidly changing information technology.

SELECTING APPROPRIATE APPLICATIONS TO DOWNSIZE

No matter what potential benefits cause an organization to consider downsizing, managers must realize that downsizing is not suitable for every organization. They should also realize that not all the applications that run on their large systems are viable candidates for downsizing. Managers must often contend with the fact that it makes no sense to try to move a particular application from a large system to a smaller one, no matter how much they would like to do so. If such managers are committed to implementing a client/server network, they are going to have to do so without "pulling the plug" on the large system and without the cost savings (hardware maintenance, software leasing, operations, overhead, and so on) that they might realize if they could remove the large system. In this scenario, the best that an organization can hope for is to "rightsize" the application and to include the existing large system in the architecture of the client/server network that it hopes to develop.

Frenzel (1992) discusses some of the types of applications that are likely candidates for downsizing as well as those that are less suitable for downsizing. Table 11-5 summarizes the characteristics of applications that may be likely targets for downsizing. As you can see in the table, if the application is a small, stable (not often modified) system with its own database that is found in a single user department, it may be a good candidate

---

TABLE 11-5
Characteristics of applications suitable for downsizing

◆ Small, stable applications

◆ Applications found in a single department

◆ Applications purchased from vendors

◆ Technically simple systems

◆ Low-risk systems easily managed by user departments

for downsizing. Applications that have been purchased from vendors are other potential candidates, particularly if the vendor has a "family" of software products with virtually equivalent functionality that are designed for different sized computing platforms. Low-risk applications (those that can be easily managed and maintained by end-user departments) and applications that do not rely on specialized or advanced technology are also possible downsizing targets, according to Frenzel.

Very large or enterprise-wide applications that are used by workers from many parts of the organization may be difficult to downsize, especially if they involve the retrieval and manipulation of data from integrated databases. In addition, Frenzel believes that evolving applications (ones being continually modified, updated, or expanded) that are customized to meet the organization's unique computing needs are probably best left on the organization's large systems. As may be observed in Table 11-6, Frenzel also believes that applications that rely on advanced, specialized technology and applications that user departments would have a difficult time managing and maintaining (high-risk systems) may be unsuitable for downsizing.

AVOIDING PITFALLS DURING THE DOWNSIZING PROCESS

Organizations that have experienced downsizing have helped to identify factors that can cause the process to go awry. Some of the lessons they have learned are summarized in Table 11-7.

As you can see from the table, organizations need to have a clear vision of the IT infrastructure (the configuration of hardware and software) that they are trying to create and that backup, security, and recovery mechanisms for the new infrastructure should not be overlooked. Further, it is important for organizations to have a plan for moving applications from larger to smaller systems and they should ensure that this plan is effectively communicated to both IS users, key managers, and users. This plan should anticipate potential difficulties in acquisition and implementation of components needed for the new network and should not overlook potential connectivity and interoperability problems that may be encountered, especially if network components are likely to be supplied by multiple vendors. In addition, the plan should include adequate provisions for IS staff,

TABLE 11-6 Characteristics of applications unsuitable for downsizing	◆ Large, enterprise-wide systems used by numerous workers
	◆ Applications using data from integrated databases
	◆ Evolving applications customized to the organization's requirements
	◆ Applications relying on specialized or advanced technologies
	◆ High-risk applications

◆ Unanticipated difficulties combining and coordinating products from multiple vendors

◆ Immature application development tools that make it difficult to develop applications for client/server networks

◆ Difficulty in changing the culture and mentality of the IS department to that needed to fully support a client/server architecture

◆ Attempting major downsizing too quickly or without using a pilot/test system to help familiarize the organization with downsizing and its potential pitfalls

◆ Inadequate management of key managers' resistance to change and of the general fear and anxiety that downsizing can instill throughout the organization

◆ Lack of a clear vision of what information technology infrastructure downsizing is supposed to create

◆ Failure to develop a clear migration path or plan to smaller systems for applications traditionally run on large systems

◆ Inadequate preparation or training of IS staff and users for the new systems being implemented (client/server systems can be perplexing to both groups)

◆ Insufficient communication to both IS employees and users about the changes being made

◆ Inadequate backup, security, and recovery mechanisms needed in client/server networks

◆ Impatience with IS staff's and users' learning curves for using client/server networks

◆ Insufficient consideration of network interconnection, interoperability, reliability, and integration issues

◆ Difficulty in acquiring and implementing the IT infrastructure necessary for the applications being downsized

◆ Unrealistic expectations about how long it can take to realize anticipated payoffs

TABLE 11-7   Factors that can derail downsizing

user manager, and user training and should not overlook the importance of managing resistance to change. Finally, planners must realize that the implementation of client/server networks can be the equivalent of a cultural change within the organization and should be prepared to exercise patience if movement along the learning curve is slower than they prefer.

Many of the organizations that have been involved in downsizing programs have learned the lessons summarized in Table 11-7 the hard way. For too many organizations, downsizing has been a "learn as you go" process. Fortunately, organizations that are planning to downsize may be able avoid the same pitfalls by studying what (and why) mistakes have been made by other organizations.

◆ IMPACTS OF DOWNSIZING ON NETWORK MANAGERS

Thus far in this chapter, we have discussed what downsizing is, the potential advantages and challenges associated with downsizing, the need for downsizing plans, and potential pitfalls that may be encountered during downsizing programs. Another important downsizing topic, and the final one that will be addressed in this chapter, is the impact of downsizing programs on LAN and network managers.

As noted in the previous section, downsizing programs should be planned efforts. Planning to downsize is especially important because it is likely to have a major impact on the user departments and systems to which the applications (and the responsibility for managing and maintaining the systems) are being moved. While the jobs of both users and user managers are likely to be altered as the result of downsizing, the work of network professionals who manage and maintain the systems in user departments is likely to be changed forever.

Several chapters in the text have noted the importance of network planning. Effective network managers are likely to follow the guidelines that have been provided in these chapters and to have long-range plans for developing, installing, maintaining, and upgrading the networks for which they are responsible. However, when the organization decides to downsize applications that have traditionally run on large systems, existing network plans will have to be revised or scrapped. In some cases, it may be possible to fold downsizing plans into existing network plans; in other cases, there may be no easy way to combine the two sets of plans and the network planners must start over.

The process of modifying network plans is likely to involve reconsideration of needed hardware and other network components. In many instances, downsizing may necessitate acquisition and installation of more powerful servers and workstations, higher-speed communications media, higher-quality (and higher-speed) output devices, more sophisticated backup and security mechanisms, and higher-capacity storage equipment. In practice, the functionality of downsized applications may be insufficient unless such hardware modifications are made. Acquisition and installation plans for new hardware (and for upgrading or removing existing hardware) are likely to be needed along with new rounds of vendor selection and contracting, and site preparation.

Existing software (both application software and system software) may also have to be modified, especially if it is supposed to interface with the applications that are being moved to network. Programmers may have to be assigned to these modification projects, and if the downsizing effort is supposed to happen quickly, the applications that they are currently working on may have to be delayed or put on the back burner until the downsizing modifications are made. In some instances, programmers may have to be trained before they can begin to work on the downsizing project and this may involve significant time and money. Some of this training and expense may result from the need to acquire middleware and client/server application development tools.

In addition, user training programs will have to be developed and implemented so that users will be able to use the application after downsizing has occurred. User managers may have to be trained in how to manage and maintain the applications that are being moved to their work areas. Training programs may also be needed for IS staff members in these work areas (including training for network managers and the technical staff that operates and maintains existing networks).

The process of converting to the new computing platform will have to be planned, especially if the organization wishes to avoid the expense of running parallel systems for an extended period. System testing and debugging plans usually must be developed, particularly if the application being moved has great importance to the organization. In addition, a new set of system maintenance plans may have to be developed and new system maintenance software may have to be acquired and installed.

Network managers are likely to be responsible (at least in part) for modifying existing plans, acquiring necessary hardware and software, developing and implementing training programs, and converting to the new system. Because of this, their jobs and those of their staff members are likely to be significantly affected by organizational downsizing initiatives. These added burdens should not be overlooked by either corporate planners or the network managers who will be affected by downsizing.

◆ SUMMARY

The term downsizing has a double meaning to network managers. For general managers, downsizing refers to layoffs, flattening the organizational hierarchy, and reductions in force. To network managers and IS professionals, downsizing also means moving applications that have traditionally been run on large, centralized systems to smaller computing platforms, including LANs.

Downsizing may be contrasted to both upsizing and rightsizing. Upsizing means moving applications from a smaller computing platform (such as a stand-alone PC) to a larger, more powerful network (such as a LAN or client/server network). Rightsizing refers to fitting an application to the best-sized computing platform in order to satisfy an organization's computing needs in the most cost-effective manner; rightsizing programs may involve both upsizing and downsizing.

Organizations are downsizing applications for a variety of reasons including to reduce hardware leasing, operations, and maintenance costs, to reduce the size of the IS staff (especially the size of the computer operations staff), to avoid hardware upgrade costs, and to reduce software leasing costs. Some organizations are downsizing in order to capitalize on the potential advantages of distributed processing systems and client/server networks, and others are downsizing in order to simplify the complex information processing infrastructures that they currently have in place. Organizations may also downsize because they wish to give user departments greater control over data and computing resources, to increase the variety of software available to users, to take advantage of commercially available application packages, and to promote end-user application development.

In spite of these potential advantages, there are many reasons why managers should be concerned about embarking on major downsizing programs. Among the potential disadvantages of downsizing are diminished centralized control of applications, systems, and application development; sizable expected hardware, software, and systems integration costs; significant impacts on the jobs and needed skills of end users and end-user managers; an enhanced need for effective data management; the likelihood of unnecessary duplication in applications and application development; the possibility of increased parochialism among user departments and the feeling that subunit performance is more important than corporate performance; new software maintenance challenges; resistance by key managers (both IS and non-IS); and the need for major retraining programs for IS professionals and users.

In order to minimize or avoid these potential downsizing problems, downsizing plans are needed. Downsizing plans are likely to include a number of elements including a "blueprint" of the infrastructure for the new network, the identification of applications to be downsized, application migration plans, new hardware and software needed for the new system, and training programs. These plans should be effectively communicated to all the areas that will be impacted and should be flexible enough to accommodate midcourse corrections.

An important part of downsizing planning is the selection of the applications that will be moved to smaller computing platforms. In general, smaller, self-contained applications with independent databases are better candidates for downsizing than large, enterprise-wide applications that utilize integrated databases. In addition, the ability of user departments to effectively manage and maintain downsized applications should be a factor in selection decisions.

Organizations that have undergone downsizing have identified several potential pitfalls. Drawing from their collective experiences, it is apparent that organizations need to have a clear vision of the IT infrastructure that they want to create as well as a good, effectively communicated migration plan for moving applications from larger to smaller systems. Organizations should anticipate potential difficulties in the acquisition and implementation of network components and that unexpected connectivity and interoperability problems are likely to occur. In addition, they should make adequate provisions for IS staff, user manager, and user training, and should realize that the implementation of client/server networks can be the equivalent of a cultural change within the organization.

Downsizing programs are likely to have significant impacts on the jobs of network managers and their staff members. Downsizing decisions are likely to result in the revision of existing network plans. These decisions will also likely require the acquisition of new hardware and software, the development of new training programs for both IS staff and users, and a variety of other short-term and long-term burdens for network managers.

✳ KEY TERMS

Computer operators   Rightsizing
Distributed processing   Upsizing
Downsizing

✳ REVIEW QUESTIONS

1. What are the differences between downsizing, upsizing, and rightsizing?
2. Discuss the reasons why organizations are downsizing.
3. Describe the advantages and disadvantages of distributed processing systems.
4. Describe the potential disadvantages of downsizing.
5. Why are downsizing plans important?
6. Describe the elements that are likely to be included in a downsizing plan.
7. Describe the characteristics of applications that are likely targets for downsizing. Also, describe the characteristics of applications that may be unsuitable for downsizing.
8. What are the potential pitfalls of downsizing programs? What can organizations do to avoid these pitfalls?
9. Describe the impacts that downsizing programs can have on network managers.

✳ DISCUSSION QUESTIONS

1. Downsizing, upsizing, and rightsizing are affecting the mix of information processing professionals that organizations need. Discuss the changes that are taking place and identify the types of staff positions that are decreasing in number as well as the types of staff positions that are increasing in number. Also discuss how these trends are affecting the relative power and clout of the different types of IS professionals.

2. Some organizations are avoiding or delaying downsizing by out-sourcing some of the noncritical applications run on large systems, by consolidating multiple data centers, and by creating "unattended" data centers. In addition, major price reductions have been implemented by large system vendors in hopes of slowing down the trend toward downsizing. Discuss how and why each of these factors can affect organizations and may lead them to reconsider whether they should downsize.

✳ PROBLEMS AND EXERCISES

1. Do a computer-assisted search for articles about organizations that have downsized to smaller computing platforms. Obtain a copy of an article that especially interests you, read it, and summarize the downsizing that took place, the reasons why downsizing was undertaken, and what happened as a result. Produce a one-page, word processed executive summary of the article and develop a brief presentation for your fellow students.

2. Invite an IS manager from a local company that has downsized (or is currently downsizing or considering downsizing) to make a presentation to your class. Ask him or her to summarize the downsizing, why it was done, and the results that the organization has experienced.

✳ REFERENCES

Ambrosio, J. "Not a Disappearing Act." *Computerworld* (August 17, 1992): pp. 1, 62.

Atre, S. and P. M. Storer. "Client/Server Tell-All." *Computerworld* (January 18, 1993): p. 73.

Betts, M. "'Old-Timers' Wrestle with Dinosaur Image." *Computerworld* (August 24, 1992): pp. 1, 12.

Booker, E. "Motel Chain Locates on Client/Server Route." *Computerworld* (July 4, 1994): p. 61.

Bozman, J. S. "Mainframe-to-PC LAN Shift Taking Hold." *Computerworld* (August 31, 1992): p. 4.

Cafasso, R. "Rethinking Re-Engineering." *Computerworld* (March 15, 1993): p. 102.

Dostert, M. "LAN Saves TV Network $1M." *Computerworld* (August 9, 1993): p. 53.

"Downsizing from Mainframes and Mid-Range Computers." *Datamation* (June 15, 1994): p. S7.

Frenzel, C. W. *Management of Information Technology*. Boston, MA: Boyd & Fraser, 1992.

Halper, M. "Sabre Eyes Downsizing." *Computerworld* (March 29, 1993): pp. 1, 10.

Hoffman, T. "Getting There From Here." *Computerworld Client/Server Journal* (May 1994): p. 21.

Johnson, M. "Aerospace Giant Commits to Workstations." *Computerworld* (January 13, 1992): pp. 41, 44.

King, J. "Re-Engineering Slammed." *Computerworld* (June 13, 1994): pp. 1, 14.

Krepchin, I. "Will Cheaper Mainframes Slow Downsizing?" *Computerworld* (October 1, 1993): p. 63.

Liebs, S. "Mastering the Migration Maze." *Information Week* (November 29, 1993): p. 36.

Lindquist, C. "County Thinks Small, Dumps 4381 for LANs." *Computerworld* (April 6, 1992): pp. 1, 20.

Littman, J. "Breaking Free." *Corporate Computing* (June 1993): p. 104.

Maglitta, J. "Utilities Plug into Client/Server." *Computerworld* (March 28, 1994): pp. 1, 8.

Miller, S. "The Downsides of Downsizing." *Chief Information Officer Journal* (May/June 1993): p. 25.

Moran, R. "Mainframes Keep Going and Going..." *Informationweek* (March 21, 1994): p. 28.

O'Leary, M. "The Upshot of Downsizing." *CIO* (June 1993): p. 42.

Panepinto, J. "Client/Server Breakdown." *Computerworld* (October 4, 1993): p. 107.

Parker, C. and T. Case. *Management Information Systems: Strategy and Action.* 2nd ed. Watsonville, CA: Mitchell/McGraw-Hill, 1993.

Radding, A. "Downsizing Without the Fuss." *Computerworld* (November 30, 1992): p. 81.

Ricciuti, M. "The Mainframe as Server: Is IBM Totally Bonkers—or Brilliant?" *Datamation* (May 15, 1994): p. 61.

Scannell, E. "IBM Hosts Downsize Via Objects." *Computerworld* (July 11, 1994): pp. 1, 12.

The, L. "Does Downsizing Add Up for Financials?" *Datamation* (March 15, 1994): p. 51.

"UNIX Servers—The Downsizing Choice." *Datamation* (March 15, 1994): p. S8.

Vizard, M. "Client/Server Caveats." *Computerworld* (February 15, 1993): pp. 1, 16.

Whiting, R. "From COBOL to Client/Server." *Client/Server Today* (April 1994): p. 34.

Case Study:
Southeastern State University

From reading the previous segments of this case, it should be apparent that upsizing has been more common than downsizing or rightsizing on the Southeastern State University campus. Upsizing has been most evident in the creation of new LANs, both student and administrative, within the various LANs. As noted, the LANs created are microcomputer-based and usually consist of various types of IBM-compatible workstations running either DOS or Windows (depending on the generation of processors in the individual machines). The servers in the LANs are usually high-end microcomputers with greater RAM and disk space than that found in workstations; adequate levels of network performance and response times are achieved.

To date, there has been no pressure to develop a LAN utilizing a minicomputer or retired mainframe as a server, though the potential for doing so exists; some of the larger systems have been replaced by newer, more powerful, and often cheaper substitutes. For example, one of the Control Data Corporation Cyber 830s in the Computer Services Department has been recently taken offline, because it was no longer needed as a node in the state university system's computer network and because most of the noncommunications-oriented applications that it had been used for had been moved to newer VAX super-minis (in the cluster of VAX machines operated and maintained by Computer Services Department employees) or to the newly installed RS/6000. The university is contemplating what to do with the Cyber 830, a machine that has served the university well for a number of years in both administrative and academic computing areas. The university is leaning toward donating it to one of the nearby two-year schools in the state university system.

The trend toward distributed processing and a client/server network architecture is quite evident at SSU, best illustrated by the acquisition and installation of the RS/6000. In addition, this trend can be observed in the recent acquisition and installation of an IBM AS/400 that will be used to implement the (Banner) student information system. The university intends to move all the database and applications associated with admissions, registration, and financial aid to this machine.

As noted, the AIX operating system (IBM's proprietary version of UNIX) is used to run the RS/6000, and several academic computing applications (such as COBOL and FORTRAN compilers and Oracle database management applications) traditionally run on a VAX have been moved to the RS/6000. This has created a cleaner breakout of academic and administrative applications (most of the latter reside on the VAXs) and has improved network performance for both student and administrative users, because they are no longer competing for time and resources on the same computers. Improved system performance is especially evident during peak processing (and network traffic) periods such as at semester's end.

The installation of the RS/6000 and the migration of academic applications to this machine still creates an imperfect system. For example, for many of the LANs in student computer labs, user access to this UNIX server is accomplished through one of the VAX minicomputers. That is, from the LAN workstation, the student must first logon to the VAX, then select the RS/6000 from among the VAX menu options available to student users. After this selection is made, the student must logon to the RS/6000 to utilize the applications available on that machine. Hence, although student users of applications on the RS/6000 are no longer competing for processing time on the VAX minicomputers (other than that needed to establish and maintain connections to the RS/6000), they are still sharing communications media with academic users. Network traffic projections indicate that there may soon be some problems as the volume of messages passed over the communications lines approaches the media's capacities.

Thanks to an external grant, one of the student LANs (the one used for CAD applications) in the College of Engineering has been linked directly to the RS/6000. Essentially, this was done to enable students using the CAD LAN to perform computer assisted engineering (CAE) tests on their designs. Student users of this LAN have GUI menu options for both the VAX and the RS/6000. If they want to access the VAX (for example, to connect to the Internet), they click on the VAX icon, which takes them directly to the VAX logon screen. To logon to the RS/6000, they click on the icon for this UNIX server and are taken immediately to the logon screen for this machine.

CASE STUDY QUESTIONS AND PROBLEMS

1. Why is upsizing to LANs from stand-alone microcomputer systems a desirable option in state-supported public universities? List the general advantages of upsizing and identify those most pertinent to this type of situation. Briefly explain each of your selections.
2. Why is the movement toward a distributed client/server computing network desirable at SSU? List reasons why this change is desirable and compare this list to the general set of advantages associated with distributed systems mentioned in the chapter. Which of the potential disadvantages of distributed systems are most likely to occur at SSU? Justify your selections.

3. Based on the information provided in this and previous segments of the SSU case, identify at least three ways in which the useful life of the Cyber 830 could be extended at SSU. Also, discuss the advantages and disadvantages of using this machine as a node or server in networks at SSU.

4. Identify several specific mechanisms that could be used to avoid the projected network traffic problems.

5. In light of the fact that different operating systems are used to run each of the new minicomputers at SSU (and that each of these is different from the OS used with the VAX machines), what role(s) do you see for middleware in the distributed client/server network that is emerging at SSU? What tasks would middleware be most likely to perform at SSU?

PART FOUR

Network Management

Managing Network Operations

CHAPTER OBJECTIVES

After completing this chapter, you will be able to:

◆ Identify the range of activities associated with managing LAN operations.

◆ Identify issues associated with data ownership in LAN environments.

◆ Distinguish between physical and logical network users.

◆ Identify the various levels of user access privileges found in LANs.

◆ Describe some of the activities of managing workgroups.

◆ Describe some of the activities of managing printing operations.

◆ Identify the major types of printer interfaces and print server configurations found in LANs.

◆ Discuss some of the challenges of supporting graphical user interfaces in networks.

◆ Identify some of the causes of workstation lockups when using applications with large memory requirements.

◆ Explain the function of refresh rates in multiuser, multitasking networked database applications.

After a LAN or other computer network has been installed, network administrators must necessarily turn their attention to managing day-to-day network operations. Some of the primary goals of managing daily network operations include ensuring that the applications, data, information, and computing resources needed by network users are accessible, correct, and provided in a timely manner. A wide variety of tasks are performed by network managers and other network personnel to achieve these operational goals. Some of the major duties likely to be performed daily are summarized in Table 12-1. The manner in which these tasks are actually carried out depends upon the type and size of LAN installed, the network operating system (NOS) used to run the network, the operating system(s) used to run the LAN's workstations, the characteristics of the workstations, the number and variety of servers included in the LAN, the types of network printers, the applications supported on the network, and a number of other factors. Because of the wide range of factors that determines the daily tasks performed for specific LANs, it is impossible to account for all the possible combinations and permutations of operational duties. Hence, this chapter will focus on several major functions associated with managing daily operations, including the management of user access privileges, managing workgroups, managing printing operations, and managing user interfaces.

- Creating, maintaining, and deleting user accounts
- Creating login scripts for users and system administrators
- Designing and implementing directory structures
- Establishing and changing security for directories and files
- Establishing and changing user and workgroup access privileges
- Adding and removing workstations and other devices from the network
- Installing, updating, and testing software applications
- Establishing and changing configurations for individual workstations
- Creating and modifying drive mappings
- Monitoring server operations

- Monitoring and managing printing operations
- Establishing and changing print server configurations
- Backing up important data and application files
- Creating and maintaining network documentation
- Troubleshooting network and user problems
- Performing regular and preventive maintenance activities
- Providing user training, support, and documentation
- Planning and implementing server, workstation, and printer upgrades
- Developing RFPs, RFQs, and RFBs

TABLE 12-1   Network managers' and staff's operational tasks

◆ MANAGING USAGE AND WORKGROUPS

The manner in which a LAN and groups of LAN users are organized depends on the characteristics of the user community. The spectrum of activities that network users perform is typically as diverse as the users themselves. However, applications on the LAN must be organized to ensure access to the network resources needed by individual users while simultaneously protecting data and resources by restricting access to only authorized users.

Payroll information is a classic example of data to which access must be restricted. For example, a clerk in the sales department typically has no need to know the data in the payroll system and therefore should not have the authorization to access it. Appropriate security mechanisms must be established (and maintained over time) to prevent the clerk and other unauthorized users from accessing payroll data. Data may also need to be protected from inexperienced users who could inadvertently alter or delete data or files that they should not be able to access in the first place.

MANAGING ACCESS PRIVILEGES

Establishing maintaining, and changing the ability of individual users to access network applications and resources (such as databases and printers) is one of the important ongoing responsibilities of network managers. Most of these tasks revolve around users with workstations attached to the network; however, in an increasing number of organizations, establishing and maintaining access to LAN resources by remote users, telecommuters, and mobile workers has become very important.

Providing remote access to network resources by authorized users can make the network vulnerable to invasion by outside individuals or groups who have no business trying to access an organization's computers and data. In addition, this can open the door to network disruptions caused by viruses or other renegade programs released by unscrupulous or malicious outsiders. For example, in 1988, a student at a major university released a renegade program that traveled telecommunication lines and entered a large number of systems through an important worldwide network. The renegade program consumed and rendered inactive the memory resources of the nodes it invaded. Thousands of researchers saw their research projects brought to a screeching halt because critical systems were not available. This student was the first person convicted under the federal Computer Virus Eradication Act.

While the danger of invasion by outsiders cannot be ignored, greater threats may be found within organizations. Research indicates that most security violations are done by employees. Some violations are willful security breaches (or intentional sabotage) related to organizational problems such as employee terminations, employee dissatisfaction, and criminal intent. Other access violations are unintentional and result from insufficient or inappropriate access controls. Because of the potential for both willful and unintended security breaches, the LAN administrator must be vigilant in ensuring that adequate security mechanisms and access controls are in place.

PROVIDING GENERAL ACCESS TO THE NETWORK

The user of every computer or terminal physically attached to the LAN has the potential to become a physical user. However, some computers on the network may be used by more than one employee in the company. Therefore, networks may also embrace the concept of logical users. A *physical user* is the person actually using the computer attached to the network. The physical user may be restricted from using other attached workstations; that is, there may be a one-to-one correspondence between a physical user and the workstation that he or she may use.

Logical user configurations enable multiple users to share the individual workstations attached to the network. For example, in a computer lab at a college or university LAN, each computer in the lab can be assigned a workstation number or identifier. However, because many students may use the same workstation in a day, they may be designated as logical users who may logon to the network using the same user identification (for example, by entering "student" or the workstation number/identifier as the user identification).

In an office setting, a different set of parameters may be needed. Office workers may use a data storage system (such as a relational database) that requires data input into the system to be available not only to the worker who actually entered the data, but to all other workers with authorization to access the database. For example, an accounts payable system may be concurrently used by four or five accounts payable clerks each day and each clerk may enter invoices and adjustments, print checks, and print reports from the same database. In such a configuration, it may be important to designate all users as physical users to distinguish who input the particular data item from who made later adjustments to the item.

Novell NetWare's Version 4.xx has altered traditional distinctions between physical and logical users. NetWare Version 4.xx has been designed for use with interconnected LANs on an enterprise-wide basis. In such situations, an individual user's workstation may be physically attached to one LAN, but he or she may need access to data or computing resources found in other LANs. To gain access to the data or resources in other networks, the user may be designated as a logical user of those networks while remaining a physical user within the LAN to which his or her workstation is attached. The logical user designation in NetWare 4.xx makes it unnecessary for the user to logon to each network containing needed resources (previous versions of NetWare required this extra step); however, he or she may still need to have password clearance to access particular applications or data.

ACCESS PRIVILEGES

Differing levels of access can be granted to users depending on the nature of their needs. For example, virtually everyone may be granted permission to logon and logoff the network. In addition, everyone may be granted access to general-purpose applications available on the network. However, only a few users may be allowed to access sensitive databases and only users with verifiable needs may be allowed to attach to other networks as logical users.

Access rights may be granted to a series of applications based on the ownership of the data. For example, if the programs and data are "owned" by a particular workgroup, all members of that workgroup may access the applications while nonmembers may not be permitted to access the programs at all. When particular departments, subunits, or workgroups own specific databases or files, formal permission may be required for a user to be designated as an authorized user. For example, most LAN administrators may view access to employee address lists, employee personnel files, payroll records, and employee benefit records by senior members of the Human Resources Department to be appropriate. However, written authorization from the highest-ranking manager in the Human Resources Department may be needed to formalize such authorization.

Although personnel data is certainly sensitive, some parts of it are more sensitive than others. For example, the company may keep records on the medical treatment of employees related to the medical reimbursement plan. Access to such highly sensitive data could be restricted to only one or two senior managers. Even employees who input the data could be restricted from viewing anything other than the one record being input to the system.

Some of the more common types of access that may be granted to users in LAN environments are summarized in Table 12-2; the implementation of these is discussed in more detail in Chapter 14. Note that access can usually be granted at either the directory or file level (within directory levels). The user may be allowed access to specific directories but not others and then may have a wider range of access rights for some files in the subdirectory than for others.

Most of these access privileges are granted when the network manager creates user and group accounts. The set of rights granted to individual users or groups is called *trustee rights*—these determine the user's ability to read, delete, and update files and directories. If a new user is designated as a member of a specific group, he or she may inherit the access rights that the group has when his or her account is created.

Access Privilege	Brief Description
Access control	Allows the user to modify the list of users who can access a directory or file
Create	Allows the user to create files and subdirectories
Erase	Allows the user to delete files or directories
File scan	User may view the contents of a file
Modify	User may rename a directory or file or change the attributes of the directory or file
Read	User may open and read existing files in a directory
Write	User may open and write to all files in a directory
Supervisor	User has all the rights listed above to all files and subdirectories in the directory

TABLE 12-2
Types of access
privileges

In Novell NetWare LANs, the SYSCON utility is the primary tool used to create new user and group accounts. However, the creation of new accounts and specification of access privileges can also be done via the MAKEUSER and USERDEF utilities. User access privileges can also be enforced by customizing logon scripts for individual users. Trustee rights can be reviewed through NetWare's RIGHTS command and modified in Novell LANs through the GRANT command; specific rights can be removed by using the REVOKE command. Similar types of access utilities and commands are found in other NOSs.

In LANs with a variety of different types of microcomputers and microprocessors, it is often necessary to restrict users from accessing certain types of applications. For example, even though general-purpose Windows applications may be supported on the network, it may be necessary to restrict access to these by physical users whose workstations are incapable of running Windows. In practice, it is often advisable to customize the user's logon script to display only directories or file options that his or her machine is capable of running (assuming, of course, that the user is authorized to access these resources).

In networks where users have access to databases, other levels of access privileges may be needed. For example, some users may be granted update privileges (for example, the ability to add, delete, or modify the contents of records in a file). Other users may be given read (or view) only access—they can look at data records but make no changes to them. Coupled with these privileges may be file-level, record-level, and field-level access. File-level access permits access to a specific file. Record-level access restricts access to a specific set of records in a file. Field-level access restricts access to specific fields within data records.

As you can see, depending on the types of applications, data, and users found in a network, the set of access privileges and access policies needed may be quite elaborate. When organizations develop sophisticated systems for assigning access to data and applications, ongoing review of access policies and the mechanisms used to enforce them is important.

MANAGING WORKGROUPS

Many organizations organize access privileges around groups of employees who work on similar tasks and who need access to the same applications and data. Novell NetWare and other NOSs allow network managers to designate specified users as workgroup managers who may add users, change trustee rights, and perform other functions for the groups and users assigned to them. In other words, network managers can delegate some of the responsibilities associated with the administration of access privileges to workgroup managers. In Novell LANs, the rights of workgroup managers are assigned through the SYSCON utility. Figure 12-1 illustrates one of NetWare's group information screens accessible by network supervisors.

Of course, workgroup managers must be adequately trained in carrying out these network-related tasks for their groups of users. Often, the burden of doing the training falls to the network manager.

As discussed in Chapter 7, in the last decade, companies increased their utilization of teams and collaborative workgroups as a means of increasing productivity and managing quality. Networks have played a significant role in enabling a group to accomplish its goals by supporting the information processing needs of the group. Many of these workgroups have some unique requirements for data and applications (such as groupware) and also have a need for their data and applications to be inaccessible by other network

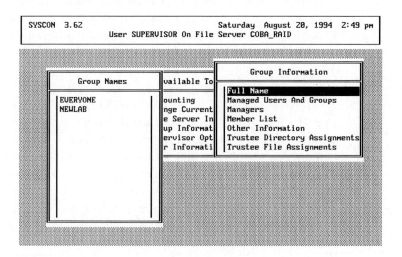

FIGURE 12-1
Group
management

managers. These special needs of workgroups and the increasing popularity of groupware and other types of workgroup computing software have made the job of managing workgroups by network administrators more challenging and complex.

Generally, a workgroup consists of a small number of users (usually two to eight) who need to be able to access common information databases, communicate online and offline, schedule meetings, and review and annotate reports. A workgroup may also prepare and review correspondence, coordinate plans, and track activities. Group members may be geographically dispersed and travel frequently. Again, software designed to facilitate such workgroup requirements is generically known as *groupware.*

Many groupware packages enable multiple users to have concurrent access to specific spreadsheets, database files, and documents. Concurrent users may also have updating privileges for the files they can access so that authoring activities can be shared; online conferencing and decision support capabilities are included in many groupware packages.

The characteristics and benefits of groupware are often difficult to explain to the inexperienced. Many magazine articles have attempted to explain how Lotus Notes (one of the most widely used groupware products) works, but most network managers confirm that the best way to convince users of the potential benefits of Notes is by providing them with hands-on experience with the product. The use of a facilitator when Notes (or other groupware) is installed can help ensure that users will be aware of the benefits of groupware.

The ability to electronically change messages is often a valuable part of workgroup support. For this reason, groupware packages generally include a mail facility. Enhanced e-mail systems are becoming more common in many packages. These electronic messaging capabilities allow senders to transmit messages at times when a conversation would not be convenient or when the sender is travelling. The software ensures that the message is routed to the appropriate destination; the recipient can make a brief annotation and return a reply.

Notes includes e-mail capabilities (see Figure 12-2), but, as may be observed in Figure 12-3, this groupware is more than e-mail. It could also be described as a flat file database intended for electronic conversations; as may be observed in Figures 12-4 and 12-5, Notes can electronically track e-mail conversations and maintain "profiles" of individuals frequently communicated with; these capabilities help ensure that specific conversations or descriptions will not be forgotten or misplaced. In addition, Notes can electronically record conversations (that is, create a conversational database) so that key word searches or reviews of conversations can be made.

PC Eudora Groupware is another package that enables workgroup members to compose, transmit, and receive e-mail online. Figures 12-6 and 12-7 illustrate some of the messaging capabilities of this groupware package.

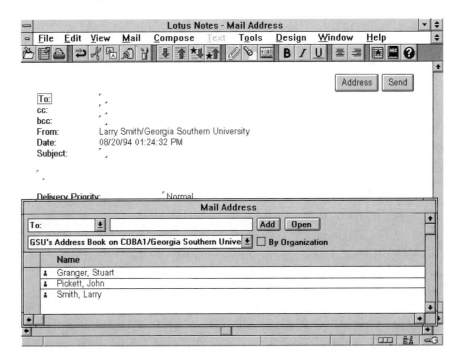

FIGURE 12-2
Notes' e-mail

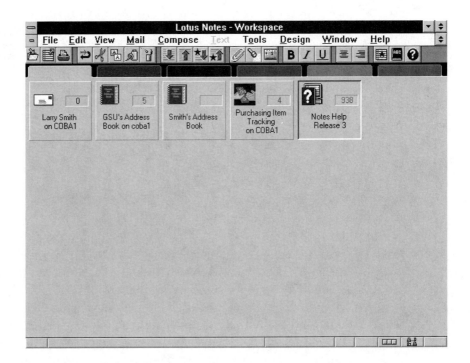

FIGURE 12-3
Some Notes
capabilities

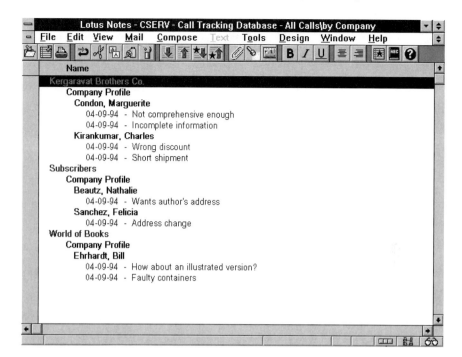

FIGURE 12-4
Using Notes
to track
conversations

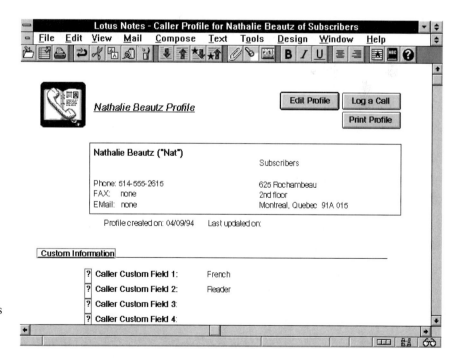

FIGURE 12-5
Using Notes to
maintain profiles
of frequent
contacts

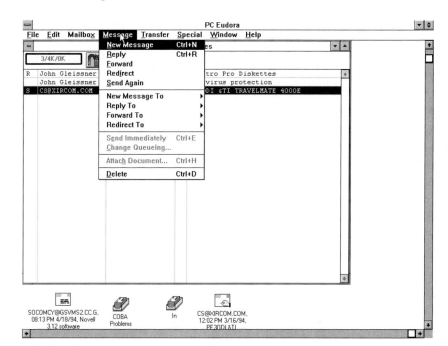

FIGURE 12-6
PC Eudora
Message menu

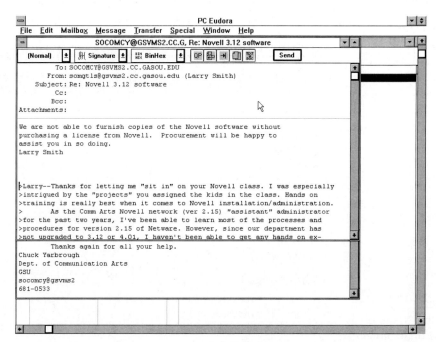

FIGURE 12-7
Sample message

Most e-mail packages include ***broadcast features*** that enable the same message to be simultaneously transmitted to a specified list of recipients. Such capabilities (illustrated in Figures 12-8 and 12-9) make it possible to distribute important messages quickly to all members of the workgroup, all

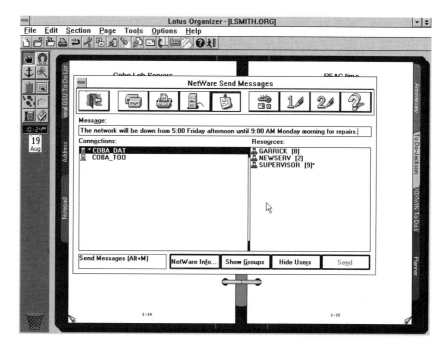

FIGURE 12-8
Sending a
message to
network users

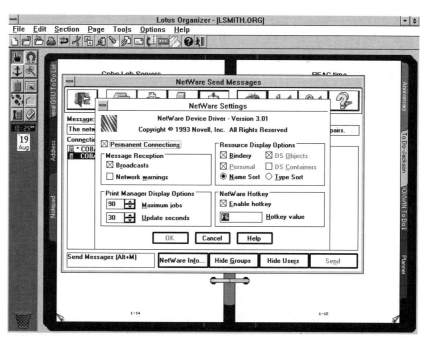

FIGURE 12-9
Windows'
network interface

users of the network, or all members of the organization. For example, messages can be sent to members of the workgroup who are travelling, in meetings, working on other projects, or engaged in other activities that make phone calls inconvenient. Notice of an important meeting can be broadcast to a stored list of recipients, each of whom is asked to acknowledge plans to attend; when using Notes or other groupware that incorporates group scheduling software, the recipient may be referred to an electronic calendar to check the time and place of the meeting.

Groupware can be used to facilitate communication and collaboration among the members of a workgroup even if their individual workstations are part of a larger, hierarchical networking system. In other words, groupware can make it possible to implement organizational structures such as project teams and matrix organizations even if the underlying computing platform has a hierarchical architecture. Generally, structuring such a system requires a specialist trained in the methods of providing peer-to-peer access in the midst of other network configurations. Some organizations have been making use of such workgroup support capabilities for several years. For example, Texas Instruments routinely assigns managers to multiple project roles—the ability to communicate across traditional organizational lines is necessary. Other organizations are seeing the benefits of such capabilities and are developing similar systems.

◆ MANAGING NETWORK OUTPUTS

One of the primary goals of implementing a network is to share resources that otherwise would have to be provided for every computer in a stand-alone mode. The inconvenience of waiting in line to use a particular computer attached to a laser printer or color printer, and the expense of providing every single computer a laser printer and a color printer, are powerful forces behind the demand for networked systems.

In terms of network outputs, there can be great diversity in the needs of network users and groups. Some users have extremely high demand for printing volume, while others have little volume but unique formatting requirements. In a network environment, the network administrator has the option of setting up a printer to act as a local printer for a particular workstation, or he or she can configure the printer to service a print queue on the network (thereby making the printer available to any user having access to that queue). In Novell NetWare LANs, a printer physically connected to a workstation can fulfill either of these roles, depending on how the network is configured.

In the NetWare architecture, a printer allows the user to print from an application as when using a stand-alone PC system. The software in the NOS directs the user's print job to the appropriate printer. Most networks allow any user with access to the printer queue to print at the user's discretion. The print server will release the jobs in the print queue to the printer in sequential order unless the network administrator has provided for a different priority.

DIFFERENCES AMONG USER PRINTING NEEDS

As has been noted, there may be many different printing needs among network users. In a network manager's ideal world, every kind of printer would use the same printer driver and accept the same electronic signals. However, such uniformity is not likely, so the network administrator must make allowances for the different needs of users. In fact, most networking systems still provide elaborate systems for standardizing printer output on a character-by-character basis controlled by escape (ESC) character code sequences embedded in the stream of characters produced by the computer. In these situations the image of the character is stored in the printer itself and activated by the introduction of a binary code specifying the character's ASCII code. In more recent times, and particularly in graphical environments, the printer drivers output characters as a part of a graphic image that can accommodate scaled sizes of characters and special images. Many of the applications producing such sophisticated output are network-aware; the character-by-character management of the printing output is no longer necessary.

Two of the first steps in administering a diverse printing environment are to identify the kind(s) of printer(s) needed by each user and provide a queue path to the printer(s). For example, a user who needs to frequently print voluminous reports on continuous green bar paper is likely to need a heavy-duty dot matrix or band printer designed for high volumes of character-based printing. However, this printer may be of no use to a worker in the public relations department who needs to print high-resolution graphics; a laser printer is likely to be needed for this task. In addition, some users will need access to more than one kind of printer; therefore, more than one print queue must be made available.

For character-based printing in a DOS or OS/2 environment, print queues can be set with batch files. DOS is not considered to be a network-aware system. Command files have to be set up to guide the user to network print queues. For example, the LAN manager may set up two batch files, PRINTER1.BAT and PRINTER2.BAT, so that the user needs only to type in PRINTER1 or PRINTER2 to route the print job to the proper printer.

Users of Windows and Windows NT applications can take advantage of the network awareness of these operating systems. These OSs allow users to connect to printers using dialog boxes. The dialog boxes are designed to logically connect the standard LPT*n* or the COM*n* designations of a stand-alone computer system to the print queue(s) named and defined for the user's work environment. Hence, if a laser printer is connected to the queue named LASER and a dot matrix printer is connected to the queue named DOTMATR, the user can simply select the one he or she needs just as he or she would when using a dialog box on the screen of a stand-alone computer. Several queues could be named, allowing an even wider choice of printers and output. Some network administrators follow the policy of naming print queues according to the physical location of the printer; for example, ACCT_LASER might be used to specify the queue for a laser printer in the accounting department.

Some network users need to print in secured environments; for example, documents such as accounts payable or payroll checks are often printed in supervised settings, because the printouts are valuable and because they are typically printed on sequentially numbered paper stock, each page of which must be accounted for. Laser printers equipped with the special toner needed to print the MICR (magnetic ink character recognition) characters found at the bottom of checks are available for these special print needs. Needless to say, access to such print queues is likely to be severely restricted.

MANAGING SPOOLING OPERATIONS

In addition to the network manager, people designated as queue operators may be involved with the management of spooling operations. *Spooling* operations place print jobs in print queues and direct them to printers as they become available. A queue operator is a user who has been given the security clearance to modify print queues and print server configurations. In Novell NetWare LANs, queue operators are able to access and use the PCONSOLE utility, which is used to manage print queues and create and set up print servers. If, for example, a print job is ruined by a printer malfunction, a queue operator may be able to reprint the job or eliminate the current print job and reprint after the printer has been repaired.

The complexity of managing spooling operations is contingent on how the print server(s) and printer queues are configured. Some NOSs allow a printer attached to a particular workstation to be accessible to other computers on the network; in client/server LANs, the workstation becomes a nondedicated print server. Other systems allow a certain number of printers to be attached to a file server so that it can be used as both a print server and a file server, although this configuration usually slows file server operations. Some smaller systems even allow one computer to be used as both a nondedicated file server and a nondedicated print server. Several printer vendors (including Hewlett-Packard) and some third-party manufacturers produce printer interfaces that emulate the appearance of a print server on the network and allow processors in the printers to manage print queues and spooling operations. However, the most common configuration found in LANs is the dedicated print server. In Novell NetWare 3.xx LANs, a single print server can serve up to 16 printers; print servers in NetWare 4.xx networks can handle up to 25 printers.

TYPES OF PRINTER INTERFACES

In most stand-alone PC systems, a parallel transmission interface is used to communicate with a printer. The most common parallel cabling uses a 25-pin plug similar in appearance to a 25-pin serial interface, and the most common interface from the cable to the printer is a Centronics interface. Most IBM-compatible PCs can have 3 parallel printers attached to expansion cards in the expansion bay.

In IBM-compatible systems, the most common serial interfaces are the 25-pin serial plug or the 9-pin interface (frequently called the AT interface). Apple-based products use serial rather than parallel interfaces in most instances. As is true for most peripherals, printers can also be accessed with the increasingly popular SCSI interface; SCSI can be used to daisy-chain several printers in succession.

In a network, the hardware interfaces for printers connected to networked PCs are the same as those found in stand-alone systems. However, the NOS includes modules to direct print output from other workstations to the proper queue so that it can be printed out on the correct printer. The print server software (loaded either in the file server or in a dedicated print server) directs the output to the designated printer. In some NetWare LANs, a small TSR (Terminate and Stay Resident) program named RPRINTER can be initialized by a workstation in the AUTOEXEC.BAT file at boot-up time. The workstation can then dynamically receive output from the network and forward it to its attached printer. In this arrangement, the printer is called a *remote printer* as it is located some distance away from the file server or print server (and to indicate that the printer is not directly connected to the file server or print server).

Most network printing systems provide for a form feed character at the end of the printer stream to position a new sheet of paper at the print head between print jobs. In addition, the LAN manager or queue operator can configure the system to print the beginning page with a large image of the user's name (or an abbreviated identifier) to separate the end of one print job from the beginning of the next. In Novell NetWare LANs, these and other print job configurations are specified via the PRINTCON utility.

Operating systems such as DOS, OS/2, UNIX, and Windows NT all provide for printing directly from the operating system. DOS offers the PRINT command for printing ASCII files stored on disk. DOS also permits the use of the COPY command to direct graphically-oriented files stored on disk to printers; of course, the printer must be able to receive and print the image. The drivers needed to ensure this usually vary among printer vendors and printer models.

In networks, workstation users can send print jobs to network printers through a function generically known as *redirection*. DOS has limited capabilities in this regard, but the NOS is designed to take what would normally be a DOS output (character or graphics) stream and divert it to a NOS print queue. For example, in NetWare LANs, the CAPTURE command may be used to take DOS output and redirect it to a designated printer queue.

Some applications that will run on a network but are not network aware can create chaos when attempting to print on network printers. Some spoolers will not initiate printing until the capture operation is completed. If the application being used is dependent on interaction with the printer during printing operations, there may be problems in attempting to get the application to print successfully on network printers. Sometimes a NOS will expect an application to terminate when activity in the printer output stops for some period of time (a *timeout interval*). When this time span (usually between 30 and 90 seconds) has expired, the spooler terminates the print job. Should printing operations for an application simply be slow to execute, print jobs can be prematurely terminated.

REPORTING OF PRINTING OPERATIONS

Most NOSs can provide LAN and queue operators with online reports concerning the number of applications currently being redirected to print queues, the number of active printers, which queue(s) are servicing which printers, and the number of print jobs awaiting their turn to be printed. Several status items related to current operations can be determined by such reports. For example, the print server modules can indicate the NOS' perception of a printer's current status. If a printer indicates "waiting for job" and the list of print jobs in the queue indicates a ready condition, there is a connection problem between the queue and the printer. In addition, if a print job is listed as being active for an unreasonably long period of time, the application may be hung up or no longer active on the workstation that initiated the print job. (Most systems will permit a queue operator to interrupt a print job—often the only solution to a job that is hung up.) Also, if the number of print jobs becomes longer over time, it may be advisable to add another printer to the queue.

Two types of information needed by LAN administrators are usually provided by the NOS or third-party software. The first, data organized by the user, includes: 1) the number of stored (spooled) print jobs for the user; 2) the number of lines printed for the user; and a list of printers accessible by the user. The second type, aggregated information, includes: 1) the average number of jobs in the queue; 2) the average amount of time jobs have been held in the queue before printing; 3) a map of which queues service which printers; 4) names of users with priority for print jobs; 5) names of users to be notified in case of aborted job; and 6) the maximum time for ready print jobs to begin printing. Such information can be used to monitor network printing operations over time and can help the network manager determine the types of changes that should be made.

PRINTER CONTROL TASKS

A large network seldom remains fixed in number of workstations, servers, and printers. The number of devices attached to large networks may change daily in response to changes within the enterprises they serve. It may be necessary to add new devices to the network to accommodate new users or to replace aging machines. In addition, malfunctioning devices often have to be removed from the network for servicing or retired altogether. Keeping up with such expansions, changes, and contractions can be time-consuming.

Printers must be serviced occasionally for such items as depleted toner, torn ribbons, paper feed problems, or failure to print characters faithfully. While these problems can often be resolved quickly, LAN administrators typically configure large networks to ensure some *redundancy* in printing operations. This frequently means having more than one printer available to service the same queue. Such configurations allow network personnel to take one printer out of operation without having to curtail printing operations. Fortunately, many network systems can retain print jobs in print queues even when the print server is reinitialized; this preserves print jobs in process (that is, in queues or being printed) at the time the printer was shut down.

In operating environments such as Windows, spooling mechanisms such as the Print Manager can be used as a front end to the NOS spooling system. The Print Manager releases the print job to the NOS' spooler when the printing operation is completed within the Print Manager itself. Thus the Print Manager itself does not continue to be burdened with the storage responsibility for the print job file even when the network printer to which the job has been directed is not immediately accessible.

JOB CONTROL TASKS

As a general rule, the LAN administrator uses a first-in, first-out priority scheme for print queues serviced by network printers; this is usually the default setting of a print spooler. However, it is possible to assign priority to the print jobs of particular users or for particular types of print jobs. For example, priority could be given to short jobs over some long, time-consuming jobs. For urgently needed print jobs, the network manager or queue operator can manually change the order of jobs in the print queue. Generally, a queue operator is empowered to halt the printing of a job, delete a job from a queue, move a job from one queue to another, and change the priority level. A queue operator can execute most of the changes to printing operations that a LAN administrator can perform except actually taking the print server out of service.

Some systems allow one queue to be given priority over another. For example, in Novell NetWare 3.xx LANs, if the priority is set as "1," the jobs in that queue will print before jobs in a queue with the priority set as "2" (assuming that both queues are served by the same printer or group of printers).

Fairness and sensitivity to user needs should govern how such print job control tasks are carried out by LAN managers and queue operators. Modifying the order of print jobs in print queues and the implementing of priority schemes should be in response to valid user needs.

◆ MANAGING NETWORK INPUTS

Between 1984 and 1994, the world of computing underwent two definitive changes: 1) commercial businesses and institutional users moved from stand-alone personal computers and mini/mainframe hardware platforms to networks containing numerous microcomputers; and 2) both casual and business users moved from drab and mundane command line interfaces to GUIs. Figure 12-10 depicts a GUI in Novell NetWare. Both of these changes have caused considerable adjustment in the manner in which input is entered in networks.

With the introduction of Novell NetWare 3.12, the NetWare DOS Requester could be installed on personal computers capable of running

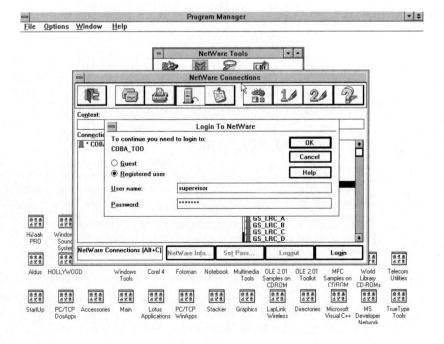

FIGURE 12-10
NetWare GUI
interface

Windows and Windows applications, including their GUIs. The Requester's automated installation program adds the files needed to cause Windows to become network aware and makes some relatively complex additions and changes to the Windows configuration found on stand-alone microcomputer systems. These additions and changes are important to the millions of Windows users who are also network users.

Windows and other graphically-oriented microcomputer operating systems can display dialog boxes, icons, and application interfaces in a variety of display resolutions. Windows supports video resolutions ranging from the old EGA standard of 640×350 to resolutions for Super VGA of $1,200 \times 1,000$ or even higher. In general, better defined and crisper screen images are possible with higher resolutions. In addition, resolution also determines how many lines of text can be displayed on a single screen by a word processor and how many columns and rows of a spreadsheet can be displayed onscreen at one time.

Several items in GUI interfaces are of concern to network managers. First, prior to the introduction of GUI interface tools by NOS software vendors, there was no support for GUI interfaces on some networks. Through trial and error, some LAN managers figured out how to accomplish some rudimentary support for applications with GUI interfaces, and seminar leaders made considerable profit demonstrating how to accomplish a marriage of GUI applications and networks. In addition, some third-party software vendors provided menuing systems for such applications. However, a comprehensive, supported method for integrating GUI applications in network environments did not exist.

Second, differences in the architectures of machines capable of running GUI applications and those incapable of doing so have presented network managers with some major challenges, especially in LANs containing a mixture of new and old microcomputers. For example, one of the fundamental challenges in introducing GUI operating systems and applications to networks can be the presence of a large number of workstations that use Microsoft DOS or another vendor's version of Microsoft DOS. Even later versions of DOS such as DOS 6.0, 6.2, 6.21, or 6.22 are limited to 640K of usable RAM. While progress has been made in memory management, such as using upper memory blocks above the 640K barrier to load drivers, there are still some hardware and network drivers that must be at least partially loaded in the 640K memory.

On a DOS machine, if there is less than 500K of RAM available after executing the CONFIG.SYS and AUTOEXEC.BAT files (which must include the network drivers that can consume 40–90K), the user already runs the risk of not being able to run some DOS programs. This risk exists even before the drivers for Windows' GUI interface have been loaded. In addition, when some network-intensive Windows applications such as FTP (a file transfer program used in TCP/IP networks, including the Internet) are loaded, the total RAM can be taken down as low as 400K. Although Windows has some capability to pool memory below 640K with memory above 640K, some programs still cannot run, or the number of programs that can be run simultaneously is reduced. In short, because of the network and GUI requirements for memory below 640K on a DOS machine, a significant number of programs cannot be run on the workstation without manual intervention and reconfiguration by network personnel.

Intervention can take several forms. With DOS 6.xx comes the capability of choosing between multiple configurations at bootup without special software. For example, DOS 6.xx will permit you to choose whether you want to run the memory-intensive FTP program to enable you to transfer files on the Internet, or to run your scanner with the TWAIN driver that consumes 50K of RAM. FTP will not run if the scanner driver is loaded, and the scanner program (which uses the scanner driver) will not load with FTP already loaded. While you cannot run the two programs simultaneously, you can run one, reboot, and run the other. As you can see, this procedure takes away one key benefit of the GUI environment: the ability to run more than one program at a time.

Another manual intervention that can be used is to increase the workstation's RAM. Many microcomputers' RAM can be increased to 16MB, 32MB, 64MB, or more. However, loading additional RAM does not cure the problem of a lack of memory below the 640K ceiling. Still, Windows and some Windows applications may run faster or more efficiently because of the additional RAM.

When Windows runs short of available memory resources, it may choose to selectively provide some features of the GUI interface. For example, if you are inputting data to a GUI-capable, network-aware, relational database program, you could get a message that the fonts you have chosen cannot be displayed because of a lack of memory. Since Windows does not always release 100 percent of the memory resources a program consumes, even unloading another program may not remedy the problem. In these cases, the only solution may be to exit and reload Windows, or to reboot under DOS and then load Windows. This can be frustrating when you are intensely inputting data into a program and must constantly switch back and forth to other programs for support information.

Windows NT no longer requires DOS. OS/2 is also capable of circumventing the infamous 640K barrier, as can UNIX GUI interfaces. However, each of these operating systems presents some disadvantages to the desktop computing user, most notably the fact that there is a greater variety of application software (especially business application software) available to Windows and DOS users.

Compounding the aforementioned challenges is the fact that a network may include many different combinations of hardware and software. Many organizations cannot afford to upgrade each workstation uniformly across the network, and sometimes there is no need for such uniformity. For example, some users on an otherwise sophisticated network may have no need for software that is more complicated than a good word processor. Other users may need to have a half-dozen or more applications running simultaneously during the day. While both types of users may need to utilize the network to access the applications required to perform their jobs, the sophistication of their workstations may be quite different.

When there are differences among the types of workstations attached to the network, each machine must be configured according to the hardware, software, and network connections required. For example, machines running applications with GUI interfaces have initialization files (long ASCII text files) that record the general categories of hardware and software installed on a machine. These initialization files tell Windows, or whatever microcomputer OS product is being used, how to load the program. Hence, if there is a high degree of diversity among the workstations on the network, each machine may need a different set of initialization files and different CONFIG.SYS and AUTOEXEC.BAT files.

In general, diversity among workstations attached to the network demands one (or more) of the following: 1) to have a boot PROM (Programmable Read Only Memory) installed on each workstation that individually configures the boot-up files for the machine; 2) to have a floppy disk for each machine used to boot individually to configure the workstation; and 3) to have a local hard disk to store the boot-up files and initialization files needed to configure the computer. Remember, too, that when a machine is taken out of service and replaced with a newer machine (or when personnel move from office to office) all of the uniquely configured hardware and software must accommodate the change being made (either temporarily or permanently).

In most cases, microcomputer OSs with graphical user interfaces will not even load if the hardware settings are not correct. However, the Windows program provides a means of changing the setup configuration from the DOS prompt in case the user makes a change that prevents Windows from loading.

Despite all the caveats given above, it is possible to administer a broadly diversified network that includes multiple hardware and operating system platforms and still accommodate the individual requirements of each machine. For example, network managers in many universities have successfully implemented and maintained networks using applications running under DOS, DOS/Windows, UNIX, System 7 (Macintosh), and OS/2. In addition, more detailed technical literature is being published that informs LAN managers of the previously undocumented challenges they may face in implementing such combinations of platforms.

MANAGING VIDEO CONTROL IN MULTITASKING OPERATIONS

Challenges to controlling the video in multiuser, multitasking operations continue to mount. One reason why this area of network operations continues to be bothersome is the huge resource requirements of some of the newer multimedia applications.

Network managers have also found that it can be difficult to stabilize the video for general-purpose application programs and some of the robust multiuser relational database packages that accommodate GUIs and image graphics. For example, available technology allows a professor to shoot student portraits with a camera designed to dump the images to discrete files on the hard disk of a laptop. These images can then be imported into a pictorial grade book, enabling the professor to associate the names and faces (and grades) of individual students more quickly than might otherwise be possible. Technology also exists to record video clips as digital images for visual playback onscreen. Such technologies are relatively new and require huge memory resources when compared to that needed to store character-based data. From a networking point of view, the resources required by these applications can be problematic; the program may attempt to use network resources (for example, disk space and server RAM for data retrieval) in ways that prevent other users from executing even the simplest of programs.

Many times a user's attempt to display a graphic, play a sound byte, or play a video clip causes an application error which can be difficult to pinpoint. In some cases, the workstation must be rebooted to recover from such errors; this requires rebooting DOS (which loads hardware and software drivers), reloading Windows (which loads additional drivers and initialization files), and reloading the application(s). This operation is time-consuming and may leave the cause of the error undetermined. Many application errors are caused by bugs in the program that can be repaired only by making changes to the source code and recompiling the program. This remedial action typically must be done by the company that developed the software, and it is rarely done quickly. The software vendor may blame the problem on the workstation's operating system or the NOS; OS and NOS vendors are likely to say that it is the software vendor's problem to fix. In short, little external help may be available for troubleshooting errors encountered with multimedia applications on the network. LAN administrators may be left to find their own solutions.

One technique that may overcome some problems with multimedia applications is to reinstall the program with a minimal configuration. Generally speaking, the more demand that the application places on computer resources, the more frequently application errors occur. One useful approach is to reload the application at the lowest possible video resolution. This level of display may not be acceptable for the long run, but it may enable you to at least run the application. Later, the video monitor or video card could be changed to see if a substitute will work more effectively. When using this approach, only the drivers needed to support the application should be loaded. For example, in a DOS 6.xx personal computer environment, you can press (F8) when DOS begins to load to give you line-by-line control over which drivers to load during the CONFIG.SYS and AUTOEXEC.BAT. This "peel the onion" approach is often useful for isolating the causes of the application errors.

The LAN administrator may also find that one workstation can run an application and another will not. At least in this situation you know that the application is not completely at fault. If it were, it would not run on any workstation at all. Again, an effective approach for solving this problem is to analyze the drivers and initial settings for the machine on which the application will not run. Trying lower video resolutions and substituting other video boards or monitors may also be useful. It is always advisable to carefully read the documentation that comes with the application to see if there is a documented problem with running the application on the platform you are using.

A nice tool available from many software vendors is a database of the known incompatibilities and errors experienced with the software. A conscientious vendor will record all problems and publish them; it may also solicit assistance from responsible users who have encountered the same problem. In all fairness to software vendors, with the broad spectrum of software and hardware in use today, the statistical probability of there being a problem somewhere on some machine even after seemingly comprehensive testing is high. Software has always evolved through trial and error, and today's complex applications are no exception.

Problem databases may be available on CD-ROM from either the vendor or a third-party firm. It may also be available through a vendor-sponsored bulletin board system or from an automatic voice response system that can be used to acquire a fax copy of documentation/solutions to problems that were detected too late to be published in the application's manual. Such information is usually available from responsible software (and hardware) vendors who make it their business to be knowledgeable about which product components work well and which do not.

REFRESH RATES IN MULTITASKING OPERATIONS

Many LANs include applications for either multitasking or multiuser/ multitasking operations. For years, Windows has experienced problems in controlling the video display in multitasking situations primarily because of the lack of power in older personal computers. In addition, Windows version 3.1 has a rudimentary system of multitasking that cannot assign one application a higher priority than another. Even stand-alone computers running Windows have experienced the same problems. In DOS machines, difficulties such as a blanked display or a frozen display can be corrected only by rebooting the workstation, causing the loss of data. Rebooting a workstation is frustrating and time-consuming even if no data is lost.

OS/2 is capable of handling multitasking operations more effectively because of its ability to implement preemptive multitasking. UNIX platforms with GUI interfaces are not especially troubled by such problems, usually because of the power of the computers on which UNIX is often run.

Multiuser database systems are becoming more common in LANs. In LANs that include such applications, one of the most important display issues addressed by LAN administrators concerns the refreshing of user video screens, particularly when changes are being made by concurrent users. The microcomputer-based relational database management systems used in multiuser LAN environments make it possible to control the time intervals for automatically refreshing the video displays of tables being used. When one user makes an update to the database, other users should be able to view the change in a reasonable amount of time, usually in 30 seconds or less.

The multiuser relational database application is not the only situation for which refresh rate is important. In workgroup support applications such as groupware packages, the refreshing of user screens to accommodate changes made by other members of the workgroup is important.

◆ SUMMARY

Some of the primary goals of managing daily network operations include ensuring that the applications, data, information, and computing resources needed by network users are accessible, correct, and provided in a timely manner. A wide range of activities may be performed daily by network managers including the management of user access privileges, workgroups, printing operations, and user interfaces.

LAN applications must be organized to ensure access to network resources needed by individual users. However, network managers must also protect data and resources by restricting access to authorized users. A system of access privileges must be created and maintained for both individual users and workgroups that simultaneously satisfies user needs for network resources and protects these same resources from both external and internal security threats.

When providing general access to the network and to designated applications, users may be categorized as either physical or logical users. Physical users are often restricted to using particular workstations in the network, while logical users may be able to use any of the workstations in the LAN. In Novell NetWare 4.xx LANs, logical users are users whose workstations are physically attached to one LAN but who have authorization to access resources in other LANs.

Access to particular data or applications may be restricted by the owners of the data—the employees who are responsible for the data. Highly sensitive data is likely to be accessible by a very limited number of employees. Data that is not sensitive may be accessible by most (or all) LAN users. In LANs with databases, access may be restricted at the file, record, and field levels. Trustee privileges are the set of access privileges possessed by the individual or group.

The increasing prevalence of groupware and workgroup support software has made the management of workgroups an important part of many network managers' jobs. This includes the installation of groupware and e-mail systems and training group members in the features and benefits of these systems. Novell NetWare enables LAN administrators to designate specific users as workgroup managers who are empowered to add users, change trustee rights, and carry out a number of other tasks for groups and group members.

Managing network outputs is another important component of the administration of network operations. In many settings, the primary thrust of this is the management of printing operations, which includes configuring individual workstations, creating and maintaining print queues, installing and maintaining print servers, and managing spooling operations. Many networks enable network administrators to designate queue operators—individuals who are empowered to create and modify print queues and print server configurations as well as to perform printer job control tasks.

During the last decade, graphical user interfaces (GUIs) have become more common in stand-alone microcomputer systems and LANs. Supporting GUI interfaces presents network managers with a number of challenges, particularly in LANs with diverse types of workstations. Often, workstations must be individually configured to handle the video output associated with GUI applications. Multimedia applications also present a special set of video output challenges in LANs. In LANs with database applications accessed by concurrent users, ensuring adequate screen refresh rates is an important concern of LAN managers.

✳ KEY TERMS

Broadcast features	Spooling
Logical user	Trustee rights
Physical user	

✳ REVIEW QUESTIONS

1. Identify the goals associated with managing network operations.
2. Identify the types of tasks that network managers engage in daily.
3. Explain the traditional distinction between a logical user and a physical user of a LAN.
4. In Novell NetWare 4.xx LANs, compare logical users with physical users?
5. Identify the different types of access privileges that can be assigned to users and workgroups.
6. What are trustee rights?
7. What are the differences among file-level, record-level, and field-level access privileges? How do view-only rights differ from update rights?
8. What are the characteristics of workgroups?
9. Identify some of the activities associated with managing workgroups.
10. What factors determine differences in printing needs and printing operations among networks?
11. What types of printer interfaces may be found in LANs?
12. What is redirection?
13. What types of printer statistics and status reports are used by LAN managers to monitor and modify printing operations?
14. Explain why redundancy is an important issue for printing in network environments.
15. Identify the tasks that may be performed by queue operators.
16. What types of printing priorities are possible in networks?
17. What is the 640K barrier? Why is it important in network environments?
18. What steps could circumvent the 640K barrier in workstations?
19. What challenges are associated with the management of multimedia applications in network environments?
20. Why are refresh rates important in concurrent-user LAN databases?

✳ DISCUSSION QUESTIONS

1. Assume that you are the new LAN administrator for a business consulting firm. Describe the process you would use to determine the nature and types of access privileges needed for users and groups in your LAN.

2. In general, what types of printing priorities do you feel should exist in a LAN? Explain the reasons behind your choices.
3. The controller of the company that employs you as the LAN manager has just called and requested that you set up a laser printer to print checks with the new software the company just purchased. What type(s) of printing configuration(s) and environment(s) would you recommend? Justify your recommendations.
4. Why may the 640K RAM ceiling in DOS be a problem if you have 16MB of memory on your Windows/DOS personal computer on a LAN? What can LAN managers do to rectify this problem?
5. How can you ensure that you will not lose the individual configurations for the different types of workstations found in a LAN with GUI application interfaces?

✳ PROBLEMS AND EXERCISES

1. Survey ten users either at your school or at a company to find out what their printing needs are and what types of printers are used. Tabulate the results and note the variations.
2. Ask a LAN manager from a local organization to come to your class to discuss how user access privileges are determined, implemented, and maintained in his or her organization.
3. Type in the following command on a DOS-based microcomputer or workstation: **MEM/C|MORE** (Enter). DOS should give you a breakdown of the usage of RAM in the computer's current configuration. Summarize the contents of this breakdown and the approaches that could be used to increase available RAM.
4. Ask a LAN manager from a local firm to come to your class to discuss the video output challenges faced in his or her organization's networks and how these are being addressed.
5. In a recent computing or networking periodical, find an article on managing multimedia in LAN environments. Summarize its contents in a paper or presentation to the class.

✳ REFERENCES

Ambrosio, J. "Tales from the Warehouse." *Computerworld* (April 26, 1993): p. 55.

Bridge, F. R. "What Is a LAN Administrator?" *Library Journal* (October 15, 1991): p. 56.

Csenger, M. "Newbridge Set to Add to Access-Device Line." *Communications Week* (April 25, 1994): p. 31.

Cuffe, S. and M. Wallack. "People: Essential to Downsizing; Education Is the Key to a Successful Move to Networking." *LAN TIMES* (January 24, 1994): p. 55.

Cummings, J. "Life Cycle Costs Critical Factor in Distributed LANs." *Network World* (May 2, 1994): p. L.1.

Cummings, S. and T. Giorgis. "Peer-to-Peering Through Windows: Windows-Based Peer-to-Peer NOSes Give 'The Little Guys' All the Networking Features They Need." *LAN TIMES* (August 23, 1993): pp. 87, 90.

Editor. "LAN Checklist: 12 Tough Questions to Ask a LAN Administrator." *Journal of Accountancy* (June 1992): p. 30.

Glazer, S. "Before and After: Three Office Makeovers." *Inc.* (Winter 1994): p. 72.

Harslem, E. "The Network Rules." *Computerworld* (May 2, 1994): p. 35.

Horwitt, E. "The Cost of Doing Business with LANs." *Computerworld* (February 7, 1994): p. 54.

Klett, S. P. "Video Networking Standard Backed." *Computerworld* (March 31, 1994): p. 16.

Mackin, K. "Printers Talk Back: New Interface Improves Network Printer Management." *LAN TIMES* (March 28, 1994): pp. 46.5, 46.6.

Maglitta, J. "Large Shops Rethink PC Focus." *Computerworld* (February 14, 1994): p. 6.

Miller, M. J. "Networked PCs: Nothing Personal?" *PC Magazine* (January 11, 1994): p. 77.

Musich, P. "LAN Costs Surprisingly High: Studies Cite Need for Better Management." *PC Week* (December 20, 1993): p. 37.

Perey, C. "Transporting Multimedia Across Local Area Networks." *Business Communications Review* (January 1994): p. 53.

Radosevich, L. "WordPerfect, Novell Consolidate Groupware Efforts." *Computerworld* (April 4, 1994): p. 12.

———. "Users Brace for LAN Mail Wars." *Computerworld* (February 7, 1994): pp. 1, 10.

———. "HP, IBM Scramble to Fill Messaging Void." *Computerworld* (May 2, 1994): p. 63.

———. "Management Void Limits LAN E-Mail." *Computerworld* (April 25, 1994): p. 15.

Schmidt, R. V. "Internetworking: Future Directions and Evolutionary Paths." *Telecommunications* (January 1994): p. 55.

Scott, K. "Multimedia Enters the Network Realm." *Open Systems Today* (May 9, 1994): p. 36.

Torgan, E. A. "Saber Menu System Puts the LAN Administrator in Control." *PC Magazine* (October 27, 1992): p. 61.

Van Name, M. L. and B. Catchings. "Getting Soaked by the Corporate-Data Hose." *PC Week* (February 14, 1994): p. N8.

Waltz, M. "Native NOS Printing: Managers Look for Better NetWare and NT Print Services." *LAN TIMES* (March 28, 1994): pp. 46.8, 46.11, 46.14.

Case Study:
Southeastern State University

LAN managers at Southeastern State University rarely describe their jobs as being a laugh a minute. While there is considerable variation in the size of LANs found on different parts of campus, the administrators of most networks perceive their jobs as being far from routine. For most, it is difficult to anticipate what any day at work will be like. However, most realize that they will spend a good part of their day helping users solve problems they are having with their applications, setting up new user accounts (including Internet accounts), and showing the uninitiated how to logon to the network, access applications, locate files that have been directed to network drives rather than to floppy disk drives, send their work to network printers, and logoff from applications and the network. As you might expect, the time spent on these tasks by network managers typically increases at the beginning (and end) of semesters, especially during the fall semester, when freshmen find their way to the computer labs for the first time.

In spite of the fact that they often find themselves answering the same questions and addressing the same problems hundreds of times during the week, most LAN administrators manage to perform these user-intensive tasks pleasantly. Like anyone, they can get testy and short-tempered when the network is experiencing stability problems or when a student fries a keyboard by spilling a can of soda in it despite the large signs plastered on every wall proclaiming that no food or drinks are allowed in the lab. However, most manage to maintain an even temper in what many other SSU employees consider to be a high-stress job.

Some LANs (such as that in the College of Business Administration—COBA) have expanded from being simply a student computer lab to incorporating all clerical, faculty, and administrative users. The responsibilities and challenges faced by managers of these LANs have increased over time. While the administrators of such networks typically have several student assistants (usually at least one is available at any time of the day), easy work days are still a rarity. Many find it easier to answer the questions posed by freshmen than to try to educate stodgy professors who resent the intrusion of computers into their lives when they are so close to retirement.

However, the real challenges faced in these expanded LANs stem from the fact that workstations are spread throughout the multiple floors and offices of buildings rather than concentrated in a single room. In addition to the large diversity of microcomputers found among faculty, clerical, and administrative offices, the members of the College of Business Administration's formal groups (that is, the college's departments) are not necessarily physically close to one another. While the personnel of the COBA's different departments tend to be grouped close to their administrative and clerical work areas, it is not uncommon to find one or two departmental personnel on each floor and wing of the building, and some even have offices in other nearby buildings.

For budgetary purposes, the Dean and Department Chairs of the COBA would like to see subnetworks created so all personnel associated with a particular department could communicate easily with one another through e-mail, and so that printing needs of department members (at least access to laser printers, which the university cannot afford to put in each staff member's office) could be satisfied at one central location. These subnetworks would include their own servers to store both the general-purpose and special-purpose applications needed by the personnel in each department. The administrator and LAN manager hope to configure these in such a manner that a file server in one department can serve as a backup server for at least one other department.

Currently, every faculty member who wants one has a dot matrix printer attached to the computer in his or her office. However, in most instances, it is necessary to copy files to a diskette and carry the diskette to departmental clerical staff members for printing on laser printers; this happens even if a member of one department has an office next door to the clerical work area for a second department. The laser printers in the clerical areas are attached to individual workstations, and some of the clerical workers' workstations have been configured as nondedicated remote print servers so that they can accept and process print jobs sent through the network by faculty members. The clerical workers whose workstations have been set up in this fashion are not very satisfied with the arrangement. They complain that it really slows down the performance of their workstation, that they are sometimes startled when the laser printer next to them starts to process a print job that they did not initiate, and that when they do initiate a print job, the wait is a lot longer than it used to be.

In addition to wanting to create departmental subnetworks so that computing and printing costs can be cleanly broken out for budgetary purposes, the Dean and Department Chairs of the COBA want to support the activities of interdisciplinary research teams and committees composed of members from the COBA's several departments. Toward this end, and at the urging of the LAN manager, the COBA has recently acquired a site license for Lotus Notes that allows all faculty, clerical, administrative, and users of the workstations in the student labs to be concurrent users.

CASE STUDY QUESTIONS AND PROBLEMS

1. As noted, there is a great diversity among the types (and ages) of workstations found in the LANs in student computer labs. Within most of these labs, both Windows and DOS versions of general-purpose applications are supported (for example, WordPerfect, Lotus 1-2-3, and Paradox). Based on the information and guidelines found in this chapter, what mechanisms should be in place to ensure that student users at particular workstations are prevented from accessing a version of an application that the machine is incapable of running? Also, explain why users of some workstations may have more applications to choose from than others.

2. Discuss the advantages and disadvantages of the manner in which printing operations are handled within the COBA departments at SSU. Describe several alternatives that could better satisfy the printing needs of departmental personnel.

3. Discuss the advantages and disadvantages of creating subnetworks around departments within the COBA as well as the tentative plan that has been devised for doing so.

4. Explain how the COBA network manager's job has changed since the LAN in the student lab was expanded to include all the microcomputers in the COBA building (and some in other buildings). What new complexities and challenges has this expansion probably created? What priority recommendations would you give the LAN manager for addressing problems experienced by COBA personnel and those experienced by students?

5. Explain why Lotus Notes may help COBA administrators support the work of interdisciplinary research teams, cross-disciplinary committees, and other groups that cross traditional departmental boundaries. Also, discuss the impact that the installation of Lotus Notes is likely to have on the operational duties of the LAN manager.

Network Monitoring and Maintenance

CHAPTER OBJECTIVES

After completing this chapter, you will be able to:

◆ Discuss why network management and monitoring is important.

◆ Identify the elements needed to ensure effective network management.

◆ Describe the characteristics of problem tracking and reporting systems.

◆ Describe the tasks performed by network management systems.

◆ Describe how network management systems help network managers detect and correct errors.

◆ Identify several LAN monitoring and management software packages.

◆ Describe the network tuning process and why it is needed.

◆ Describe the capacity planning process and the approaches and tools used in this process.

◆ Identify some of the diagnostic tools and troubleshooting tools used by network personnel.

After a network has been installed, tested, and is operational, network managers are faced with the challenges of day-to-day network management. In Chapter 12, it was noted that the daily activities of LAN managers may include creating new user accounts, directories, and logon scripts; providing training, support, and documentation to users; assigning and modifying access privileges; backing up critical network files, databases, and applications; installing new software and software upgrades; adding, removing, or moving workstations and other devices attached to the network; monitoring the network for problems and fixing them; and performing regular maintenance activities. In this chapter, we will be focusing on the last two sets of functions: network monitoring and maintenance.

◆ WHY ARE NETWORK MONITORING AND MANAGEMENT IMPORTANT?

One of the major responsibilities of network managers is to help users solve problems and keep them productive. This may include creating passwords for new users, recovering or recreating damaged files that users need to perform their jobs, and ensuring that the network is available and reliable.

ENSURING USER ACCESS

One of the priorities of network managers is to ensure that users have access to the network when they need it. Workers are increasingly dependent on computers and computer networks to perform their jobs. If users are not able to access the computing resources needed to do their jobs, their frustration is likely to rise and their productivity to drop. As neither of these outcomes is desirable, network managers work hard to ensure network availability, reliability, and effectiveness.

Network availability and reliability are key determinants of user perceptions of the network. *Network availability* means that all needed network components are operational when a user needs them. As you might expect, users are likely to be frustrated when the hardware or software needed to perform their jobs is not working properly. *Network reliability* refers to the extent that the network functions consistently across time. Needless to say, users are not likely to consider a network to be reliable if it is down too often or for extended periods of time. Users are also not likely to perceive the network as being reliable if response times are inconsistent (vary greatly from one time period to another) or if the system returns different responses to the same user command.

Network availability and reliability are often dependent on the number and quality of network components. As may be observed in Table 13-1, the availability of a single network component is a function of how long, on the average, it can be expected to operate before failing (mean time

Component Availability

$A = MTBF/(MTBF + MTTR)$

where **A** is component availability; **MTBF** is Mean Time Between Failure (the average time period that a component is expected to operate before failing—a measure typically available from the component's manufacturer or vendor); **MTTR** is Mean Time To Repair (the average amount of time required to place a failed component back in service)

Network Availability

$A = a_1 \times a_2 \ldots \times a_i$

where **A** is network availability for a network made of **i** components; a_1 is availability of component 1; a_2 is availability of component 2; a_i is availability of component **i**

Component Reliability

$R(t) = e^{-bt}$

where **R(t)** is the probability that the component will not fail during time period **t**; **e** is the mathematical root *e* (a constant); **b** is the inverse of MTBF for the component

Network Reliability

$R = r_1 \times r_2 \times \ldots r_i$

where **R** is network reliability for a network made up of **i** components; r_1 is the reliability of component 1; r_2 is the reliability of component 2; r_i is the reliability of component **i**

Network Effectiveness

$E = A \times R$

where **E** is overall network effectiveness; **A** is network availability; **R** is network reliability

TABLE 13-1   Calculations used in assessing data communication networks

between failure—MTBF) and how long it takes to repair the component, on the average, after it has failed (mean time to repair—MTTR). Of course, overall network availability assessments should take into account the availability of all network components.

Network reliability is directly related to the availability of its components (both hardware and software availability). As may be seen in Table 13-1, MTBF is used to calculate both network availability and network reliability.

Network availability and reliability can be enhanced in a number of ways, including building fault-tolerance into the system; having spare components available to quickly replace those likely to fail; regularly backing up critical systems; having a well-trained and well-equipped technical support and troubleshooting staff; and through effective network performance monitoring and maintenance programs. By having mechanisms such as these in place, users are likely to find that the network is available when they need it and that they can count on it to function consistently. In short, they are likely to be happy (or at least satisfied) with the network.

A concept closely related to network availability and reliability is *network effectiveness*. Users are likely to perceive the network as being effective (and useful) when it is both available and reliable. A mathematical expression for network effectiveness is depicted in Table 13-1.

One of the goals of network managers should be to maximize user perceptions that the network is effective, useful, and usable. If users do not see the network as having these characteristics, there are likely to be high levels of user frustration with the system, and network managers are likely to be the targets of barrages of user complaints. In addition, without these characteristics, network utilization may be inhibited; users will find other ways to perform their jobs by working around rather than with the network. This, of course, is just the opposite of the effect that network developers intended. High levels of user frustration with a network are likely to reduce workers' use of it.

Users are likely to have acceptable perceptions of the availability, reliability, and effectiveness of networks when network managers have good network monitoring systems that alert them to problems and help them quickly diagnose the causes.

SATISFYING COMPUTING NEEDS COST-EFFECTIVELY

Network managers also strive to provide cost-effective solutions to the telecommunications needs of users. As noted, network managers often have many alternative network options to select from. It is important for them to select wisely, cost-effectively, and, whenever possible, to make selections that will help the organization reduce (or avoid) costs in the long run.

Effective network planning can play an important role in ensuring cost-effective computing and networking solutions. In some instances, the budget may enable network managers to buy components with greater power and capacity than are currently needed; this can accommodate expected growth and may delay the expense of upgrading. The network may also be designed for modular expansion; this may make it possible for the organization to make (less expensive) incremental, rather than wholesale, upgrades to the network. In addition, through good planning, the migration of computing equipment from one location to another may be possible; such planned migration can prolong the useful life of the equipment and allow the organization to get the most from its investment. Examples of planned migration were provided in Chapter 11; these included organizations that transformed their once-centralized mainframes or minicomputers into servers in the client/server networks they developed.

◆ ELEMENTS OF EFFECTIVE NETWORK MONITORING AND MAINTENANCE SYSTEMS

Effective network monitoring and maintenance systems are typically composed of a combination of proactive and reactive elements. Proactive elements include careful planning; the development of standards and procedures; the maintenance of network documentation; the hiring, training, and development of a competent staff; and user training programs. Reactive network management elements may include the use of network monitors (and network management software) to identify problems; well-designed problem-reporting procedures; effective problem diagnosis and analysis procedures; and the availability of appropriate diagnostic and troubleshooting tools.

Before proceeding, it should be noted that network monitoring and maintenance systems are likely to be partially automated and partially manual. For example, network management software often makes it possible to automatically monitor network performance and even to pinpoint where problems are occurring. However, many technical problems (such as replacing defective network interface cards) can be rectified only manually.

NETWORK PLANS, STANDARDS, PROCEDURES, AND DOCUMENTATION

We have repeatedly emphasized the importance of network planning. This is truly one of the keys to successful networks. However, we realize that it is very easy for network managers to get caught up in day-to-day network operations; this can leave little time to develop long-range plans. Also, the dynamic nature of information technology (especially that found in client/server networks) makes long-range network planning difficult. Yet, network plans cannot be overemphasized, because so many things revolve around these plans (including budgeting, downsizing activities, vendor selection, staffing, training, budgeting, and network maintenance and enhancement activities).

Network standards and procedures should be an outgrowth of planning. Standards help to ensure consistency and interoperability among network components; these help the organization by specifying minimally acceptable implementation guidelines and performance levels. Procedures help ensure consistency in operating and maintaining the network; procedures are also helpful in resolving network problems.

Network documentation can also play an important role in effective network management. Good, up-to-date network documentation can facilitate the jobs of network managers and technicians. As you may observe in Table 13-2, the documentation for a network is likely to include wiring diagrams, information on hardware and software components, vendor contracts, network standards, diagnostic procedures, disaster recovery plans, and a variety of other items.

In short, effective network management includes ensuring that an appropriate mix of both automated and manual proactive/reactive elements are in place. It is also important to have network plans, appropriate

◆ Network graphs and maps for each WAN, backbone network, or LAN

◆ Wiring diagrams for all computer networks (including locations of wiring closets, conduits, and so on)

◆ A listing of all internetwork connections

◆ Backup, redundancy, and security mechanisms (and their locations)

◆ Serial number inventory of all network components

◆ The organization's software and hardware standards

◆ Preventive maintenance guidelines and schedules

◆ Equipment and application migration and movement plans

◆ Disaster and recovery plans and procedures (including contingency plans)

◆ Vendor contracts and license agreements

◆ Diagnostic procedures and guidelines for each component and type of problem (including troubleshooting manuals)

◆ Operations manuals for network personnel

◆ Copies of all local, state, and federal telecommunications regulations with which the organization must comply

◆ Hardware operation manuals

◆ Software documentation manuals

◆ Maintenance history logs

◆ Problem reporting and tracking equipment and procedures

◆ Vendor maintenance records (including MTBF and MTTR)

◆ List of all dial-in connections (including telephone numbers)

◆ List of all network management software and diagnostic tools

◆ User access rights policies

◆ User passwords, e-mail addresses, workstation addresses, and so on

◆ List of all hardware and software resources in each network

TABLE 13-2   Items included in network documentation

standards and procedures, and up-to-date network documentation. An appropriately trained and equipped staff that possesses good diagnostic and technical skills will also help to ensure network availability, reliability, and effectiveness.

ORGANIZING NETWORK MONITORING AND MAINTENANCE FUNCTIONS

A network manager's job is likely to be easier when the elements of network management and maintenance systems are appropriately organized. This often involves implementing a network control center and developing appropriate problem-reporting and tracking systems.

Network Control Centers

A *network control center* is responsible for monitoring the network and initiating corrective actions after problems have been detected. In a LAN environment, the network control center may be the LAN manager's office. In distributed processing systems, there may be multiple control centers.

Network control centers are typically equipped with one or more dedicated computers running network control software. This software collects information such as error rates, data transfer rates, and retransmission rates. The data captured can be stored in a network database and later analyzed; for example, trend analyses can be performed that enable network managers to identify correctable degradation in network performance. This data is often gathered via network monitors capable of probing individual network components for problems and utilization statistics.

A network control center may also possess an online problem reporting system used to retrieve information about similar problems encountered in the past and through which information about a current problem can be saved. Such systems can help network managers and technicians avoid solving the same problem repeatedly. To help resolve problems, network documentation is typically stored at the control center; this is also where the documentation is maintained.

Problem Reporting and Tracking Systems

One of the most important functions performed by the control center involves recording the incidence of problems and their resolution. This may require creating and implementing problem reporting procedures.

Network managers often develop and publish problem reporting procedures and distribute them to users. These procedures typically specify the information that users should be ready to provide when reporting a problem and to whom the problem should be reported. When users are familiar with these procedures, network managers and technicians are likely to be able to rectify problems faster and more easily. Hence, it is important to write these procedures in language that users can understand.

Table 13-3 lists the information likely to be collected when users report network problems. After the problem is resolved, a description of what was done to solve the problem is added to the original problem report. All the information about the problem and its resolution should be stored in a database; the information is then readily available to network managers and technicians when similar problems are reported by users in the future. Collecting such information also assists in *problem tracking* (monitoring the status of problem resolutions) and in assembling statistics about network problems (including network availability, network reliability, and network effectiveness).

In general, all the information gathered while solving network problems should be documented. Problem reporting and tracking software can automate the gathering and storage of problem data for network managers. Such software also provides network managers and technicians with a fast, efficient mechanism for identifying similar problems that have occurred in the past (and how they were solved) and for reviewing the status

TABLE 13-3
Information
collected about
network
problems

◆   The person who reported the problem

◆   The time and date the problem was first noticed

◆   The location of the problem

◆   A detailed description of the problem

◆   Why the problem occurred

◆   How the problem occurred

◆   Other users involved in or affected by the problem

◆   The severity of the problem

of problems yet to be fully resolved. This type of software can be a valuable network administration tool and an important component of an overall network monitoring and maintenance system.

A final point that must be made about control centers and problem reporting and tracking systems is the fact that not all of the identified problems can be fixed by network personnel. Network managers and technicians must be able to determine when reported problems are beyond their ability to rectify. Being able to make this determination quickly will facilitate the resolution of the problem.

Most often, external help with problem resolution comes from vendors. You may recall from Chapter 9 that good contracts and agreements with vendors usually specify that their network components are expected to work properly; if they don't, it is the responsibility of the vendors to correct the problem and to replace unusable components with ones that work correctly. If a reported problem is part of the area that the vendor is responsible for, information about the problem is relayed to them. When the organization has a service contract with the vendor, the terms of the service contract will dictate the degree to which the vendor will be involved in the resolution of the problem (and often, how rapidly the vendor will get a service technician to the site).

◆ NETWORK MANAGEMENT SYSTEMS

Problem reporting and tracking systems are important because they provide network managers with a way to document network problems discovered and reported by users. However, reacting to user complaints is less desirable than avoiding such complaints by having systems in place that alert network managers and technicians to potential problems before users become aware of them. Correcting problems before they affect users is likely to have a positive impact on the extent to which workers perceive the network to be available, reliable, and effective.

A ***network management system (NMS)*** can be described as a combination of hardware and software used to monitor and administer the network. In essence, an NMS monitors network performance, captures data needed to calculate network statistics, identifies network performance areas that exceed normal or desirable levels, and alerts network managers and technicians when problems are detected. Some NMSs can help network managers troubleshoot and correct problems by suggesting what diagnostic tests or checks should be performed or by recommending how problems should be corrected. A summary of the features and functions of NMSs is provided in Table 13-4. Because NMSs can help network personnel detect and correct potential problems before they become critical, they can be important network administration tools.

Figure 13-1 illustrates how a network management system may be configured. As may be observed in the figure, the key components of an NMS include monitors, filters, alerts/alarms, a report generator, a network database, a network management workstation/console, and the network management software itself.

The monitors included in NMSs are dispersed among network components. In some NMSs, monitors are called *agents*. Monitors may be found in dedicated hardware and software, in intelligent microcomputer device controllers, and in intelligent network devices such as intelligent hubs, bridges, or routers. A monitor continually gathers information from the devices and components for which it is responsible; the data collected will vary from one type of device or component to another. For example, the monitor for an

---

- Network maps and topology graphs
- Workstation status monitoring
- Server status monitoring
- Shared equipment monitoring (printers, disk drives, and so on)
- Communications media monitoring
- Virus detection and eradication
- Hacker/unauthorized intruder detection
- User logon statistics
- Software utilization statistics
- Metered use of software to ensure that license agreements are not violated
- Message traffic statistics
- Error statistics
- Event logging
- Alerts and alarms
- Trend analysis of network statistics
- Graphical presentation of network statistics and trend lines
- Problem diagnosis and solution recommendations
- Account use and maintenance activities
- Workstation configuration information
- User security and access levels
- Software configuration software
- Menu generators for developing application and function menus

TABLE 13-4 Features and functions included in network management systems (NMSs)

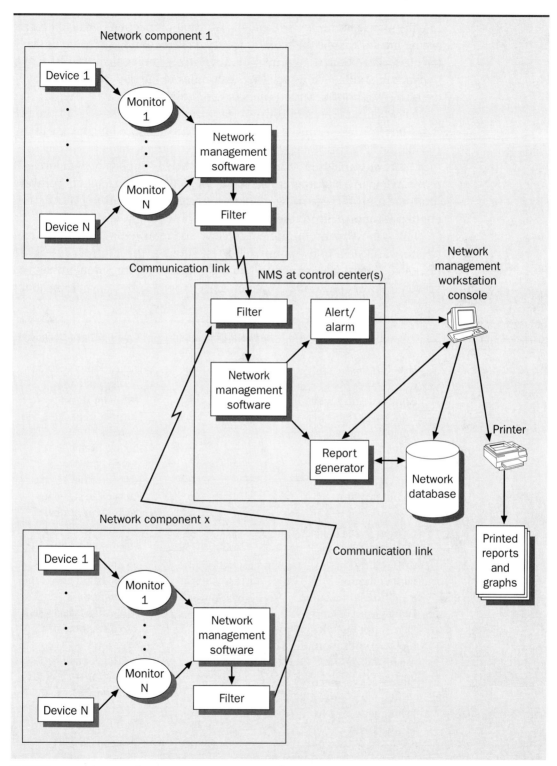

FIGURE 13-1 A network management system configuration

intelligent disk controller will collect information needed to calculate disk access statistics, while the monitor for an intelligent bridge or router is likely to collect data about the number of message packets transmitted from one network to another. Table 13-5 lists examples of the data that may be collected by monitors for various network components.

Network management software captures the data passed to it from the monitors; the data captured is often stored in a database for later use in the calculation of network statistics. As this data comes in, the network management software analyzes it for irregularities such as unusually high levels of errors. When irregularities are detected, the network management software modules found in the network components send warning information about the abnormality to the network control center by way of a filter.

Filters in network components receive and reformat warning messages and forward them to the control centers. Filters at the control centers are responsible for detecting redundant messages about the same problem to prevent network personnel from being overwhelmed by information about

Network Component	Types of Data Collected	Network Component	Types of Data Collected

Workstations	User logons
	Current user
	Current status (active/
	    inactive)
	Current application(s)
	    in use
	Messages generated
	Messages received
	CPU utilization
	Disk utilization
		Communications
		    media
			Utilization percentage
			Number of packets
			    transmitted
			Number of packets with
			    errors
			Busiest nodes/
			    workstations
			Highest transfer rate
			Packet size statistics
			Packet type statistics

File servers	Current status (active/
	    inactive)
	CPU utilization
	Disk utilization
	Application utilization
	Disk space available
	Disk swapping rates
	Memory utilization
	Cache utilization
		Internetworking
		    devices (such as
		    bridges and
		    routers)
			Collision rates
			Number of lost tokens
			Local traffic volumes
			Traffic volumes to other
			    networks
			Types of messages sent
			    to other networks
			Error rates
			Routing statistics
			Path failures
			Path changes

Printer servers	Status (active/inactive) Print queue status Print queue lengths Print activity		

TABLE 13-5   Data collected for NMSs by various monitors

the same problem. As today's client/server networks often consist of diverse set of hardware and software from multiple vendors, it is often necessary for filters to reformat warning messages to the forms needed by the control centers.

The network management software modules at the control centers analyze incoming data and display it on the network manager's workstation/console monitor. Most NMSs make it possible to display the data in a variety of formats, including text-based tables and graphs (including network maps). Report generator modules of the NMSs play an important role in these display options. Other report generator modules enable network managers to analyze data stored in the network database; these enable a variety of network statistics to be calculated and to graphically display the results of trend analyses for network statistics. Such statistics and graphs can play an important role in network tuning and capacity planning (these concepts are discussed a little later in the chapter).

Alerts and alarms inform network personnel when serious irregularities in network performance occur. Alerts are used to inform network administrators of important, but less critical, network performance problems than alarms. An *alert* is used to indicate a potential problem situation (for example, an excessive number of transmission errors). Text messages onscreen or changes in screen colors are often used to signify that important abnormalities are occurring; for example, green may be used when performance is within normal ranges, yellow for when normal levels are being slightly exceeded (or exceeded frequently within a specific time frame), and red when performance levels significantly exceed normal limits for extended periods of time.

An *alarm* is used to inform network personnel of very important performance irregularities (such as a server crash) that demand immediate attention and correction. When such conditions exist, the NMS may start to issue audio signals through the network management workstation or console, may cause the workstation monitor or console screen to flash, and may even initiate a call to network personnel pagers and the transmission of fax messages. All of these mechanisms are intended to attract immediate human attention to the problem so that it can be corrected as quickly as possible.

EXAMPLES OF NETWORK MANAGEMENT SYSTEM PRODUCTS

Table 13-6 includes examples of NMS products used in both LANs and larger networks. NMSs for larger networks (such as client/server networks consisting of diverse computing platforms, backbone networks, WANs, and interconnected LANs and WANs) will be discussed more fully in Chapter 15.

LAN Product	Vendor
ARC Monitor	Brightworks Development, Incorporated
Close-up/LAN	Norton-Lambert Corporation
LANalyzer	Excelan
LAN Probe	Hewlett-Packard
LANVision	Triticom
NetWare Management System	Novell, Incorporated
SMT for FDDI	Vendors with FDDI token ring LAN products
Watchdog	Network General Corporation

Larger Network Products	Vendor
ACCUMASTER Integrator	AT&T
CMIP	Several major vendors including IBM
DECmcc	Digital Equipment Corporation
Net/Master	Cincom
NetView	IBM
SNMP; SNMP.2	IBM, Hewlett-Packard, Sun Microsystems, and others
Spectrum Services	Pacific Bell and other Bell operating companies

TABLE 13-6
NMS products for LANs and larger networks

LANalyzer is one of the more widely used LAN NMSs; this product is often used to monitor and manage LANs using Novell NetWare. Figure 13-2 is a representative LANalyzer screen for monitoring network performance;

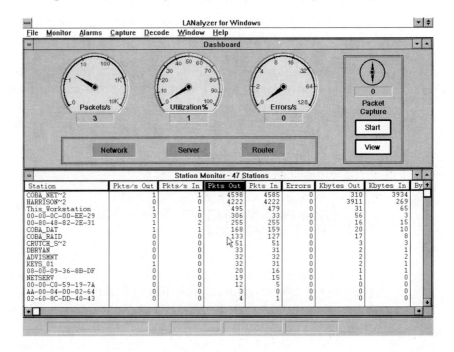

FIGURE 13-2
Monitoring network performance

as you can see, LANalyzer makes it possible to monitor (in real time) the number of data packets being transmitted over the medium, the number of transmission errors, and the overall network utilization. The lower part of the screen in Figure 13-2 shows that the activity of individual workstations can also be monitored and even sorted from most to least busy. Figures 13-3a and 13-3b illustrate the process of setting alarm thresholds in LANalyzer, and Figure 13-4 shows one of the menus used in LANalyzer to view long-term network trends.

The screens depicted in Figures 13-2 through 13-4 really only scratch the surface of all the capabilities of a LANalyzer and other LAN NMSs. Of course, LANalyzer has very good alarm and alert capabilities including color changes, sound output, and so on. Also, LANalyzer can be used to locate individual workstations involved in unusually high numbers of error transmissions; this facilitates troubleshooting and problem correction. In addition, LANalyzer (like most other NMSs) can generate a wide array of network statistics to help network managers in network tuning and capacity planning activities.

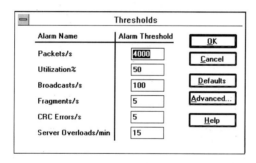

FIGURE 13-3a
Setting alarm
thresholds

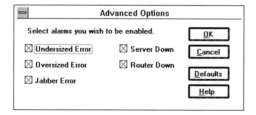

FIGURE 13-3b
Advanced alarm
options

FIGURE 13-4
Monitoring
trends

Prices for LANalyzer and other LAN NMS products vary according to the size of the LAN (for a moderately sized LAN, the price may be in the $1,000 range). Network managers should also be aware that an NMS (especially one with graphical interfaces such as LANalyzer) may run efficiently only on dedicated high-end microcomputers (such as those with 8 or more megabytes of RAM, 50 MHz or greater clock speeds, SVGA (or higher) video, and high capacity hard-disk systems). These hardware costs should be factored into the equation for an NMS.

LAN MONITORING AND MANAGEMENT CAPABILITIES OF LAN OSs

While many early versions of LAN operating systems included few, if any, LAN monitoring and management capabilities, some of the NMS functions mentioned previously are now being built into some of the popular LAN NOSs. For example, IBM's LAN Server contains modules for monitoring IBM Token Ring Networks. This product is capable of monitoring specified nodes including workstations, file servers, print servers, and gateways, and it will alert network managers when problems occur. In internetworking environments, LAN Server can interface with NetView, IBM's integrated network management program (NetView is discussed more fully in Chapter 15).

TRENDS IN NETWORK MANAGEMENT SYSTEMS

One of the important trends in NMSs is toward user-friendly GUIs. Such systems can represent individual workstations and network nodes as icons that are labeled with network addresses or descriptive names (including user names). Such interfaces allow network managers to view network utilization and traffic in real time with actual packet rates and utilization statistics for any location on the network. In addition, several packages enable network managers to set thresholds for network statistics; whenever these are exceeded, the event is recorded with a time stamp in an *event log file*—all recorded events can be displayed (or printed) on network statistics graphs.

Table 13-7, adapted from an article by Barbara Bochenski (see references), lists some of the features that network managers see as being desirable in NMSs. This table is somewhat of a wish list, as it includes capabilities not yet found in most NMSs; however, it may give you some idea as to how NMSs may evolve and add functionality in the years ahead. You should note that Table 13-7 applies to both LAN-oriented and larger network NMSs such as NetView, Net/Master, and ACCUMASTER Integrator.

- ◆ The ability to store network management data in relational databases with the enhanced reporting capabilities that relational data structures can provide

- ◆ The ability to view an entire client/server network from one console, not just pieces of it or just the part monitored by that console

- ◆ The ability to easily distribute network management functions in a client/server network across a variety of machines without sacrificing performance speed and without additional licensing fees

- ◆ The ability to schedule and perform simultaneous back ups on diverse servers and hosts in client/server networks

- ◆ The ability to share information easily among network management tools

- ◆ The development of a standard for network management databases

- ◆ The ability to manage wiring hubs and routers from one NMS console or workstation

- ◆ The ability to monitor the status of all workstations and servers in a diverse client/server network

- ◆ New functions such as the ability to electronically distribute software, to configure software and change its management, and to manage software licensing

TABLE 13-7
Ideal NMS
features

◆ NETWORK TUNING

Network tuning is concerned with keeping the performance level of a network at acceptable levels. This involves analyzing network statistics (usually those captured by NMSs) and utilizing these statistics to ensure that the network maintains satisfactory performance levels.

Network tuning is necessary because network configurations and use typically change over time. For example, new workstations or higher resolution monitors may be added, new (or more sophisticated) versions of application software that demand more memory may be installed, and internetworking devices (such as bridges or routers) may be added. Any of these changes has the potential to degrade normal network performance levels by placing greater demands on servers, increasing the volume of message packets transmitted over the communication medium, and so on. If multiple changes are made quickly, users may experience reduced response time. And, if excessive demands are placed on the system, network availability levels may decrease.

By analyzing network statistics, network managers may be able to identify trends that indicate that network performance levels are declining toward unacceptable levels. For example, if there are sharp rises in error rates (relative to those observed in the past) or in the number of alarms or alerts in a given time period, network managers may conclude that potential problems are brewing and that corrective intervention is needed.

Balancing a network is one of the approaches that network managers use to tune a network; balancing often involves using existing network resources in a different configuration. For example, if bottlenecks are occurring as users request applications from a server, moving some of the applications to another server might ease the bottleneck and improve response time. A similar approach can be used if bottlenecks occur in printing operations (perhaps by changing the printer defaults for some users so that their work is routed to alternative printers rather than the ones experiencing the bottlenecks). Another example of tuning via reconfiguration is the installation of copies of a new version of an application software package with enhanced GUIs on workstation hard disks to reduce the traffic on the communication medium. Since newer, graphically-oriented versions of application packages are often larger (in terms of megabytes of storage requirements), by installing copies of these on at least some of the workstations, traffic on the communication medium may once again fall into acceptable ranges.

Another approach to tuning the network is to upgrade the hardware. For example, if a new version of an application software package recently installed on the network runs more slowly than the previous version, it may be necessary to increase the RAM in workstations, servers, or both. Adding more memory to printers can also address printer performance problems caused by slow processing of large print files.

Table 13-8 summarizes the steps in the network tuning process. An important thing to remember about this process is that once potential solutions to a problem have been identified, they should be implemented one at a time rather than in combination. If performance improves after making several changes at once, it may be difficult to determine which change is most responsible for enhancing network performance. Also, if system performance deteriorates after making a combination of changes, it can be equally difficult to determine which change was the main culprit. By implementing solutions one at a time, network managers and technicians can more accurately assess the performance impacts of each, positive or negative.

Network managers and experienced technicians should play key roles in network tuning. These network personnel are likely to have the greatest and most technically detailed knowledge of the network's computing equipment and software (and how network, hardware, and software interact). Such in-depth understanding of the network's computing resources is needed to accurately interpret trends in network statistics and changes in network performance levels. Over time, experienced network managers

1. Monitor system performance; capture network performance data.

2. Analyze network statistics; identify trends and potential problems.

3. When a (potential) problem is identified, determine its likely causes and the different ways it can be solved.

4. Choose the best solution for the problem. "Best" may mean the one with the greatest potential gain in network performance; the most cost-effective solution (biggest gain per dollar spent); or the quickest/easiest solution to implement.

5. Implement the chosen solution; measure its impact on system performance; document its effectiveness.

6. Repeat steps 4 and 5 (as needed) until acceptable performance levels are obtained.

7. Repeat steps 1–6 as needed to maintain network performance within acceptable ranges, especially after any significant change in network resources or utilization.

TABLE 13-8
The network tuning process

and technicians become more adept at using NMSs and at identifying and implementing solutions that will maintain network performance within acceptable levels.

◆ CAPACITY PLANNING

Capacity planning involves forecasting network utilization and workloads and developing plans to ensure that the network will be able to support anticipated performance demands. Analyzing network statistics captured by NMSs can help network managers forecast future workloads for a single LAN. However, this value of NMS data is likely to be limited if the organization decides to downsize applications from mainframes to existing LANs, to interconnect LANs, to add a significant number of new users to an existing LAN, or to otherwise make major changes to existing networks.

A number of tools and approaches are available for assisting network managers in capacity planning. These include the development of forecasts derived from network utilization statistics, the use of simulation models, and the use of workload generators.

FORECASTING

As mentioned, NMSs and network management modules included in the later versions of LAN operating systems often can perform trend analyses on network statistics and graphically display the results of these analyses. The trend lines that are generated can help network managers see historical trends in LAN utilization and performance. Some network management systems include forecasting modules that use the historical data to project trends out into the future; others enable network managers to transfer network data to spreadsheets or statistical packages so that forecasts can be done. Forecasting network utilization trends enables the network manager to anticipate when user performance demands are likely to reach (and exceed) current computing capacity levels.

NETWORK SIMULATION SOFTWARE

Another capacity planning approach involves the use of simulation models. Simulation generally involves modeling an object or system and then using the model to anticipate how it would behave in different operating conditions. In brief, simulation makes it possible to perform rigorous tests on a model (rather than on a real object or system) to test its ability to withstand the conditions that the real object or system is likely to face. The models may be physical (as is often the case when a prototype is built for a new car or airplane design) or logical (for example, a car or plane can be designed with a CAD package). Similarly, the tests performed on the models can be either physical (such as braking or crash tests for cars, aerodynamic tests in a wind tunnel for planes) or logical (such as computer simulation or the use of CAE test modules).

The majority of models and simulation tests used by network managers are logical rather than physical. The *network simulation software* used for capacity planning enables network managers to change network configurations of hardware or software, modify resource utilization rates and message traffic, and run tests to determine how network performance would be affected by these and other changes. Network simulation software enables network managers to estimate server utilization, disk utilization, traffic over the communication medium, message and print queue lengths, message bottlenecks, and response times to user requests for access to computing resources under varying operating conditions. As with forecasting approaches, network simulation software can help network managers determine the types of performance demands that can exceed current network computing capacities. In addition, it can help network managers to identify components that may be under- or over-utilized for a particular configuration.

Network simulation software can vary in functionality and sophistication. Low-end packages may require users to enter a set of data for a particular configuration in order to generate a report concerning expected performance levels under those conditions; to generate a report for a second configuration, a new set of network parameters will have to be entered. While this can be a time-consuming approach, it does allow network managers to see the estimated performance effect(s) of making one change at a time. More sophisticated network simulation packages are likely to draw the data used to develop performance estimates for alternative network configurations from the NMS's database of network statistics.

The value of network simulation tools can be seen in their ability to anticipate how a new network configuration is likely to perform before implementing it. They also allow network managers to compare anticipated performance levels of alternative configurations; this may help managers select configurations that are most likely to satisfy future computing needs.

WORKLOAD GENERATORS

Workload generators are used to estimate transaction loads (workloads) for an existing (or proposed) network configuration. As was noted in Chapter 8, like capacity planning tools, workload generators can play an important role in network planning. In addition, these can be used in conjunction with a performance monitor to perform stress tests. In network environments, *stress testing* typically involves placing increasingly higher performance demands on network components until the network crashes; workload generators generate the messages and network traffic needed to imitate the increasing performance demands. Workload generators can also be used to determine if an NMS's performance monitors, alarms, and alerts are functioning properly. As workloads increase, the performance monitor should progress from alarm status to alert status. In addition, if a test network for a proposed configuration is built, workload generators can be used to determine the validity of the network performance estimates made by network simulation software.

Workload generators can only be used on an existing network. Hence, to use it to determine the performance of a proposed network configuration, a test network has to be implemented. However, for either current or test networks, workload generators enable network managers to see first hand how network performance will change as workloads increase; they also make it possible for network managers to estimate when network computing capacities will be exceeded.

◆ DIAGNOSTIC TOOLS

Network management software can be used to help identify the likely locations and sources of network problems. Such diagnostic and troubleshooting capabilities are valuable aspects of NMSs. However, there is a variety of other hardware and software tools that can help network personnel identify and correct network problems and maintain desirable network performance levels. Some of the more important tools used by LAN managers are breakout boxes, bit error rate testers, protocol analyzers, remote control software, pocket LAN adapters, and cable testers.

BREAKOUT BOXES

A *breakout box* helps LAN managers and technicians determine whether two data communications devices are correctly connected. It is often used for testing the functionality of the individual pins in connector plugs (for example, RS232 interfaces) and for determining which pins do what.

BIT ERROR RATE TESTERS

Bit error rate testers (BERTs) are used to determine if data is being transmitted reliably over a communications link. This is done by sending and receiving specific bit patterns and data packets over the link.

PROTOCOL ANALYZERS

Protocol analyzers capture and analyze individual data packets sent over the network; they also provide network managers with a variety of valuable statistics. Protocol analyzers can be used to test individual workstations and communications channels. They can also be used to detect different types of errors, to generate network traffic pattern statistics, and to analyze the contents of individual message packets.

The Sniffer, a Network General Corporation product, is an example of a protocol analyzer found in some LANs. It is essentially a microcomputer system attached to the network which monitors traffic on the network. The Sniffer can graphically display total traffic in terms of kilobytes per second, frames per second, or as a percentage of available bandwidth. This program can graphically display the contents of individual frames transmitted from a single workstation and can keep count of the extent to which individual workstations are contributing to network traffic and errors.

Some network management system products (including LANalyzer) also include protocol analysis utilities. Figure 13-5 shows a screen from LANalyzer used to analyze the format and content of data packets.

The statistical information that can be captured and displayed by protocol analyzers such as the Sniffer can be invaluable to network managers. By helping network managers monitor network traffic trends, protocol analyzers can play an important role in capacity and network planning.

REMOTE CONTROL SOFTWARE

Remote control software can be used by network managers to help diagnose and solve user problems from a remote workstation. Many software

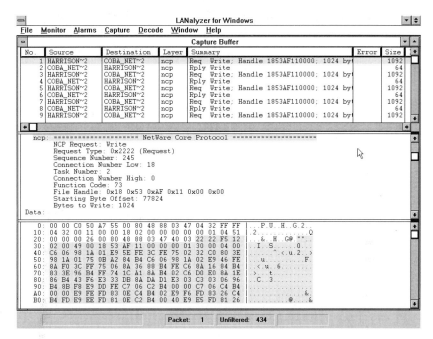

FIGURE 13-5
Analysis of a data packet

vendors use such software to troubleshoot user problems over the telephone and to reduce the expense involved with traveling to user locations to solve routine problems.

The key characteristics of remote control software are its abilities to view screens displayed on remote monitors and to enter commands remotely. Essentially, a network manager can remotely take control of a particular workstation (perhaps from the control center or from a laptop while traveling to a distant location). By being able to take control, the network manager can run diagnostics and even correct problems without having physical contact with the user's workstation. By watching the screen changes, attentive users may be able to watch what the network manager does and learn how to correct similar problems in the future on their own.

POCKET LAN ADAPTERS

A pocket LAN adapter makes it possible to establish a network interface for a workstation through the workstation's serial or parallel communication port(s). This essentially bypasses the workstation's existing adapter and helps the network staff member determine if the existing adapter is the cause of the problems. If problems still exist after establishing a connection to the network through the pocket LAN adapter, it is possible that the workstation's network interface card (NIC) is the source of the problem.

CABLE TESTERS

Wiring problems are a common cause of resource request failures in LAN environments. Because of this, the network management toolkit often contains a cable tester. Essentially, a cable tester generates a signal on the medium and monitors its transmission over the communications medium. Cable testers can detect short circuits or breaks in the cable and often can determine the location of the problem to within a few inches. Needless to say, this can be a valuable tool in identifying and quickly correcting wiring problems.

OTHER DIAGNOSTIC AND TEST EQUIPMENT

Table 13-9 summarizes some of the other diagnostic and test equipment likely to be found in the network management and maintenance toolbox. These, the previously discussed types of diagnostic and testing tools, and a variety of other tools and equipment beyond the scope of this book are available from numerous vendors including the Specialized Products Company, Jensen Tools, Inc., and the Blackbox Corporation.

◆ SUMMARY

One of the major responsibilities of network managers is to help users solve problems and to keep them productive. A priority is to ensure that users have access to the network when they need it. When users are not

Tool	Function
Patch panel	Provides an electrical connection to all parts of the network and centralized access to each network communication circuit; allows a spare circuit to be substituted for a failed circuit
Handheld test sets	Can be inserted between two network devices to test voltages and to send/receive test patterns of bits in order to isolate errors; can also be used to determine problems with RS232 connector cable
BKERT	A block error rate tester (BKERT) calculates the percentage of blocks received by a network device that contain at least one error. It can help locate upstream sources of transmission errors.
Fiber identifier	Used to locate nonworking fibers in fiber optic networks
Self-testing modems	Used to ensure that a workstation's (or remote laptop's) modem is functioning correctly

TABLE 13-9

Some diagnostic tools used to troubleshoot and correct problems

able to access the computing resources that they need to do their jobs, they are likely to be frustrated and nonproductive. To avoid these outcomes, network managers work hard to ensure network availability, reliability, and effectiveness.

Network availability means that network components needed by users are operational when they need them. Network reliability refers to the extent that the network functions consistently over time. Both network availability and reliability are dependent on the number and quality of network components. Both can be enhanced in a number of ways, including building fault-tolerance into the system; having spare components available to replace those that fail; regularly backing up critical systems; having a well-trained and well-equipped network staff; and through effective network performance monitoring and maintenance programs. Users are likely to feel that the network is effective if it is available and reliable.

Effective network monitoring and maintenance systems are a combination of both automated and manual proactive and reactive elements. Some of the proactive elements are careful planning; developing standards and procedures; maintaining network documentation; the hiring, training, and development of competent network personnel; and user training programs. Reactive network management elements include the use of network management software to identify problems; well-designed problem reporting procedures; effective problem diagnosis and analysis procedures; and readily available diagnostic and troubleshooting tools.

Network standards and procedures are an outgrowth of planning. Standards help ensure consistency and interoperability among network components by specifying minimally acceptable implementation guidelines and performance levels. Procedures help ensure consistency in network operations and maintenance; these also are helpful in resolving network problems. Good, up-to-date network documentation can also facilitate the job of a network manager.

A network manager's job is likely to be easier when the elements of network management and maintenance systems are appropriately organized. This often involves implementing a network control center and developing appropriate problem reporting and tracking systems.

A network control center is responsible for monitoring the network and initiating corrective actions after problems have been detected; in a LAN environment, the network control center may be the LAN manager's office or desk. A network control center is usually equipped with one or more dedicated computers running network management/control software. This software collects information such as error rates, data transfer rates, and retransmission rates, and stores the data in a network database for analysis. Network monitors capable of probing individual network components for problems and utilization statistics are used to capture the data. A network control center may also possess an online problem reporting system than can be used to retrieve information about similar problems encountered in the past so that current problems can be resolved quickly.

One of the most important types of software used by network managers is known generically as a network management system (NMS). An NMS is a combination of hardware and software used to monitor and administer the network. An NMS monitors network performance, captures data needed to calculate network statistics, identifies network performance areas that exceed normal or desirable levels, and alerts network managers and technicians when problems are detected. Some NMSs further help network managers troubleshoot and correct problems by suggesting what diagnostic tests or checks should be performed and recommending how problems should be corrected.

Alerts and alarms are very important parts of a network management system, because they inform network personnel when serious irregularities in network performance occur. NMS software packages for both LANs and larger networks are available from a variety of vendors. In addition, network management functions are incorporated into many of the recent versions of network operating systems.

Network management software can play an important role in both network tuning and capacity planning. Network tuning is concerned with keeping the performance level of a network at acceptable levels; this involves analyzing network statistics to ensure that the network maintains satisfactory performance levels. Network tuning is necessary because network configurations and use typically change over time. Balancing is one of the approaches that network managers use to tune a network; this often involves using existing network resources in a different configuration. Capacity planning involves forecasting network utilization and workloads and developing plans to ensure that the network will be able to support future performance demands.

Analyzing network statistics captured by NMSs can assist network managers in forecasting future workloads. Network simulation software and workload generators are two other tools that may be used by network managers for capacity planning.

To ensure network availability, reliability, and effectiveness, it is also important for network personnel to have ready access to appropriate diagnostic and troubleshooting tools such as breakout boxes, bit error rate testers, protocol analyzers, remote control software, pocket LAN adapters, and cable testers. A breakout box helps determine whether two data communication devices are correctly connected. Bit error rate testers (BERTs) are used to determine if data is being transmitted reliably over a communications link. Protocol analyzers capture and analyze individual data packets sent over the network; they also provide network managers with a variety of valuable statistics. Remote control software can be used by network managers to help diagnose and solve user problems from a remote workstation. A pocket LAN adapter makes it possible to establish a network interface for a workstation through the workstation's serial or parallel communications port(s) that bypasses the workstation's existing adapter and helps determine if the existing adapter is the cause of the problems. A cable tester generates a signal on the communications medium and monitors its transmission over the medium; these can detect short circuits or breaks in the cable.

✳ KEY TERMS

Balancing

Bit error rate tester (BERT)

Breakout box

Capacity planning

Network availability

Network control center

Network effectiveness

Network management system (NMS)

Network reliability

Network simulation software

Network tuning

Problem tracking

Protocol analyzer

Remote control software

Workload generator

✳ REVIEW QUESTIONS

1. Why are network monitoring and maintenance important?
2. What are the differences among network availability, network reliability, and network effectiveness?
3. Identify examples of both proactive and reactive network management practices.
4. Identify examples of automated and manual network management processes.
5. What are the differences between network standards and procedures? How are standards and procedures related to network planning?
6. What is network documentation? What is likely to be included in the documentation for a network?
7. What is a network control center? What equipment/software is likely to be found in a network control center?
8. What characterizes problem reporting and tracking systems?
9. Briefly describe the functions performed by each of the major components of a network management system.
10. What are the differences between alerts and alarms?

11. How can network management systems help network personnel identify and resolve network problems?
12. Identify some of the network management software products for LANs.
13. What types of network management functions are likely to be included in current versions of network operating systems?
14. What are some of the trends in network management software packages?
15. What is network tuning? Why is network tuning needed? How are balancing and hardware upgrades related to network tuning?
16. Identify each of the steps in the network tuning process.
17. What is capacity planning? Why is it important? Briefly discuss each of the major tools that may be used in the capacity planning process.
18. Why is it important for network managers to have diagnostic and troubleshooting tools readily available?
19. What are the differences among breakout boxes, bit error rate testers, protocol analyzers, remote control software, pocket LAN adapters, and cable testers? Identify some other troubleshooting and diagnostic tools that network personnel may use.

✳ DISCUSSION QUESTIONS

1. How may day-to-day network management and operations make it difficult for network administrators to engage in network planning? Describe several approaches that you feel would help ensure that network managers have sufficient time to both plan for the future and manage daily network operations.
2. Why is it important for network managers and technicians to have up-to-date network documentation? How can having up-to-date network documentation help identify and resolve problems? What steps can be taken to ensure that network documentation is properly maintained?
3. Discuss how a database containing reported network problems and their resolution can help network managers rectify current problems in a timely manner.
4. Why is it important for network managers to be able to determine who (a vendor or the network staff) is responsible for solving a problem reported by users? How is this determination related to license agreements? Should network managers attempt to resolve problems for which a vendor is responsible? Why or why not?
5. Suppose your organization has standardized on network hardware and software. Also, suppose that a particular user insists on using a type of workstation and software that your organization does not support. If the user has a problem, how much support should the network staff provide?

6. Discuss why the most important factor in network tuning is the competence and experience of network personnel.

✳ PROBLEMS AND EXERCISES

1. If a network interface card (NIC) has an MTBF of 30,000 hours and an MTTR of 2 hours, what is the availability of this component?
2. Suppose a LAN is made up of the following four key components along with their corresponding MTBFs and MTTRs:

Component	MTBF (in hours)	MTTR (in hours)
File/Print Server	40,000	10 hours
Workstation	75,000	12 hours
NIC	30,000	2 hours
Cable	90,000	24 hours

   Calculate the network availability.
3. If network availability is .99 and if network reliability is .87, what is network effectiveness?
4. Use *Faulkner Retrieval Services*, McGraw-Hill's *Data Pro Reports*, or information supplied directly from vendors to find the MTBFs for three different network components.
5. Use *Faulkner Retrieval Services*, McGraw-Hill's *Data Pro Reports*, or information supplied directly from vendors to find the MTBFs for three different Ethernet NICs (from three different vendors).
6. Find a current article comparing network management systems for LANs in trade publications such as *Network World, Communications Week, LAN Magazine,* or *Computerworld*. Summarize the article's contents in a written report or presentation to your class.
7. Invite a local network administrator to speak to your class about the network management system used in his or her organization. Be sure to ask why that system was chosen, what its strengths are, what its weaknesses are, the types of monitors it includes, and its alert/alarm and report generation capabilities.
8. Invite a local network administrator to speak to your class about the approaches that his or her organization uses for network tuning and capacity planning.
9. Invite a local network administrator to speak to your class about the diagnostic and troubleshooting tools used by network staff in his or her organization.

✳ REFERENCES

Bates, R. J. "Bud." *Disaster Recovery for LANS*. New York: McGraw-Hill, 1994.

Bochenski, B. "Network Management Platforms." *Client/Server Computing* (July 1994): p. 88.

Cope, P. "Buyer's Guide: Design and Optimization Packages; Building a Better Network." *Network World* (January 4, 1993): p. 34.

Frank, A. "How Much Technical Support Do You Get?" *LAN Magazine* (September 1994): p. 42.

Fitzgerald, J. *Business Data Communications: Basic Concepts, Security, and Design.* 4th ed. New York: John Wiley & Sons, 1993.

Goldman, J. E. *Applied Data Communications: A Business-Oriented Approach.* New York: John Wiley & Sons, 1995.

Hayes, S. "Analyzing Network Performance Management." *IEEE Communications* (May 1993): p. 52.

Henderson, T. "Novell Approach to Management." *LAN Magazine* (June 1994): p. 166.

Jander, M. "Proactive LAN Management." *Data Communications* (March 21, 1994): p. 49.

———. "Simulation Made Simpler." *Data Communications* (August 1992): p. 69.

Juliano, M. "Managing Private Networks: Portable Test Instruments vs. Network Management Systems." *Telecommunications* (August 1992): p. 19.

Kalman, S. "Managing from the Desk." *LAN Magazine* (April 1994): p. 141.

Kim, D. "The Internetwork Manager's Survival Guide." *LAN Magazine* (April 1994): p. 111.

Lieberman, R. W. "A User's Guide to LAN Performance Modeling." *Telecommunications* (October 1992): p. 58.

Liebing, E. "Network Simulation: Look Before You Leap." *LAN TIMES* (May 11, 1992): p. 37.

Nassar, D. "How to Become a Protocol Analysis Pro." *LAN TIMES* (July 20, 1992): p. 36.

Panza, R. "Tales of Troubleshooting." *LAN Magazine* (September 1994): p. 107.

Pickering, W. "How Much Help Is Enough?" *Datamation* (April 1, 1994): p. 49.

Schott, S. *Data Communications for Business.* Englewood Cliffs, NJ: Prentice Hall, 1994.

Stamper, D. A. *Business Data Communications.* 4th ed. Redwood City, CA: Benjamin/Cummings, 1995.

———. *Local Area Networks.* Redwood City, CA: Benjamin/Cummings, 1994.

Strauss, P. "Third NetView Not the Charm." *Datamation* (April 1, 1994): p. 55.

Case Study:
Southeastern State University

Every LAN at SSU that serves more than 50 users utilizes some type of network management system (NMS) software; only smaller LANs are likely to rely solely on NOS' monitoring and management utilities. Because Novell NetWare is the most commonly used NOS at SSU, NMS products designed for use in NetWare LANs are typically used. LANalyzer is the most prevalent.

In some of the networks, LANalyzer or a competing NMS product is installed on the file server and the server's monitor doubles as the network control console when the LAN manager runs the NMS software. However, because LANalyzer is a large program (in terms of megabytes of RAM), when it is run in such configurations, it degrades both server and overall network performance. For this reason, most LAN managers using LANalyzer have moved it to a dedicated workstation, usually one with a great deal of RAM and disk storage and a very fast clock speed.

In most of SSU's LANs, the workstation with the NMS software becomes the de facto network control center. It is often located in the same room as the student LAN it services or in an adjacent room (which may have formerly been a closet). For some of the larger LANs at SSU, the network administrator's office is the network control center.

An extreme example of this latter arrangement is the office of the College of Business Administration's LAN manager. It is located several doors down the hall from the LAN in the largest student computer lab, but still within convenient walking distance. (You may recall from previous segments of the case that this LAN was recently expanded to incorporate the microcomputers in all the faculty, clerical, and administrative offices of the COBA.)

Upon first entering this LAN manager's office, a newcomer is often struck by its appearance. Its bookshelves are filled with software packages (after all, the installation disks and documentation for the applications supported on the network need to be stored somewhere), which gives the room the image of a retail software or microcomputer store. Since the room also contains the network's file server, RAID storage system, and backup servers, there are computers and monitors scattered among several desks, tables, and work areas; this helps create the impression of a somewhat loosely organized version of a military command center or NASA control center. This latter effect is undermined somewhat by wires coming down from the ceiling (the main conduit for network cabling within the COBA building) that attach to the servers and to the high-end laptop computer that the LAN manager uses to run LANalyzer and monitor network operations. However, the newcomer can't help but be impressed.

In a typical workday, the LAN manager's office is a beehive of activity. In addition to the LAN manager, at least one student assistant is always working there; frequently there is a line of students spilling out into the hall waiting to set up new user accounts or to enlist the help of the student assistant in modifying access privileges. It is not at all uncommon to find the LAN manager on the phone to the help desk of a software vendor, keeping an eye on a server monitor and both hands on the keyboard, stepping through a diagnostic routine as guided by someone at the other end of the line. The are no idle hands anywhere in the room and there are always roving eyes, especially to the laptop's LANalyzer screen; both the LAN manager and the student assistants are alert to background color changes on the laptop's monitor.

The LAN manager's office is both the network control center and help desk. It is also the place from which all network troubleshooting starts. Because of this, a variety of diagnostic toolkits and pocket adapters can be found on the tables and desktops. Occasionally, network interface cards, workstation expansion cards, printheads, and circuit boards can be found on the desks and tables, attesting to the fact that they are used as workbenches for setting dip switches, testing interfaces and cabling, maintaining printers, and carrying out a variety of other troubleshooting and installation activities.

This LAN manager's office is far from what most people would consider to be a stress-free work environment. Many colleagues comment that this manager must be glad to leave the hectic office setting behind at the end of the day. Little do they know that since the LAN has been expanded, LANalyzer has been integrated with the office phone and will page the manager, 24 hours a day, when alarm status conditions are detected.

CASE STUDY QUESTIONS AND PROBLEMS

1. In addition to the documentation for the software applications on the COBA LAN, what other types of network documentation should be stored and maintained in the network control center?
2. How would you organize the network documentation for the COBA LAN? What approach(es) would you use to maintain the network documentation in order to ensure that it is up to date? Justify your recommendations.
3. How would you set up a problem tracking system for the COBA's network control center? How would you implement such a system?

4. Currently, there is only a single phone line for the COBA LAN manager's office. Discuss why this might be a problem if you wanted to create a true help desk and problem tracking system, especially now that the LAN has been expanded.

5. What types of network statistics and trends are likely to be most important to the COBA LAN manager? What types of capacity planning and network tuning approaches are most likely to be used in this environment? Justify your selections.

Managing Network Security

14

CHAPTER OBJECTIVES

After completing this chapter, you will be able to:

◆ Describe the importance of network security.

◆ Identify the most prevalent LAN security threats.

◆ Identify the types of security measures LAN managers can implement.

◆ Describe the role of encryption in LAN security.

◆ Identify the types of security mechanisms used to protect data and files in LANs.

◆ Explain the importance of backup procedures.

◆ Describe the characteristics of data recovery procedures.

◆ Identify the characteristics of disaster recovery plans.

◆ Provide an overview of emerging backup/data recovery methods and techniques.

◆ THE IMPORTANCE OF NETWORK SECURITY

The importance of protecting a computer system's data system is a product of the amount of time, effort, creativity, and money spent in creating the system and the amount of time, effort, creativity, and money that would be needed to reproduce the system. Replacement costs, especially for large centralized systems, can be staggering. Network managers and other information systems personnel must be concerned with threats to the security and physical integrity of their systems.

People often appreciate the full value of something only after it is lost. In business, regret for losses is measured in terms of replacement costs: the greater the costs, the greater the regret. Such a view of loss assumes that what has been lost can be replaced. It may be impossible to fully replace information systems, however, because the business may lack the resources to do so, because the loss of the system makes it unable to continue in business, or because essential data is either temporarily or permanently lost.

For an increasing number of organizations, networks have become the central computing platform, and data and information have become the lifeblood of their business. In response, vendors of network products have created tools to keep data secure and to assure the continuing operation of systems. In addition, the sophistication of security and backup mechanisms has continued to increase in response to the increasing complexity of networks.

SECURITY THREATS

As mentioned in Chapter 12, many organizations are often more concerned with security breaches from external than from internal sources. However, research on network security indicates that most security breaches are "inside jobs." This should not be surprising, because the people most knowledgeable about the architecture of the system and how it is used are employees who have built, maintained, or used the system. In short, networks should be designed to thwart security breaches from both outsiders and insiders (employees).

Numerous threats exist to the security of networks and other computing systems, more than can be adequately addressed herein. Some of the major categories of security threats are listed in Table 14-1 and are briefly discussed in the following paragraphs.

TABLE 14-1
Threats to
network security

- ◆ Unauthorized access to data and applications
- ◆ Unauthorized alterations to data and applications
- ◆ The introduction of viruses and other malicious software
- ◆ Eavesdropping (wiretapping) on network traffic

If unauthorized users can gain access to the network, valuable data and applications may be laid open to potential theft, fraud, and malicious mischief. In its more sinister forms, unauthorized access could be a part of industrial or other types of espionage.

The managers of most large network systems guard seriously against intrusions from unauthorized outside users. For example, network managers may implement mechanisms to lock out external users after the third unsuccessful login attempt. Usually only the network administrator can deliberately deactivate this and other intruder-detection mechanisms. Figure 14-1 illustrates intruder detection/lockout mechanisms available in Novell NetWare; note that in this example, only users with supervisory status may modify the parameters.

Unauthorized modifications to systems, applications, or data are cousins to unauthorized entry. An employee may secretly change the hourly rate of a payroll record, pad the hours of a relative or close friend, or enter a payroll record of a fictitious person. While such changes may be made by unscrupulous authorized employees, most of the control mechanisms focus on preventing unauthorized users from making such modifications. Often these involve the development and monitoring of user access privileges such as those discussed in Chapter 12.

Network interruptions are also of concern to LAN managers. Causes of network interruption include damage to a disk (particularly the server disk), a virus that disables the resource management system of the NOS, and the severing of the media connecting two or more parts of the network. While some of these interruptions may be inadvertent, others are deliberate.

Of particular concern is the introduction of viruses and other malicious software. While some viruses cause only minor (yet annoying) disruptions (such as screen messages and workstation lockup), others can

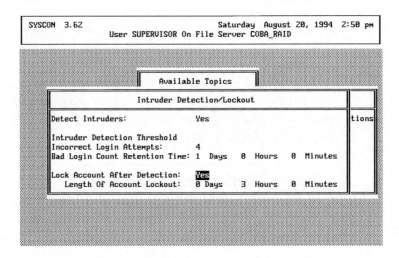

FIGURE 14-1
Intruder detection

cause serious problems such as system crashes and the destruction or corruption of data and application files. In most instances, viruses result in loss of user productivity; in addition, users often lose confidence in the reliability of the network. Because viruses can be spread from machine to machine in networks (and from network to network), many organizations have developed virus prevention and protection guidelines such as those listed in Table 14-2.

Eavesdropping or wiretapping is often another security concern among network managers. Eavesdropping is essentially the ability to monitor ("hear") network traffic. While eavesdropping does not damage the system, it can result in the leakage of confidential information to unauthorized persons. Wireless data networks (particularly cellular data networks) are especially vulnerable to this type of threat. Encryption—the scrambling of data messages—is one of the primary mechanisms used to inhibit eavesdropping.

OTHER SECURITY CONCERNS

In general, any threat to free, open, and dependable communication among network users may be a threat to the entire system. For this reason, network managers must be concerned with creating and maintaining mechanisms which avert or reduce threats to the security of the network.

Regular users are among the individuals most likely to breach the security of a network. By giving access only to those individuals who have ownership of the data and have a valid need to access particular data or applications, LAN managers are able to address many of the major security concerns. However, LAN managers must also keep up with changes to

---

TABLE 14-2

Steps to prevent virus infections

◆ Check all new software (especially public domain software downloaded from bulletin board systems) and user files on disk by using a virus-detection program before permitting them to be introduced to the network.

◆ Keep original application disks in a secure location so that they can't be infected.

◆ Make frequent backups of important files and data.

◆ Limit the exchange of disks containing executable code (such as .EXE files).

◆ Put write-protect tabs on all floppy disks.

◆ Never boot a hard disk system from any floppy disk exept the original, write-protected system disk.

◆ Treat virus security as an extension of overall network security.

◆ Educate users about the symptoms and problems of viruses and the steps they can take to prevent the network from being infected.

the status of users, especially those leaving the organization holding grudges against their former employers. Many disgruntled employees have destroyed critical data or have sabotaged the network upon departing the organization. As a result, many network systems, especially those in banking and financial institutions, have official procedures to delete users from the system minutes after the human resources department is informed of the employee's imminent departure.

In addition, authorized users may attempt to access data that they have no legitimate reason to access. For example, not long ago, the news media carried stories of how the Internal Revenue Service punishes employees who electronically "snoop" through taxpayer records. Even if the person doing the "snooping" makes no attempt to change any of the data, access without valid reason is a violation of federally protected privacy rights.

Serious threats to a network are posed by individuals with criminal intent who would intentionally change data to defraud the organization. Financial institutions and accounting departments are constantly prepared for those who would perpetrate fraud. However, fraud is not always easy to detect. For example, it may be difficult to detect the creation of dummy accounts by accounts payable clerks. Often, only more than a normal amount of transactions flowing to a particular account or address alerts someone's attention to the possibility of fraud. To deal with such problems, many accounting systems keep audit logs which record the user names of anyone making an entry or deleting an entry to the accounting system. Network operating systems may supplement such systems with *server error logs* (see Figure 14-2), especially those that affect files such as bindery files, which contain information about user passwords and access privileges. In this example, once intruder detection is turned on, notations are

```
SYSCON   3.62                           Saturday  August 20, 1994  2:51 pm
                     User SUPERVISOR On File Server COBA_RAID

┌──────────────────────────────────────────────────────────────────────────┐
│                          File Server Error Log                             │
├──────────────────────────────────────────────────────────────────────────┤
│ 8/3/94 4:00:51 pm  Severity = 0.                                           │
│ 1.1.60 Bindery open requested by the SERVER                                │
│                                                                            │
│ 8/3/94 4:36:51 pm  Severity = 0.                                           │
│ 1.1.62 Bindery close requested by the SERVER                               │
│                                                                            │
│ 8/3/94 4:36:51 pm  Severity = 4.                                           │
│ 1.1.72 COBA_RAID TTS shut down                                             │
│   because backout volume SYS was dismounted                                │
│                                                                            │
│ 8/3/94 4:53:35 pm  Severity = 0.                                           │
│ 1.1.60 Bindery open requested by the SERVER                                │
│                                                                            │
│ 8/3/94 4:54:02 pm  Severity = 0.                                           │
│ 1.1.62 Bindery close requested by the SERVER                               │
│                                                                            │
│ 8/3/94 4:54:03 pm  Severity = 4.                                           │
```

FIGURE 14-2
File server error log

made in the server error log; if accounting applications are involved, notations are also made in the audit log.

Any network that allows access to the system via the telephone has a special set of security concerns. If remote users can access network resources, they may determine, at leisure, how to damage or defraud the system, perhaps in the evening when no one is around to interrupt the plot. Peer-to-peer networks are particularly vulnerable to security problems because many lack sophisticated control systems; many essentially assume that network users are honest. Server-based networks are usually better at controlling user network access, such as by controlling the time of the day when a user (even a remote user) can access the system. Figure 14-3 illustrates that NOSs (in this case Novell NetWare) often include utilities for controlling the times that individual users can logon to the network.

Many federal and state laws address network security threats and concerns. Some of the more important federal laws are listed and briefly summarized in Table 14-3. In general, prosecutions of computer criminals have become more aggressive, and penalties for convictions are becoming stiffer.

◆ PHYSICAL SECURITY

Two major types of protection mechanisms for computer networks are: protection mechanisms for hardware and the physical environment, and mechanisms for protecting software and data. The second set includes constructing secure electronic pathways to the data so that only authorized users are able to access the network resources that they need. Some of the major types of both physical and electronic security approaches are listed in Table 14-4.

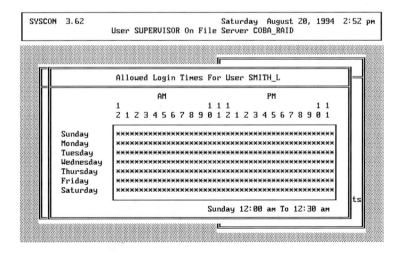

FIGURE 14-3

Login time control

Federal Law	Brief Description
Electronic Communications Privacy Act of 1986	Makes the interception of electronic communications such as data communications or e-mail a federal crime; also covers tampering with computers in a data network.
Computer Fraud and Abuse Act of 1986	Makes it a federal offense to access a computer in a fraudulent scheme to steal, alter, or destroy data, hardware, or software without authorization. This law also makes it illegal to traffic in computer passwords with the clear intent to defraud.
Computer Security Act of 1987	Gives total control of the data encryption standard (DES) to the National Institute of Standards and Technology (NIST). DES is widely used by both private industry and the federal government.
Computer Virus Eradication Act of 1988	Makes it illegal for anyone to knowingly install a program having information or commands that may cause loss, expense, or risk to health or welfare. This law also makes it illegal to give such programs to others.

TABLE 14-3　Laws addressing network security threats

TABLE 14-4
Physical and electronic security mechanisms

Physical Mechanisms	
Access cards	Fire-resistant storage containers
Badge readers	Security guards
Biometric verification systems	Sprinkler systems
Burglar alarms	Surge protectors
Closed-circuit television	Theft-proof safes and vaults
Electronic security systems	Workstation locks
Fire detectors	Uninterruptible power supplies

Electronic Mechanisms	
Call-back systems	Pseudo-logons
Call-tracing systems	System monitors
Encryption	Terminal profiles
File and database security mechanisms	User profiles
Identification and authentication systems	

Physical security includes protecting network and computer resources with such devices as locked doors, workstation locks, badge readers, biometric verification systems (such as retina scanners, fingerprint readers, hand geometry scanners, and voice verification systems), fire detectors, fire extinguishers, sprinklers, burglar alarms, closed-circuit television, and security guards. Theft-proof safes or vaults and fire-resistant containers may be used to protect backup tapes and disks. Physical security may also include methods for minimizing the impact of natural disasters by positioning hardware resources in a certain way, following housekeeping policies that minimize exposure of data and I/O media to contaminants, and having meetings with users to clarify security policies. The most critical locations for physical protection usually are server sites, the data library (where backup tapes and disks are stored), and the network's communications lines.

◆ ELECTRONIC SECURITY APPROACHES

A number of electronic security mechanisms may be used to protect network resources from unauthorized intrusions. User profiles are related to establishing user access privileges and trustee rights; these are discussed in Chapter 12. *User profiles* essentially specify the restrictions that apply to the use of network resources by individual users. These may even specify the workstations individual workers are authorized to use. *Terminal profiles* specify the types of applications that can be performed at particular workstations.

Callback systems are used to keep unauthorized users from being able to logon to networks. In such systems, the user briefly logs on to the system; the system breaks the connection and calls the user back at the number in the user profile. Pseudo-logons and call-tracing systems are designed to mislead unauthorized users into believing that they have logged onto the network; the call is traced and the police are notified.

System monitors keep a log of all network activity. This includes data such as who logged on to and off of the network, where, and when; the files accessed; and what data were added, deleted, or modified.

IDENTIFICATION AND AUTHENTICATION PROCEDURES

Virtually all well-managed and properly secured LANs require users to identify themselves and to authenticate identification through passwords. A **password** is essentially a string of characters that, when input to the computer system, allows access to certain hardware, software, or data.

In many networks, unique passwords are mandatory. In systems without passwords or that allow numerous users to logon using the same password, network users are generally allowed access to a limited set of network resources and may not be allowed to make changes to the resources they can access. Even when sophisticated identification and authentication procedures (such as biometric verification systems) are used, network managers should not be overly confident that these mechanisms can prevent security breaches by knowledgeable users.

Several guidelines that network managers use to develop password policies follow.

♦ Passwords should be frequently changed, particularly for access to critical applications. While users resent overcontrol, if the information is of a critical nature, password changes on a monthly, weekly, or even daily basis may be desirable.

♦ Longer passwords are better than shorter passwords. LAN managers often set a five- to seven-character minimum password length, because longer passwords are more difficult to guess.

♦ Layers of passwords (multiple passwords) should be used to access important applications or data. For example, one password may be needed to logon to the network, a second to access a particular application, and a third to access data used by the application.

♦ Users must be informed that they cannot write down their passwords and keep any real degree of security. Such written records may be found by the unscrupulous in desk drawers, filing cabinets, or safes.

♦ Users should be advised not to use initials, common abbreviations, key dates, or other familiar material for passwords. Such passwords make the guessing game that much easier, especially for associates who know the user well.

♦ Users should change passwords immediately if they suspect that their passwords have been learned or stolen by unauthorized users.

♦ Users should not leave workstations unattended without logging off the network. Even a brief trip down the hall can be sufficient time for someone to change the user's password and be able to later gain access to the network.

While other policies may be needed for a particular network, these are likely to be used in most LANs.

As you might expect, password utilities are included in most NOSs, and many of these have become more sophisticated over time. For example, in earlier versions of Novell NetWare, passwords are transmitted unchanged over network media; this makes them susceptible to being picked up by eavesdropping technologies. However, in the 3.xx versions of NetWare, passwords are encrypted (scrambled) before they are transmitted from workstation to server, and in NetWare 4.xx, logon keystrokes are not transmitted over network media at all.

As noted, NetWare enables network managers to create and enforce time controls for user logons that restrict a user to logging on only during specified time periods. In addition, logon scripts (such as that depicted in Figure 14-4) can be individualized to enforce access privileges, including access to specific directories (Figure 14-5). NetWare also includes a security

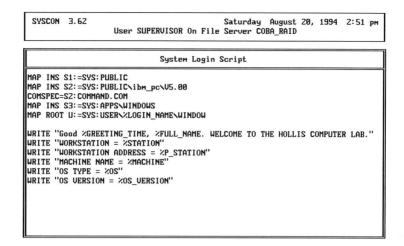

FIGURE 14-4
Logon script

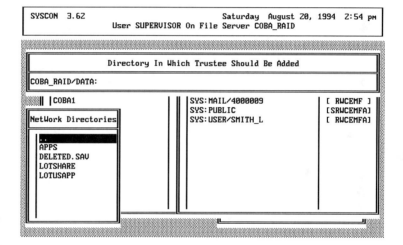

FIGURE 14-5
Directory access
privileges

evaluation utility that enables the LAN manager to review the password requirements and access rights of individual users (Figure 14-6).

In certain situations, LAN administrators may retain sole control over password changes. Either the LAN manager or a designated employee may be allowed to change passwords. This approach usually obligates users to follow preestablished password control procedures and helps to ensure that proper passwords are assigned, that they are changed at proper intervals, and that the choice of passwords is appropriate.

If users are permitted to change their own passwords, network managers often allow **grace logons**. These allow the user to briefly logon to the network using an old (and, technically, expired password), but require the password to be changed before the user is allowed to access any of the network resources.

```
SECURITY EVALUATION UTILITY, Version 2.23

User COTTON (Full Name: TRAVIS COTTON)
  Does not require a password


User SUPERZ (Full Name: supervisor equlivelent)
  Has [SRWCEMFA] rights in SYS:MAIL/1 (maximum should be [    C     ])
  Is security equivalent to user SUPERVISOR
  Does not require a password


User COBA02
  Does not require a password
  No Full Name specified


Group NEWLAB
  Has [ R     F ] rights in SYS:SYSTEM (maximum should be [         ])
  Has [ R     F ] rights in DATA: (maximum should be [        ])
  Has [ R     F ] rights in SYS: (maximum should be [        ])
  No Full Name specified

Press any key to continue ... ('C' for continuous)
```

FIGURE 14-6
Security
evaluation

Other important management considerations related to password mechanisms are whether they will permit **concurrent logons** and whether a user will be restricted to logon to a specified terminal or computer. While concurrent logons are useful and almost essential for network administrators and some other network personnel, allowing users to do so can cause security problems. For example, a user could logon to one terminal, leave the room without logging out, and logon to another terminal unless the system prevents this. Network security risks are high in such an environment, because another person could use the first workstation to access network resources that are normally restricted for that individual. In general, the network administrator must be convinced that there is adequate reason to permit a user to logon from multiple terminals, particularly if the user has a history of not logging out when finished using the network.

To ensure that appropriate logoff procedures are followed, several mechanisms may be used. For example, some DOS workstations running Windows invoke a screen saver graphic after a specified number of minutes. When a key or mouse button is pressed to restore the screen, a password input field appears and the user cannot reenter the application without entering the proper password. More restrictive mechanisms may include a screen message that the workstation must be rebooted and logged on to the network. In addition, utilities that automatically logoff workstations after a certain period of inactivity are often considered to be mandatory for systems with sensitive data or application. Such utilities eliminate some of the potential harm that could be done if a user simply forgets to logoff.

Network managers must remember that users may become indignant at strict security procedures. However, as a network administrator, you must determine whether the degree of security is appropriate for the value of the data in the system. If critical data exists, you must enforce proper security procedures even if these are unpopular.

ENCRYPTION

Encryption is as old as the military cyphers (codes) used to send messages behind enemy lines before electronic communications. Encrypting data is equivalent to scrambling video channels in cable television systems. As encryption requires decryption (unscrambling) capabilities on the part of the receiver, such schemes help ensure that sensitive data or information does not fall into the wrong hands, at least not in a usable format.

In electronic communications, encryption can be used to prevent users from seeing portions of a message, particularly logons, passwords, or other sensitive information within the message itself. Commercially, encryption has been widely used to secure satellite, microwave, and other types of wireless voice and data communications. The military and foreign service also use encryption to ensure the confidentiality and security of voice and data messages, especially those destined for unsecured areas. Encryption is also becoming more common in LANs and other computer networks. As noted, Novell NetWare 3.xx versions utilize encryption to inhibit eavesdropping on passwords transmitted on network media.

Two of the most common encryption schemes used in data communications networks are the data encryption standard (DES) and RSA (named after the developers of this scheme—Rivest, Shamir, and Adleman). The Escrowed Encryption Standard (ESS) is also expected to become widely used for data communications among federal agencies.

◆ FILE AND DATABASE SECURITY

Most networking systems incorporate mechanisms for protecting directories, files, and data from unauthorized users. Such systems have been in place in LANs for some time. For example, such a scheme has always been available in Arcnet LANs, one of the early LAN systems developed by the Datapoint Corporation. The NOS in early Arcnet LANs (called the Resource Management System) provided for almost full resource sharing. However, each user on the system had to have access rights assigned either at the directory, volume, or file level. Access was further broken down into levels to segregate those needing read only access and those needing permission to make modifications to the files. These features are now common in most NOSs, especially those used in client/server LANs.

Networks evolve. As the network continues to grow and change, its managers must try to develop preventive techniques, detection techniques, and sets of remedial actions for security breaches. All techniques and controls will involve some degree of cost. These costs must be matched to the organization's investment in the system and to the costs of restoring the system in the event of its destruction.

The remainder of this section focuses on how file and database security can be implemented in Novell NetWare and UNIX LANs. You will see that file and data access schemes can vary in complexity and ease of maintenance, factors that may be important in determining what type of LAN to implement in the first place.

In a LAN, files are protected by both the local operating system and by the NOS. File Allocation Tables are some of the primary means of controlling the placement of files on the disk as well as the addition, deletion, and modification of files. On a workstation's local disk, few mechanisms can be used at directory and file levels to prevent knowledgeable coworkers from accessing data.

Peer-to-peer networks such as NetWare Personal and Windows for Workgroups offer little file and database security. While such systems enable users to share resources, they typically offer little protection for either shared or private data. Other peer-to-peer systems (such as LANtastic) possess limited password control utilities, but their security features are not as sophisticated as those found in most client/server NOSs. However, peer-to-peer systems can often be used in conjunction with server-based systems to achieve adequate levels of security for mission-critical applications.

FILE SECURITY IN NOVELL ENVIRONMENTS

One of the most recent versions of NetWare is version 4.01. The security mechanisms in this version are similar in many ways to those found in the 3.xx versions. File and database security is built around users, groups, trustees, and trustee rights to files and directories. However, in NetWare 4.01, another layer of security is available with its NetWare Directory Services (NDS). Whereas versions 3.xx were primarily concerned with the welfare of server security, NetWare version 4.01 is concerned with security for the entire network.

In this NOS, the security is implemented in three levels: the user logon level, the NDS level, and the file system level (at the server). Version 4.01 includes user name verification, password verification, account restrictions, station logon restrictions, time restrictions, and intruder lockout mechanisms.

NDS security provisions are made through an object-oriented database. The user must have access rights to resources listed as objects in the NDS database in order to even be able to see the object. Special access rights are needed to see or modify the information associated with the object.

User rights to specific directories flow down the NDS' tree structure. As in previous versions of NetWare, as branches of the tree are followed, user rights can be blocked when the Inherited Rights Mask filters out whatever user rights the network manager wants to control. The residual, or effective, rights implemented in NetWare are a product of the user rights assigned specifically to a directory or file by the supervisor, or they can be rights inherited from the parent directory in a modified form determined by the LAN administrator.

Table 14-5 summarizes NDS rights and what they enable the user to accomplish. As noted, the NDS system is organized as is an object-oriented relational database. If you are familiar with such databases, you will not be surprised to see references to objects, properties, and containers.

The default in NetWare Version 4.01 automatically provides the user with Browse rights, so that the user can view his or her location in the NDS tree. In the majority of cases, the user will require the Browse right to an object only to use the object and the corresponding resource. Depending on the system, the other rights listed in Table 14-5 are restricted to specific users or to only the network manager.

Each of the objects in the NDS tree has properties. Each type of property has certain access rights associated with it. The rights associated with the property are designed to control how the users' and other objects' properties relate to the object. These additional properties are listed in Table 14-6.

For example, if you as a user have the Compare right, you will have sufficient security to determine if you have a valid logon script. However, you will not be able to read it, because the Compare command does not carry with it the permission to read the object's (in this case, the logon script's) properties.

As was true for the 2.xx and 3.xx versions of NetWare, user rights such as Read and Write rights are granted as part of the logon script and the

Object Right	Description
Browse	The right to see an object in the NDS tree. Similar to the read access in previous versions. This right does not assure that property values of an object can be seen.
Create	Allows the creation of a new object below the current object in the NDS tree. This applies only to a special *container* object.
Delete	You must have this right to delete any object from the NDS tree. Restrictions apply as to which objects can be deleted once populated with subordinate objects.
Rename	Allows modification of an object's name. Restrictions apply here as to which objects can be modified.
Supervisor	As in earlier versions, the supervisor has full rights to an object and its properties. Unlike previous versions, there is an auditor function over which the supervisor may not have full rights.

TABLE 14-5

NetWare 4.01 rights to objects

Property Right	Description
Compare	The right to compare any value to the value of the property. Although this procedure can return True or False, the user still will not be allowed to see the property value.
Read	The right to read the value of the property. This right implies the Compare right.
Add or Delete Self	The right to add or delete itself, without interfering with other values of the property.
Write	The right to make modifications, allowing the user to add, delete, or modify. The acquisition of this right implies that the Add or Delete Self right is present.
Supervisor	All rights are granted to the property.

TABLE 14-6

Rights assigned to properties

user's Print Job Configuration. In NetWare 4.01, these rights are granted using the Access Control List (ACL). This command is similar to the rights command in version 3.xx. If a user has the Browse right to a particular volume (such as COBASYS), but has Write rights to the ACL of this volume, that user can make changes to the ACL. The user can modify the ACL and grant more rights to the volume.

As noted, the NDS system provides for rights related to objects and properties to flow down the NDS tree just as file system rights do in previous NetWare versions. Such rights are known as *inherited rights*. If the user is granted a set of rights to a container (an object in Novell 4.xx which *contains* other objects), the same user inherits those rights for all objects below it in the tree. In an environment where there is a need to ensure that certain rights are not inherited through the NDS system, the LAN manager can accomplish this through the Inherited Rights Mask (IRM). This filter specifies which rights can be inherited. If it appears on the list, the right can be inherited.

NetWare also incorporates a concept called *security equivalence* (Figure 14-7). The network administrator can create users with equivalent rights to another user. The most important of these options is that of making a user a *supervisor equivalent*—this user has all the rights of the supervisor (these cannot be blocked with the IRM). This user becomes a "super user" with few restrictions on the ability to change the system.

Supervisory equivalence has important security implications. Novell keeps information related to users, passwords, and rights in three files known collectively as the *bindery*. If for any reason the password(s) for the supervisor or supervisor equivalents are forgotten or unavailable, the only way to recover control of the system is to reinstall NetWare or to rename the bindery files. If the bindery files can be renamed, NetWare will assume that the installation is new and allow passwords to be initialized.

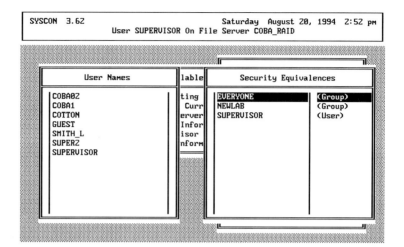

```
SYSCON  3.62                           Saturday  August 20, 1994  2:52 pm
                    User SUPERVISOR On File Server COBA_RAID
```

FIGURE 14-7
Security
equivalence

One way to rename the bindery files is through a supervisor equivalent that has the same rights and privileges as the supervisor. At best, this technique is a desperate recovery procedure that compromises the security parameters for other users. Unfortunately, security equivalence is also useful to a person who intends to break the network administrator's security scheme.

FILE SECURITY IN UNIX ENVIRONMENTS

Similar to NetWare, UNIX grants control access to files and directories through a system of permissions. Users may access those files and directories for which they have the appropriate permission; all other volumes, files, and directories are inaccessible.

The exact procedures used vary somewhat among the different dialects of UNIX such as AIX (IBM's proprietary version) and Xenix. The discussion in this section will be generic.

Permissions can be displayed for all files in a given directory with the "ls -l" command. This is the *list directory* command, and it additionally lists the name of the file's owner, the size of the file in bytes, and the date and time stamp for the last change made to the file. Permissions are generally shown as a sequence of ten characters at the beginning of each item in the display. The first field or column at the beginning of the sequence is reserved for the type of object (Table 14-7).

The rights indicated by the special characters are given in Table 14-8.

The following are a couple of examples to help you see how the files are defined in terms of access.

"-rwxrwxrwx" indicates that a file has complete read, write, and execute access for everyone.

Indicator	Definition
-	Ordinary file
d	Directory object
b	Special block I/O file
c	Special device special character I/O file

TABLE 14-7
Object types in
UNIX

Indicator	Definition
r	Read permission. The user can copy or display the file. This permission may be granted to an entire directory.
w	Write permission. The user can change or modify the file. This permission given for a directory or to subdirectories or files within that directory.
x	Execute permission is given for files and search permission for directories. For a file, the user can execute the file itself; for the directory, the user can enter the directory with the cd (change directory) command.
-	No permission.

TABLE 14-8
Permission
indicators in
UNIX

"-rw-------" indicates an ordinary file with read and write access limited to the user.

"drwxr-x--x" indicates a directory with search access for everyone, read access for the user and group, and write access for the user only.

When a file is created in UNIX, the operating system automatically assigns the following permissions: "-rw-r--r--." This designation means that everyone may read the file, but only the user may write to it. When you create a directory, the system assigns the permissions; for example, "drwxr-xr-x" indicates that every user can search and read the directory, but only the user may create and remove files and directories within it. Permissions can be changed with the "change mode" command; however, this command requires that you designate how to change the permissions for a specific file or directory.

In UNIX, when a file is created by a user, the operating system automatically designates the creator as the owner and permits the creator of a file to access the file according to the user permission system. The operating system also designates a group as the owner of the file. Group ownership allows a designated group to access the file according to the group permission scheme. The default is to assign access privileges to the user's group.

UNIX permits the administrator to prevent ordinary users from gaining access to the data and program files on the system and to thereby protect the system's vulnerable files. UNIX uses the system files, usually kept in a "/dev" directory, to transfer data to and from the computer's devices, but the files can also be used to gain direct access to networked devices. The network administrator can logon as a superuser of the system and use the "chmod" (change mode) command to set the correct permissions. Directories containing vulnerable files can be set so that read and write access is granted only to file creators and the network manager.

UNIX includes a set of features—the *process accounting system*—that allows the network administrator to track the amount of time that each user spends on the system. The process accounting system records the number of processes (applications) started by a user, how long each process lasts, and additional information, such as how often the process accesses I/O devices and the size of the process in bytes. Process accounting was originally intended for monitoring time sharing approaches to data processing, but is now more commonly used as a sort of auditing function that keeps a detailed record of system, command, and system resource usage.

UNIX is particularly sensitive to an aborted system or a system that has been shut down hastily. Lost data and damaged files often result from aborts or shutdowns. Damage to files can also be caused by hardware errors in the disk drives and worn disks. Occasionally damage causes the entire file system to become inaccessible, but usually it affects only a few files.

UNIX permits the user to repair a file system with a "file check" command. The following is an example of the procedures used to perform file checks and repairs.

Phase 1—Check blocks and sizes

Phase 2—Path names

Phase 3—Connectivity

Phase 4—Reference counts

Phase 5—Check free list

If UNIX finds a damaged file during any of these phases, the system asks if it should be repaired or salvaged. Answering yes initiates the file repair process. The file check command deletes any file that it considers to be damaged beyond repair. In general, the file check procedure is used as a last resort after other file restoration and repair remedies have been exhausted.

UNIX generally will request that you check the file system at boot time if the system has been shut down improperly. During this procedure, the NOS will check and repair files affected by the abrupt shutdown.

UNIX offers at least two methods of backing up file systems. One uses the "sysadmin" program and the other the "tar" command. The former is a formal maintenance program for systems that require regularly scheduled file backups. These applications characteristically have many users and a large number of files that are modified daily. Sysadmin automatically identifies modified files, copies them to the backup media, and can, at the user's option, list these files. The sysadmin command is restricted to the superuser (typically the network administrator). The scheduling of backups in UNIX systems parallels that of Novell systems.

The "tar" command is useful on systems with one or two users, or on any system where an individual user may want a copy of a directory. Tar lets the system manager or user choose the files and directories to be copied. It does not, however, go through the identification procedure mentioned above.

◆ BACKUP AND RECOVERY PROCEDURES

Organizations, no matter how physically secure their systems, are always vulnerable to disaster. Thus, a key issue is backup. Most large organizations have a backup plan for data, hardware, and software resources. Data and software backup involves creating duplicate copies of important data and software, and often, storing the originals at a distant site. Hardware backup often involves establishing a contingency plan that will be enacted if a major problem befalls some key hardware element. One solution is to keep duplicate hardware available for backup purposes, but for hardware other than microcomputers, low-cost peripherals, and dumb terminals, this alternative may be prohibitively expensive. In such instances, arrangements may be made with another party, such as a firm specializing in disaster recovery, for hardware backup support.

In network environments, most people consider backing up data on disk to be the primary backup procedure; however, there should also be redundancy plans for hardware and software. A portion of the strategy for making data backups is contingent on the media used for actually archiving the data. If the media lack portability, durability, or easy access, they may not be desirable as backup media.

Tape has been the traditional mainstay for backing up data and software for stand-alone PCs and PC networks. However, as the price of hardware continues to drop, more and more companies are relying on other media (such as disks, including optical disks) for backup as well as on duplicated hardware and software.

THE NEED FOR BACKUP AND RECOVERY PROCEDURES

One of the most important aspects of backup and restoration of data and applications is developing and implementing a backup plan (schedule).

While creating a dependable and practical strategy for backing up the data and software can be a time-consuming process that requires constant evaluation, such a plan can prevent a lot of grief in the long run. In addition to having a backup plan, network managers must ensure that the policies and procedures are carried out.

Some of the questions the network administrator must answer when devising a backup plan include the following.

1. Will the backup hardware be server-based or workstation-based?
2. Who will perform the backups?
3. Will full backups or incremental backups be used?
4. Where will the backups be stored or located?

However, before developing such a plan, the need for regular backups must be weighed against the value of the data and applications. Generally, the more valuable (critical) the data and applications are, the greater the need for regular backups. For example, some companies cannot afford the cost of restoring the data without archives; other companies cannot afford the time that would be required to manually restore the data; and some companies cannot afford either.

Specific backup procedures must be established—the need to back up data can arise unexpectedly. Although some systems operate for years without loss of data, backup procedures cannot be ignored. Backup and restoration procedures are much like life insurance, in that the network administrator hopes that such procedures will never be needed. Yet if you don't have them when you need them, you will sincerely regret not having taken the time to put them in place.

HARDWARE BACKUPS

In the days of multimillion dollar centralized, host-based systems, few companies could afford to consider the duplication of hardware as a viable backup alternative. However, in today's LANs, the idea of duplicating the hardware is more feasible. The hardware is not nearly so expensive and it consists of modular parts, not all of which need to be duplicated. As a result, a variety of hardware backup systems may be used in LANs; some of these are summarized in Table 14-9. In addition, today's networking software allows different levels of *fault tolerance* (resistance to network crashes) including the ability to *duplex servers* (run two or more servers in tandem; each is a backup to at least one other server).

Duplexing servers involves electronically connecting them so that one trails the other by a few milliseconds and can take over the lead processing should something happen to the lead server. Sometimes duplexing even increases network throughput, because it increases both the number

◆ Duplexed servers

◆ Disk duplexing

◆ Disk mirroring

◆ Redundant arrays of independent disks (RAID)

◆ Optical backup systems

TABLE 14-9
LAN backup
systems

◆ Virtual disk systems

◆ Tape backup systems

of disk read/write heads that can be utilized and the overall efficiency of the network's disk storage system. A server may be duplexed by another server for a few thousand dollars.

A junior version of server duplexing is the duplexing of disk drives within the same server. This approach may include two separate disk controllers, often SCSI (Small Computer System Interface), which allow several disk drives to be daisy-chained together. If one disk/controller fails, the second will take over. The primary vulnerability of this approach is that the failure of the central processing unit renders even a second disk/controller useless unless mounted in another CPU. This is why server duplexing is often a superior solution. Novell NetWare and other NOSs offer a third duplexing alternative—disk mirroring—in which two disks are duplicated, but use the same disk controller. The danger in this approach, however, is that if the disk controller fails, both disks fail.

Another hardware alternative is RAID (Redundant Array of Independent Disks); RAID is a relatively new type of secondary storage system whereby multiple disks, disk drives, and disk controllers are combined in an array. Often, one of the disks is designated to keep error correction codes derived from the actual data bits stored on the disks; these can be used to reconstruct damaged or missing data in the event of a disk or disk drive failure.

RAID systems vary in complexity, the degree of redundancy, data protection (see Table 14-10), and cost. Most include disk mirroring, and some include disk duplexing. In addition, many vendors now offer interchangeable RAID configurations—drives can be removed from and replaced in the server without having to power down the server.

OPTICAL BACKUP SYSTEMS

Optical disk backup systems are increasing in use in LANs. Systems, applications, and data are backed up by using either write-once, read-many (WORM) disks or erasable optical disks (such as magneto-optical disks). The costs of such systems vary considerably.

LAN Sweep is an example of an online optical backup system. It is a menu-driven system that provides unattended backups for LAN servers.

Level	Brief Description
0	Data is striped (spread) across all drives in the array.
1	Disk mirroring is provided; important data is written to two disks.
2	Data distribution is done on a bit-by-bit basis; bits in each byte are spread across disks in the array.
3	One drive stores addressing, parity, and error correction information files; data is striped across the other drives in the array.
4	Similar to level 3 except that blocks of data are striped across the other drives in the array.
5	Blocks of data are striped across all drives in the array; parity and error correction codes are also distributed across all drives.

TABLE 14-10
Levels of redundancy in RAID systems

Full backups (backing up all software and data stored in all servers), incremental backups (backing up specific servers at specific times), and differential backups (backing up only the files that have been used or modified since the last backup) can all be performed with this product. All the LAN manager has to do is to program the time, date, and type of backup desired (in a backup "profile"), and the system will take care of the rest. If the optical disk fills to capacity during the backup, the system will alert the LAN manager (or other designated personnel).

VIRTUAL DISK SYSTEMS

Virtual disk systems provide online, realtime continuous backup services. This is one of the most sophisticated and expensive backup systems available for LANs. Vortex Systems is the best known vendor of such products; to date, it has developed virtual disk systems for both NetWare and Banyan Vines LANs.

Virtual disk systems eliminate the need for daily backups as they perform continuous backups automatically and transparently. If the network crashes, the LAN manager can simply turn back the clock to just before the crash and the virtual disk system will recreate all the data as it existed at that time.

SOFTWARE AND DATA BACKUPS

Companies sometimes decide that protecting software is more important than providing hardware or data backups. They assume that hardware can be replaced relatively quickly, which in many cases is true. Of course, the situation is much more serious when servers or workstations fail, especially when servers fail. Whether server or workstation hardware will be used for backups becomes crucial in the debate about how to protect software.

Workstation-based backup devices have been used for some time on networks, and now server-based software protection systems are available. In Novell NetWare LANs, the backup software is implemented as a NetWare Loadable Module (NLM). Security considerations are moot as to whether the user is logged on as supervisor or has the proper security to see all files in all directories. Programs can be written as NLMs that will allow the administrator to perform the backup in an unattended mode. Therefore, provided that the files selected to be duplicated can be backed up by the software, it matters little whether a server or a workstation is used to actually perform the task.

Software backup strategy begins with a determination of how often files will be backed up. For a very small system, it may be possible to back up all important software every day. However, in medium- and large-sized operations, there may not be sufficient time to back up each file and application every day. Many companies choose to back up the entire set of software (and data) either weekly or monthly and to back up only the files modified since the last daily backup; most tape and disk backup systems can determine by flags or attributes (for example, date/time stamps, byte counts) which files have been modified since the last backup.

No matter what, backup procedures must be carefully thought out. If you back up the entire network weekly and perform differential backups daily (on only those files changed since the last backup), you still must save the entire set of tapes or disks until the next complete backup. In fact, many organizations save at least two and sometimes three generations of backups in case of media failure or an undetectable disk error.

Another issue related to software and data backup is whether your backup device can archive an open file. If it cannot, you must wait to back up until all users have logged off the system. For many organizations, this is only likely at night. Even if a software/data backup system vendor claims that its product can save open files, be careful. In LANs with large database files, changes can be made to the file while the backup is in progress, and the data in the backed up version may not be the most current.

If the software/data from several servers must be archived, the backup schedule can become hectic. Depending on time constraints, additional hardware or faster backup hardware may be needed to ensure that all software and data can be saved during the available backup period. In such instances, more than one network staff member may be needed to perform the backup. All staff performing the backup should precisely document backup procedures and the actual tasks performed should be recorded. After the backup is finished, the backups should be checked for completeness and correctness of format(s).

In many instances, the network administrator may decide not to back up (or to infrequently back up) the actual program files for the applications available on the system, as program files should be readily available and these are usually relatively easy to restore. However, more microcomputer-based networks are using graphically-oriented applications, and restoring the configurations for the server and individual workstations can be time consuming. Hence, it is often advisable to regularly back up the configuration files for such applications.

Another complication is that the company will receive updates for an application that requires server and workstation reconfiguration. In such cases, a log book should be kept to ensure that the reinstallation procedure does in fact store the new version of the program on the disk(s) in the same form and sequence in which they were initially created.

ESTABLISHING BACKUP PLANS AND PROCEDURES

Backing up data and software is an essential but unexciting network task. It is often viewed as being a necessary evil of network administration, but the fact remains that no one really wants to do it. Some companies waste human resources by requiring highly compensated managers, programmers, and analysts to take turns in the backup process. A much better management strategy is to make one person responsible for performing backup tasks (under the supervision of the network manager).

A successful backup plan is really a list of logistical tactics. It should list what is to be backed up, when backups are to be made, the steps involved in each scheduled backup, and what is to be done with the backup copies of data and software.

Because of the possibility of natural disasters, backup copies should not be concentrated in one location. Some security businesses specialize in backup storage, and it is often wise to learn about their services. In addition, it may be desirable to store multiple generations of backups. Reasons include the fact that the storage media may deteriorate and that some backups are just plain faulty. Some organizations have faithfully conducted backups, only to learn the hard way that the procedures used were faulty and the data could not be restored. Hence, it is important to regularly test the effectiveness of backup procedures to ensure that accessible data is being stored on the backup media.

Many businesses rotate a set of four or five backups and then permanently store month-end backups. These multiple generations provide the organization with the flexibility of restoring data from a particular time. The approaches used to store predetermined sets of backup generations are often referred to as *retention policies.*

Backup policies and procedures must be thorough. The more frequent the backup, the more orderly the backup storage, and the more comprehensively data files are reproduced, the greater the effectiveness of backup policies and procedures. In general, a thorough set of backup policies and procedures ensures that network recovery procedures will be successful.

◆ NETWORK RECOVERY PROCEDURES

Recovering from LAN failures is different from recovering from the failure of a centralized, host-based system. In client/server LANs, the application processing is shared among servers and client workstations. Because of this, when failures occur, they can be more difficult to diagnose and correct. For example, a power failure may cause a workstation to lose the data it is processing, but may not affect the server. Needless to say, such problems need to be anticipated by network managers, and the network must be able to recover from this type of failure.

RECOVERING FROM SERVER FAILURES

Serious problems occur when servers lose power. While most LAN managers ensure that servers are equipped with an uninterruptible power supply (UPS), if the UPS does not function properly or does not allow sufficient time to power down the server in an orderly fashion, far-reaching problems can occur.

Recovering from server failures can be complex. In NetWare LANs, one of the first recovery tactics often involves the use of the VREPAIR (volume repair) utility. This utility systematically scans the organization of the data on the server disk(s), looking for imperfections. If VREPAIR is not successful in restoring network software, data, and the security setups for network users, another utility—BINDFIX (bindery fix)—can be used to attempt to repair damage to the object databases. The bindery includes information related to users, their passwords, their trustee rights to access various directories and subdirectories, and other controls maintained by the NOS. If these operations are not successful in allowing the network to run as it did prior to the failure, the network administrator may have no other choice but to restore the network from backup copies of software and data.

DATABASE RECOVERY

In a large number of networks, users have concurrent access to multiuser databases. Recovering from such failures involves ensuring that the data being processed at the time of the failure is not lost. NetWare has an online system called the *Transaction Tracking System (TTS)* designed to assist in such recovery operations by reconciling modifications to the database.

If the TTS does not function properly, restoration of the database from backups may be the only way to repair the database. However, even backups made a few hours before (or a day before) the failure can result in the loss of data, and paper copies or other audit records may be needed to manually bring the database up to date. This is one of the reasons why virtual disk systems and other continuous backup systems are popular in online transaction processing LANs.

LOW-LEVEL RECOVERY

While restoring a database from backups will sometimes solve data recovery problems, in the case of lost or damaged files, more drastic measures may be needed. For disks in DOS-based workstations, there are tools such as Norton Utilities, PC Tools, and DOS utilities such as CHKDSK and SCANDISK; however, these tools often do not work properly for NOS-controlled disks. For server disks, NOS utilities such as NetWare's VREPAIR may provide a system-level solution to lost or damaged data files, but in some instances this may not do the trick and a low-level format (reformatting of the server disks) may be necessary.

Low-level formats of server disks may also be needed to use data with new NOSs. For example, one network manager reported that after installing Windows NT, a low-level format had to be done before the disk could be used with another operating system.

BINDERY FILES

In NetWare, the bindery files are pivotal. If they are deleted, damaged, or otherwise rendered useless, a significant amount of manual repair work will have to be done. This is one area for which hardcopy documentation of trustee assignments and other key network controls is valuable.

In NetWare LANs, it may be possible to correct problems in bindery files using the BINDFIX utility. However, sometimes the only possible recovery for bindery files is to rename them and to allow the NOS to recreate them. Backups for bindery files can be used only when there has been a complete backup of the server disk and when the contents of the bindery files match those on the disk.

ESTABLISHING RECOVERY PROCEDURES

The creation of written and tested recovery procedures is critical to protecting data. Recovery procedures are best developed in the calm of preparation for trouble rather than in the traumatic moments immediately following file server disk destruction. LAN administrators have been known to panic after a server disk failure and have initiated incorrect recovery procedures, making a bad situation even worse. Conversely, well-designed and faithfully executed backup procedures can give network administrators confidence that most data can be restored.

The availability of multiple backups can also give you the leeway to make mistakes in the heat of the moment and still be able to recover. However, network managers should be aware that some backup systems are not intended for NOS systems and therefore do not properly back up some hidden NOS or bindery files. Of late, most backup system vendors have constructed their software to run as a NetWare Loadable Module, thereby making certain of its capability of running under the NOS in most cases.

As is true for backup procedures, recovery procedures should be in writing and should be systematically executed following failures. While actual procedures are likely to differ from network to network, many are likely to include the elements listed in Table 14-11.

DISASTER RECOVERY

Disaster recovery describes the contingency measures that organizations have adopted to recover from, or to prevent, any monumentally bad event or disaster. A disaster may result from natural causes (such as floods, hurricanes, or earthquakes) or from human sources (such as willful or accidental destruction of equipment). The two main strategies for disaster recovery are physical security and adequate backup.

DISASTER RECOVERY PLANS

Although disaster recovery has long been considered important for organizations dependent on large mainframe or minicomputers, it is also advisable to devise disaster recovery plans for LANs. A ***disaster recovery plan*** is a strategy for quickly reestablishing the organization's computing capabilities following a major system failure or natural disaster. Some companies maintain specially-designed facilities for this purpose and go immediately to the disaster recovery mode. This type of recovery may include restoring backups, movement and diversion of telecommunications facilities, and bringing the

---

1. Determine the type of problem that has occurred and its cause.
2. Attempt to recover damaged or corrupted data; disk utilities and NOS utilities such as NetWare's VREPAIR and BINDFIX may be needed.
3. If necessary, restore the most recent backup of the lost data.
4. If necessary, enter data collected since the last backup until the database is current.
5. Test to ensure that the recovery is complete.
6. Document the problem and the steps needed to correct it; develop procedures to avoid future occurrences of the same problem.

TABLE 14-11
Network recovery procedures

---

system online at a remote facility having computing capabilities equivalent to the capabilities of the ordinary facility.

Network managers must know that disaster recovery plans may include assistance from third-party firms specializing in disaster recovery services. Two prominent firms in disaster recovery are SunGard Recovery Systems and Comdisco Disaster Recovery Services. Each of these firms operates a handful of "hot sites" equipped with standby mainframes, telecommunications equipment, and other hardware resources. Depending on the extent of their needs, clients may pay anywhere from $2,000 to $100,000 monthly for the assurance that they will be bailed out during a disaster—and these sums do not count the extra charges that are levied if a disaster does, in fact, strike.

Disaster recovery plans help the organization systematically reestablish computing capabilities following disasters. Because of this, off-site, secure storage of backups is usually a focal point of such plans. Obviously, if the current facility is burned or flooded, the on-site backups may become useless.

Another important part of a disaster recovery plan may include an insurance policy that will pay immediately in case of disaster. This could allow a company to quickly acquire the hardware needed to reconstruct the network.

◆ TRENDS IN NETWORK SECURITY

LAN backup and recovery is becoming a much more complex topic because of the increased value of the information stored in LAN databases. Equipment such as automatic tape changers, diagnostic routines that can be run prior to backup (to ensure that backup mechanisms are working correctly), optical disk backup systems, and continuous backup systems such as virtual disk systems are all a part of this progression.

Much of the software that accompanies backup and restoration hardware allows for the programming of unattended backups with multiple tapes or multiple disks of one kind or another. These systems automatically back up a system at specified times, often at night when some systems are not being used.

In addition, companies are learning the value of well-designed, written backup procedures. These and systematic recovery procedures can minimize the work and effort needed to recover from system failures.

An ounce of prevention is worth a pound of cure. The acquisition of reliable hardware and software designed to run in network environments can minimize the threat of system failures, as can uninterruptible power supplies and good virus detection/eradication software (and policies).

Many newer network management systems, especially those designed for WANs, allow for the central backup of key network systems. Such archiving of remote systems under the control of a central network management system can, if properly implemented, be simpler than developing and implementing plans for each different location.

Disk technologies such as the RAID scheme with "hot swappable" disks may prevent the need for conventional backup facilities. Because of such systems' ability to make limited recoveries without taking the system down, most failures can be forestalled.

◆ SUMMARY

The importance of network security is often determined by the value of the network's data, software, and other computing resources. In general, organizations are concerned with both external and internal security threats; however, the latter are the most common. Among the security threats that network managers are most concerned with, unauthorized access to data and applications, unauthorized alterations of data and applications, the introduction of viruses and other malicious software, and eavesdropping (wiretapping) on network traffic are the most serious. Security breaches have the potential to cause network interruptions that make network resources unavailable to users.

Both physical security and electronic security mechanisms may be utilized by network managers to protect network resources. Physical security mechanisms include locating the network in areas secured by access card systems, badge readers, and biometric verification devices. It may also include fire detection systems, fire-resistant storage containers, closed-circuit television systems, and burglar alarms.

Electronic security mechanisms include user profiles, terminal profiles, callback systems, system monitors, passwords and other verification/authentication schemes, logon scripts that enforce user access privileges, automatic logoff mechanisms, encryption, and file and database security mechanisms. In Novell NetWare LANs, trustee rights assignments play a prominent role in electronic security. The NOS' bindery files contain user passwords and security parameters. In NetWare 4.01, the NetWare Directory Services (NDS) is a focal point for network security. Similar security provisions are found in UNIX networks. In any of these networks, the ability to make major changes in access rights is restricted to the network manager and specially designated network staff members.

Backup and recovery procedures are also important aspects of network security. Backup mechanisms may include hardware backups, for example, server duplexing, disk duplexing, disk mirroring, tape backup systems, optical disk backup systems, RAID (redundant arrays of independent disks) systems, and virtual disk systems. Backup plans and procedures should be documented and faithfully executed.

Recovery is concerned with restoring the network to operational status after a major network failure has occurred (for example, a server failure or the failure of an important server disk). Recovery is also concerned with restoring data files and databases to their most current state. This involves identifying and repairing damaged data and system files using disk and other data recovery utilities, restoring data and system files from the most recent backups, and manually reentering lost data. Written and tested recovery plans and procedures can make recovering from failures a lot easier.

Disaster recovery is concerned with restoring computing capability in the wake of a monumentally bad event or natural disaster. Disasters may result from natural causes (such as floods, hurricanes, or earthquakes) or from human sources. Disaster recovery plans should be developed to ensure that the organization can systematically restore computing capabilities after a disaster. These plans may include temporarily shifting computing operations to the computers of a third-party firm that specializes in disaster recovery services.

Among the important trends in LAN security are the growing popularity of continuous backup systems for critical data and applications and the ability to perform unattended backups. Virus detection and eradication software is also being more widely used. Server duplexing and disk redundancy schemes such as RAID are becoming more commonplace. And more companies have developed written backup, recovery, and disaster recovery plans.

✳ KEY TERMS

Bindery
Concurrent logons
Disaster recovery plan
Duplexed servers
Encryption

Fault tolerance
Grace logons
Password
Physical security
Security equivalence

✳ REVIEW QUESTIONS

1. Why is network security important? In general, how do you gauge the relative importance of security in one LAN over another?
2. What are the major types of security threats?
3. Why must network managers be concerned with both external and internal security threats? Identify several examples of each.

4. Why are network managers concerned with viruses? What approaches may be used to prevent virus infections?

5. What is an intruder detection system? What role do server error logs play in such systems?

6. Describe how security differs between peer-to-peer and server-based networks?

7. Identify and briefly describe some of the federal laws related to network security.

8. Identify the mechanisms that may be used to enhance the physical security of a network.

9. Identify and briefly describe the types of electronic security mechanisms that may be found in computer networks.

10. What is a password? How are passwords used in network security?

11. Identify the guidelines that network managers often follow when developing password policies for network users.

12. Identify and briefly describe the various types of electronic security mechanisms (utilities) found in Novell NetWare LANs.

13. What are the differences between grace logons and concurrent logons? Why may each be needed?

14. Why may automatic logoff mechanisms be needed? Identify and briefly describe some automatic logoff mechanisms.

15. What is encryption and why is it important?

16. Identify and briefly describe the types of file and database security mechanisms found in NetWare LANs.

17. Identify and briefly describe the types of file and database security mechanisms found in UNIX networks.

18. What is "security equivalence" and why is it important?

19. What are bindery files and what role do they play in network security?

20. In network environments, what is meant by the term *backup*? Why are backups important?

21. What decisions must network managers make as they devise backup plans? Why are backup plans important?

22. Identify and briefly describe the various types of hardware backups that are used in LANs.

23. What differences may be observed in RAID systems?

24. What are the differences between optical disk backup systems and virtual disk systems?

25. What types of software backups are used for networks?

26. What are the characteristics of effective backup plans?

27. What are network recovery procedures and why are they important?

28. What approaches may be used to recover from server failures?

29. What approaches may be used to recover from disk failures?

30. What elements are likely to be included in recovery plans?

31. What are disaster recovery plans? What elements are likely to be included in disaster recovery plans?
32. What trends may be observed in network security?

✳ DISCUSSION QUESTIONS

1. Why have internetworking and the development of distributed client/server networks increased the security challenges faced by network managers?
2. Why are network managers increasingly concerned with network fault tolerance? What do you feel are the primary approaches that should be used to enhance the fault tolerance of networks? Justify your recommendations.
3. Discuss the issues that should be considered when determining whether differential, incremental, or full backups are appropriate. Also, discuss the advantages and disadvantages of storing backups off-site.
4. Discuss why effective network security strategies are likely to include hardware, software, and data backups. Also discuss why these backups are crucial for network recovery and disaster recovery.
5. Discuss the issues associated with the decision to hire an external firm for network/disaster recovery. What factors are most important in deciding whether or not to contract with such a firm?

✳ PROBLEMS AND EXERCISES

1. Consider the following scenario: A new system has been in use for 3 months. The old system has been dismantled. An average of 5 accounts has been added each day for 65 days. Each account takes approximately 1 hour to enter and the average hourly labor rate, adjusted for benefits, is $8.50 per hour. Weekly backups are made, but no daily backups are made. The business must have access to the system each day in order to remain in business. Give an estimated value of this system, assuming that the program files can be easily reproduced from backups at the vendor's location. List the potential security weaknesses that you see in the description of this system.
2. John's business involves accessing drivers' records from the state system and checking these records for moving violations for insurance companies. His contract with the state provides that he is responsible for the privacy of the data he accesses and that any security problems with his system may be a cause for termination of the contract. John's business makes a net profit of $250,000 per year, and this company is

his sole source of income. John is considering adding dial-up telephone lines for the convenience of two or three insurance companies, which are his best customers. The new addition will cost $15,000, and the sophisticated security package that comes with it will cost an additional $10,000. The security package will use a system of passwords to secure the system. Should John buy the additional security package? Why? What other security mechanisms should John consider?

3. Jason works for a large financial institution as manager of front office operations. He overhears a conversation between two bank tellers; one teller discloses that she has found another job and will be leaving next week. She continues by saying that she thinks she can remove $100 from her drawer for personal use without anyone detecting the loss. Jason's boss is responsible for the security of the entire computer system. What should he do?

4. Karen has just accepted a new position with a small company that uses computers to create graphics for advertising. Karen's position is that of computer support person. Karen has just found out that the storage of these graphics on the computer is vital to this small business, but that no copies of the graphics files exist except on the main computer, which is networked to three desktop workstations. Karen investigates; streaming tape backup hardware costs about $350, and the tapes cost approximately $15 each. When Karen approaches the owner of the business about establishing a backup system, the owner objects on the basis of cost. She asks Karen, "Why do we need backups? That is certainly an expensive addition." Karen is convinced that the system should be installed, and she is also concerned that she might be held responsible for data loss. How can she try to convince the owner to purchase a backup system?

5. Contact an officer from the local police or FBI office assigned to deal with data security crime issues. Ask how the officer would recommend that data be secured in financial institutions. Ask the officer for information related to the techniques used to access unauthorized data in your local area.

6. In your data processing department, you have only 3 hours each evening between the closing of one day and the opening of business for the next day. The LAN system must be online during the hours of business. There has been such an accumulation of data in the past few months that 4 hours are required to make a complete backup of the entire system. The business is closed for 24 hours on Sunday. What suggestions could you make to both accommodate the time restrictions of the business and make an adequate backup of the LAN once each day? Also, make suggestions as to where the backups should be kept.

7. Invite a local network manager to come to your class to discuss the network security mechanisms used in his or her organization. Be sure to ask about backup, recovery, and disaster recovery procedures.

8. In a networking or computing periodical, find a recent article on network security mechanisms for peer-to-peer networks. Summarize the article's contents in a paper or class presentation.

9. Find a recent article on either optical disk backup systems or continuous backup systems (such as virtual disk systems) for LANs. Summarize the article's contents in a paper or class presentation.

10. Find a recent article on computer crimes committed in network environments. Summarize the article's contents in a paper or class presentation.

✳ REFERENCES

Anthes, G. H. "Piracy on the Rise; Companies Fear Liability." *Computerworld* (April 18, 1994): p. 1.

Anthes, G. H. and J. Daly. "Internet Users Go on the Alert." *Computerworld* (February 14, 1994): p. 14.

———. "Intruders Stalk Internet Accounts." *Computerworld* (February 7, 1994): p. 14.

Bates, R. J. *Disaster Recovery for LANs.* New York: McGraw-Hill, 1994.

Berg, A. "Network Backup Strategies and Tactics: Plan Ahead by Assessing Your Needs." *LAN TIMES* (October 18, 1993): pp. 50.6, 50.7.

Bowden, E. J. "StationLock Physically Secures Net Workstations." *LAN TIMES* (November 1, 1993): p. 23.

Cooper, E. B. "Taming the Beast: Backup Challenges Administrators." *LAN TIMES* (August 8, 1994): p. 54.

Daly, J. "PC Identification System May Sink Software Pirates." *Computerworld* (January 31, 1994): p. 37.

Donahoo, M. and B. Homer. "Quarantined Area: Tackling Viruses NLM-Style." *LAN TIMES* (August 9, 1993): p. 71.

Duncan, T. "The Long Arm of the Network." *LAN TIMES* (August 8, 1994): pp. 59, 66.

Krzemien, R. "The Disaster-Recovery Plan: Your Bridge Over Troubled Water." *LAN TIMES* (October 18, 1993): pp. 50.14, 50.16.

McConnell, J. "Getting a Management Handle on Virtual LAN Infrastructures." *Network World* (May 23, 1994): p. L.10.

Stephenson, P. "Negotiating a Cure for Viruses: What the Law Cannot Do." *LAN TIMES* (October 18, 1993): p. 72.

Ubois, J. "Safe and Secure." *LAN TIMES* (August 8, 1994): pp. 58, 65.

Case Study:
Southeastern State University

Southeastern State University has had its share of problems with computer viruses in recent years. These programs, with their built-in codes to copy themselves or to transmit copies to other networked machines, have become an ongoing problem. As a result, virus alerts are regularly published in the student newspapers, distributed in flyers, sent out in memos to faculty and staff, and posted on bulletin boards in almost every building on campus. The university's administration has realized that it can never truly eliminate the viruses that find their way to SSU, but can only take preventive measures and react to problems as they occur.

Needless to say, viruses have added a new dimension to the operational duties of LAN managers at SSU. While most LANs now incorporate virus detection and eradication software that is automatically run every time a server or workstation is booted up, it is a rare day that passes without the software detecting something. Once a virus is detected, network staff has to move quickly to isolate it and to keep it from spreading.

Some viruses have attached themselves to application software transmitted over the network. Others have infected the boot sectors of hard disks, making it impossible to make the workstation operational, and others have even infected network drivers. Viruses have caused server crashes, workstation lockups, and a variety of screen messages ranging from semi-humorous (such as letters of a word processing document sprouting wings and flying off the screen) to social commentaries (such as "Legalize Marijuana") to obscenities. Viruses have overwritten data, applications, and system files stored on disk, effectively destroying them and requiring their restoration from backups or reinstallation. In every instance, virus infections in LANs have caused disruptions in network operations and made it difficult for students to complete their assignments, often at the busiest times of the semester. Viruses have permanently driven some students away from the labs; they have lost all faith in the reliability of the networks and refuse to do any work on LAN workstations.

Recently, encrypted viruses have taken up residence at SSU. These have been designed to avoid leaving the "signatures" that virus detection and eradication software looks for when scanning disks and software.

In some instances, infections have gotten so bad that low-level formats of all the server and workstation disks in a network have been needed to get rid of them. Of course, this meant that all the software and files had to be reinstalled before users were allowed back on the network. Doing so was a time-consuming effort which added to the considerable amount of network personnel time and effort devoted to reformatting the hard disks in the first place. As a result of this, several LAN managers have considered removing the disk drives on workstations and/or replacing them with diskless workstations.

No long ago, an accounting instructor asked his 35-student class to turn in a spreadsheet assignment in both hardcopy and disk form. Having experienced virus problems in the past, he scanned each disk for viruses before retrieving the student files. He found that 23 of the 35 disks were infected with viruses; 13 of these were infected with at least 2 viruses.

A computer science instructor asked his class to do presentations on "hot" topics in computing. One student did a presentation on viruses and brought in a copy of a book, *The Little Black Book of Computer Viruses*, that contained detailed instructions for writing the code needed for viruses. He also brought in disk copies of virus logic that he had gotten through coupons that came with the book.

Viruses on student disks have infected stand-alone computers found in faculty members' offices and homes. This usually occurs because faculty members access files stored on student disks without scanning them. Needless to say, this has enhanced faculty interest in virus detection and eradication software.

Now that SSU is moving toward internetworking LANs and connecting stand-alone LANs to university networks and the Internet, concerns about virus infection are increasing. The university knows that the network plans are important, but it also is aware that virus problems may mushroom because there will be new avenues for viruses to come to SSU and new channels through which they can spread. Because of this, no one at SSU expects to see a decrease in the all-too-common virus alerts, memos, and flyers.

CASE STUDY QUESTIONS AND PROBLEMS

1. From this and the previous segments of the case study, identify at least five different avenues via which computer viruses can enter LANs at SSU. Describe how SSU's movement toward internetworking and establishing connections between traditionally stand-alone LANs and the VAX network and Internet has created new avenues for viruses to enter and spread at SSU.
2. If you were a LAN manager at SSU, what policies would you implement in computer labs to minimize the possibility of virus infections?

3. What provisions for viruses should be incorporated in backup and recovery plans? Develop and justify specific guidelines and policies that should be included.
4. How do you feel about *The Little Black Book of Computer Viruses*? What are the advantages and disadvantages of the existence of such a book?

PART FIVE

Internetworking

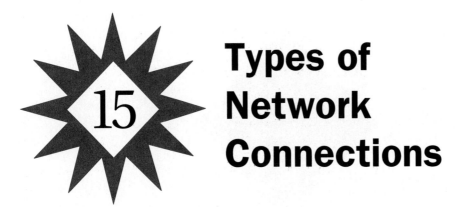

Types of Network Connections

CHAPTER OBJECTIVES

After completing this chapter, you will be able to:

◆ Describe the approaches used to establish microcomputer-to-mainframe connections.

◆ Describe how users can connect to LANs from remote locations.

◆ Describe the differences among the major types of LAN interconnection hardware.

◆ Describe how backbone networks can be used to interconnect networks.

◆ Briefly describe the interfaces that may exist between LANs and private branch exchanges (PBXs).

◆ Briefly describe the hardware, software, and application interfaces found between LANs and WANs.

◆ Identify some of the major network management software and protocols for interconnected networks.

As we have noted throughout this book, one of the important trends of the 1990s is enabling users to share computing resources. Many LANs have been created for this purpose. This trend is a driving force in the interconnection of LANs, the establishment of interfaces between LANs and WANs, the creation of client/server networks, the burgeoning popularity of the Internet, and the increasing popularity of telecommuting and mobile computing, which enable workers to establish remote connections (sometimes wireless connections) to corporate networks. This chapter discusses network connections.

In this chapter, we will begin by focusing on micro-to-mainframe connections; these were the first types of remote connections and are still widely used today. Next, we will cover how remote users can connect to LANs through dial-in ports. After that, we will address how LANs can be interconnected and LAN-to-WAN interconnections. Finally, we will discuss the network management software and protocols used to monitor and manage message traffic in interconnected networks.

◆ ESTABLISHING MICRO-TO-MAINFRAME CONNECTIONS

Prior to the rise of the microcomputer in the 1970s, most computer systems were centralized, host-based systems. Their popularity has continued into the 1990s, but such systems are giving way to client/server networks in most organizations. In centralized systems, the processing power for an entire company is concentrated in a single large processor (the host), usually a mainframe or a minicomputer.

Throughout the 1950s and early 1960s, these systems typically operated in a batch-processing mode and used punched cards and tapes for data input and storage. All input, processing, and output was done at the host's central location with the assistance of the computer operations staff. Later, it became possible to have online access to the host. This made it possible for users to carry out processing tasks from distant terminals. Online access and remote job entry made it possible for users in different parts of the organization to use the centralized computer.

With online access, all input went directly from the users' terminals to the central computer for processing; after processing, all outputs to users came directly from the host. The first terminals used in these online systems were called *dumb terminals* because they could do no processing on their own, serving only as input/output mechanisms linking users to the central CPU.

Over time, centralized systems and their support equipment have grown more efficient and sophisticated. In many instances, dumb terminals have given way to smart terminals or intelligent terminals, which have sufficient processing capabilities to relieve the central CPU of some of its processing tasks.

It has also become quite common to use microcomputers in place of terminals in centralized, host-based systems. Microcomputers equipped with modems and communications software packages (such as ProComm and CrossTalk) can interact with a host as do terminals. The communication software used includes terminal emulation modules that cause the host to recognize the microcomputer as a terminal. Terminal emulation can also be accomplished by installing a terminal emulation board in one of the microcomputer's expansion slots.

When users can access computing resources from remote locations, the geographic scope of both host-based and distributed networks expands. Workers can telecommute via connections between home computers and the organization's network. In addition, remote access enables salespeople and other professionals in the field to stay in contact with their main offices.

In many instances, the computers to which users at remote locations connect have some type of dial-in, or direct dial, capability. The media used to carry the message are usually communications circuits (both local and long distance) provided by telephone companies. However, they may also be wireless connections such as those available in cellular telephone and data networks. Because the host in a host-based system is often a mainframe (or minicomputer), the connection between it and a computer at a remote location is often called a ***micro-to-mainframe link***. This type of remote access is illustrated in Figure 15-1.

◆ COMMUNICATIONS PROTOCOLS FOR MICROCOMPUTERS

A variety of microcomputer protocols may be used to establish micro-to-mainframe connections. Some are quite simple; others are more sophisticated. Like any protocol, the protocols that enable data and files to be

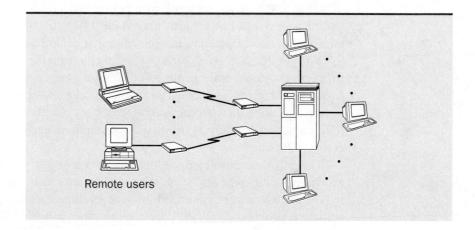

FIGURE 15-1
Micro-to-mainframe connections

Remote users

transferred between microcomputers and larger computers (minicomputers and mainframes) ensure that both sender and receiver agree on details such as the size and format of message frames, the number of start and stop bits (or characters), the error checking methodology that will be used, the location of control characters, half or full duplex transmission, and so on.

A number of microcomputer communication protocols are listed and briefly described in Table 15-1. Most of those listed are often classified as file transfer protocols that allow two computers to exchange files. KERMIT is among the best known and most widely used file transfer protocols. The many versions of KERMIT allow microcomputers to connect to a variety of mainframes and minicomputers; typically, KERMIT uses asynchronous links to transfer ASCII files. Along with XMODEM and YMODEM, KERMIT is often included in terminal emulation programs and communications software packages for microcomputers.

IBM's APPC/PC protocol allows files to be transferred between microcomputers and larger systems through high-level program-to-program links. This is a peer-level type of communication that makes it unnecessary for the microcomputer to emulate a terminal.

◆ COMMUNICATIONS SOFTWARE PACKAGES

To connect to a mainframe, network, or commercial databases—such as Prodigy, CompuServe, America Online, Dialog Information Services, or the Dow Jones News Retrieval Service—a remote microcomputer typically utilizes communications software. Many communications software packages can be purchased (such as ProComm and CrossTalk) or one can acquire public domain communications software. In addition, communications modules may be bundled with microcomputer operating systems such as Windows.

Most communications software packages incorporate terminal emulation modules. Because IBM mainframes are the most common types of larger systems found in businesses, IBM-3270 terminal emulation is a very common type of terminal emulation module; this emulates how the IBM 327x family of terminal devices—terminals, printers, and cluster controllers—communicate with IBM mainframes. Other commonly emulated terminals are the IBM 3101, the DEC VT 100, the DEC VT 220, and the DEC VT 320 graphics terminal. The two generic terminal types most commonly found in communications software programs are ANSI and TTY.

Terminal emulation is necessary because microcomputer keyboards are laid out (mapped) differently from terminals designed to be connected to the mainframe or minicomputer. The modules used for terminal emulation cause the larger system to recognize keystrokes on the microcomputer as terminal keystrokes. As the type of terminal emulation needed

Protocol	Description
X-ON/X-OFF	Among the oldest protocols; first used on teletype machines. Can be used to transfer text files. If idle, receiver transmits X-ON signal to sender; if busy, receiver sends X-OFF signal. Today, it is most commonly used in bulletin board systems (BBSs).
XMODEM	One of the first public domain protocols designed for micro-to-micro communications; often used in micro-to-mainframe connections. Divides data into 128-byte blocks. Uses stop and wait ARQ and includes an effective checksum error detection scheme.
YMODEM	Uses message frames containing 1024 bytes and CRC-16 error checking capable of detecting 99.99 percent of errors. Has greater throughput than XMODEM.
KERMIT	Perhaps the most widely used protocol for microcomputers. A public domain family of protocols. Can be used for both micro-to-micro and micro-to-mainframe connections. Requires only an asynchronous serial connection and KERMIT running on each of the communicating computers.
X.PC	A protocol developed by Tymnet that constructs packets on microcomputers capable of being transmitted over X.25 (packet switching) networks. Can also be used to transfer files between microcomputers.
PC-BLAST II	A recent version of BLAST (Blocked Asynchronous Transmission) that gives asynchronous transmission the efficiency and throughput of synchronous transmission. It can be used to establish connections between microcomputers, mainframes, and midrange systems and automatically converts file formats between computers (for example, from DOS to UNIX). Can be operated in both half- and full duplex.
MNP	Microcom Systems' networking protocol (MNP) is a set of protocols (arranged in classes of sophistication) supported by many of the modems used with today's microcomputers. Supports both asynchronous and synchronous data transfers up to 38,400 bps.
ISDN	Integrated Services Digital Network (ISDN) is a service provided by common carriers that permits data, text, audio, and visual images to be transmitted over ISDN circuits at 64,000 bps. Usually implemented in microcomputers through ISDN terminal adapter cards.
APPC/PC	Advanced Program-to-Program Communications/PC (APPC/PC) can link microcomputers and mainframes to allow cooperative applications to exchange data between machines running different operating systems. Developed by IBM, this protocol is said to be well-suited for diverse client/server networks.

TABLE 15-1 Protocols used in micro-to-mainframe connections

depends on the host type (mainframe or minicomputer) to be connected to, it is important to select communications software supporting the needed type of terminal emulation.

In addition to the type(s) of terminal emulation supported, a number of other factors should be considered in selecting appropriate communications software. Some of these are briefly described in the following sections.

OPERATING SYSTEM SUPPORT

Communications software packages vary in the extent to which they support multiple mainframe and microcomputer operating systems. For example, not all communications packages work with Windows; if Windows support is needed, the choice of communications software may be limited.

REMOTE END MICROCOMPUTER SUPPORT

If micro-to-micro connections are desired, it is important to consider the types of microcomputers (for example, IBM, IBM-compatible, Apple) supported at the remote end.

COMMUNICATIONS PROTOCOL SUPPORT

The variety of communications protocols supported (such as those listed in Table 15-1) differs among communications software. It is important to select communications software that supports the needed protocols.

INFORMATION SERVICES SUPPORT

If users need access to online information services such as CompuServe, Dow Jones News Retrieval, Dialog, and so on, communications software with prewritten logon scripts may be desirable. Such logon scripts are typically easy to use, but communications packages vary in the number of automatic logon scripts included.

REMOTE CONTROL CAPABILITIES

Some communications packages make it possible to control or configure a microcomputer remotely; others do not. As noted in Chapter 13, remote control capabilities can enable network managers to troubleshoot and correct user problems from the network control center. If these capabilities are desirable, communications packages with remote control functions should be selected.

SECURITY FEATURES

If other microcomputers will have dial-in access to the microcomputer on which the communications software is installed, security modules are desirable. Password security, dial-back security, and call logging modules are included in some communications packages, but not others.

SPECIAL USE CAPABILITIES

Some communications software packages support special communications applications such as the ability to transmit data over cellular data and telephone networks, and the ability to send and receive fax transmissions.

In addition, some packages enable users to support multiple, simultaneous communications sessions with other computers.

SCRIPTING LANGUAGES

Scripting languages enable users to write, store, and use their own communications scripts (such as those used for unattended file transfers over long-distance common carrier lines when rates are lowest). These languages make it possible to develop and maintain customized communications routines that run automatically at predetermined times. Communications packages vary in the number of commands available in the scripting language (and whether a scripting language is available). If scripting language capabilities are desirable, the sophistication of a package's scripting language and the applications for which it will be used may be key selection criteria.

GENERAL EASE OF USE

The overall user-friendliness of the communications package is an important factor in selecting communications software. For example, communications packages with graphical user interfaces (GUIs) are often perceived to be more user-friendly than those that require users to enter text commands. Menu-driven packages, especially those with online help, are also generally considered to be user friendly. Other desirable features include keyboard dialing (the ability to use the keyboard to enter phone numbers), auto-redial (the ability to redial previously dialed numbers—sometimes at regular intervals without additional user intervention), auto-answer (answers incoming calls and manages security features), and stored number/setup parameters (which allow phone numbers and setup parameters—such as baud rate, parity, terminal emulation type, and so on—to be stored, recalled, and modified). Generally, the easier software is to use, the better.

TECHNICAL SUPPORT

Another factor that may be an important consideration in the selection of communications software is the availability, quality, and cost of technical support provided by the vendor or developer.

OTHER POPULAR FEATURES

Some of the other popular features and capabilities of communications software include the ability to use a mouse to make menu or directory selections (or to generally use the software), support for e-mail applications, the ability to capture and store incoming data on a disk, and text editor capabilities (especially important for packages with scripting languages).

Users and network managers must consider the factors listed to make appropriate selections. Some of the packages that are most likely to be encountered include Bitcom, CrossTalk, Freeway, MicroPhone, Relay Gold,

PCDIAL LOG, PC-Dial, ProComm, and WATSON. Many of these have added features and functionality unavailable in previous versions. In addition, some packages provide support beyond data communications. For example, both PCDIAL LOG and WATSON combine data communication features, telephone (voice) communications, and voice mail on the same microcomputer. Voice mail support is rapidly becoming another popular special use feature in communications software packages designed for both voice and data communications.

◆ CONNECTING TO LANs FROM REMOTE LOCATIONS

The increasing popularity of telecommuting and mobile computing has prompted many organizations to enable workers to establish connections to LANs from remote locations. Such remote access to LANs allows workers to check and exchange e-mail messages, upload and download files, and to generally utilize LAN resources.

COMMUNICATIONS SERVERS

One of the most common mechanisms used to allow remote users to connect to a LAN is through *communications servers* (sometimes called *access servers, network resource servers,* or *telecommuting servers*). This approach is illustrated in Figure 15-2. The communications server manages interfaces between modems as well as the exchange of messages and files. When this approach is used, a special type of remote control software is found in the access server. This software answers the modem, validates user IDs and passwords, and logs the remote user onto the LAN; sometimes it allows the remote user to access and remotely control a particular workstation attached to the network (usually the one in his or her office). Two of the most popular remote control software programs used in communications servers are Close Up/LAN (from Norton/Lambert) and Remote LAN Node (from Digital Communications Associates). Communications servers typically cost from $10,000 to $20,000 (including the software); one of the primary determinants of the cost is number of users permitted simultaneous access to the LAN through the communications server from remote locations.

DIAL-IN ACCESS THROUGH A LAN WORKSTATION

A second approach used to allow a remote user to connect to a LAN is by using communications software to provide a dial-in connection to a LAN workstation. For example, a manager at a home computer may be able to dial in directly to the workstation in the office (which, in turn, is attached to the LAN). In this instance, the communications software typically must include remote control capabilities. From the remote location, the user

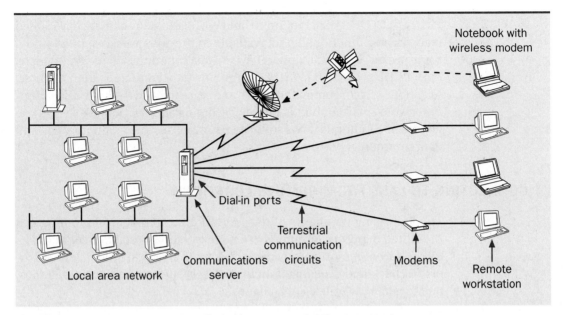

FIGURE 15-2　Remote connections through a communications server

essentially seizes control of the LAN workstation to which he or she has dialed in. By doing so, the remote user is able to access and utilize the computing resources available on the LAN. Some of the popular remote control software packages for establishing this type of connection include Carbon Copy (from Microcom), Close Up (from Norton/Lambert), and PCAnywhere (from Symantec/Peter Norton). Prices for these packages range from approximately $100 to $300.

LAN MODEMS

A third approach used to allow remote users to connect to a LAN is through devices known as *LAN modems* or *dial-in servers*. These are usually designed to provide remote users with access to a particular type of LAN (for example, Ethernet or Token Ring), or through a particular network operating system protocol such as IP, IPX (for Novell LANs), or AppleTalk. A LAN modem is essentially a modem with a network adapter card attached to the LAN; a remote user's computer is recognized as another network workstation when accessed through the LAN modem. Some LAN modems can support multiple simultaneous remote users. Examples of dial-in servers include Gatorlink (from Cayman), LANMODEM (from Microtest), and LANRover/E and NetModem/E (both from the Shiva Corporation). Prices for these are comparable to those for moderate to high-end microcomputers—from $1,500 to $3,000.

◆ INTERCONNECTING LANs

Although stand-alone LANs are still being implemented, the predominant trend is toward connecting two or more LANs, as well as toward connecting LANs and hosts, and LANs and WANs. By the turn of the century, stand-alone LANs are expected to be a rarity.

Several approaches may be used to interconnect LANs. These include providing interconnections through special-purpose hardware (repeaters, bridges, routers, brouters, switches, and so on), as well as establishing interconnections via backbone networks or PBXs. We will focus on each of these approaches in this section.

LAN INTERCONNECTION HARDWARE

The several important types of LAN interconnection hardware include repeaters, bridges, brouters, routers, and switches. Gateways can also be used to interconnect two or more different types of LANs. These interconnection technologies were introduced in Chapter 2, but we will be taking a closer look at them in this section.

Most discussions of interconnection technologies are based on the OSI reference model to assist in understanding their functionality. As may be observed in Figure 15-3, our discussion is no exception.

Repeaters

While generally considered to be a type of interconnection hardware, *repeaters* are best described as LAN technologies that geographically broaden LAN resources and connect LAN segments. Repeaters are not used to connect two or more LANs.

Repeaters are considered to be at the physical layer of the OSI model (Figure 15-3). This is because repeaters do not modify the data packets that pass through them; their only job is to ensure that transmitted signals remain strong enough to be accurately decoded by receivers.

Recall from Chapter 5 that one of the specifications included in the IEEE 802 standards for LANs is the maximum length of a LAN segment. Essentially, this maximum length indicates how far apart two network devices can be before a repeater or other signal regeneration hardware (such as an amplifier) is needed. Through the use of repeaters, distances beyond those specified in the IEEE 802 standards can be achieved; that is, repeaters make it possible to extend the geographic scope of the LAN. Figure 15-4 illustrates the use of repeaters.

Signal regeneration hardware is necessary because the signals transmitted by network devices attenuate as they travel along the communications medium. The weakened signals are a frequent cause of transmission

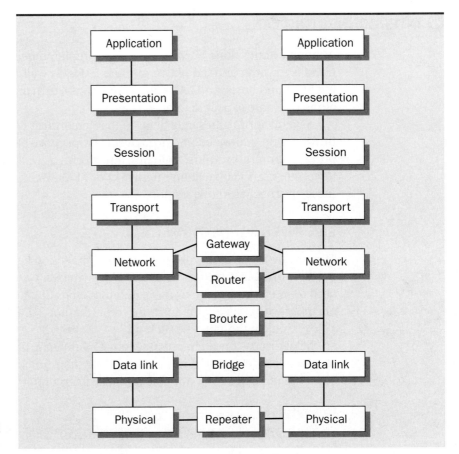

FIGURE 15-3
Interconnection
technologies and
the OSI model

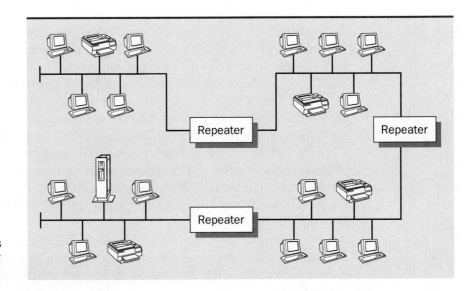

FIGURE 15-4
Using repeaters
to connect LAN
segments

errors detected by receivers (and hence, retransmissions and extra traffic over the medium). Repeaters correct for attenuation by regenerating (restoring to original full strength) the signals received before passing the signals on down the medium.

Because the primary task of repeaters is to renew the signals that pass through them, they are relatively simple devices. In most instances, the entry and exit media to the repeater are the same (for example, twisted pair to twisted pair, coax to coax). However, some repeaters have different entry and exit media (for example, coax to fiber optic or twisted pair to coax) and also serve as media transfer boxes. Also, some repeaters have more than two ports; this makes it possible to connect three or more LAN segments through the same device.

Bridges

Traditionally, *bridges* have been described as technologies used to connect two separate LANs that each used the same media access protocol (for example, both LANs employ CSMA/CD or token passing for media access control). Figure 15-5 illustrates the use of a bridge to connect two Ethernet LANs with bus topologies. Today, however, some bridge technologies are used to connect LANs that employ different media access protocols. These newer bridges incorporate protocol conversion capabilities—they are capable of reformatting data packets from one media access protocol (for example, CSMA/CD frames) to another (such as that used in an IBM Token Ring). These expanded capabilities of bridges have blurred the traditional distinctions among bridges, routers, and gateways.

In light of the expanded definition of a bridge, perhaps the best way to describe it is as a device used to connect two networks at the data link level of the OSI model (see Figure 15-3). However, network managers must remember that references to bridges may encompass the traditional or the newer definition of bridging technologies.

Because a bridge can be used to pass messages and data packets between two LANs, it is more sophisticated than a repeater. A bridge must know the addresses of the workstations or devices attached to each network. It must also be able to determine when a particular packet sent by a workstation in one LAN should be transferred to the other LAN for delivery to its intended destination. In addition, if the bridge has protocol conversion capabilities, it must be able to locate the destination addresses in the different types of packets and to reformat the packet if the intended receiver is located in the other network.

Bridges may be implemented in several ways. For example, they may be cards that plug into an expansion slot on the microcomputer; this may be sufficient for connecting two small LANs, especially two small LANs using the same protocol (in this case, the bridge may pass all messages to

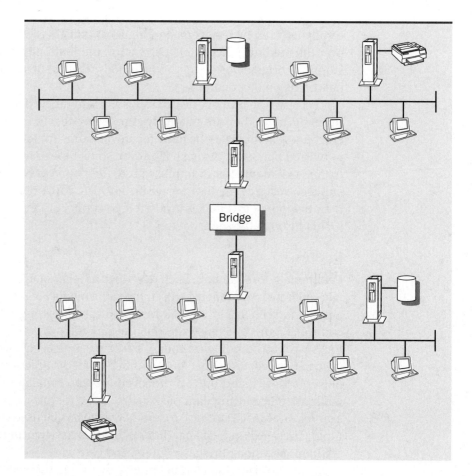

FIGURE 15-5
A bridge con-
necting two
Ethernet LANs

both networks regardless of the intended recipient). Today, however,
bridges are most frequently stand-alone devices and may possess special
characteristics that depend on the features of the two LANs being connected. For example, if the two LANs operate at different speeds (say, one
is a token-ring LAN running at 4Mbps and the other is a token-ring LAN
running at 16Mbps), the bridge must be capable of speed conversion—this
is usually implemented through buffers. Also, if the two LANs are implemented using different media (say, one uses twisted pair and the other
coax), media conversion capabilities are needed.

Many of the bridges on the market today are called *learning bridges* (or
transparent bridges). The name is derived from the fact that they build automatically their own routing table from the messages on and across the
networks (unlike older bridges that required network personnel to manually input the data needed to build the routing table).

It is also important to note that bridges can be used to interconnect
LANs that are geographically distant. These are often called *remote bridges,*
in contrast to local bridges used to connect networks near one another.

Remote bridges include connections needed to directly interface with high-speed digital data circuits such as Switched 56 (56,000 bps) and T-1 lines (1.544 Mbps). Remote bridges may be used by organizations to build private WANs.

Bridge performance is generally compared in terms of filtering rate and forwarding rate. Both of these rates are measured in frames or packets per second. *Filtering* is the process used by the bridge to examine a packet and determine if it is addressed to a workstation in the other network. Filtering rates typically range between 10,000 and 60,000 packets per second. *Forwarding* is the process that enables a packet to physically cross the bridge to the other network; forwarding rates can range from less than 1,000 to more than 30,000 frames per second.

Some bridge software (such as BridgeIT! from Triticom) enables network managers to monitor bridge operations in real time and to capture data for later analysis. Bridge statistics can play important roles in network tuning and capacity planning (discussed in Chapter 13). For example, if forwarding rates significantly exceed the number of packets transmitted between workstations within each of the networks, network performance may be improved by combining the two networks into a larger network without a bridge (this assumes that the two networks are geographically near).

Routers

As may be observed in Figure 15-3, **routers** are located at the network layer of the OSI model. The responsibility of the network layer is to establish, maintain, and terminate links between network nodes (for example, between two workstations in two different networks). Technologies at the network layer must know the source and destination addresses of messages to be able to carry out these functions.

Routers are responsible for delivering message packets across multiple LANs. Two or more of these LANs may be separated by wide area communications links. Hence, like remote bridges, routers may play a crucial role in the development of private WANs. Figure 15-6 illustrates the use of routers to interconnect an Ethernet LAN and an FDDI LAN over an X.25 network. Note that each of the routers in the figure could be used interconnect either of the two LANs with multiple LANs and WANs.

Similar to bridges, routers examine the message packages received and forward them to the appropriate network for delivery. However, unlike bridges, routers do not filter (examine) all message packets transmitted by the workstations within the networks they interconnect. A router filters only the packets addressed specifically to it.

The forwarding process also differs between bridges and routers. While bridges forward all messages intended for devices in the other network, a router will confirm the existence of the network address before forwarding.

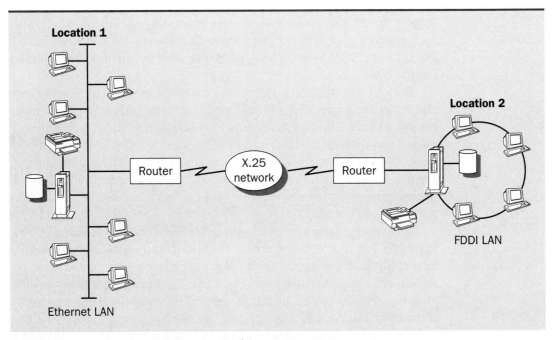

FIGURE 15-6   Routers interconnecting LANs through an X.25 network

This minimizes mistaken traffic over the communications links connecting the networks and generally improves network performance. In addition, prior to forwarding a message, a router will examine current information about available paths to the destination and network traffic in order to select the best route from the router to the message's destination (note that this means that a router must "know" its own location in a network). Once the route selection is made, the router sends the message on its way.

Like bridges, routers consult routing tables to determine the destination to which the message should be sent. The routing tables used by routers are more sophisticated than those used by bridges, because they contain destinations of devices (including other routers) in multiple networks as well as information on possible communications paths between the router and potential destinations.

Routing tables are used in the selection of the "best" path for the message from the router to its destination. Several heuristics used by routers to select the best path are summarized in Table 15-2. As may be observed in the table, the overall goal behind most routing heuristics is to minimize the amount of time it takes for the message to be sent from the router to its destination. A particular router may use a combination of these fundamental routing heuristics in selecting the best routes for the messages that it handles. One or more of these heuristics are likely to be used at each of the routers that a message passes through on its way from sender to

Heuristic	Brief Description
Minimize number of hops	This routing heuristic considers the number of other routers that the message will have to pass through to reach its final destination. As the message will have to be filtered and forwarded by each router (which takes time at each stop), minimizing the number of hops often means that the message will get to its destination in the least amount of time.
Fastest route	This routing heuristic considers the speed of the communications links between the router and the message's destination. By selecting the fastest available communications links, the message should get to its destination in the least amount of time.
Avoid heavy network traffic	This heuristic considers the amount of current network traffic on the possible paths between the router and its destination. Routers can do this because their routing tables are continuously updated with the latest (minute-by-minute) information on network traffic. By avoiding busy links, the message is less likely to be delayed in message queues at other intermediate nodes between the router and the destination. Once again, the message should arrive at its destination in the least amount of time.
Minimize protocol conversions	This heuristic considers how many times the message will have to be reformatted from one protocol to another. A message may have to flow across different networks to reach its destination, and its packets may have to be converted from one format to another in the process. By minimizing the number of times it has to be reformatted at intermediate hops, the message should reach its destination in the least amount of time.

TABLE 15-2   Heuristics used to select the best routes between routers and destinations

receiver. Hence, the router is often most concerned with determining the best route between it and the next router through which the message will pass.

The two most common ways to implement routers are as cards installed in microcomputer expansion slots (this essentially turns the microcomputer into a router) or, more commonly, as self-contained units. Each type of router is designed to work with specific network layer protocols such as those summarized in Table 15-3. Recall from earlier chapters that IPX is the internetworking component of Novell NetWare's SPX/IPX protocol; it is IPX that makes it possible to route messages among devices in two or more NetWare LANs even if they utilize different data link protocols (for example, CSMA/CD and token passing). IP (internetworking protocol) is the internetworking component of TCP/IP; it is responsible for formatting message packets for transmission over the network and for routing packets across the

Network Layer Protocol	Network Type(s)
AFP	AppleTalk
IP	Internet (and other networks using TCP/IP)
VIP	Vines
IPX	NetWare

TABLE 15-3
OSI network
layer protocols

network to their destinations. The TCP (transmission control protocol) portion of TCP/IP is responsible for end-to-end message delivery including the acknowledgement of packets received, the retransmission of lost packets, and ensuring that packets are received in the correct order.

Because routers are more sophisticated than bridges, it should not be surprising to learn that router prices are typically two to three times greater than those for bridges. In addition, because of the additional processing that routers perform (such as determining the validity of destination addresses and selecting the best route to the destination), forwarding rates in packets per second may not be as high as those for bridges.

Brouters

A *brouter* is a cross between a bridge and a router. Brouters combine the functions of bridges and routers and are typically able to handle multiple protocols. For example, a brouter may serve as a router for messages transmitted over the Internet using TCP/IP, but may bridge all other traffic (such as that between two Ethernet LANs). Some brouters enable network managers to enhance network performance by temporarily suppressing the brouter's routing functions in order to speed up its bridging operations; of course, this would only be done at times when users don't need the brouter's routing capabilities.

Gateways

Like routers and brouters, *gateways* are more complex than bridges because they interface between two dissimilar networks. In addition, they are considered to be more complex because they provide interfaces above the network layer in the OSI model (see Figure 15-3). Gateway technologies are necessary when two networks do not share the same network layer protocol.

The major function performed by a gateway is protocol conversion; however, speed and media conversion may also be necessary. For example, a gateway may translate one network protocol (say, IP) to another (say, AFP). Also, a gateway may be used to translate from one data format to another (such as from ASCII to EBCDIC) or to open program-to-program sessions (such as APPC sessions).

To carry out such functions, the gateway must understand the protocols of the two networks being connected and must translate from one protocol to another. Such protocol conversion involves changing the format of data packets received from one network to the format used in the other network. Figure 15-7 illustrates the use of a gateway to interconnect an Ethernet LAN using NetWare (the IPX/SPX) protocol to an FDDI LAN.

Gateway technologies include stand-alone units, circuit cards in network servers, and front end processors (discussed in Chapter 16) connected to a mainframe. Each type is designed for operational transparency—users do not need to know the protocol of the network from which they are requesting or to which they are sending data.

LAN to SNA (IBM's System Network Architecture) gateways are quite common due to the large number of SNA installations in the U.S. and worldwide. These are usually implemented as a card in the LAN device that acts as the communications server. This card handles all the terminal emulation functions needed to enable a LAN workstation to connect to a mainframe in the SNA network. It also makes the installation (and expense) of installing 3270 emulation cards in all the LAN workstations unnecessary.

Like bridges, gateways can be remote or local. Remote gateways make it possible to interconnect dissimilar networks over wide area communications circuits. For example, it may be possible for two dissimilar LANs to be

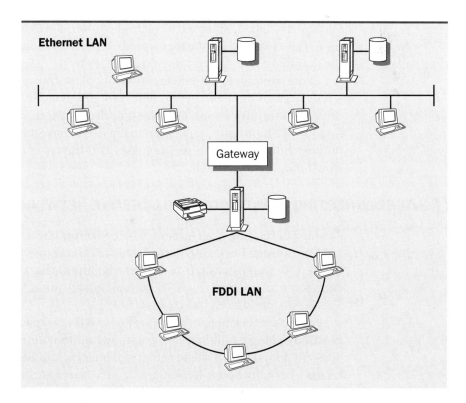

FIGURE 15-7
Gateway to interconnect Ethernet and FDDI LANs

interconnected through X.25 network circuits. In this case, the gateways at each LAN interface would have to convert incoming messages from X.25 message formats to the format used by the LAN protocol and to convert outgoing messages from the format used by the LAN protocol to X.25 format.

LAN Switches

LAN switches are among the newest LAN-to-LAN and LAN-to-WAN internetworking technologies. The range of LAN switch technologies is quite large, and the number of products available from vendors is increasing. Many of these are considered hubs for routing traffic between different LANs, for example, NetWare, DECnet, and AppleTalk LANs. These are often designed to be used as alternatives to bridges and routers; some carry out the same functions as collapsed backbones (discussed a little later in the chapter).

Organizations are often interested in LAN switches because they see them as avenues for increasing bandwidth, as a means for overcoming or bypassing bottlenecks created by the message processing done by bridges and routers, and as stepping stones for moving toward Fast Ethernet and ATM networks. In fact, many of the latest LAN switches are designed to connect existing LANs to Fast Ethernet or ATM networks. LAN switches are also being used to connect FDDI LANs with copper-based networks.

Network managers will be hearing more and more about LAN switches in the years ahead. Many exciting products and developments are emerging in this area, and most experts predict that LAN switches will be among the dominant internetworking technologies by the year 2000.

Large companies with diverse network platforms may need all the different types of interconnection hardware discussed in this chapter to establish communications links among their various networks. For this reason, it is important for network managers to have a working knowledge of these interconnection technologies and the types of connections for which they are best suited.

◆ INTERCONNECTING LANs THROUGH BACKBONE NETWORKS

As noted in Chapter 3, an approach being commonly used to interconnect LANs is through backbone networks (BNs). Backbone networks may also be used to establish micro-to-mainframe interfaces, LAN-to-mainframe connections, and LAN-to-WAN connections through a shared medium. This is illustrated in Figure 15-8.

Most backbone networks are located in a limited geographic area, such as within a single building (for example, to interconnect LANs found on different floors) or in a single campus of buildings. Many colleges and universities have fiber optic backbones to interconnect networks found within different buildings.

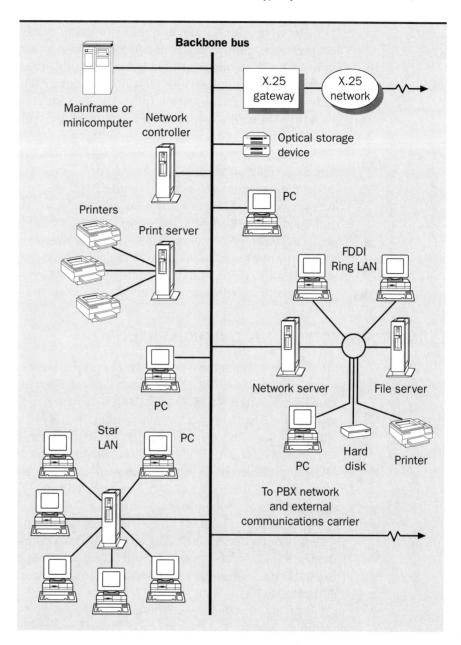

FIGURE 15-8
A backbone
network

In addition to providing a variety of network interconnections, backbone networks are often capable of supporting multiple network layer protocols (such as TCP/IP and IPX), and many carry several types of message
traffic (such as voice, data, video, and fax). No matter what medium is used
for the backbone, most backbone networks support connections to other
types of media; as noted in Chapter 3, even wireless LANs can be connected
to backbones. Backbone networks generally operate reliably and securely,
usually at high enough speeds to transparently support business operations.

A relatively recent development in using backbones to establish internetwork connections is generically known as a *collapsed backbone* or *superhub*. A collapsed backbone or superhub is most commonly found in a backbone network for a single building. The superhub essentially consists of a connection hub with slots for cabling to all LANs to plug into, as well as slots for internetworking technologies such as bridges, routers, and gateways.

Without built-in fault tolerance (which many have), the use of a superhub creates a single point of network failure. However, many experts expect the use of superhubs and collapsed backbones to increase. The growing popularity of ATM (asynchronous transfer mode), which can transmit multimedia files at speeds in excess of 100 Mbps, is likely to make switched media backbones (such as superhubs) more popular than shared backbones such as the one depicted in Figure 15-8. Because of its potential to effectively transmit voice, video, and data simultaneously, ATM is likely to emerge as the most common protocol used in enterprise network backbones.

◆ INTERCONNECTING LANs THROUGH PBXs

Most organizations have their own telephone switchboards, or *private branch exchanges (PBXs)*. The earliest PBXs—which one can still observe in old movies from the 1930s and 1940s—consisted of telephone operators stationed at board devices, switching plugs in and out of the board to provide physical connections for callers. Today, of course, the vast majority of such connections are software-controlled and handled by computers. Many include services such as voice mail, call accounting systems, caller identification, and call waiting. Interfaces with cellular telephone (and data) services are becoming more common.

Besides phones, PBXs can be used to establish connections among a variety of other devices including terminals, microcomputers, LANs, mainframes, minicomputers, printers, and fax machines. They may also be used to establish connections with wide area communications services. This is illustrated in Figure 15-9.

Virtually all of the newer PBXs use digital signals for both voice and data communications, and most provide *voice-over-data* capabilities, allowing users to simultaneously be engaged in voice and data communications over the same link. For example, with voice-over-data, a user can be transferring a large computer file while talking to a coworker on the phone.

Because a PBX can be used to interconnect workstations and other information technologies within an organization, it may be used to create a LAN. In addition, as it can for voice communications, the PBX can serve as a sophisticated (data) switch capable of supporting numerous concurrent connections among workstations. Similar to conference calls, most PBXs

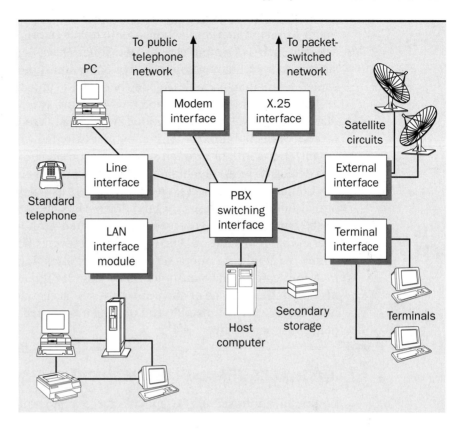

FIGURE 15-9
A PBX

can provide three or more users with realtime computer conferencing capabilities. Because most PBXs use digital signals, they can also provide the media needed for modemless micro-to-mainframe links. Some PBXs (especially those that support ISDN) are also being used to support desktop-to-desktop videoconferencing.

◆ LAN-TO-HOST CONNECTIONS

We have mentioned several ways in which LANs can be connected to hosts. These include connections through PBXs, connections through backbone networks, and connections through gateways (for example, LAN-to-SNA gateways). In this section we will briefly discuss several other types of LAN-to-host connections including modem sharing, interfaces through communications servers, LU.6 connections, and LAN connections with UNIX servers.

MODEM SHARING

Modem sharing implies dial-out capabilities for LAN users. Dial-out capabilities enable LAN users to establish connections to a host—establish a micro-to-mainframe connections—or to online information and database services such as CompuServe or Dow Jones News Retrieval. In most instances, modem sharing is managed by communications servers. However, some LAN modem packages include both dial-in and dial-out capabilities.

Communications servers (and LAN modems) can provide one or more dial-out lines. The communications lines available to hosts may be dedicated (connected directly to the host) or switched (may go through the organization's PBX or through switched common carrier circuits); generally, dedicated lines offer higher transmission speeds.

Communications servers tend to get more expensive as their number of dial-out lines increases. To hold down costs, most network managers install communications servers with a limited number of dial-up lines; user demand for dial-out services may not always be fully met. For example, if a communications server has four connections to the host, up to four users can simultaneously access the host. If a fifth user requests a connection to the host, he or she will receive an "access denied" or "all ports busy" message. A connection can only be established when one or more of the host ports is free.

INTERFACES THROUGH COMMUNICATIONS SERVERS

Communications servers can provide access to hosts if they have modem sharing and dial-out capabilities. Communications servers that include modem sharing and dial-out functions are sometimes called *modem servers*. In a previous section, we noted that IBM terminal emulation can be accomplished through LAN communications servers; this can provide LAN users with access to IBM hosts and to SNA network resources.

Note that some of the more sophisticated communications servers may include internetworking technologies such as bridges, routers, and gateways (which may be needed for LAN users to connect to hosts). In addition, communications servers may link wide area services such as X.25 packet-switched networks and T-1 digital circuits.

LU.6 CONNECTIONS

As with modem sharing, communications servers in LANs may be responsible for managing LU.6 connections. LU.6 connections enable users (logical units—LUs) to establish program-to-program communications sessions. This is done within the APPC framework discussed previously. Such connections enable microcomputers and hosts to exchange files and data even if they are in different networks with different computing platforms.

CONNECTIONS AMONG LANs AND UNIX SERVERS

UNIX is often described as an open-system, multiplatform operating system capable of running on computers of all sizes manufactured by different vendors. These characteristics have increased its popularity in client/server networks. Downsizing and the general trend toward internetworking have also increased UNIX's popularity.

In the diverse, distributed computing networks currently emerging, network managers are increasingly likely to establish connections between LANs and servers running UNIX (which may include mainframes, minicomputers, and LAN servers). This is often done through LAN/UNIX gateways (usually via TCP/IP) or through LAN communications servers that support the logon of LAN workstations to UNIX servers and UNIX workstation emulation. Routers, bridges, and brouters capable of protocol conversion between UNIX and the protocol used in the LAN may also be used to establish LAN-to-UNIX server connections.

HOSTS AS LAN SERVERS

In some situations, hosts are able to connect to a LAN as a node and to function as a server. This is one of the most effective ways to connect LAN workstations and hosts, but it is not always an alternative.

Downsizing has made the use of large computers as LAN nodes more popular. While some companies completely remove large computers as part of their downsizing, others try to leverage their investments (and prolong the useful life of the large system) by folding it into the new network. Many companies, for example, have found that mainframes make excellent database servers.

◆ LAN-TO-WAN CONNECTIONS

We have covered all of the major ways in which interconnections between LANs and WANs may be established. To quickly summarize, LAN-to-WAN interconnections can be achieved through interconnection hardware such as bridges, routers, brouters, switches, and gateways. You may recall, for example, that remote bridges and remote gateways make it possible to interconnect networks via wide area communications circuits.

LANs may also be able to connect to WANs and WAN services through PBXs, backbone networks, and superhubs. In addition, communications servers with modem sharing and dial-out capabilities may be used to establish LAN-to-WAN connections.

Wireless connections between LANs and WANs are one of today's major issues concerning LAN/WAN interfaces. Another issue concerns the transfer of multimedia files between LANs and WANs.

◆ MANAGING INTERCONNECTED NETWORKS

The networks that exist in today's large organizations are complex. Often, the hardware used is produced by a variety of vendors. For example, an organization's network may include one or more IBM mainframes, several DEC minicomputers, LANs built around both IBM-compatible and Apple Macintosh computers, and a PBX manufactured by Northern Telecomm. In addition, the organization may have information flowing among its multiple operating locations over X.25 networks, T-1 lines, cellular data networks, and a variety of leased and dial-up circuits. Needless to say, sophisticated network management systems (NMSs) may be needed to manage and control such diverse and complex networks.

In Chapter 13, our discussion of network management mentioned that managing network operations and performing tasks such as network tuning and capacity planning can be very challenging, even for smaller, standalone networks. Interconnecting LANs and establishing connections between LANs and WANs can make the task more difficult. For example, it may be hard to simultaneously monitor the operations (and interactions) of the various subnetworks. Problem tracking and troubleshooting may also be more complicated with interconnected networks.

NMSs for diverse, interconnected systems can ease these tasks by giving managers access to network configuration information, cost accounting information, and a variety of network performance and utilization statistics. Such systems can help network personnel detect and correct problems, rectify network security problems, and forecast network workloads. Most of the products on the market today utilize standards that enable the capture and exchange of network operations and control data in a common format; the network management information captured is typically stored in a database generically known as a *management information base (MIB)*. Many current NMSs include user-friendly interfaces including GUIs, color mapping systems, pull-down menus, and the ability to summarize specific network data by clicking on an icon.

In the following sections we will briefly discuss some of the major NMSs developed to manage large, distributed, interconnected networks. Our discussion will focus on NetView, ACCUMASTER Integrator, DECmcc, and NetWare Management System. Our discussion will also address SNMP, SNMP.2, and CMIP; these are the main transport protocols used to structure and transfer network management data in diverse, distributed networks.

NETVIEW

NetView is an established network management program designed by IBM for managing multivendor voice and data networks. It was designed for managing host-based network management services in SNA networks. The main NetView program runs on a mainframe. However, IBM and IBM-compatible microcomputers running NetView/PC are able to monitor network operations in LANs, non-IBM computing equipment, and some PBXs and to provide the mainframe running NetView with network statistics, alerts, and alarms.

Because NetView was developed to manage integrated voice/data networks, it incorporates a number of voice-oriented features and services. For example, it contains a network billing system that generates call-detail records captured by the PBX for billing individual departments. It also contains least-cost routing modules for long-distance calls.

General network management capabilities permit users to write macros for automated diagnostic routines. NetView also contains help desk modules that help control center personnel isolate malfunctioning network components; these modules also provide recommendations for overcoming these problems. It includes monitors for session failures caused by terminal malfunctions, hardware monitors with alarm and alert features, and overall network status monitors that can reactivate devices downstream from a failed network resource.

NetView is evolving; IBM expects future versions of NetView to provide enhanced financial administration functions, service-level tracking, improved network management support for heterogeneous LANs, and superior capacity planning capabilities.

ACCUMASTER INTEGRATOR

AT&T has developed its own distributed processing architecture (the United Network Management Architecture [UNMA]) and network management and control program—ACCUMASTER Integrator. These provide organizations with end-to-end management and control of distributed, multivendor voice and data networks.

ACCUMASTER Integrator maintains a comprehensive inventory of all network components. Its mapping system contains schematics of the connectivity between network elements. The program's alarm and alert features enable network managers to isolate network problems and to determine who is responsible for solving the problem.

The ACCUMASTER Integrator collects information on whatever data communication equipment and media the organization uses. It will collect, coordinate, and integrate information from LANs, WANs, PBXs, T-1 lines, and hosts. It also supports transmissions over both physical (twisted pair, coaxial cable, fiber optic cable) and wireless media. If SNA networks are parts of the enterprise network, ACCUMASTER Integrator's SNA Management Application enables the two NMSs to work together. ACCUMASTER can also interface with Cincom's Net/Master NMS.

DECmcc

Digital Equipment Corporation's network management architecture is called the Enterprise Management Architecture (EMA). Like ACCUMASTER Integrator, it assumes the presence of a variety of network components manufactured by different vendors. EMA is built around a relational network database, called the DECmcc Director, containing information on network components, interfaces, and user profiles.

Through DECmcc and EMA, DEC has attempted to achieve a balance between distributed and centralized network management. It allows network failure and configuration management to be distributed, while functions such as network security and performance accounting are handled at a single control center. Recent versions of DECmcc and EMA include remote bridge management software, terminal server management modules, remote system management software, and LAN traffic monitoring programs. Like NetView and ACCUMASTER Integrator, DECmcc is an evolving product that has added functionality and additional automated network management features.

NETWARE MANAGEMENT SYSTEM

Novell's NetWare Management System allows network managers to manage interconnected NetWare LANs from a single control console, enterprise-wide if necessary. The NetWare Management System includes modules for configuration management, security, performance management, accounting management, and fault detection.

The usual implementation involves the installation of a NetWare Management Agent (NMA) on each server in the network; NMA consists of a set of NetWare Loadable Modules (NLMs) that transmit statistics about the operations of server hardware, software, and data resources to the network control center. These statistics can alert network personnel of potential (or actual) resource allocation problems. NMAs can respond to polls from other network management systems such as IBM's NetView.

The reporting system of NetWare Management System includes mapping modules to display and print out schematics of the entire network (or subnetworks). It can also map network resources onto building floor plans.

SNMP

One of the major differences among distributed network management systems is the manner in which the network management information is formatted and transported. The protocols used by network management systems can differ, and if two or more NMSs are used to manage a network, protocol conversion may be necessary.

Simple Network Management Protocol (SNMP) was originally developed to control and monitor the operations of network components on networks using the TCP/IP protocol. Now, however, it can be used with other network-level protocols. It has been endorsed by a variety of vendors including Hewlett-Packard, IBM, and Sun Microsystems.

SNMP is an application-layer protocol that outlines the format of data packets used to transfer network management information among network components. It also specifies how network data is structured and stored in the management information base (MIB). Because of its close ties to TCP/IP, SNMP data formats are structured for efficient transfer over TCP/IP networks.

SNMP was developed to be a simple protocol. The agents used to monitor the operations of network components are less sophisticated than those in other network management protocols. Agents are able to respond to the NMS's requests for information and to set parameters for the devices they monitor, but not much more. For example, SNMP agents are not intended to filter or analyze the data they collect; without filtering, the data packets transmitted from agents to network control centers can account for a significant percentage of network traffic (especially in large networks).

Some of SNMP's other shortcomings include a limited command set and limited provisions for security. To overcome such shortcomings, a second version of SNMP has been developed.

SNMP.2

SNMP.2 is sometimes called *SMP (simple management protocol)*. This network management protocol was developed to overcome the shortcomings of SNMP. Specifically, it was designed to enable the implementation of multiple agents in the same network device, reduce the number of data packets containing network management information transmitted across the network, support multiple transport protocols (not only TCP/IP), enable the distribution of network management functions and control centers, and to have enhanced security capabilities.

One way in which the limited functionality of SNMP agents has been overcome by SNMP.2 is by making it possible to include multiple agents in key network components. For example, a database server in a distributed client/server network might include one agent to monitor networking activities, a second to monitor disk access operations, and a third to monitor query processing operations. Each of these might transfer data to different network management programs or control centers.

Network traffic is reduced by SNMP.2 through bulk retrieval and transfer mechanisms that allow an agent to transfer more network management information at once. Agents also have more intelligence and sophistication, including some filtering and analysis functions.

Support for multiple transport protocols enables SNMP.2 to be deployed on more than just TCP/IP networks. SNMP.2 can now work transparently with other network-layer protocols including IPX (used in NetWare LANs) and AppleTalk.

Distributed network management is possible with SNMP.2 because it allows for the creation of multiple network management entities within the same network. The data collected by agents can be passed to a local network manager entity, where it can be filtered, analyzed, and stored. Only important network data collected by agents needs to be passed from local manager entities to the central network management entity. The segmentation of network management functions can lead to the creation of a hierarchy of network management entities. It can also help to reduce network information traffic. Such network management segmentation capabilities, along with remote configuration capabilities, help in making SNMP.2 networks more secure than SNMP networks.

CMIP

Common management information protocol (CMIP) is generally considered to be more sophisticated than SNMP. In fact, SNMP was first developed as predecessor to CMIP. However, because SNMP and now SNMP.2 have taken on lives of their own, CMIP may be years away from being a serious challenger. Indeed, SNMP and SNMP.2 are presently being used ten times as often as CMIP.

Relative to SNMP, CMIP is a more complex protocol for exchanging network management information among network components and management entities. CMIP possesses a richer command structure and a more extensive management information base (MIB). In addition, CMIP agents typically possess filtering and analysis capabilities. These attributes give CMIP the potential to overcome SNMP's shortcomings and to provide more effective network control.

◆ SUMMARY

One of today's major networking trends is the interconnection of LANs and other networks. Factors such as the movement toward client/server networks, the increased popularity of the Internet, and the increasing prevalence of telecommuting and mobile computing have become driving forces to interconnect networks.

Historically, microcomputer-to-mainframe connections increased in importance during the 1970s and 1980s as organizations realized the cost-effectiveness of using microcomputers in place of terminals in host-based systems. Through modems and communications software, remote users could access host resources over common carrier lines.

To support micro-to-mainframe links, a number of microcomputer-oriented communications protocols have been developed such as KERMIT, XMODEM, and YMODEM. Most include file transfer utilities to facilitate the uploading and downloading of files and data between microcomputers and larger computers. In addition, many communications software packages (such as ProComm and CrossTalk) support these protocols. Communications software packages also include terminal emulation capabilities, which enable microcomputers to imitate host terminals.

When selecting communications software, network managers should take into account a number of factors. Factors include the types of operating systems and microcomputers supported; the communications protocols supported; built-in logon routines for needed information services (such as CompuServe, Dow Jones News Retrieval Services, and so on); remote control capabilities; security features; the sophistication of the scripting language; the general user-friendliness of the software; the level and quality of available technical support; and special features such as mouse support, fax transmission capabilities, and voice mail.

The prevalence of LANs has made the ability for remote users to access LAN resources more common. Remote users typically gain access to LANs through communications servers, by dialing in and seizing control of LAN workstations, and by dialing into LAN modems—modems with network interface cards that are attached to a LAN.

Another important networking trend is the interconnection of LANs. Such interconnections allow users of one LAN to access computing resources in other LANs.

A variety of interconnection technologies can be used to connect networks and network segments. These include repeaters, bridges, routers, brouters, and gateways.

A repeater connects two or more segments of the same network. This device is found at the physical layer of the OSI model. The main function of a repeater is to renew weakened network signals; repeaters can thus make it possible to extend a network beyond its normal geographic limitations. Some repeaters include media and speed conversion capabilities.

A bridge is used to connect two networks at the data link level. Traditionally, bridges have been used to interconnect two networks running the same data link protocol. However, the definition of bridges has expanded to include devices that interconnect two networks using different data link protocols. Bridges are more sophisticated than repeaters because they must examine (filter) data packets to determine which packets need to be forwarded to the other network. In addition to speed and media conversion capabilities, bridges may also be able to convert messages from one protocol format to another. Many of the bridges used today are called *learning bridges* because they automatically build their own routing tables.

Local bridges are used to connect networks near one another, while remote bridges connect networks over wide area communications circuits.

Routers are found at the network layer of the OSI model. To operate at this level, they must have their own network address and must be able to identify the source and destination addresses of the messages that they handle. Like bridges, routers examine the message packets they receive and forward them to destinations in other networks, either through local connections or over wide area communications links. Prior to forwarding, routers verify the existence of the destination address and determine the best route for the message to take from the router to its destination. In determining the best route for the message, the router consults routing tables containing information on alternative routes to the destination as well as current network traffic information. Best route selection also involves the use of one or more routing heuristics such as minimizing the number of intermediate hops, using the fastest or least congested routes, and minimizing the number of protocol conversions.

A brouter is a cross between a bridge and a router. These are typically used to interconnect three or more networks; the traffic between two of these networks may be bridged, while that involving the other network(s) is routed. Some brouters enable network managers to increase bridging operations by suppressing routing activities when they are not needed.

A gateway is more complex than a bridge because it enables two dissimilar networks using different network-layer protocols to be connected. Because of this, gateways are found above the network layer in the OSI model. The major function performed by gateways is protocol conversion. However, a gateway may also have to convert data formats and have speed and media conversion capabilities. Like bridges, gateways can be remote or local.

LAN switches are relatively new internetworking technologies. These can be used to interconnect different types of LANs as well as to connect LANs and WANs. They can also be used as collapsed backbones and superhubs. Many experts expect LAN switches to be one of the dominant internetworking technologies by the late 1990s.

In addition to using interconnection technologies such as bridges, routers, and gateways, networks may be interconnected through backbone networks or private branch exchanges (PBXs). Backbone networks can support interconnections between hosts, LANs, and WANs through shared or switched media. Superhubs (collapsed backbones) may combine a variety of local internetworking technologies as well as provide access to WANs. Interconnecting networks through PBXs enables both voice and data networks to be integrated.

LANs may be connected to hosts in several ways. For example, LAN-to-host connections can be established through communications servers that provide modem sharing and dial-out capabilities. Communications servers may also support program-to-program LU.6 connections as well as connections to UNIX servers. Some hosts can be used as nodes in LANs; downsizing has made this an attractive option.

LAN-to-WAN connections can be established through interconnection technologies such as routers, brouters, and gateways. LANs and WANs may also be interconnected via backbone networks, superhubs, and PBXs. Wireless LAN-to-WAN connections are growing in popularity and importance as is the ability to transfer multimedia files from LANs to WANs.

Interconnecting networks has made network management more complicated. The network management systems (NMSs) needed for diverse, distributed networks are more sophisticated than those used to manage stand-alone LANs. Some of the best known NMSs for large, complex networks are IBM's NetView, AT&T's ACCUMASTER Integrator, Digital Equipment Corporation's DECmcc, and Novell's NetWare Management System.

Three of the most important network management protocols are SNMP, SNMP.2, and CMIP. SNMP is the least sophisticated and most widely used of the three. However, because of its simplicity, SNMP has relatively unsophisticated monitoring, security, and network control capabilities. To overcome these and other SNMP shortcomings, SNMP.2 was developed; it provides greater network control, more sophisticated monitoring, and better security provisions than its predecessor. CMIP is the most sophisticated network management protocol; while having the potential to overcome SNMP shortcomings, to date, it is used significantly less than SNMP.

✳ KEY TERMS

Bridge	Micro-to-mainframe link
Brouter	Repeater
Common Management	Router
Information Protocol (CMIP)	Simple Network Management
Communications servers	Protocol (SNMP)
Gateway	Superhub
LAN modem	Voice-over-data
LAN switches	

✳ REVIEW QUESTIONS

1. Briefly describe how microcomputers can connect to mainframes and minicomputers.
2. Identify and briefly describe the major types of microcomputer communications protocols.

3. What is terminal emulation? Why is it needed?
4. Identify some of the popular terminal emulation schemes.
5. Identify and briefly describe the criteria that may be used in selecting communications software.
6. Identify and briefly describe some of the factors that contribute to the user-friendliness of communications software.
7. Describe the different ways in which remote users can connect to LANs.
8. What is a repeater? What functions are performed by repeaters?
9. What is a bridge? What functions are performed by bridges? How are bridges implemented?
10. What are the differences among learning bridges, local bridges, and remote bridges?
11. What are the differences between a bridge's filtering and forwarding processes?
12. What is a router? What functions are performed by routers? How are routers implemented?
13. What are the differences between the forwarding processes used by bridges and routers?
14. What are the differences between the routing tables used by bridges and those used by routers?
15. Identify and briefly describe the routing heuristics used by routers to determine the best routes for message packets.
16. What is a brouter? What functions are performed by brouters?
17. What is a gateway and what functions does it perform? How are gateways implemented?
18. What are LAN switches? Why are these growing in popularity?
19. What are backbone networks? How may networks be interconnected through backbones?
20. What are superhubs? Why are superhubs becoming popular?
21. What is a PBX? How may networks be interconnected through PBXs?
22. What are voice-over-data capabilities?
23. Identify the different ways in which LANs may connect to hosts.
24. What are the differences among modem sharing, LAN-to-host access through communications servers, LU.6 connections, and LAN connections with UNIX servers?
25. Why are some mainframes used as nodes in LANs?
26. Identify the different ways in which LANs can be connected to WANs.
27. Why is managing interconnected networks more challenging than managing stand-alone networks?
28. What is a management information base (MIB)? What is the role of MIBs in network management?
29. What are the differences among NetView, ACCUMASTER Integrator, DECmcc, and NetWare Management System?

30. What are the differences among SNMP, SNMP.2, and CMIP? In what ways are SNMP.2 and CMIP superior to SNMP?

✳ DISCUSSION QUESTIONS

1. What types of equipment and communications software should network managers recommend for telecommuters? What features should the communications software for telecommuters possess? Why should these features be included? If the telecommuter works half the time at home and half the time in the office (which is equipped with a LAN-attached workstation) what type of remote LAN access would you recommend? Why?

2. What are the benefits of having high-level program-to-program interfaces such as those provided by APPC/PC and LU.6 connections? Where would such connections be located on the OSI reference model? What types of underlying protocol conversion might be necessary for such interfaces, and how are these related to middleware?

3. Suppose network performance is deteriorating because of substantial increases in the forwarding rate for a bridge connecting two Novell NetWare LANs. One of the LANs is a 4 MBps token ring using twisted pair cable and the other is a 10 MBps Ethernet LAN using coaxial cable; they are located on separate floors of the same building. Describe several specific changes that could enhance network performance.

4. When could a repeater replace a bridge? When couldn't a repeater replace a bridge? When could a bridge replace a router? When couldn't a bridge replace a router? Why do many LAN managers prefer bridges over routers for network security reasons?

✳ PROBLEMS AND EXERCISES

1. Develop a hierarchy of communications and network management protocols that maps into the OSI reference model (that is, list data link protocols, network-layer protocols, protocols above the network layer, and so on).

2. Find a recent article comparing communications software packages in a trade publication such as *Communications Week, Computerworld, Data Communications, LAN Magazine, Mobile Office, Network Week, PC Magazine,* and *PC Week.* Identify the criteria used to compare the packages and summarize the strengths and weaknesses of the different packages in a paper or presentation.

3. Find a recent article comparing a particular interconnection technology (bridges, gateways, routers, LAN switches, superhubs) in a trade publication. Identify the criteria used to compare the different products and summarize the differences among the products.

4. Find a recent article about one or more of the network management systems discussed in this chapter (NetView, ACCUMASTER Integrator, DECmcc, NetWare Management System) in a trade publication. Summarize the content of the article in a paper or class presentation.

5. Find a recent article about one or more of the network management protocols discussed in this chapter (SNMP, SNMP.2, CMIP) in a trade publication. Summarize the content of the article in a paper or class presentation.

6. Invite a local network manager to discuss the types of interconnection technologies used in his or her organization and how they are used. Be sure to cover the types of changes he or she expects to make in interconnection technologies in the years ahead.

7. Invite a local network manager to discuss the type(s) of remote access provided to LAN resources. Ask why these are being used and how he or she expects remote LAN access to change in the future.

8. Invite a local network manager to discuss the network management system in place in his or her organization. Ask how the current system could be improved.

✳ **REFERENCES**

Bachus, K. and E. Longsworth. "Road Nodes." *Corporate Computing* (March 1993): p. 55.

Barnhart, S. "Myths and Facts About Network Management Interoperability." *Telecommunications* (June 1993): p. 47.

Biery, R. "Collapsed Backbones: Next Step in Premises Networks." *LAN TIMES* (September 14, 1992): p. 47.

Bradner, S. and D. Greenfield. "Routers: Building the Highway." *PC Magazine* (March 30, 1993): p. 221.

Briscoe, P. "ATM: Will It Live Up to User Expectations?" *Telecommunications* (June 1993): p. 25.

Cikoski, T. R. and J. S. Whitehill. "Integrated Network Management Systems: Understanding the Basics." *Telecommunications* (June 1993): p. 41.

Dostert, M. "Integrating PC LANs with UNIX Systems." *LAN TIMES* (May 25, 1992): p. 23.

Ellison, C. "Productivity from Afar: Modem Remote Control Software." *PC Magazine* (February 25, 1992): p. 189.

Flanagan, P. "Multiprotocol Routers: An Overview." *Telecommunications* (April 1993): p. 19.

Goldman, J. E. *Applied Data Communications.* New York: John Wiley & Sons, 1995.

Jander, M. "Proactive LAN Management: Tools that Look for Trouble to Keep LANs Out of Danger." *Data Communications* (March 21, 1993): p. 49.

Jeffries, R. "ATM to the Desktop: Prospects and Probabilities." *Telecommunications* (April 1994): p. 25.

Krepchin, I. "These Hubs Ease LAN Changes." *Datamation* (August 1, 1994): p. 49.

Layland, R. "The Superbox: A Cure for this Old LAN." *Data Communications* (September 1992): p. 103.

Lundardoni, M. "Network Simplification Using ATM." *Telecommunications* (June 1993): p. 33.

McCullough, T. "Don't Forget Repeaters for Connectivity." *LAN TIMES* (June 22, 1992): p. 69.

McCusker, T. "HP Opens Up Network Management." *Datamation* (February 15, 1993): p. 22.

Merry, R. C. "Bridges vs. Routers: Comparing Functionality." *Telecommunications* (April 1993): p. 29.

Molloy, M. "ATM LANs: A Marketplace Overview." *Telecommunications* (April 1994): p. 35.

Moore, S. "AT&T Licenses OpenView for Integrated Net Management Scheme." *Computerworld* (July 4, 1994): p. 15.

Saunders, S. "PBXs and Data: The Second Time Around." *Data Communications* (June 1993): p. 69.

Stallings, W. "SNMP-Based Network Management: Where Is It Headed?" *Telecommunications* (June 1993): p. 57.

Strauss, P. "Welcome to Client-Server PBX Computing." *Datamation* (June 1, 1994): p. 49.

St. Clair, M. "Broadening the Scope of Bridges." *LAN TIMES* (April 6, 1992): p. 49.

Thyfault, M. E., S. Stahl, and J. C. Panettieri. "Breaking Down the Walls." *Informationweek* (April 19, 1993): p. 22.

Tissot, A. F. "The Changing Role of the PBX in Today's Office Environment." *Telecommunications* (November 1993): p. 51.

Tolley, K. "Grading Smart Hubs for Corporate Networking." *Data Communications* (November 21, 1992): p. 56.

Valovic, T. "Network Management: A Progress Report." *Telecommunications* (August 1992): p. 23.

Violino, B. and S. Stahl. "No Place Like Home." *Informationweek* (February 8, 1993): p. 23.

Wexler, J. M. "LAN Switch Gains Foothold: Hubs Used to Segment Nets Rather than Increase Bandwidth." *Computerworld* (March 22, 1993): p. 47

———. "LAN Switches Unsnag Jammed Nets." *Computerworld* (August 12, 1991): p. 43.

Case Study:
Southeastern State University

Before 1990, faculty members at Southeastern State University used telephone circuits to connect to SSU's mainframes and minicomputers. In the 1970s and early 1980s, teleprinters with acoustic couplers were the primary devices that enabled users to interact with SSU's hosts. As microcomputers began to proliferate, modems replaced acoustic couplers, but telephone lines remained the media used to establish micro-to-mainframe connections. Telephone lines are still used today to provide dial-in connections to SSU's minicomputers from remote locations; both faculty and students take advantage of the available ports. Remote users can also connect to some of the LANs on campus through dial-in connections. For example, remote users (both faculty and students) can dial in to the CD-ROM server in the library to do computer-assisted searches of the databases and periodical indexes stored on that LAN. In addition, by connecting to one of SSU's VAX computers, remote users can gain online access to the library's card catalog; the card catalogs of the other colleges and universities in the state's system can also be accessed. By connecting to the Internet (through the VAX), remote users can access library card catalogs worldwide.

When working from their homes, remote users first load communications software, then establish a connection to the host (in most instances, the communications software automatically dials the number for them). To facilitate such connections, the Computer Services Department has developed extensive documentation for remote users so that they can set the necessary parameters in the software.

Several forms of terminal emulation are supported for SSU's micro-to-mainframe connections, however, because the majority of the connections are through the VAX computers. DEC VT 100, DEC VT 220, and DEC VT 320 emulations are the most commonly used (the latter is most common for microcomputers running Windows). KERMIT is the primary file transfer protocol used at SSU; Computer Services gives a free copy of KERMIT to all remote users. KERMIT and the previously mentioned terminal emulations are also used to establish connections between LAN workstations and VAX resources.

The manner in which the computers in faculty, administrative, and clerical offices connect to SSU's VAX resources has evolved. Prior to 1985, telephone services at SSU were provided by the local telephone company, an independent family-owned company that was not part of the regional Bell system. The telephone in each office at SSU was essentially connected to the telephone company's main switching center. Microcomputers (or teleprinters) with modems (or acoustic couplers) connected to SSU's host computers in the same manner as for remote users. In fact, faculty in their offices competed with remote users for dial-in ports. This system lacked voice-over-data capabilities. Hence, if a faculty member was connected to the host from an office workstation, the office telephone was unavailable to make or receive calls.

In 1985, the university purchased a PBX. The administrator determined that by paying for the PBX over the next seven years, its monthly telephone costs would be almost exactly the same as what it was paying on a monthly basis to the local telephone company. In addition, the PBX offered additional functionality unavailable from the local telephone system at that time, including voice-over-data, conference calling, speakerphone capabilities, call forwarding, call waiting, and a sophisticated call accounting system. For micro-to-mainframe connections, speeds up to 19,200 bps were supported—this was 4 times higher than the maximum possible from the previous system on a good day (2400 was more common for campus users; remote users often had to settle for 1200 bps with the old system). This system was not without its problems, but its implementation represented a significant step forward in both voice and data communications at SSU.

In 1993, SSU replaced its PBX with CENTREX services from the local telephone company. The reason for the change was that as the result of growth, the PBX had reached its capacity and had to be replaced by either a larger PBX or an alternative system. A committee of administrators hastily created an extremely flexible RFP (request for proposal) with minimal input from SSU's user community and put it out for public bid. The local phone company won the bid. It was only after the contract had been awarded that anyone asked users at SSU what kinds of voice and data services they used and needed. This information was collected by phone company employees. In the interviews, users were often told that some of the usual services (such as voice-over-data) would simply not be available with the new system. As you might expect, this news did not sit well with many users. They viewed the change (and still do) as a step backward for telecommunications at SSU.

Largely because of the RFP's wording and the university system's requirement that the lowest bid be accepted, the CENTREX system now in place lacks the range of functionality of the PBX that it replaced. For example, voice-over-data capabilities are no longer available, and the maximum data communications speed supported by the CENTREX system is 9600 bps. On the plus side, voice mail is supported by the new system, and this has been implemented by most departments at SSU.

For faculty members whose office microcomputers are attached to LANs that are, in turn, connected to the VAX computers and other networks on campus (including the RS/6000 AIX server), the reduced functionality of the CENTREX system is not considered to be a problem. These users basically have direct, modemless access to these larger computers at typical LAN speeds, so the loss of voice-over-data capabilities does not upset them. They use the CENTREX system only for voice communications. However, users in colleges and departments that have not implemented LANs (or have implemented only stand-alone LANs) continue to feel shortchanged by the loss of voice-over-data capabilities and reduced transmission speeds.

Computer Services has pointed out another shortcoming: There is no longer a backup micro-to-mainframe connection for the existing LANs. When the PBX was in place, LAN workstations in faculty, administrative, and clerical offices were attached to both the PBX and the LAN. If users were unable to access VAX resources through the LAN, they were usually able to get to the VAX through the PBX. However, when the PBX was removed, all the phone units were replaced and the previous connections between microcomputers attached to LANs and the PBX were removed. New connections between phone units and microcomputers were installed only in offices that were not part of an existing LAN.

Administration members responsible for developing the RFP that brought CENTREX services to SSU argued that this implementation plan saved the university a lot of money and that everyone still has a means to access VAX resources from his or her office. However, the fact remains that a backup system for micro-to-mainframe connections by LAN users was lost when the PBX was replaced by CENTREX services.

In spite of its criticism of the move to the CENTREX system, Computer Services has utilized the PBX's replacement as leverage to push the university toward a fiber optic backbone. All the minicomputers and servers for which Computer Services is responsible are directly connected to the backbone, and, to date, trenches have been dug and fiber optic cable has been brought to each of the buildings on campus, including dormitories. The catch is that the groups occupying the buildings are budgetarily responsible for wiring their facility so that they can connect to the fiber optic backbone. While everyone would like to see fiber replace copper throughout all the buildings on campus, the fact remains that doing so is expensive. As a result, only new buildings have fiber running through them, and the occupants of older buildings feel like second-class users. Even in older buildings with preinstalled LANs, transceivers are used to convert the light waves passed over the fiber to electric signals that can be passed over the copper-based media used to create the LANs. This approach, while being cost-effective in the short run, means that the full capabilities of fiber optic telecommunications cannot be realized in many parts of campus.

Partly in response to the internal friction caused by the replacement of the PBX by CENTREX services, the administration recently voiced its intention to eventually use the fiber optic backbone to integrate both voice and data communications at SSU. Phone units will be integrated with LAN workstations, and the proposed system will be able to transfer multimedia files between networked devices; desktop-to-desktop videoconferencing will also be supported. While this proposal has evoked positive responses among users, many of the skeptics hope that this plan will be more thoroughly thought out and better executed than some of the other changes that have recently taken place at SSU.

CASE STUDY QUESTIONS AND PROBLEMS

1. Identify and briefly describe the approaches that SSU could use to provide remote users with dial-in access to LANs (such as the CD-ROM LAN in the library). Which of these do you feel is most likely to be in place? Justify your selection.

2. Identify and briefly describe the approaches that SSU has in place to provide users (both on campus and remotely) with connections to host computers such as the VAXs and the RS/6000. What types of hardware do you suspect are in place at the host computers to support such access? Also, because of the variety of LANs and operating systems found on campus, what types of interconnection technologies are likely to be in place?

3. Critique the RFP process that led to the selection of CENTREX services to replace the PBX. How should this have been handled? (Review Chapter 9 prior to developing your answer.)

4. Comment on Computer Services' complaint that the replacement of the PBX has left LAN users without an alternative method of accessing hosts at SSU. What types of backups or alternative access mechanisms would you recommend for SSU to ensure that LAN users will be able to access hosts?

5. Discuss how superhubs could be used at SSU to integrate the different types of networks found at SSU.

6. What types of media and protocols will be needed to support the transfer of multimedia files and desktop-to-desktop videoconferencing at SSU?

Wide Area Networks

CHAPTER OBJECTIVES

After completing this chapter, you will be able to:

◆ Describe the differences between LANs, WANs, and MANs.

◆ Identify the different types of services and digital lines available from common carriers.

◆ Briefly describe the characteristics of value-added networks (VANs) and packet distribution networks.

◆ Identify and briefly discuss the functions of some of the important hardware devices found in WANs.

◆ Briefly describe the differences between teleprocessing access methods and teleprocessing monitors.

◆ Identify and briefly describe the functions performed by data link and network-layer WAN protocols.

◆ Briefly describe the differences between frame relay and cell relay systems.

◆ Identify and briefly discuss the major issues involved in building global networks.

◆ Discuss some of the major trends in WANs and their management.

The focus of this book has been on local area networks and how these are managed. However, while stand-alone LANs are still being created, the trends are toward interconnecting LANs and interconnecting LANs to wide area networks (WANs). Because of these trends, it is important for network managers to be aware of the differences between LANs and WANs and of the management challenges that WANs present.

◆ WANs, LANs, AND MANs

The primary criterion used to differentiate between WANs and LANs is the geographic expanse of the networked computers. However, as you will read in the following descriptions, WANs and LANs also differ in terms of the importance of the roles of larger computers, as well as in data transmission speeds.

WIDE AREA NETWORKS

Wide area networks (WANs) are distributed networks that can cover broad geographic areas. An example of a WAN is illustrated in Figure 16-1.

As in many distributed networks, mainframes and minicomputers often play a prominent role in WANs. While servers and microcomputers are starting to share the limelight with larger systems in WANs, mainframes and minicomputers continue to play the pivotal roles; they are frequently the focal points for network interconnection. Because of this, they are often called *nodes*.

WANs may cover a significant region of a state, an entire state, several states, all the U.S., and even the entire world. Generally, a distributed

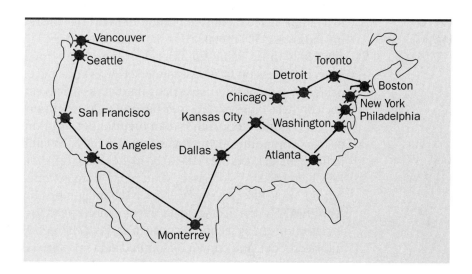

FIGURE 16-1
A wide area network (WAN)

network is considered a WAN if a user at one site would normally have to use long-distance services to call users at one or more of the other network locations. This rule of thumb might not apply in large cities with very large local calling areas, but it is still a useful guideline for distinguishing WANs from more geographically limited networks.

Because WANs can cover very broad areas, the communications media used to interconnect different locations can be slower than those used in more geographically restricted networks. However, some of the fastest data communications services available from long-distance phone companies may be used in WANs, especially when high-volume data transfers between large computers are needed. In general, media speeds for WANs are increasing, and these increases are driven by new and faster services offered by long-distance service providers such as AT&T, MCI, and Sprint.

LOCAL AREA NETWORKS

Relative to a WAN, a LAN covers a much smaller geographic area. A LAN may be restricted to a single room (such as a LAN in a university computer lab), a set of rooms or offices in the same building, or the set of buildings that makes up a single campus of a school or business. LANs rarely involve common carrier services other than to dial out on a switched circuit for voice, data, video, or fax.

Some experts feel that the term *LAN* should cover only networks confined to a fixed distance (for example, a radius of five miles). However, there is really no hard-and-fast rule. As noted, the geographic area covered by a LAN is more likely to be determined by the networking technologies used than anything else. For example, Apple Computer recommends that its AppleTalk networks (used to connect Macintosh computers) be limited to a radius of 300 meters. For Ethernet networks, the recommended working radius is 2,500 meters (less than 2 miles) or less. Fiber optic networks, especially those using the fiber distributed data interface (FDDI), may cover distances up to 200 kilometers. However, as noted in Chapter 15, repeaters and other internetworking hardware devices can be used to extend each of these types of networks beyond their recommended limits.

Microcomputers, rather than mainframes or minicomputers, are the primary computing technologies found in LANs. In addition, media speeds in full-fledged LANs are likely to be higher than those found in many WANs; LAN data transmission speeds can range from several hundred thousand bps (in AppleTalk LANs) to well over 100 million bps (in fiber optic LANs). As with WANs, the average data transmission speed for LANs continues to increase, driven by technologies such as Fast Ethernet, ATM (asynchronous transfer mode), and copper distributed data interface (CDDI).

Many organizations have both LANs and WANs. In fact, there may be stand-alone LANs unconnected to any other type of network. However, as noted throughout this book, there is a growing trend to interconnect LANs and to merge LANs and WANs. By the turn of the century, stand-alone LANs are expected to be rare.

METROPOLITAN AREA NETWORKS

Another geographically defined network is a ***metropolitan area network (MAN)***. A MAN, as its name implies, is a distributed network that spans a metropolitan area (a city and its major suburbs). In geographic scope, a MAN is considered to fall between a LAN and a WAN. Figure 16-2 depicts an example of a MAN.

As noted in Chapter 5, MANs are covered in IEEE 802.6 standards. These standards define MANs as covering distances up to 200 miles and possessing speeds on the order of 100 Mbps. These standards also recognize that MANs may transmit data, voice, and video signals. Partly in response to this standard, many of the Regional Bell Operating Companies (RBOCs) offer multimegabit digital services in metropolitan areas.

The American National Standards Institute (ANSI) FDDI standard is not far removed from IEEE's MAN standards. The FDDI specifications call for a token-ring LAN that uses fiber optic cable to interconnect up to 1,000 nodes. Transmission speed is 100 Mbps (or higher), and distances up to 200 kilometers may be covered.

Microcomputers and servers are likely to be the central computing technologies in MANs. Also, the data transmission speeds found in MANs are faster than the averages found in either LANs or WANs. Because of these factors, MANs are considered distinct from either LANs or WANs.

◆ WAN APPLICATIONS

Many of the day-to-day telecommunications applications discussed in this text are commonly found in WANs. In fact, many were initially designed for use in WANs to overcome geographical distances between locations.

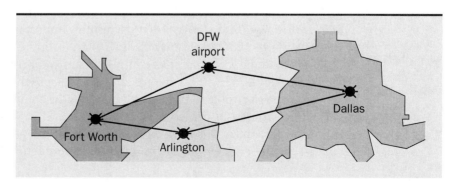

FIGURE 16-2

A metropolitan area network (MAN)

E-mail systems, the computer-age equivalents of traditional mail-boxes, are commonplace today. While e-mail is supported in LANs, many of the first e-mail systems were created for WANs. For example, Bitnet is a worldwide e-mail system that was built upon interconnected IBM hosts. In addition, many subscribers to the e-mail systems available through information utilities such as CompuServe, Prodigy, and America Online access their e-mail through WANs.

Voice mail systems are being increasingly used to digitize and store voice messages. These high-tech equivalents of answering machines are especially important for capturing messages from long-distance callers (that is, those paying to use the wide area services provided by telephone companies).

Fax machines have become standard fixtures in most offices for sending hardcopy document images from one location to another over ordinary phone lines. These are especially important for sending copies of documents to distant locations.

Videoconferencing (video teleconferencing) is also becoming more widely used in large organizations; it allows organizations to hold electronic committee meetings in real time with participants never having to leave their buildings, even if they are scattered around the globe.

Electronic bulletin board systems (BBSs) are often used to distribute information to workers and people outside the organization. Remote users of the BBS can "modem-in" to the bulletin board to post new information or see what other messages have been posted. As noted, many hardware and software vendors have established BBSs (often using an 800 number) to provide users with help and as a means to exchange tips and advice.

Cellular phone systems are also being used for both voice and data communications. Portable computers can be used in conjunction with cellular phone systems to allow users to gain access to organizational computing resources from remote locations and while driving or flying between locations. Cellular distributed packet data (CDPD) networks are also being implemented to provide wireless wide-area data communication capabilities.

ELECTRONIC DATA INTERCHANGE

Electronic data interchange (EDI) is another important data communications use of wide area networks. EDI is an example of an interorganizational system (IOS) that links computers in two or more organizations. It allows standard business documents, such as purchase orders and invoices, to be electronically exchanged between the companies. An example of an EDI system is provided in Figure 16-3.

Many organizations have created EDI links with customers and suppliers. It has been adopted by many U.S. manufacturers, retailers, and

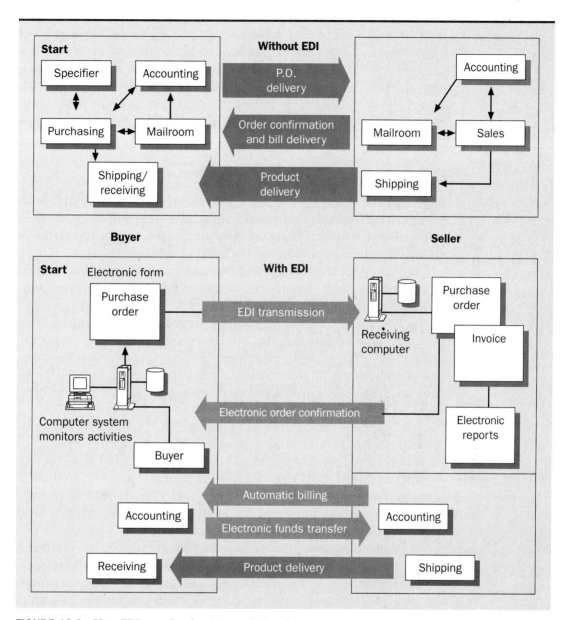

FIGURE 16-3   How EDI can alter business relationships

financial institutions including General Electric, General Foods, General Mills, General Motors, Procter & Gamble, K Mart, J. C. Penney, Sears, Chase Manhattan Bank, and Citicorp. Also, nearly 80 percent of all orders placed by pharmacies are done via EDI.

EDI software is available from a variety of vendors. In addition, EDI services are provided by third-party firms (including IBM) and value-added networks (described a little later in this chapter). Several EDI protocols have been developed; the CCITT's X.12 protocol and EDIFACT are two of the most widely used EDI protocols.

EDI is widely used because it saves time and money. Transaction documents can be exchanged instantly, and this can minimize printing and paper handling costs as well as postage costs. EDI also serves as a means of "locking in" customers and suppliers by making it easier to do business with the company.

USING WANs FOR COMPETITIVE ADVANTAGE

In general, the variety of wide area applications is increasing. From a business perspective, WANs are being used to save time and overcome geographical restrictions. They are also helping organizations operate more efficiently, enhance decision making, and be more innovative.

Through WAN links, data can be instantaneously transmitted between locations. Work can be done faster; for instance, through WANs, stock brokers can execute many more trades than they could through voice networks. WANs also benefit organizations by reducing *information float*, the time it takes to get information from the source into the hands of decision makers. For example, e-mail and fax technologies enable messages and documents to be transmitted around the globe almost instantaneously, and the timely receipt of such information facilitates rapid decision making.

Better service can be another by-product of WANs. The interorganizational systems that some companies have set up with their suppliers or customers can help ensure that materials are always on hand when needed.

WANs also help publishers such as the Gannett Corporation (*USA Today*) distribute page images to widely dispersed printing plants. In addition, WANs support overnight package delivery companies such as Federal Express and United Parcel Service.

WANs enable geographically dispersed organizations to function as if they were all located at a single site. This may enable dispersed organizations to realize some of the economies of scale enjoyed by single-site firms. For example, by linking inventory databases together among sites, firms can maintain less aggregate inventory. Also, through WANs, organizations can often exert better control over their dispersed units, ensuring that they are following organizational operating procedures.

WANs have allowed some organizations to create an "electronic presence" in geographical areas where they have no (significant) physical presence. For example, through a WAN, a financial services organization can provide its full range of products to remote field offices staffed by skeleton crews. In some cases, the electronic presence has the effect of extending business hours across time zones.

◆ TYPES OF WANs

Organizations have a number of options with respect to WAN services and facilities. Many WANs are built around the services available from common carriers, value-added network (VAN) vendors, and communications media vendors who provide facilities that may be acquired for private use. Figure 16-4 illustrates some of the major types of WAN services.

COMMON CARRIERS

Common carriers are firms licensed and regulated by the government to provide wide area communications services to the general public. These companies offer the use of the wide area telecommunications infrastructures that they have developed.

Common carriers include Bell system phone companies, which provide in-state and regional service, long-distance carriers such as AT&T, MCI, and Sprint, and locally owned telephone companies. Common carriers offer many types of channels; the most widely used are ***dial-up lines*** and ***leased lines***. Dial-up telephone lines are phone lines rented from local telephone companies; this is the telephone service that most people have in their homes. After installation, there is typically a monthly fee for keeping the line in service and additional charges for long-distance usage. Traditionally, dial-up lines have been switched lines, which use analog signalling and provide data transmission rates from 300 bps to 9600 bps.

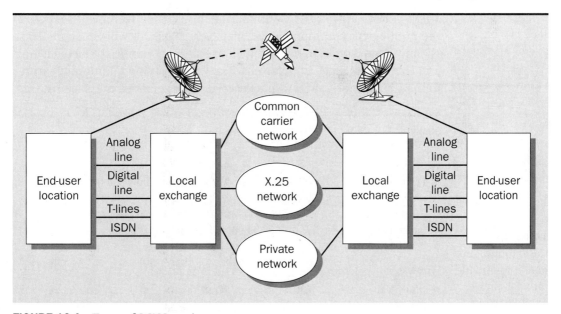

FIGURE 16-4 Types of WAN services

However, the Bell system companies and their competitors are making other, higher-speed digital switching services available in many areas. Among these are "384" (38.4 kbs) and "switched 56" (56 kbs) services. As digital lines and communications equipment become more common, additional, higher-speed services are expected.

Leased lines can be rented from common carriers. Lines are often leased by organizations when network traffic volumes make leasing cheaper than the long-distance toll charges incurred through dial-up circuits. Typically, leased lines provide more error-free communications and higher data transmission speeds than dial-up lines. Speeds on leased lines may range from 9600 bps to 1.544 Mbps provided by T-1 lines and the even higher speeds available on T-2, T-3, and T-4 lines (see Table 16-1).

Dedicated lines can be leased or purchased from common carriers; these can also be obtained from other vendors that sell communications media for private use. Dedicated lines are "private" lines in that they do not have to be shared with others. In many instances, an organization may share a nondedicated leased line with one or more other organizations. Dedicated lines are economical when one needs speed above that offered on switched lines, and when large volumes of data (especially sensitive data) are sent between locations. Dedicated lines provide the infrastructure on which private networks—networks that the organization has exclusive use of—are built.

VALUE-ADDED NETWORKS (VANs)

Value-added network (VAN) vendors are businesses that provide services to the public over common carrier facilities. These networks usually lease channels from common carriers, then re-lease the channels to their customers. They "add value" to these channels by providing additional features to their customers, such as packet switching services, message routing,

TABLE 16-1
Communications media commonly found in public common carrier networks

Medium	Common Transfer Rate (bps)
Physical Wires	
◆ Private line	300, 1200, 2400, 4800, 9600, 19,200, 38,400, 56,000, 64,000, 80,000
◆ Switched line	300, 1200, 2400, 4800, 9600, 19,200, 38,400, 56,000
◆ Leased line	2400, 4800, 9600, 19,200, 56,000, 64,000
◆ T-1, T-2, T-3, T-4	1.5M, 6.3M, 46M, 281M
Wireless	
◆ Microwave	To 45M
◆ Satellite	To 50M

network management, store and forward capabilities, error checking, and protocol conversion.

Value-added carriers offer their customers or "subscribers" relatively high-quality and inexpensive service in return for a membership fee and usage charges. Unlike long-distance dial-up charges, which are based on "connect time" (how long the circuit is used), value-added usage charges are often based on the volume of data actually transmitted over the network. This makes it possible for VANs to offer subscribers attractive prices for their services.

Many VANs are examples of **packet distribution networks (PDNs)**. The name is derived from the manner in which data is transmitted over this type of WAN—in fixed-length packets (depending on the network, these may be 128, 256, 512, or 1024 data bytes in length). PDNs are also commonly called X.25 networks after the Consultative Committee on International Telegraph and Telephony (CCITT) X.25 recommendation that defines the interface between terminal equipment and data circuit-terminating equipment (DCE) for devices that operate in the data mode on public data networks. Figure 16-5 provides an example of a PDN.

Note that private packet switching networks have been created by some organizations. In fact, private PDNs are growing in popularity. However, publicly available PDNs handle packetized data traffic for most organizations.

Because subscribers to publicly available PDNs are charged for the amount of data they transmit rather than for how long they are connected to the network, subscribing to PDN services can often be cheaper than using

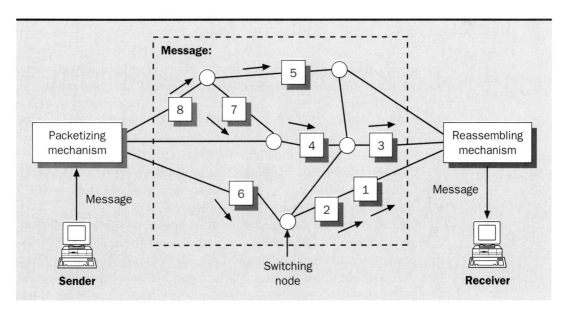

FIGURE 16-5 A packet distribution network (PDN)

either dial-up long-distance connections or leased lines. Many of the VANs provide wide geographic coverage, and most can be accessed through a local telephone call (or 800 number). However, because subscribers share PDN circuits, these networks may bog down when network traffic is very high. In addition, if an organization consistently has very high data volumes to transmit, leased lines may be cheaper than PDN services.

Within the United States, public PDNs include GTE's Telenet, General Electric's Mark Net, and Tymnet from Tymeshare. Packet switched networks are also available in most countries around the world. Some examples of these are Canada's Datapac, Australia's Austpac, and Euronet, which connects major European cities. The development of global packet switching services and VANs is currently underway.

The variety of VAN services is constantly growing. Some VANs are designed to serve varied organizations while others serve only particular industries or organizations interested in special types of telecommunications services. In general, VAN services have become so popular that common carriers, Western Union, and large corporations such as IBM, Boeing, and RCA are providing them. For example, in response to widespread (and growing) interest in videoconferencing, AT&T and Sprint have introduced public videoconferencing services. Also, many of the information utilities and online database services such CompuServe, Dialog, Dow Jones News Retrieval Service, and The Source connect their computers to public packet switching networks. A well-known industry-specific VAN is the cooperatively owned SWIFT network, which connects over 2,000 banks worldwide for funds transfers and other financial clearing purposes.

ISDN

An ongoing effort to provide integrated, all-digital communications service to users is the *Integrated Services Digital Network (ISDN)* gradually being introduced around the world. Although these services are already available in some areas, global ISDN in not expected before the turn of the century. ISDN is a digital network with built-in intelligence that enables all types of signals—fax, video, voice, and data—to be sent over a single network.

The common form of ISDN is also known as *narrowband ISDN*. *Broadband ISDN* (or *B-ISDN* or *ISDN-2*) is a newer version of ISDN which, at 155 Mbps, is about 10 times faster than narrowband ISDN.

Narrowband ISDN is a switched digital network service available from common carriers. It offers two different service levels or interfaces. The *basic rate interface (BRI)* is also referred to as *2B+D;* this is because the interface includes 2 (64 Kbps) "bearer" channels and 1 "data" channel (16 Kbps). The bearer channels carry communications services such as voice, video, and data transfers; the data channel is used for network management functions such as initiating and terminating "calls." The second service level is known as *PRI* (for *primary rate interface*). This interface is also known as *23B+D* because it provides 23 bearer channels (64 Kbps) and 1 data channel (64 Kbps); PRI fits neatly into a 1.54 Mbps T-1 line—the main medium used to connect two PRI locations.

In order for a workstation to transmit data over the ISDN network, an ISDN terminal adapter is necessary. ISDN adapters are available as expansion cards that plug into one of the microcomputer's expansion slots and as stand-alone units; most include a phone jack so that voice communications can be accommodated through the adapter. ISDN adapters remain expensive, typically costing more than $1,000.

In addition to ISDN terminal adapters, technologies exist to connect an entire LAN to the ISDN network. The three most popular approaches used to do this are through ISDN gateways, ISDN bridges or routers, and ISDN servers. Superhubs may also be used to interconnect an organization's networks with the ISDN network. ISDN and X.25 interconnection technologies are also available.

◆ WAN HARDWARE

Throughout the book, we have focused on the hardware and software components of LANs. We have also discussed some of the internetworking technologies found in organizational networks. However, our discussion of data communication technologies would be incomplete without identifying and briefly describing the major types of hardware found in WANs.

Like LANs, WANs typically include servers, workstations, and peripherals such as printers. However, other devices are usually used only in WANs. These devices include front end processors, controllers, multiplexers, and concentrators. The major hardware devices used to support WANs are depicted in Figure 16-6.

HOSTS

In WANs, hosts are usually mainframe or minicomputers. In large WANs, several mainframes or minicomputers may be used as hosts; when these are interconnected by high-speed communications lines, they are commonly called *network nodes*. Hosts typically provide WAN users with access to application programs and database management systems; in this way, they function as "servers" in WANs. In addition, communications functions are carried out at the nodes so that users at one location can access network resources at other locations.

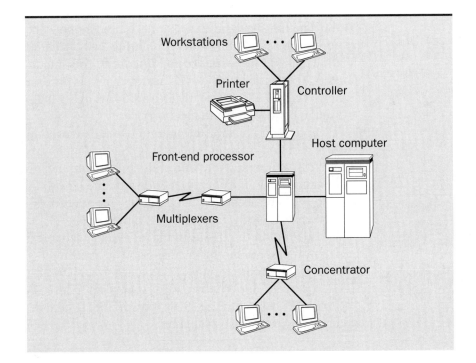

FIGURE 16-6
Major types of
WAN hardware

FRONT END PROCESSORS

A *front end processor* is a computer added to a host to relieve it of having to perform both communications and processing operations. Communications related tasks, such as accepting data from terminals, sending processing results to requesters, and providing output to terminals, printers, and other output devices are some of the more time-consuming chores that computers perform. If a host is responsible for both communications-related tasks *and* processing tasks, network performance can be negatively impacted. By offloading communications-related tasks to a front end processor (FEP), the host can concentrate on processing tasks, improving network performance.

FEPs vary in sophistication and the variety of communications activities they are asked to perform. They may be called upon to perform communications line control functions (such as polling), protocol conversion, error checking, encryption, data compression, message queuing, and message logging. In addition, FEPs may be used to manage continuous backup systems and to carry out network management functions; they may even be used for some preliminary processing of data. For instance, the airlines' passenger reservations systems use front end processors to edit data coming in from agents.

CONTROLLERS

Controllers are specialized computers whose purpose is to relieve the host CPU of the burden of communicating with a large number of terminals and peripheral devices. These are known by many names including *terminal controllers, cluster controllers,* and *communications controllers.* Like multiplexers, controllers make it possible for several terminals or peripheral devices to be connected to a single high-speed line to the host. Controllers are often found at the remote ends of high-speed lines and frequently carry out tasks performed by front end processors at these locations. Controllers may be found in almost any type of network topology: plex, hierarchical, star, bus, ring, or hybrid.

Controllers perform almost the same role that an appointment secretary does for a busy executive. The controller collects messages from terminal devices that are destined for the host CPU, packages them appropriately, and sends them on to the host for processing. After the host processes the messages, they are sent back to the controller, and it is the controller's task to ensure that they are sent to the correct devices.

MULTIPLEXERS

A *multiplexer* is a hardware device that enables several terminal devices to share the same high-speed line. Multiplexers can be very useful in WANs because the communications lines between locations often have a greater capacity than a single terminal device can use. Thus, if a company is to be charged for using a communications line, it makes good sense to pack as much data as possible over it.

A multiplexer accepts data streams from several terminals (typically in multiples of four: 4, 8, 16, and 32) and transmits them over a high-speed communications channel to the host. This enables multiple remote terminals to be connected to the host through a single line.

Like modems, multiplexers are typically used in pairs so that signals combined at one end can be sorted out—demultiplexed—at the other. The three major types of multiplexers are frequency division (which assigns incoming data streams to distinct frequency ranges—subchannels), time division (which allocates time slices to incoming data streams), and statistical (which includes sufficient intelligence to minimize the number of empty time slices by allocating the time slices that would normally go to "idle" terminals to "busy" ones).

Recently, T-1 and optical multiplexers have been receiving a great deal of attention in the networking literature. These are used on the high-capacity digital lines available from common carriers.

CONCENTRATORS

A *concentrator* is a hardware device that combines the functions of a controller and a multiplexer. Most concentrators work by using a store-and-forward messaging approach—storing messages from many remote devices until the quantity of data is sufficient to make it worthwhile to send over a high-speed line.

In the airlines' passenger reservations systems, concentrators are commonly used at key nodes to gather messages sent in from airline agents. When enough messages are collected to make forwarding worthwhile, the messages are sent to the centralized reservations system for processing. All of this happens so fast that most agents are not even aware that their messages are being held up by the concentrator.

OTHER WAN HARDWARE

Because an organization's WAN has typically evolved over a number of years, it is often composed of a mixture of many types of computers and communications channels; in addition, parts of it are likely to use different transmission modes and data codes. To make it possible for diverse systems components to communicate with one another and to operate as a functional unit, internetworking technologies such as gateways, bridges, routers, superhubs, and so on are often required.

Increasingly, WANs are being expanded through mobile computing and telecommuting. More organizations are making it possible for employees to access WAN resources from remote locations. Such remote access may even be accomplished through wireless media such as cellular data networks.

WAN TOPOLOGIES

WANs can use various topologies just as can LANs (LAN topologies were discussed in Chapter 2). A hierarchical system can be organized in layers much like an organizational chart; in such systems, messages often travel through the system in hierarchical paths. However, most WANs incorporate several topologies, making the overall topology a hybrid, such as that shown in Figure 16-7.

◆ WAN SOFTWARE AND PROTOCOLS

In any large communications system, there are a variety of communications software programs making it possible for users to carry out their tasks. For example, multiuser network operating systems are quite commonly used in WANs to enable users to share resources at nodes. In addition,

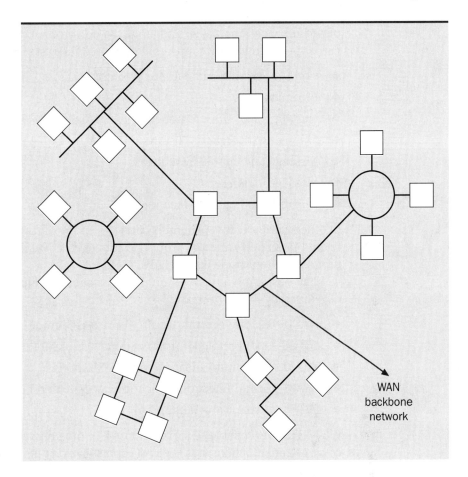

FIGURE 16-7
A WAN with
hybrid topology

the different types of communications devices found in WANs require software to carry out their functions. For example, software enables controllers to poll terminals and FEP software enables front end processors to carry out their tasks. And, of course, remote users are likely to use communications software (such as CrossTalk and ProComm) in conjunction with modes to connect to WAN resources.

In addition to these, other important types of WAN software include telecommunications access programs and teleprocessing monitors. These are briefly discussed in the following sections.

TELEPROCESSING ACCESS PROGRAMS

In WANs, *teleprocessing access programs* provide users access to applications and databases. Access programs may be located in hosts, front end processors, nodes in packet distribution networks, controllers, servers, and microcomputers. These are used to fulfill user requests to retrieve data, perform processing tasks, and so on. In addition, access programs may carry out the following communications-related tasks:

◆ Code conversion (such as ASCII to EBCDIC)

◆ Error checking

◆ Handling calls from remote users over dial-up lines

◆ Managing the network's store-and-forward capabilities

◆ Message and circuit switching

◆ Message logging

◆ Terminal polling and selecting

In other words, access programs carry out many of the communications-related tasks performed by front end processors. Not surprisingly, access programs often reside within FEPs.

The best known telecommunications access programs were initially developed by IBM. These are:

◆ BTAM (Basic Telecommunications Access Method)

◆ QTAM (Queued Telecommunications Access Method)

◆ TCAM (Telecommunications Access Method)

◆ VTAM (Virtual Telecommunications Access Method)

◆ NCP (Network Control Program)

While BTAM is capable of carrying out all of the previously mentioned access program functions, it is the least sophisticated of the listed access methods. It is designed to reside in the host and to serve as the interface between the FEP and application programs. It primarily supports input operations.

QTAM is an extension of BTAM, but supports only asynchronous communications (BTAM supports both asynchronous and synchronous). In addition to the tasks performed by BTAM, QTAM can edit messages, manage message queues, and route (and reroute) messages. In most WANs, QTAM has been replaced by TCAM or more sophisticated access programs.

TCAM is capable of performing all the functions of BTAM and QTAM in addition to handling a number of network management tasks such as reporting the status of terminals, communications lines, and message queues. It also incorporates a number of recovery and security features that enhance the fault tolerance of the network. Because of these features, TCAM has replaced BTAM and QTAM in many WANs. TCAM is also capable of managing IBM's SNA (Systems Network Architecture) networks.

VTAM can also be used to manage SNA networks. It resides in the host and controls the establishment, maintenance, and termination of access to the host's application programs by network terminals. VTAM permits application programs to share communications lines, controllers, and terminals. It is sometimes used in conjunction with the Network Control Program (NCP).

NCP resides in front end processors and is capable of carrying out the tasks commonly associated with FEPs. It routes data and controls the flow of data between the FEP and other network resources including the host and controllers. It works with rather than replaces either TCAM or VTAM and moves some of their functions to the FEP.

TELEPROCESSING MONITORS

Teleprocessing monitor software handles a number of important communications and network management tasks. Some of these are:

◆   Error checking

◆   File and database management

◆   Managing access privileges

◆   Message logging

◆   Message queue management

◆   Password checking and authentication

◆   Recording network statistics

◆   System recovery and restart

In contrast to access programs, which essentially move data into and out of the host, teleprocessing monitors interface between access programs and the rest of the host's software (including its operating system, database management systems, security software, and application programs). These are especially important in online transaction processing systems. Teleprocessing monitors are often considered to be "mini" operating systems with data communication interfaces.

IBM's Customer Information Control System (CICS) is one of the most widely used teleprocessing monitors. It was designed for use in high-volume transaction processing environments. CICS runs in conjunction with the host's operating system as a table-driven program that can service up to 64 programs (layers). It essentially performs communications-related tasks for the operating system so that the OS can process tasks associated with application programs. While originally designed for mainframes, minicomputer versions of CICS have been developed, especially for IBM's AS/400 family of processors. Server (LAN-based) versions of CICS are also available.

WAN DATA LINK PROTOCOLS

The two primary media access control methods used in LANs, token passing and CSMA/CD (carrier sense multiple access/collision detection), are rarely used in WANs. However, WAN data link protocols carry out the same types of functions. WAN data link protocols accept message packets from the network layer of the OSI (Open Systems Interconnection) reference model and add data link control data to them. The process of adding this data link control data is often called *enveloping*, because the message packet is enclosed in the data link control data much as a handwritten letter is enclosed in an envelope.

The two main categories of WAN data link control methods are asynchronous and synchronous. Synchronous protocols are most common in today's WANs. Some of the most widely used are identified and briefly described below.

♦ Synchronous Data Link Control (SDLC) is one of the oldest synchronous data link protocols. It was developed by IBM in the early 1970s and is the primary protocol used in SNA WANs (of which there are more than 50,000 worldwide). It is used in a majority of the WANs that use synchronous protocols.

♦ High-Level Data Link Control (HDLC) is a synchronous protocol standard developed by the International Standards Organization (ISO).

♦ Advanced Data Communications Control Procedure (ADCCP) is a synchronous data link protocol standard endorsed by ANSI (American National Standards Institute).

♦ Link Access Procedure—Balanced (LAPB) is the standard data link protocol used in X.25 packet distribution networks.

All of these protocols operate in a similar manner. In addition, their frame formats are similar enough to enable fairly easy conversion from one protocol to another.

WAN NETWORK-LAYER PROTOCOLS

Within WANs, network-layer protocols perform three major functions: message routing, network control, and congestion control. Message routing gets data from the message originator to its final destination. In many WANs, the message can take alternate paths through the network; network-layer protocols must be aware of these alternatives and must select the route the message is to take. As noted in the discussion of routers in Chapter 15, the path selected for a particular message may be determined by a number of factors including network traffic, number of intermediate nodes (hops), and communications link speeds. Network routing tables play a significant role in message routing in WANs.

Several routing techniques may be used in WANs including broadcast routing, static routing, weighted routing, and adaptive routing. *Broadcast routing* is most similar to the routing schemes used in LANs; the message is sent (broadcast) to all devices, but is only accepted by the device(s) to which it is addressed. In *static routing*, messages always use the same path between two nodes unless the link is unavailable. In *weighted routing*, the path that a message will take from among the available alternatives is randomly selected by using weights determined by the utilization of the different links. For example, if two alternative links exist, and one is used 60 percent of the time, the random selection is set up so that there are 6 chances in 10 that the most-utilized path will be chosen. In *adaptive* (or *dynamic*) *routing*, the best or fastest alternative route currently available (based on network routing tables and current network traffic) is selected; it is called adaptive because it is done on the fly and repeated at each node to which the message is passed.

The network control functions carried out by WAN network layer protocols include enforcing message priority schemes. They also involve sending out and aggregating node status information for updating routing tables and determining the best paths for messages to use through the WAN.

Congestion control functions are concerned with minimizing transmission delays caused by very busy links or bogged down nodes that are unable to process messages in a timely manner. By monitoring such bottlenecks, WAN network-layer protocols should be able to route messages around the network congestion.

Routing, network control, and congestion control functions are key activities carried out by network-layer protocols such as TCP/IP. As noted, TCP/IP (Transmission Control Protocol/Internet Protocol) is one of the oldest networking standards. It was developed for the U.S. Department of Defense's Advanced Research Project Agency Network (ARPANET). ARPANET has since evolved into the Internet.

The two parts of TCP/IP are IP (the internetworking protocol) and TCP (the transmission control protocol). IP is the internetworking component of TCP/IP; it is responsible for formatting message packets for transmission over the network and for routing packets to their destinations. The TCP portion of TCP/IP is responsible for end-to-end message delivery; this includes acknowledging received packets, retransmitting lost packets, and ensuring that packets are received in the correct order.

In addition to TCP/IP, some of the network management protocols discussed in Chapter 15 play important roles in data communications in WANs. This includes SNMP, SNMP.2, and CMIP, especially when these are used in distributed client/server networks that span large geographic areas.

OTHER IMPORTANT WAN PROTOCOLS

A number of other important protocols exist in WAN environments, too many to adequately cover in an introductory book such as this. For example, several protocols have been developed for electronic data interchange (EDI); two of the most important are the X.12 and EDIFACT protocols inasmuch as these are the most widely used.

PDN Standards

A number of important standards (protocols) are associated with the X.25 standard used in packet distribution networks. Like the X.25 standard, these recommendations have been developed by the CCITT. Some of these are:

◆ X.28—this standard governs dial-up asynchronous access to PDNs.

◆ X.32—this standard governs dial-up synchronous access to PDNs.

◆ X.75—this standard defines the interface between two different public packet switching networks, both national and international; it is often considered the packet switched network "gateway" protocol.

◆ X.121—a global addressing scheme for PDNs that includes a four-digit code signifying the global zone, country code, and VAN codes within countries as well as a ten-digit address for the destination node.

Frame and Cell Relay Standards

Closely related to X.25 standards are *frame relay* and *cell relay* technologies. Both of these technologies have been developed for high-speed fiber optic links (such as T-1, T-2, T-3, and T-4 lines) available from common carriers and PDNs; such communications services were not available when many of the public PDNs were first implemented.

Frame relay (also called *fast packet*) systems are built around a data link protocol (LAP-D, for Link Access Procedure-D Channel) that defines how frames of data are assembled and routed through a PDN. In frame relay systems, frames are formatted by a FRAD (frame relay assembly device) rather than a PAD (packet assembly device—the computer and software used to assemble data packets in traditional PDNs). Frame relay systems are designed to provide faster data transfer rates than are possible with traditional X.25 implementations. Such systems provide for variable length packets capable of transporting up to 8,000 characters at once; sophisticated switches, along with advanced network control and congestion control features, enable frames to be transmitted at speeds up to 1.544 Mbps.

Cell relay systems are essentially the same as frame relay systems, except that the frames are fixed in length (53 bytes total and 48 data characters). The use of fixed-size frames makes it possible to transmit over the highest-capacity communications links available from common carriers (such as T-3 and T-4 lines) with speeds ranging from 45 Mbps to 2.4 gigabits per second. Such speeds, along with predictable processing, forwarding, and delivery rates, make cell relay systems ideally suited for the simultaneous transmission of voice, data, fax, image, and video. The best known cell relay system is ATM (asynchronous transfer mode). In the years ahead, ATM is expected to emerge as one of the most important WAN protocols. In addition, LAN-ATM hubs are being created to interconnect LANs and ATM WANs.

E-Mail Standards

A final set of standards (protocols) found in WANs focuses on worldwide e-mail networks. The "X" prefix marks these standards as developed by the CCITT.

◆ X.400—this standard defines the operations of global e-mail services.

◆ X.435—this standard defines the transmission of electronic data interchange (EDI) documents over X.400 e-mail networks; it is known as the EDI Messaging and Services standard.

◆ X.500—this standard defines global directories (and directory services) of e-mail users in X.400 networks.

◆ CHALLENGES TO THE DEVELOPMENT OF GLOBAL NETWORKS

From the foregoing discussion, it should be evident that worldwide WANs and WAN services (such as PDNs and e-mail) are becoming increasingly important. Today, more than ever before, a corporation's decision to go international is almost certainly tied to its ability to develop a global telecommunications network capable of supporting its international operations. Global telecommunications technologies are enabling companies to establish both physical and electronic presence in parts of the world that would not have been feasible in the past.

For example, the creation of the 1992 European Community (EC) has resulted in a growing number of U.S. firms exploring the possibility of expanding to Europe. A study performed by KPMG Peat Marwick indicates that the number of U.S.-based companies planning to expand into Europe has risen sharply since 1990. Some plan to open European plants and offices. Others are planning to acquire European companies, mergers, or joint ventures. Regardless of the approach, information systems will play a key role. Organizations are already spearheading many projects including enhancing and expanding existing networks, building new telecommunications centers, and interconnecting new subsidiaries' systems with that of the parent company.

Building global networks is not an easy task. There are a number of obstacles to building or expanding networks in the EC and other parts of the world, including:

◆ unreliable and inconsistent telecommunications services among countries and even among cities within countries

♦ conflicting standards for different network components (such as modems, cabling, ISDN-connections, electrical connections) and protocols (such as EDI)

♦ monopolistic, government-controlled communications carriers that limit vendor service and equipment options

♦ unpredictable installation and implementation (network "turn-on") target dates

♦ widely varying government regulations and tariff structures

♦ a bewildering proliferation of alliances, coalitions, joint ventures, and projects among vendors and government services

In spite of these obstacles, expanding to Europe or other parts of the world is not as difficult as it once was. For example, the reforms sweeping Europe along with relaxed regulations are beginning to result in more open markets, better service quality, and an environment conducive to building sophisticated networks. By 1996, basic and primary access to Integrated Services Digital Network (ISDN) is expected to be readily available to commercial customers throughout Europe; virtual private networks (VPNs), based on public packet switching networks are also expected to be more widespread.

Many U.S. companies are finding that outsourcing international network development can be a cost-effective way to maneuver through the obstacles to global networking. By outsourcing network services or network development to firms that already know the region, international expansion can often be accomplished more rapidly. For example, BT North America, Inc. has helped a number of other firms (including J.P. Morgan and Company and Gillette) develop international networks. Contracting with other firms to provide network services and outsourcing network development tasks are likely to remain popular until many of the obstacles to global networking have been removed.

The changes under way in Eastern Europe and the former Soviet Union have also created unprecedented expansion opportunities and networking headaches. Since the fall of the Berlin Wall in 1989, the number of companies that have initiated efforts to expand to Eastern Europe has grown dramatically. Experts agree that it will take billions of dollars to bring Eastern Europe's telecommunications infrastructure up to that of the rest of the continent. However, most feel that this will be done quite rapidly despite the tremendous investment required to bring state-of-the-art WAN technologies to this part of the world. Wireless systems, especially satellite-based cellular voice and data networks, hold special promise in opening up the markets in Eastern Europe to the rest of the world.

◆ TRENDS AND FUTURE DEVELOPMENTS

Telecommunications and networking are bringing us closer to global information systems, both public and private. Advancing technology in this area is contributing to the globalization of the business environment. This has generated a growing demand for higher-speed communications links and international "connectivity" standards, which make it easier to develop and expand WANs.

There is an ongoing evolution of common carrier services toward digital equipment and away from analog-based systems. ISDN and ATM are on the horizon and will be among the major services provided both by common carriers and VANs by the turn of the century.

Fiber optic media are beginning to dominate terrestrial communications, while microwaves and satellite-based communications are expected to be used even more widely than they are now. Wireless networks, most using radio or microwave frequencies, are expected to increase in use, and the percentage of workers routinely using some type of wireless communications will continue to increase.

To ensure connectivity, telecommunications hardware and software vendors are expected to conform to OSI and other internationally accepted standards. Increased compliance with such standards should promote the development of enterprise-wide information systems, and standards will make it easier to establish effective LAN/WAN interfaces. Simple Network Management Protocol (SNMP) is often cited as a protocol which can be used to manage enterprise-wide networks and, for this reason, it is expected to become more popular.

UNIX is expected to grow in importance, especially in large networks. IBM's network architecture, SNA, is expected to continue to play a key role in the ongoing evolution of business telecommunications. SNA's inclusion of AIX, IBM's proprietary version of UNIX, demonstrates Big Blue's commitment to open up its network architecture and to allow it evolve.

Electronic Data Interchange (EDI) systems are already affecting the flow of documents between organizations and their suppliers and customers and is likely to proliferate by the turn of the century. Such systems are already being used by more than 40 percent of businesses in the U.S.

Many experts feel that the continuing development of sophisticated wide area telecommunications infrastructures is essential for future competitiveness in a number of industries. A consortium of U.S. companies, universities, and government agencies is working on the blueprint for a fiber optic based "data highway"—NREN (the National Research and Education Network)—which will allow data transmission rates exceeding 1 billion bps. This is expected to facilitate changes in a number of sectors, including research and development, education, medicine, and information services to the home.

Consumers are generally becoming more aware of wide area data networks and network services. Computer-literate consumers are routinely finding information on BBSs and through the Internet that previously was available only in certain geographical areas. Telecommuting is also becoming more common and is expected to increase in the years ahead.

In brief, it is important for network managers and networking professionals to stay abreast of developments that impact wide-area telecommunications. Such developments are especially important when making interconnection decisions.

◆ SUMMARY

WANs are telecommunications networks that span large geographic areas such as a major region of a state, an entire state, several states, all states, several countries, or the world. WANs can be private (that is, created and used by a single organization or group of organizations) or public. The Internet's World Wide Web is an example of a public data communications WAN that spans the globe. A private WAN may connect an organization's various operating locations; such WANs often provide the infrastructures needed to create enterprise-wide information systems. In addition, networks (such as EDI networks) that connect two or more organizations may be classified as WANs; these are often called interorganizational networks.

Relative to a WAN, a LAN covers a much smaller geographic area; a LAN may be restricted to a single room, a set of rooms or offices in the same building, or the set of buildings that make up a single campus of a school or business. LANs rarely involve common carrier services other than to dial out on a switched circuit for voice, data, video, or fax.

A metropolitan area network (MAN) is a distributed network that spans a metropolitan area. In terms of geographic scope, a MAN falls between a LAN and a WAN. Both IEEE standard 802.6 and the ANSI FDDI standard are often used to define MANs and to differentiate them from LANs and WANs.

Many popular data communication services are found in WANs. These include e-mail, voice mail, fax, videoconferencing, BBSs, cellular data networks, and electronic data interchange. WANs help businesses overcome geographic restrictions and reduce information float.

Many WANs are built around the services available from common carriers, value-added network (VAN) vendors, and communications media vendors who provide facilities that may be acquired for private use. Common carriers include Bell system phone companies, locally owned telephone companies, and long-distance carriers such as AT&T, MCI, and Sprint; while they offer many types of channels and services, the most widely used are dial-up lines and leased lines.

Value-added network (VAN) vendors provide services to the public over common carrier facilities; they usually lease channels from common carriers and then re-lease the channels to their customers. VANs "add value" to these channels by providing additional features to their customers, such as packet switching services, message routing, network management, store-and-forward, error checking, and protocol conversion. They offer their subscribers high-quality and inexpensive service in return for a membership fee and usage charges. PDNs (packet distribution networks) are probably the best known VANs. Both domestic and international PDNs are available.

ISDN (Integrated Services Digital Network) is a WAN service available through common carriers. This network enables users to transmit voice, data, image, and video signals simultaneously over the network.

Like LANs, WANs typically include servers, workstations, and peripherals such as printers. However, other devices are usually found only in WANs; these include front end processors, controllers, multiplexers, and concentrators. Host computers (usually mainframes or minicomputers) and the interconnecting technologies discussed in Chapter 15 also play important roles in WANs. Hybrid and hierarchical topologies are most common in WANs.

Communications software is important in WANs. Telecommunications access programs (such as BTAM, QTAM, TCAM, VTAM, and NCP) and teleprocessing monitors (such as IBM's CICS) are also important, especially in wide area online transaction processing systems.

Some of the more important WAN data link protocols are SDLC, HDLC, ADCCP, and LAPB. The most important tasks performed by network-layer protocols in WANs (such as TCP/IP) are message routing, network control, and congestion control. Other important protocols found in WANs include those for EDI systems (such as X.12 and EDIFACT), those related to X.25 PDNs, those related to frame relay and cell relay systems (such as LAP-D and ATM), and those related to global e-mail systems (such as X.400 and X.500).

Interest is increasing in developing global data communication networks. Among the challenges faced in doing so are unreliable services, conflicting standards, government regulations, installation difficulties, differing tariff structures, and a confusing set of alliances between businesses and governments. To address such problems, many organizations are outsourcing international network development tasks.

The development of international networks is just one of the trends for WANs. Others include increased availability of high-speed, fiber optic, cable-based links and services, the increasing importance of UXIX, the proliferation of EDI, the evolution of the Internet and development of NREN, telecommuting, and increased utilization of WAN data services by consumers.

✳ KEY TERMS

Adaptive routing
Broadcast routing
Cell relay
Common carrier
Concentrator
Controller
Dedicated line
Dial-up line
Electronic data interchange (EDI)
Enveloping
Frame relay
Front end processor
Information float

Integrated Services Digital Network (ISDN)
Leased line
Metropolitan area networks (MANs)
Multiplexer
Packet distribution network (PDN)
Telecommunications access program
Teleprocessing monitor
Static routing
Value-added networks (VANs)
Weighted routing
Wide area networks (WANs)

✳ REVIEW QUESTIONS

1. Explain the distinction among WANs, LANs, and MANs.
2. What internationally recognized standards exist for MANs? What are the distinctions between these?
3. Identify and briefly describe some of the major types of applications found in LANs.
4. What is electronic data interchange (EDI)? Why are these systems important? Identify the major protocols found in EDI systems.
5. What is information float? How may WANs help reduce information float?
6. Provide several examples of how organizations are utilizing WANs to overcome geographic distances.
7. What are common carriers? Identify several examples of common carriers.
8. What are the differences between dial-up and leased lines? What are dedicated lines?
9. Identify some of the digital lines available from common carriers.
10. What are value-added networks? Provide several domestic and international examples of VAN vendors.
11. What are packet distribution networks? Why are these also called X.25 networks?

12. How do PDN charges differ from common carrier charges?
13. What is ISDN? Why is ISDN important? How does B-ISDN differ from narrowband ISDN?
14. How do basic rate interfaces differ from primary rate interfaces in ISDN networks?
15. What are the differences among the characteristics and functions performed by front end processors, controllers, multiplexers, and concentrators?
16. Identify some of the different types of multiplexers that may be found in WANs.
17. What are the differences between teleprocessing access programs and teleprocessing monitors? Identify examples of each and briefly describe the differences among them.
18. What are the functions of WAN data link protocols? How is enveloping related to these functions?
19. Identify and briefly describe some of the most important WAN data link protocols in use today.
20. Identify and briefly describe the major functions performed by WAN network layer protocols.
21. What are the differences among broadcast, static, weighted, and adaptive routing?
22. Briefly describe the differences among the following standards: X.28, X.32, X.75, and X.121.
23. What are the differences between frame relay and cell relay networks? Why are cell relay systems such as ATM well-suited for transmitting multimedia files between nodes?
24. Briefly describe the differences among the following standards: X.400, X.435, and X.500.
25. Discuss the challenges faced by organizations that desire to develop international data communication networks.
26. Discuss the observable trends in WANs and WAN technologies.

✳ **DISCUSSION QUESTIONS**

1. Assume that your company has used a multiuser shared database in its operations in a single office for five years with measurable success. Now, two satellite offices are to be located five hundred miles away in opposite directions. Each office will handle its own territory and report to the home office. Should this database be shared on a WAN with each of the new satellite offices? Discuss the reasons behind your recommendation.
2. A customer is calling from New York City and wants to know how to electronically connect his pawn shops scattered throughout Manhattan, Queens, Bronx, and northern New Jersey. Each office needs to be able

to send e-mail to the other offices. Identify several specific ways in which this might be accomplished. Be specific about the types of links that could be used to connect the different locations.

3. A friend who manages the marketing for a diamond production firm called and asked if you could help design a satellite transmission network between the diamond mines in South Africa and marketing offices in Japan, Singapore, and New York City. The primary applications are for lengthy data transmission describing thousands of diamonds and occasional voice traffic. From the information given in this chapter, describe the types of issues that may have to be addressed during the development of such a network.

4. Discuss why network security is often of greater concern in WANs than in LANs. Identify several security mechanisms that you feel are more important for WANs than for LANs and explain why you feel they are more important (you might want to review Chapter 14 prior to answering this question).

✳ PROBLEMS AND EXERCISES

1. Obtain a map of the closest metropolitan area to your school with at least 200,000 in population. Draw a diagram connecting the downtown post office of every incorporated town or city in the metropolitan area to the downtown post office of the central city in the metropolitan area. Optimize the network so that no more than five nodes are on an individual circuit and so that the overall mileage of the interconnections is minimized.

    Using the scale of the map, measure the point-to-point distance of each network span (that is, for each major circuit to the main downtown post office). Assume that the pricing for the monthly charges for this network is:

    ◆ First mile of each circuit segment: $150 per month

    ◆ Each 1.4 mile per segment after first mile: $40 per month

    Price each circuit in your plan and estimate what the monthly charges will be. Assume that there is a $30 per termination station charge at each node.

2. Your cousin, who works at a major metropolitan bank, has just called. The top management of the bank is anxious to reduce their overall investment in computing equipment. The operations office head has heard that if she converted all their teller terminals (which require instant updating of some items) to a MAN system, several hundred thousand dollars could be saved over a five-year period. Your cousin wants to hire you as the primary consultant to analyze the situation.

Your cousin adds at the last minute that the terminal stations must remain operative with accurate information even if a power outage strikes. And, even if the system goes down for some unknown reason, it must be restored within five minutes. These are the parameters under which the bank currently operates using an SNA network running with a mainframe host. Discuss the strategies you would use to investigate the situation and the recommendations you would propose. Prepare a two-page executive summary of the issues that must be addressed by the bank and your recommendations.

3. Find a recent article on ISDN in a computing or networking periodical. Summarize the contents of the article in a paper or class presentation.

4. Find a recent article on ATM in WANs in a computing or networking periodical. Summarize the contents of the article in a paper or class presentation.

5. Find a recent article on EDI applications in a computing or networking periodical. Summarize the contents of the article in a paper or class presentation.

6. Invite a network manager from a local firm to come to your class to discuss the WANs in his or her organization. Be sure to ask about what WAN technologies, topologies, and protocols are used. Also ask how he or she expects the WAN to change in the years ahead.

7. Find a recent article on digital switching services offered by common carriers in a computing or networking periodical. Summarize the contents of the article in a paper or class presentation.

8. Find a recent article on SONET (Synchronous Optical Network) offered by common carriers in a computing or networking periodical. Summarize the contents of the article in a paper or class presentation.

✳ REFERENCES

Aspen, J., B. E. Budish, and R. Eastman. "What Is Telepresence and How Can It Enhance Business Operations?" *Telecommunications* (November 1993): p. 57.

Beasley, B. and C. Hiestand. "Adapting LANs For Wide-Area Document Imaging." *Communications News* (July 1994): p. 17.

Briere, D. "Buyer's Guide: Digital Private Line Services," *Network World* (June 15, 1992): p. 47.
———. "Frame Relay Selection Is No Picnic." *Network World* (February 1, 1993): p. 45.

Briscoe, P. "ATM: Will It Live Up to User Expectations?" *Telecommunications* (June 1993): p. 25.

Davidson, R. and N. Muller. *The Guide to SONET*. New York: Telecomm Library, 1991.

Duncanson, J. "Standing the Multiplexer on Its Head." *LAN TIMES* (September 14, 1992): p. 47.

Eckerson, W. "Federal Express Uses EDI to Streamline Its Billing." *Network World* (May 25, 1992): p. 29.

Fahey, M. "From Local to Global: Surveying the Fiber Landscape." *Telecommunications* (November 1993): p. 33.

Falk, G. "An Approach to LAN Interconnection Over X.25." *Telecommunications* (December 1992): p. 53.

Farkouh, S. "Managing ATM-Based Broadband Networks." *IEEE Communications* (May 1993): p. 82.

Fitzgerald, J. *Business Data Communications: Basic Concepts, Security, and Design.* 4th ed. New York: John Wiley & Sons, 1993.

Fitzgerald, S. "Global Warming: ISDN Heats Up." *LAN Magazine* (June 1994): p. 50.

Flanagan, P. "Videoconferencing: A Status Report." *Telecommunications* (April 1993): p. 41.

Flanagan, W. *Frames, Packets, and Cells in Broadband Networking.* New York: Telecomm Library, 1991.

Glossbrenner, K. "Availability and Reliability of Switched Services." *IEEE Communications* (June 1993): p. 28.

Goldman, J. E. *Applied Data Communications: A Business-Oriented Approach.* New York: John Wiley & Sons, 1995.

Hamilton, R. "The Multimedia Revolution and Broadband ISDN—A Call for Transparency." *Telecommunications* (December 1992): p. 57.

Harler, C. "BN's WAN Engineer Keeps Train Engineers Rolling." *Communications News* (July 1993): p. 28.

———. "How Ron Tipton Keeps His Phone From Ringing." *Communications News* (July 1993): p. 30.

Held, G. "Selecting a WAN Operating Rate." *Network Administrator* (September/October 1994): p. 4.

Heywood, P. "Global Frame Relay: Risky Business." *Data Communications* (November 1992): p. 85.

Hurwicz, M. "Frame Relay or Leased Lines—Which Is Cheaper?" *Datamation* (April 1, 1994): p. 39.

Illingworth, M. M. "Virtual Managers." *InformationWeek* (June 13, 1994): p. 42.

Johnson, J. "A Low Cost Switch with SONET in Its Future." *Data Communications* (March 1992): p. 39.

Keen, P. G. W. and J. M. Cummins. *Networks in Action.* Belmont, CA: Wadsworth, 1994.

Lippis, N. "A First Look at ATM Means Second Thoughts About FDDI." *Data Communications* (March 1992): p. 39.

McMullen, M. "Wireless Lands." *LAN Magazine* (June 1994): p. 40.

Parker, C. and T. Case. *Management Information Systems: Strategy and Action.* 2nd ed. New York: McGraw-Hill, 1993.

Schatt, S. *Data Communications for Business.* Englewood Cliffs, NJ: Prentice Hall, 1994.

Schlar, S. *Inside X.25: A Manager's Guide.* New York: McGraw-Hill, 1990.

Stamper, D. A. *Business Data Communications.* 4th ed. Redwood City, CA: Benjamin/Cummings, 1995.

———. *Local Area Networks.* Redwood City, CA: Benjamin/Cummings, 1994.

Stephenson, P. "The Orient Express." *LAN Magazine* (June 1994): p. 57.

Strauss, P. "ATM WANs: Rocket Science or Lost in Space?" *Datamation* (January 21, 1994): p. 20.

Thyfault, M. E., S. Stahl, and J. C. Panettierri. "Breaking Down the Walls." *InformationWeek* (April 19, 1993): p. 22.

Tremblay, R. and R. Darois. "Textile Firm Weaves Tight Network Fabric." *Communications News* (July 1994): p. 18.

Violino, B. and S. Stahl. "No Place Like Home." *InformationWeek* (February 8, 1993): p. 22.

Wallace, B. "AT&T Adds Array of High-Speed Net Links." *Network World* (November 30, 1992): p. 1.

Whitten, E. "E-Mail in Europe: A Corporate Perspective." *Telecommunications* (April 1993): p. 50.

Williams, J. "ISDN in the Small Business World." *Telecommunications* (December 1992): p. 60.

Zegura, E. "Architectures for ATM Switching Systems." *IEEE Communications* (February 1993): p. 28.

Zenisek, T. "Videoconferencing Over the Public Network." *Telecommunications* (December 1992): p. 19.

Case Study:
Southeastern State University

The main computer systems at SSU have been part of a wide area data communication network for more than two decades. University users statewide have been able to dial in to their local host and through that host access computing resources at other colleges and universities.

In the past, many of the connections between the hosts (nodes) at the different public colleges and universities in the state were 9600 bps leased lines. As the number of users and network traffic has increased, so has the topology of the network and the speed of the communications lines. Up until the mid-1980s, all of the leased lines used to run directly to the state's land grant (flagship) university—making this the physical center of the WAN. During the late 1980s, the topology changed. Regional hubs (switching nodes) were created, typically at the largest college or university in that geographic location. All of the colleges and universities in the region (including the state's junior colleges and technical schools) were connected to the regional hub; the regional hub was, in turn, connected to the physical center of the WAN (which, incidently, was no longer at the flagship university). In fact, the flagship university became the hub in its region of the state.

This change was made for several reasons. First, the change reduced the number of communications links coming into the computers at the physical center of the WAN. This reduced the university system's total lease charges and made operations and maintenance at the central hub. Second, network traffic studies indicated that the majority of messages transmitted from a particular college or university were intended for another school in the same region. Hence, a regional switching center made more sense than doing all switching at the central hub. Third, the university system thought that such a configuration would make it easier to develop and house computer systems and applications in the regions of the state that utilize them the most.

SSU became one of five regional hubs in the state's new data communications WAN. Seven colleges and junior colleges in the region are connected to the VAXs at SSU by leased lines (these vary from 9600 bps to 38,400 bps lines, depending on the size and nature of the school). SSU is connected to the central hub by twin 56,000 bps lines. Similar configurations are found at the other four regional hubs.

TCP/IP is the network protocol used within the state's data communications WAN. The WAN's gateways to the Internet and to information utilities such as CompuServe are located at the central hub.

For most of each semester, this configuration handles network traffic quite well and network performance levels are adequate. However, the system still bogs down at the end of each semester as the number of network users mushrooms. It can be very frustrating for interactive users to experience five to ten minute response times when the usual response is less than two seconds. Such problems occur during much of the day and often long into the evening; it is only during the wee hours of the morning that response times return to near normal levels. Still, this problem is more transient that it was during the early 1980s when all lines ran directly to the central hub.

Recently, another type of WAN has been implemented at SSU that is not yet integrated with the state's data communication WAN. This is the WAN used to support SSU's distance learning programs. You may recall from earlier segments of the case study that SSU's distance learning program involves the use of interactive television links between the university and remote sites. There are currently 6 remote sites to which undergraduate and graduate courses are delivered through distance learning technology; most are within a 100-mile radius of SSU.

T-1 lines connect SSU with the remote distance learning sites. Half of each T-1 line's capacity is used to actually transmit the video and audio signals; the other half is reserved for system control purposes. There are also dial-up lines in each classroom independent of the T-1 line. These handle conversations between technicians (who assist instructors by controlling the cameras and so on during the class) at the different sites and faxing documents (such as handouts) between the locations.

The video signal quality is acceptable, but the compressed signals from the remote sites lack the resolution needed to clearly transmit character-based computer images between locations. Hence, limited computer-oriented applications can currently be used in the distance learning classrooms at SSU.

The T-1 lines are currently leased for the duration of scheduled classes. Such classes are typically scheduled during evening hours (6 to 10 P.M.) and on weekends. Several remote sites receive two or more distance learning classes each night of the week; hence, multiple T-1 lines are being used simultaneously to connect the two sites. In addition, demand for distance learning classes is growing at the remote sites. As a result, late afternoon classes are now being transmitted, and program coordinators feel that it won't be long before classes will have to be scheduled all day long in each of SSU's five distance learning classrooms.

In the years ahead, SSU and the state university system hope to build the infrastructure needed to efficiently transmit multimedia files between the state's colleges and universities. They would also like to make desktop-to-desktop videoconferencing between users at different schools a reality. Currently, videoconferencing is possible only in specially equipped rooms in a few buildings on campus. All videoconferences are delivered via satellite hook-ups utilizing VANs that specialize in videoconferencing services. University officials expect that one day, all videoconferencing will be delivered over terrestrial fiber optic cabling.

CASE STUDY QUESTIONS AND PROBLEMS

1. Using the information provided, compose "before" and "after" drawings for the data communication WAN connecting colleges and universities in the state. Assume a rectangular state and a total of 40 different colleges and universities; in the "after" drawing, assume 7 schools connecting to each regional hub.

2. Why do you think the state's data communication WAN bogs down at the end of each semester? What could the university system do to address this problem?

3. What types of routing are likely to be used in the state's data communication WAN? Justify your selections.

4. Could packet distribution networks be used to interconnect the colleges and universities in the state? What factors would have to be taken into account to determine if PDNs could be more cost-effective than leased lines?

5. Identify several specific changes that will have to be made to the state's data communication WAN to enable the transmission of multimedia files and desktop-to-desktop videoconferencing between locations. What kinds of communications services and protocols should be utilized?

6. What changes could be made to improve the current distance learning configuration? Be specific. What specific changes would be needed to integrate SSU's distance learning WAN with the state's data communication WAN?

PART SIX

Networking Careers

Networking Careers

CHAPTER OBJECTIVES

After completing this chapter, you will be able to:

◆ Identify some of the traditional and nontraditional networking jobs and careers.

◆ Identify some of the career options available to networking professionals.

◆ Identify several areas of niche expertise valuable in today's job market.

◆ Briefly describe the ways in which networking professionals can keep their knowledge up to date.

◆ Identify some of the types of certifications available to networking professionals.

◆ Identify some of the major types of literature and publications for networking professionals.

◆ Identify and briefly describe some of the skills needed by network managers.

Never before in the history of information technology have there been more opportunities for network-related careers. Although something akin to chaos has resulted from rapidly changing networking technologies and network-related business changes such as downsizing and business process reengineering, these turbulent times have created a number of professional opportunities for individuals interested in data communications and computer networks. With organizations depending heavily on information technology networks to play a key strategic role, who is in a better position to profit than networking professionals—especially those with both strong technical and business backgrounds?

◆ TRADITIONAL AND NONTRADITIONAL NETWORKING JOBS

In this section, we consider both traditional networking careers and nontraditional ones and the educational paths that support them. Traditional telecommunications jobs in firms include positions such as network manager, network designer, help desk technician, LAN manager, and technical support specialist. These and some other traditional job titles are summarized in Table 17-1. Many organizations have such positions and career paths leading to such positions; many will continue to have similar positions and career paths. Nontraditional career paths involve both entrepreneurial and other opportunities—for example, independent consultant on systems integration, turnkey solutions, e-mail systems, network security, or desktop-to-desktop videoconferencing, or, perhaps, in-house company consultant in these or other areas (see Table 17-2).

◆ Network manager	◆ Help desk manager
◆ Network planning manager	◆ Help desk specialist/technician
◆ Network design manager	◆ Network operations manager
◆ Network designer	◆ Network operations specialist/technician
◆ Network application development manager	◆ Network security manager
◆ Network application development specialist	◆ Network security specialist/technician
◆ Voice communications manager	◆ Network data administrator
◆ Voice communications specialist	◆ LAN service/support manager
◆ Voice communications technician	◆ LAN manager/administrator
◆ Data communications manager	◆ LAN service/support technician
◆ Data communications specialist	◆ LAN installation/maintenance manager
◆ Data communications technician	◆ LAN installation/maintenance specialist/ technician
◆ Network services/support manager	◆ WAN manager/administrator
◆ Control center manager	

TABLE 17-1 Some traditional telecommunications job titles

TABLE 17-2

Some non-traditional telecommunications careers

- ◆ Consultant on specific network products
- ◆ Fault tolerance specialist/consultant
- ◆ Independent network planner/designer/installer
- ◆ Independent technical support specialist
- ◆ Independent trainer for specific network products
- ◆ Network cabling consultant/installer
- ◆ Network security consultant
- ◆ Systems integration consultant
- ◆ Turnkey networking solutions vendor
- ◆ Wireless communications specialist/consultant

Risks are taken by those who choose nontraditional career paths; sacrifices may have to be made, and success is not guaranteed. However, there are also risks in joining a firm and choosing a traditional career path, especially in this era of rapid network technology changes and downsizing. For instance, an operations specialist for a centralized, host-based network with a decade's worth of experience might discover that skills learned, although good for the company, have become obsolete in the marketplace. This could be a rude awakening for individuals whose employers have decided to remove mainframe hosts and replace them with client/server LANs; as noted in Chapter 11, operations staff members are often laid off when such downsizing occurs.

As mentioned, in addition to downsizing, outsourcing has also changed the job security and career options available to many networking professionals. When network operations and management are outsourced, companies often reduce the size of their networking staffs. While many of the displaced workers are employed by the outsourcing service provider, their long-term job security is seldom guaranteed.

In spite of the increased prevalence of downsizing and outsourcing, there continues to be a heavy demand for networking professionals. A quality graduate in a high-tech field such as computer information systems or computer science who has had several networking courses and hands-on network administration experience can usually expect several job offers and a respectable starting salary. Over time, the average salaries of networking professionals have risen. Organizations have learned that to attract and retain good networking personnel, they must be prepared to offer competitive salaries.

◆ NETWORKING CAREER OPTIONS

Numerous career options are available for individuals interested in networking and network management. Several of them are discussed in the following sections.

CAREER PATHS

A work option frequently chosen by people is the *career path*. This involves looking at employing organizations in terms of advancement opportunity and considering how networking personnel move from entry-level positions into higher-level and management positions. For instance, a possible career path in some firms is to move from LAN technician to LAN manager to network manager.

Companies vary considerably with respect to career paths. Some firms, for instance, may follow the practice of promoting top-level network managers from within. However, if things are not going well, these same companies may hire outsiders for their high-level networking positions.

In the past, the career path was considered a relatively safe form of employment and advancement. It is now seen as a riskier proposition for network professionals. With networking technology and the general business environment in a state of flux, firms cannot guarantee that current career paths will still exist in the future.

NICHE EXPERTISE

Developing a career around a particular *niche expertise* area has become especially popular in recent years. With this option, the network professional becomes an expert in some well-defined area—for instance, groupware applications, desktop-to-desktop videoconferencing, Novell NetWare, electronic data interchange (EDI) applications, superhubs, middleware, FDDI LANs, and so on. Some network professionals develop niche careers of their own (as entrepreneurs) or within the companies for which they work.

Niche expertise can be especially valuable for employees of organizations that are downsizing or outsourcing network operations and management. If a networking professional with valuable niche expertise is terminated by a company, his or her niche skills can be valuable in finding another job quickly. In fact, the niche expertise demanded by many organizations may cause entrepreneurially-oriented individuals to start their own businesses. Table 17-3 lists some of the niche areas likely to be valuable for network professionals during the 1990s.

The chaos caused by sweeping change in information technology and the business environment has made niche careers more popular among networking professionals. Part of this increased popularity is due to the fact that it is getting more difficult to be a jack of all trades. However, it should also be noted that new products and technologies are creating new niche expertise areas on which networking professionals can capitalize. As a result, an increasing number of organizations are finding it beneficial to promote the development of niche expertise among their networking staff, even realizing that doing so may threaten their ability to retain these employees.

- ◆ ATM applications and switching
- ◆ Business process reengineering
- ◆ CASE tools for network applications
- ◆ Cellular digital packet data (CDPD) applications
- ◆ Client/server application development tools
- ◆ Data compression
- ◆ Disaster recovery planning
- ◆ Electronic data interchange (EDI) applications
- ◆ Encryption systems
- ◆ FDDI networks
- ◆ Distance learning applications
- ◆ Distributed database applications
- ◆ Fast Ethernet networks and applications
- ◆ Groupware applications
- ◆ Image processing and document management systems
- ◆ Intelligent hubs
- ◆ ISDN applications
- ◆ Interactive television applications
- ◆ Internet applications
- ◆ Kiosk applications and networking

- ◆ LAN operating systems
- ◆ LAN switches
- ◆ LAN/voice integration
- ◆ Lotus Notes application development
- ◆ Message-embedded applications
- ◆ Middleware products
- ◆ Multimedia networking
- ◆ Network security
- ◆ Object oriented programming (OOP) in distributed environments
- ◆ Optical switching
- ◆ Personal communications systems
- ◆ Personal digital assistants
- ◆ Software reengineering and restructuring
- ◆ SQL servers and other database servers
- ◆ Superservers
- ◆ Systems integration
- ◆ TCP/IP applications
- ◆ Videoconferencing
- ◆ Voice recognition and voice processing
- ◆ Wide area wireless communications
- ◆ Wireless LANs

TABLE 17-3 Areas of niche expertise for network professionals

LIFESTYLE

Another option used to build a networking career is to mold that career around a preferred lifestyle. For instance, with computer networking technologies creating an increasing number of opportunities for telecommuting, a career option that many computer professionals are selecting involves finding the place they want to live and setting up careers that enable them to live there. This strategy works best by developing a niche skill—one that is both portable and in great enough demand—that permits the individual to control where and when he or she works. Application development (such as being expert in developing applications using one or more client/server application development tools) or software restructuring (being able to rewrite/reformat existing mainframe programs so that they will run on a LAN) are two areas particularly attractive to people for whom lifestyle is a major career concern.

◆ KEEPING UP TO DATE

Virtually everyone working in a fast-paced high-tech area has to constantly retrain himself or herself. New technologies and new products are constantly coming into the market. Business systems are constantly being altered entirely or fitted with new enhancements. All of this makes training, retraining, and continuing education a requirement rather than a luxury.

ADDITIONAL TRAINING

Network professionals can retool themselves in many ways, including taking classes, going to seminars or exhibitions, reading, and joining a professional association, interest group, or club. Managers and users, especially those in organizations distributing their computing systems and giving more responsibility for managing distributed systems to user departments, may also find these approaches effective for staying on top of changes.

CLASSES

A wide variety of options exists for people who want to take classes in networking related areas. Many companies, for instance, hold their own classes on networking technologies for their employees (both users and IT professionals). Classes can also be taken at a university, college, community college, or technical school either for a degree or for continuing education.

In-depth training courses are often offered by the vendors of networking products. For example, Powersoft offers courses on its products, especially its Powerbuilder product; these typically vary in length from one to several weeks. Similar training is available from vendors in areas such as object-oriented development in client/server networks, multimedia applications in networks, rightsizing, remote computing, and a variety of other areas. Such vendor-sponsored training can be an excellent way for network professionals to develop and maintain niche expertise.

SEMINARS

In these fast-paced times, going to seminars provides an excellent opportunity for networking professionals to keep abreast of the latest technologies, especially in niche areas. Whereas courses taken at a university or a college are likely to be broad, and sometimes limited by the equipment available, a one-, two-, or three-day seminar can provide a quick, in-depth look at a specific networking technology, perhaps one so state-of-the-art that it has not yet been fully integrated into academic curricula.

Often, one can find out which seminars are being given in an area by checking the trade publications, checking with local hotels, or getting on the mailing list of a company (such as James Martin and Associates) that gives seminars around the country. Increasingly, seminars are being publicized in newspapers and trade publications such as *Communications*

Week, Computerworld, Data Communications, LAN Magazine, Network Week, and *Network World.*

CERTIFICATION PROGRAMS FOR NETWORKING PROFESSIONALS

As noted in Chapter 1, networking professionals can become certified in some area. For example, Novell has developed *certification programs* for network professionals; these are briefly summarized in Table 17-4.

NetWare Certification

Novell's certifications are granted upon successfully passing certification exams. Novell has developed a comprehensive program for preparing network professionals in a wide area of expertise. The Certified NetWare Administrator (CNA) is designed to prepare the entry-level administrator for managing NetWare LANs. The examination for this certification is objective rather than experientially-oriented and is administered at testing centers scattered around the world.

The next level of certification at Novell is the Certified NetWare Engineer. There has been some controversy over the use of the word *engineer;* however, the title persists. This is a comprehensive program involving at least seven courses and examinations. A CNE is expected to be able to

Certification	Brief Description
Certified NetWare Administrator (CNA)	An entry-level certification program for administrators of NetWare LANs. Currently, the training materials focus on NetWare 2.xx and 3.xx LANs, but coverage of NetWare 2.xx will be discontinued.
Certified NetWare Engineer (CNE)	At least seven specified exams must be passed to achieve this certification. There is more emphasis on the technical nature of NetWare LANs, on troubleshooting problems, and on the network management system.
Enterprise Certified NetWare Engineer (ECNE)	An advanced course of study for professional NetWare troubleshooters and network managers. ECNEs are expected to be able to correct problems and to manage all types of NetWare LANs, especially interconnected ones running NetWare 4.xx.
Certified NetWare Instructor (CNI)	High grades must be received on the other certification exams to be eligible for this program. To achieve this certification, an exam must be passed and an acceptable peer rating for teaching another certification program course is required. This certification enables the individual to serve as the instructor in the courses that comprise the other certification programs.

TABLE 17-4 Levels of certification available through Novell

solve most irregularities on a Novell network, although most CNEs rely on various troubleshooting materials available through a bulletin board or the Novell Support Encyclopedia for unusual circumstances. CNEs are specified in many government procurement contracts as being required personnel for large networks.

Beyond the CNE course is the Enterprise Certified NetWare Engineer (ECNE) advanced course. The ECNE certification is usually pursued by highly trained professional network troubleshooters and managers. ECNEs are expected to be experts in the major bodies of knowledge related to Novell networking. ECNEs are usually found working for firms with large NetWare LANs and interconnected NetWare LANs. They may also be employed by companies that regularly install large Novell networks.

Another branch of the NetWare certification is the Certified NetWare Instructor (CNI), which requires the highest of all grades on the examinations. CNIs are certified course-by-course by testing as well as by an experiential review while actually instructing a group of students in the course for which certification is requested. All authorized training centers affiliated with Novell are required to use CNIs as the instructors for networking classes. There is a significant demand for graduates of all of these certification programs worldwide.

Novell has set up a worldwide group of Novell Authorized Education Centers (NAECs) that must maintain certain levels of expertise and classroom facilities to assure a predictable quality of instruction. Additionally, groups of accredited schools, colleges, and universities are members of the Novell Education Academic Partners (NEAPs). These teach the Novell curriculum as part of the standard curriculum or intermingled with other academic material.

Other independent training groups (firms) teach the materials that must be mastered by students to pass the Novell certification exams. These groups have no official ties to Novell itself.

Fees are charged for taking the certification exams and for exam preparation courses. Tutorial materials can also be purchased to help individuals prepare for the exams. The costs associated with achieving certification can be significant. For example, many companies have invested more than $10,000 per employee for Novell certification.

Other Network-Related Certifications

While Novell's certification programs for network professionals are among the best known, certifications can also be achieved in other areas. For example, Banyan sponsors comparable programs leading to the Certified Banyan Specialist and Certified Banyan Engineer titles. Also, Lotus Development Corporation sponsors certification in Lotus Notes application development, and Microsoft offers two fairly well-known certifications (the

Microsoft Certified Product Specialist and the Microsoft Certified Systems Engineer designations).

Certification in niche areas can be very useful for network professionals and can help ensure employability or job security. If the number of certified professionals is limited and there is substance to the certification, individuals certified are likely to command good salaries and above-average levels of job security. People completing certification programs have often found that they are hot commodities and receive numerous job offers. Many companies that have invested the time and money in helping their employees become certified often have to significantly increase the salaries of certified employees if they don't want to lose them to competitors.

EXHIBITIONS

Exhibitions are events at which one vendor or several vendors exhibit their latest information technology offerings. Some shows, such as COMDEX (Computer Dealer Expo), involve a wide variety of vendors, and are given at different locations throughout the country, a few times a year. Other types of shows involve specialty areas. For instance, a show covering presentation graphics would attract exhibitors featuring such products as commercial art systems, desktop publishing packages, laser printers, plotters, and film recorders. Exhibitions such as Object World focus on object-oriented products and applications such as OOP projects, object-oriented development tools, object databases and database management systems, and object request broker (ORB) technologies for object applications in client/server networks. Sending managers and network professionals to such trade shows benefits organizations that want to stay on top of technological developments.

READING

Reading is perhaps one of the easiest ways to learn about new technologies. It is also one of the least expensive. Many organizations have extensive in-house IT libraries, but even if they don't, it is a wise practice to encourage network professionals to keep up with the literature in this highly dynamic area. Many options are available to those who spend a lot of their time reading, including those listed in the following paragraphs.

Books

The latest books, especially those on microcomputers and local area networks, can usually be found in bookstore chains such as B. Dalton, Waldenbooks, and the like. Large cities often have bookstores with an even wider selection. For instance, New Yorkers can go to places like Barnes & Noble and the McGraw-Hill bookstore, both of which have relatively large high-technology offerings. For learning general computer networking or communications concepts, the best source is usually a local college or university library.

Periodicals and Journals

Books can take a long time to get into print, and some of the information in them is likely to be dated by the time they become available. Thus, the freshest sources of information about technology are the daily, weekly, monthly, and quarterly periodicals and journals. Useful publications for network professionals and other IT professionals are listed in Table 17-5. Some of these publications are targeted to wide audiences (for example, to all business professionals), whereas others are targeted to narrow audiences (for example, to Macintosh microcomputer users or UNIX users).

Networking professionals are likely to qualify for free subscriptions to some of these periodicals. Others, however, can only be obtained through paid subscription.

Topical Reports

In a rapidly changing area such as computer networking, professionals often need up-to-date information on vendor products. Such reports are a mainstay of publications such as *Faulkner Retrieval Services* and McGraw-Hill's *DataPro Reports*. The reports may cover a variety of networking hardware or software products, a group of related products, a technology tutorial, networking management practices, and so forth. Compared to other sources of information on networking and network products, these publications are relatively expensive.

Electronic Media

As we are living in an electronic age, it should not be surprising that a great deal of technology information is obtainable from videotapes, video disks, diskettes, and so on. For instance, one of the most familiar ways to learn about any software package is to work through the tutorial (demonstration) diskette that the software vendor typically provides. Network product vendors also commonly have demonstration disks and videos for prospective customers, and multimedia demos are becoming more common.

Increasingly, prospective clients are able to modem into vendor computers in order to download demonstration files or information files on vendor products or recent upgrades. In addition, potential clients may be able to request such information via vendor e-mail on information services such as CompuServe or the Internet. Bulletin board systems and Internet nodes for user groups of particular products are also becoming popular ways to exchange tidbits of advice and information on new developments or upgrades.

Area	Title	Frequency
General Management	*Business Horizons*	Quarterly
	California Business Review	Quarterly
	Harvard Business Review	Bimonthly
	Sloan Management Review	Quarterly
	The Executive	Monthly
General Business	*Business Week*	Weekly
	Forbes	Biweekly
	Fortune	Biweekly
	Inc.	Monthly
	Industry Week	Weekly
General Computing	*Byte*	Monthly
	Datamation	Monthly
	Computerworld	Weekly
	Communications of the ACM	Monthly
	Infoworld	Weekly
	Software Magazine	Monthly
Microcomputing	*PC Magazine*	Monthly
	PC World	Monthly
	PC Week	Weekly
	PC Tech Journal	Monthly
Macintosh Computing	*MacUser*	Monthly
	MacWeek	Weekly
	MacWorld	Monthly
Managing Information Technology	*CIO*	Monthly
	Corporate Computing	Monthly
	Infosystems	Monthly
	Information Systems Research	Quarterly
	InformationWeek	Weekly
	Journal of Management Information Systems	Quarterly
	Journal of Systems Management	Monthly
	MIS Quarterly	Bimonthly
Remote Computing	*Mobile Office*	Monthly
	Portable Computing	Monthly
Specialty Computing	*Digital News & Review*	Bimonthly
	Systems Integration Age	Monthly
	UNIX Review	Monthly
Telecommunications	*Communications Week*	Weekly
	Client/Server Computing	Monthly
	Data Communications	Monthly
	Network Week	Weekly
	Network World	Weekly
	LAN Magazine	Monthly
	LAN TIMES	Weekly
	Telecommunications	Monthly

TABLE 17-5
Some major periodicals that include articles on computer networks

Vendor Literature

One of the most straightforward ways to learn about a product is to look at the vendor's literature. This literature is, of course, almost always designed to put the vendor's product in the most favorable light, but it is frequently both free and useful.

PROFESSIONAL ASSOCIATIONS

People with common goals often join *professional associations*, clubs, or user groups in order to share knowledge and experiences. Some associations that deal with general IT-related issues are the Association for Computing Machinery (ACM), the Association for Systems Management (ASM), the Data Processing Management Association (DPMA), the Society for Information Management (SIM), and the EDP Auditor's Association. Most of these have divisions or interest groups for networking professionals. The Institute of Electrical and Electronics Engineers (IEEE), which has developed (and continues to develop) the 802.X standards for LANs, is another group with a strong interest in data communications. Most of these associations hold annual conferences or shows at which professional papers and demonstrations are presented. At most conferences, networking and other information technology product vendors set up booths to feature some of their latest wares.

Vendor consortiums and networking product user groups provide network personnel with additional ways to stay current. While the meetings and conferences of these groups are typically much more focused than those of professional associations, a great deal of in-depth information may be available.

Organizations should encourage their networking professionals to participate in these organizations and conferences. Such participation promotes the social networking of professionals and can help them and their organizations stay abreast of important developments in computer networking.

◆ SKILL AREAS FOR NETWORKING PROFESSIONALS

From the foregoing discussion, you might conclude that one of the most important attributes of the network professional is well-honed technical skills. Indeed, most niche expertise involves strong technical skills and knowledge, at least about a particular tool or product. Much of the information imparted through additional training is technical in nature as is much of the networking literature and the information shared at professional meetings. However, it would be wrong for us to leave you with the impression that strong *technical skills* are all that is needed to ensure success in computer networking. While strong technical skills can help people get jobs, other factors, such as their ability to communicate and get along with others, are often primary determinants of their long-term success.

INTERPERSONAL AND COMMUNICATION SKILLS

Employees who desire to progress along traditional career paths often find that *interpersonal and communication skills* are important factors in whether they are able keep their jobs or get promoted. Similarly, individuals in business for themselves are likely to find that interpersonal and communication skills are key to attracting and retaining clients.

By its nature, network management involves linking users with one another and with computing resources. To accomplish these tasks, network professionals need to be able to understand and communicate with users.

Interpersonal and communication skills are used by network managers daily. Network managers interact frequently with users. They also utilize interpersonal skills when talking with vendors and trying to persuade superiors to purchase new networking components.

In general, good communication between network personnel and users makes for easier and more effective network management. When users view network personnel as being approachable and responsive, they are more likely to feel comfortable communicating with them. For example, network problems are likely to be detected and corrected quickly when users feel that communication channels are open between them and the network staff. In addition, network planning and network design and development projects are more likely to be successful when user input and communication levels are high.

For individuals with a great deal of technical knowledge, users can come across as being some of the most ignorant and frustrating creatures on the planet. Many do not read the manuals and documentation, follow simple directions, remember their passwords, make backup copies, or use logon scripts or print routines—even after you have shown them how for the twenty-second time. Needless to say, these all-too-common user characteristics can try the patience of network professionals. While the temptation may be to let loose with an angry barrage of epithets and curses, such reactions are likely to damage the relationships between network personnel and users; users will not see the staff as being approachable and responsive.

Besides having the ability to communicate and get along with users, network managers often must be able to communicate and establish good working relationships with vendors. This communication is likely to be both in oral and written form; in addition to being able to talk to vendors, network managers should be able to clearly communicate in written documents such as requests for quotations, requests for proposals, and contracts.

Being able to communicate effectively with superiors and other managers can also be an important attribute for network managers. In addition to good oral and writing skills, communication with peers and superiors may demand being able to effectively present ideas through formal, stand-up presentations or impressively formatted reports. Such presentation skills may be essential for ensuring that networking needs are clearly communicated throughout the organization and responded to.

A variety of ways can be used to improve or enhance interpersonal and communication skills. These include additional training in areas such as interviewing, effective listening, technical writing, giving feedback, and doing stand-up presentations.

CONCEPTUAL SKILLS

As network managers advance in organizations, there is typically a greater need for them to possess conceptual skills. *Conceptual skills* include the ability the see the "big picture" and how all the different parts of the organization fit together to work toward achieving long-term goals. Conceptual skills also include the ability to understand how the organization fits into the business environment. As managers near the top of the management hierarchy, they are more likely to be involved in strategic planning and long-term decision making. Both of these activities demand the ability to predict how specific plans and decisions are likely to affect the different parts of the organization.

SKILLS NEEDED AT DIFFERENT MANAGEMENT LEVELS

Technical skills tend to be more important than conceptual skills at lower levels in the management hierarchy, while conceptual skills are generally more important than technical skills for upper-level managers. Interpersonal and communication skills, however, are very important at all management levels.

The skill mix varies because of differences in the roles and responsibilities of lower- and upper-level managers. As noted in Chapter 1, lower-level managers are much more likely to be focused on *operational network management*. For example, lower-level network managers are likely to be most concerned with managing the day-to-day operations of networks. For LAN managers, operational management typically involves troubleshooting (finding and correcting problems experienced by network users), monitoring LAN performance, adding and deleting users, adding new workstations or devices to the LAN, installing new software, backing up servers, maintaining the security of the LAN, and so on.

In contrast to lower-level managers, upper-level managers are much more concerned with strategic management. *Strategic network management* considers the role of networks and networking over the long term. It ensures that the organization has network infrastructure in place needed to help it achieve its long-term goals and objectives. This involves the creation of network plans that specify the networks (including interorganizational networks) to be implemented or interconnected. These plans also specify the new technologies to be included in the networks, how data will be distributed among the networks, and how network security will be enhanced.

The time frames for most of the strategic network plans being developed during the 1990s typically range from three to five years. However, some extend to ten years or more. Usually, the highest-ranking network managers in the organization are the network personnel most directly involved in strategic network management. And, from this description of strategic network management, it should be apparent why conceptual skills are needed by top-level network managers.

Mid-level network managers are typically responsible for *tactical network management*, which involves translating strategic network plans into more detailed "doable" actions. Essentially, tactical network management makes strategic network plans become realities. For example, it involves the development of implementation plans for the new networks specified in the strategic network plans. It also involves coming up with a specific set of plans and timetables for interconnecting networks, for adding new technologies, for actually distributing data, and for making the networks more secure. The time frames for tactical network plans are typically one to three years.

LAN and WAN managers at all levels may be involved in tactical network management, especially those responsible for the operations of two or more interconnected networks. A good balance of technical, interpersonal and communication, and conceptual skills is often needed for effectively carrying out tactical network management activities.

LEADERSHIP AND MANAGEMENT SKILLS

As managers advance in their organizations, their need for *leadership* and *management skills* generally increases. This is also true for network managers. In terms of leadership, they must be able to create (and sell) a vision for networks, motivate network personnel toward long-term goals, and exert the influence needed to turn the vision into reality. Good leaders can inspire others to do their best to achieve long-term goals. Network managers who are effective leaders are likely to progress faster and further than are network managers who are not.

Being an effective manager is not necessarily the same as being an effective leader. Management consists of planning (developing plans and budgets; establishing goals, objectives, standards, policies, procedures, rules), organizing (setting up the organizational structures—divisions, departments, subunits, work teams—needed to carry out plans and support operations), directing (overseeing day-to-day operations and the daily management of organizational personnel), and controlling (monitoring performance to ensure that plans are being carried out; taking corrective actions when there are deviations from plans and performance expectations). A person who is quite good at these tasks may be considered to be an effective manager without being an effective leader. This is because leadership also demands the abilities to inspire, motivate, and influence at levels beyond those needed to be an effective manager.

In sum, network managers who wish to advance in their organizations should strive to develop or enhance their management and leadership skills. While on-the-job experience may be the best way to hone such skills, network managers should also consider taking advantage of in-house management and leadership training programs or those offered by network consulting firms, colleges and universities, and technical schools and community colleges.

POLITICAL SKILLS

In addition to leadership and management skills, it is often important for managers to understand organizational politics. Like leaders, managers with *political skills* often have the power and clout to ensure that things get done. Politically aware managers understand what peers and superiors are really interested in, and this knowledge can be used to influence them and gain their support. For example, a network manager who knows that a marketing manager wants to significantly increase sales revenues might solicit the marketing manager's backing for a wireless network that will allow the sales force to remotely submit orders. This could be a win-win situation for both the network and the marketing managers; the network manager gets the network and the marketing manager gets sales force automation technology that can help increase sales revenues.

An individual's understanding of the organizational politics typically increase over time. Network managers find that investments in information technology and networks are often affected by organizational politics. They also find that changes in networks are likely to be resisted for political reasons. For example, as noted in Chapter 11, downsizing to LANs and the creation of distributed processing systems may be resisted by both user managers and IS personnel. Such changes often mean that more responsibility for network management is shifted to user departments. User managers may resist change because they don't want additional responsibilities in their subunits or they don't want to see another subunit get any more power than it already has. IS personnel may resist these same types of changes because they realize that their power and control over organizational computing resources may be diminished.

Network managers should also realize that networking and the development of distributed client/server systems have generally redistributed power within IS areas. As LANs and internetworking have become more important, the importance of centralized, host-based systems has decreased. This has often resulted in increased power and political clout for networking personnel at the expense of IS personnel associated with the operation and management of centralized systems. Redistribution of power and clout has caused bitter infighting in some organizations.

BUSINESS KNOWLEDGE

A final area that should not be overlooked by individuals interested in networking careers is **business knowledge**. To make career progress, network personnel must understand the organization's business and how its various subunits contribute to this business. To develop networks for a user area, it is essential to understand the work that users in that area perform and how it is related to the work performed in other areas of the organization. In addition, by knowing the nature of the work and interactions of the various subunits, network managers are in a better position to know what interconnections are needed to build an effective enterprise-wide network. Such knowledge should also enable them to design appropriate interorganizational systems.

Business knowledge can be gained and enhanced through practical experience. However, business courses are taught by many universities, colleges, community colleges, and technical schools as part of degree or continuing education programs. Business courses can provide students with a fundamental understanding of the different functional areas of a firms (such as accounting and finance, marketing, operations, human resources, and so on). Some programs can help students develop specific work skills in areas such as bookkeeping, budgeting, purchasing, scheduling, negotiating, interviewing, training, evaluating performance, giving feedback, and so on. There may even be in-house programs available to network personnel to help them develop the business skills and knowledge needed to effectively manage networks.

◆ SUMMARY

Career opportunities for individuals with an interest in networking have never been better. There are a variety of traditional and nontraditional networking jobs available.

Traditional jobs include positions such as network planners, help desk personnel, LAN managers, and technical support specialists. Nontraditional jobs include consulting, providing complete business systems or turnkey solutions, and developing expertise in one or more specific networking areas. The increased prevalence of LANs and internetworking has boosted the demand for both traditional and nontraditional network specialists.

Three popular career options are career paths, niche expertise, and lifestyle. Individuals on traditional network career paths may advance from being a LAN support technician, to LAN manager, to network manager. Developing niche expertise in areas such as groupware, network operating systems, interconnection technologies, middleware, and downsizing, for example, can also help ensure long-term employability. Some individuals build networking careers around preferred lifestyles; they may telecommute or develop high-demand niche expertise that affords them some flexibility in when and where they work.

To keep from becoming obsolete, it is important for career-minded networking personnel to keep their networking knowledge and skills up to date. Some of the ways to do so are getting additional training (such as attending seminars, taking classes, and completing certification programs), attending exhibitions and trade shows, reading (books, periodicals, topical reports, vendor literature, as well as the information available from electronic bulletin boards and Internet nodes), and participating in professional associations.

Network managers need to possess an appropriate combination of technical, interpersonal/communication, and conceptual skills. Technical skills are most important for entry-level networking personnel and lower-level network managers. Conceptual skills (the ability to see the big picture and how all the different parts of the organization fit together) are most needed by top-level network managers. Interpersonal and communication skills are needed by all types of network managers.

To progress along many of the traditional career paths in organizations, network managers often need leadership and management skills, political skills, and business knowledge. While such skills and knowledge can be learned through on-the-job experience, formal training and education programs can often provide the general background and skills that network managers need.

✳ KEY TERMS

Business knowledge
Career path
Certification programs
Conceptual skills
Interpersonal and communication skills
Leadership skills
Management skills

Niche expertise
Operational network management
Political skills
Professional association
Strategic network management
Tactical network management
Technical skills

✳ REVIEW QUESTIONS

1. What are the differences between traditional and nontraditional networking jobs? Identify several examples of each.
2. What are the differences among career paths, niche expertise, and lifestyle-oriented careers? Provide examples of each.
3. Identify and briefly describe the different types of additional training opportunities available for networking professionals.
4. Identify and briefly describe some of the certification programs available to networking professionals.
5. What are the differences among CNAs, CNEs, ECNEs, and CNIs?
6. What are computer exhibitions? How do these help networking professionals stay current?
7. Identify and briefly describe the different types of literature available to help network professionals keep their knowledge up to date.
8. Identify the sources of online information available to network professionals.
9. How can participating in professional associations help networking professionals keep up to date?
10. What are the differences among technical, interpersonal and communication, and conceptual skills? How can each be developed or enhanced?
11. Why are technical skills more important for entry-level jobs than for upper-level network management jobs?
12. What types of skills should upper-level network managers possess?
13. How can good communication make a network manager's job easier?
14. What are the differences among strategic network management, tactical network management, and operational network management?
15. What are the differences between leaders and managers?
16. What are the differences among the following managerial activities: planning, organizing, directing, and controlling?
17. Why should network managers possess political skills?
18. How has the emergence of computer networks changed the distribution of power among IS professionals in organizations?
19. Why is business knowledge important for network managers?

✳ DISCUSSION QUESTIONS

1. Network managers' tasks have been described throughout this book. In light of these managers' duties and responsibilities, what types of skills should network managers possess? Why are these skills needed? What are the differences between the skills needed by enterprise-wide

network managers and those needed by LAN managers in charge of smaller LANs?

2. Why do organizations that promote the development of niche expertise among their network staff run greater risks of losing their employees than organizations that do not?

3. Why do you think organizations are interested in certification programs for their networking professionals? What does certification signify? What dangers are associated with overemphasis on certification?

4. How have computer networks changed the distribution of power and political clout among subunits within organizations? Why is such redistribution of power resisted by subunit managers? What can they do to prevent or slow down the redistribution of power caused by the development of computer networks?

✳ PROBLEMS AND EXERCISES

1. Invite a network manager from a local company to discuss the networking positions in his or her organization. Ask how the responsibilities associated with these positions are changing and the types of positions he or she sees as being most important in the years ahead.

2. Invite an individual with a nontraditional networking job to come to class to describe what he or she does for a living, focussing on why and how he or she decided to pursue this career, how to get clients, and about the types of nontraditional networking positions he or she feels will be lucrative in the future.

3. Find a recent article on networking careers and networking job opportunities and summarize its content in a presentation or written report.

4. Invite a local network manager to speak to your class on the skills and knowledge he or she feels network managers should possess. Be sure to ask how he or she expects the skills needed by network managers to change in the years ahead.

5. Obtain information from Novell (or another source) related to its certification programs. Summarize the different types of knowledge associated with each of the different certifications in a paper or presentation.

6. Obtain information about the certification programs offered by Lotus, Microsoft, and Banyan. Summarize the different types of knowledge associated with each of the different certifications in a paper or presentation.

7. In light of the information given in this chapter, assess your own academic preparation for a networking career. In which areas are you being sufficiently educated or trained? Which areas are not being addressed in the curriculum for your degree? How will you fill in any gaps in the preparation for your career in the years ahead?

✳ REFERENCES

Baum, D. "Developing Serious Apps with Notes." *Datamation* (April 15, 1994): p. 28.

Bredin, A. "LAN Administrators Picking Up New Roles." *Computerworld* (May 31, 1993): p. 85.

Carlini, J. "Net Managers Need Instruction in Many Disciplines." *Network World* (January 27, 1992): p. 37.

Dostert, M. "High Demand for LAN Personnel." *Computerworld* (October 26, 1992): p. 47.

———. "Wanted: LAN Personnel." *LAN TIMES* (March 18, 1991): p. 53.

Fahey, M. "From Local to Global: Surveying the Fiber Landscape." *Telecommunications* (November 1993): p. 33.

Flanagan, P. "Future Industry Directions: The 10 Hottest Technologies in Telecom." *Telecommunications* (May 1994): pp. 31, 70.

Francis, B. "MIS Meets Multimedia on the Network." *Datamation* (July 15, 1993): p. 32.

Frazier, D. and K. Herbst. "Get Ready to Profit from the InfoBahn." *Datamation* (May 15, 1994): p. 50.

Goff, L. "Network Operating Systems: A Must for Most." *Computerworld* (July 19, 1993): p. 93.

Goldman, J. E. *Applied Data Communications: A Business-Oriented Approach*. New York: John Wiley & Sons, 1995.

Hart, J. "Top Jobs." *Computerworld* (August 29, 1994): p. 113.

Keen, P. G. W. and J. M. Cummins. *Networks in Action: Business Choices and Telecommunications Decisions*. Belmont, CA: Wadsworth, 1994.

Leinfuss, E. "Network Managers: Beyond Fire Fighting." *Computerworld* (March 30, 1992): p. 101.

Louznon, M. "In Pursuit of Troubleshooters." *Computerworld* (June 21, 1993): p. 107.

Marks, K. "Summa Cum CNE." *LAN Magazine* (April 1994): p. 99.

McCusker, T. and P. Strauss. "Managing the Document Management Explosion." *Datamation* (July 1, 1993): p. 41.

Moad, J. "How to Certify the Training Works." *Datamation* (September 1, 1993): p. 67.

Parker, C. and T. Case. *Management Information Systems: Strategy and Action*. 2nd ed. New York: Mitchell/McGraw-Hill, 1993.

Pastore, R. "Paths of Least Resistance." *CIO* (April 1, 1993): p. 37.

Snell, N. "Where's UNIX Headed?" *Datamation* (April 1, 1994): p. 25 .

St. Clair, M. "Special Report: Network Managers." *LAN TIMES* (June 8, 1992): p. 51.

Stephenson, P. "Beyond the Server: Good Network Administrators Require Good Training." *LAN TIMES* (August 5, 1991): p. 37.

Case Study:
Southeastern State University

Over the past few years, students at SSU have learned that working with LANs can give them an edge in the job market. They have witnessed peers a year or two ahead of them step into networking-related jobs immediately upon graduation, often simply because they have had *some* experience working in a LAN environment. They are also aware that several recent SSU graduates who worked as lab assistants while they were in school have risen quickly in their employing organizations, are now in charge of very sizable networks, and are being paid well.

The experience of these SSU alums has led currently enrolled students to question whether certification is needed for an entry-level networking job. After all, former students have gotten good jobs without being certified, and if they are certified now (which, in fact, many of them are) they became so only after graduating and entering the workforce.

Doubts about the need for certification are also raised when students look at the qualifications and backgrounds possessed by the LAN managers that they work for (or with) at SSU. The vast majority of SSU's LAN managers have had no formal training in computers and data communications, and only two of the few with degrees in computing-related fields have achieved any level of certification (both are CNAs). These individuals seem to the students to be functioning effectively, and from working with them, the students feel that SSU's LAN managers are doing a good job of managing their networks. By working with the LAN administrators, the students feel that they are learning a lot about LANs and LAN management, at least enough to handle themselves adequately in a job interview for a networking position.

In spite of the questions that the experiences of former students and the backgrounds of SSU's LAN managers have raised in the minds of current students about the necessity of certification, student computer lab workers at SSU are becoming more interested in certification. Ideally, they would like to take (and pass) certification exams prior to graduation so that they can have an even greater edge in the job market. They would also like to have the materials covered in certification training programs incorporated into the computer-related degree programs offered by SSU. In this best of all possible worlds, SSU students would be exposed to everything they would have to know to pass the certification exams in their regular courses. They would have to pay only a little more to take the exams and could avoid all the fees associated with the vendor-sanctioned or independent certification exam preparation programs that practicing network professionals (or their employing organizations) normally have to pay.

Student interest in networking certification is particularly high among students in SSU's MIS (management information systems) program. This program is found in SSU's College of Business Administration—graduates receive a BBA (Bachelor of Business Administration) degree in MIS. Majors take courses in all the functional areas of business, along with courses in the MIS major including structured programming, systems analysis and design, and database applications. Most MIS majors also take at least one data communication course that exposes them to information about network certification programs and about the relative scarcity of qualified networking professionals. This information often whets their appetite to acquire networking experience by working in one or more of the computer labs with LANs on campus. In fact, interest has been great enough for SSU's LAN managers to be very selective in the students they hire. Most computer labs with LANs are staffed with motivated students having a sincere interest in getting all the hands-on experience and knowledge they can about networking and network management while they are in school.

Student interest in networking and network management has resulted in modifications in the course offerings available to MIS majors. Until two years ago, only a general data communication course was offered. Now, a new course focuses on LANs and LAN management, providing students with hands-on experience in LAN installation (including installing network interface cards, network drivers, NOS software, network utilities, and network applications). It also gives them hands-on experience in setting up a print server, in generating alerts and alarms in network management systems, and in troubleshooting LAN problems. Different groups of students are responsible for bringing up different types of LANs including NetWare 3.xx, NetWare 4.xx, Windows for Workgroups, and Windows NT Advanced Server.

Because the MIS program at SSU has become involved in the Novell Education Academic Partners (NEAP) program, students now use Novell-sanctioned training materials in this new LAN course. Involvement in this program also has meant additional training for the COBA's LAN manager (and main instructor of the course) toward the CNI certification. Soon, at least one other MIS faculty member will be following the LAN manager's path toward this certification, because at least two of SSU's faculty members have to be CNIs before SSU can be designated as a site at which Novell-sponsored training programs can occur. SSU hopes to offer CNA and CNE training through its continuing education programs to help meet the demand for certified network professionals in this region of the state. Eventually, SSU also hopes to offer ECNE training.

CASE STUDY QUESTIONS AND PROBLEMS

1. Do you feel that students' doubts about the value of networking certification are justified? Why? Should they expect entry-level networking jobs to be available to those with *some* experience in networking environments? Why or why not?

2. Should students be allowed to avoid the fees that practicing managers (or their employers) must pay to receive certification training? Why or why not?

3. Based on the information provided in this chapter, is SSU's MIS curriculum adequate to prepare students for careers as network managers? Why? If you could make one change to the curriculum, what would it be?

4. Critique SSU's LAN course for MIS majors. What are the course's strengths? What are the course's weaknesses? If you could make changes to the course, what would you do?

5. Should SSU continue to pursue its involvement in the NEAP program? Explain your reasons.

6. Based on the information provided in this chapter, what specific steps do you feel should be taken by SSU to ensure that its LAN managers are well-trained and up to date? Justify each of your recommendations.

Glossary

A/B Switch A hand- or software-controlled mechanical switch that enables two or more microcomputers to share a peripheral device.

Access privileges The rights of users to access data and computing resources. These are a type of security control that essentially determines which users are authorized to access and use specific data and network resources.

Acoustical coupler An early modem with which the handset of the telephone is placed in a rubber cradle and the audible tones are translated into electrical pulses by the modem.

Active hub One of the two main types of hubs found in Arcnet LANs, used to establish workstation-to-workstation communications; these have signal regeneration capabilities enabling workstations to be located up to 2,000 feet apart.

Adaptive routing (or dynamic routing) A routing algorithm in which existing routes are evaluated and the best or fastest alternative route to the next destination is selected for a message.

Analog transmission Analog transmission occurs by changing amplitude, frequency, or phase of the signal to represent data values; it is characterized by a continuously variable signal (a sine wave).

Application program interface (API) The interface between application and network software. The job of an API is to appropriately format user requests for network services so that they can be transmitted over the network and routed to the appropriate destination.

Application software Application software includes both general- and special-purpose user-oriented programs that enable users to perform their work.

Archiving Copying or backing up data or software and storing the copies in a secure location; archiving can be an essential component in network security and recovery.

ASCII Acronym for American Standard Code for Information Interchange; this is a 7- or 8-bit code for representing characters that has been adopted by the American National Standards Institute to promote uniformity in the transmission of data.

Asynchronous One of the oldest and most common data link protocols. Each data character is transmitted individually. Typically, a *start bit* is transmitted before the bits representing the character and a parity bit (or other error detection scheme). A *stop bit* follows the data bits. The start and stop bits serve as separators for the character and enable the receiving machine to know which bits are associated with the character.

Asynchronous transfer mode A high-speed data transmission protocol in which data blocks are divided into equal size cells (frames) and are routed from sender to receiver in a manner similar to packet distribution (packet switching) networks.

Attenuation Attenuation refers to the decrease in the strength (weakening) of an electrical signal during data transmission.

Backing up The process of making a duplicate of all or part of the data or program for the purposes of redundancy and network recovery (also see Archiving).

Balancing One of the approaches used by network managers to improve network performance; this often involves using existing network resources but setting them up in a different configuration.

Bandwidth The amount of data that can be transmitted per unit of time over a medium; it is the difference between the highest and lowest frequencies that may be transmitted over the medium. Generally, higher bandwidth means higher data transmission rates.

Benchmark test A test in which one or more application programs (or data sets) supplied by the potential buyer are processed on the LAN hardware and software included in the vendor proposal.

Bindery files These store information about user passwords and access privileges in Novell 3.xx networks.

Bit A binary digit, either 0 or 1.

Bit error rate tester (BERT) BERTs determine if data is being transmitted reliably over a communications link by transmitting a known bit pattern and checking to see if it is correctly received.

Block (or frames, cells) A series of contiguous bits or bytes (data characters) that are transmitted as a group.

BNC (Bayonet-Neill-Corcèlman) connector A type of network interface technology; a BNC connector is connected to the network interface card with a T-connector, permitting the workstation connection to be severed without interfering with the network circuit.

Breakout box A diagnostic tool that helps determine whether two data communication devices are correctly connected. It is often used for evaluating the functionality of the individual pins in connector plugs (such as RS232 interfaces) and for cable testing.

Bridge An internetworking technology that connects two networks at the data link layer of the OSI model.

Broadcast routing A message routing algorithm whereby the message is sent (broadcast) to all devices; this is common in LANs utilizing CSMA/CD, but messages are only accepted by the device(s) to which they are addressed.

Brouters An internetworking technology that combines features of bridges and routers; for example, a brouter can establish a bridge between two networks as well as route some messages from the bridged networks to other networks.

Bulletin board system (BBS) A BBS is essentially a type of dial-up connection that can be accessed by microcomputer users through communications software and modems. A BBS may be used to post electronic messages and can serve as a clearing house for software and information exchange.

Bus topology LANs with a bus topology use a common medium to which all of the nodes or workstations are directly connected.

Capacity planning This involves forecasting network utilization and workloads and developing plans to ensure that the network will be able to support anticipated performance demands.

Career path The progression of employees from entry-level positions into higher-level and management positions.

Cell relay Data transmission systems that are closely related to X.25 standards and frame relay technologies; typically, data are transmitted in equal-sized blocks (cells) over high-speed fiber optic links (such as T-1, T-2, T-3, and T-4 lines) available from common carriers and packet distribution networks.

Cellular data transmission A type of data transmission service that provides remote access to network resources via cellular telephone and cellular data networks.

Centrex services These services can be leased from local telephone companies and can provide subscribing organizations with most of the functions found in a PBX.

Certification programs Typically, these are a combination of training and testing programs that enable networking professionals to become certified in specific areas. These are often sponsored by network operating system vendors (such as Novell) and application program vendors (such as Lotus).

Circuit The electronic path over which voice, data, image, and video transmission travels. Circuits can consist of physical cabling (such as twisted pair, coaxial cable, fiber optic cable) or wireless media (such as microwave transmission, satellite transmission, radio waves, cellular transmission).

Client/server computing A processing environment in which the hardware, software, and data resources of two or more computers (servers and clients) are combined to solve a problem or to process a single application.

Client/server networks Networks of servers and client workstations that enable client/server computing to occur; these may be distributed networks in which diverse hardware platforms and operating systems are used.

Coaxial cable An insulated filament of wire within an insulated cable. A second braided conductor surrounds the insulation. This type of cable is commonly used for video and for LAN messaging.

Common carrier A business organization that provides regulated telephone, telegraph, telex, and data communication circuits and services; long-distance, regional, and local telephone companies are examples of common carriers.

Common Management Information Protocol (CMIP) CMIP is a relatively complex protocol for exchanging network management information among network components and management entities.

CORBA (Common Object Request Broker Architecture) CORBA is a middleware standard developed by the Object Management Group (OMG) for object-oriented applications in client/server environments.

Communications servers (or access servers, network resource servers, and telecommuting servers) One of the mechanisms used to allow remote users to connect to a LAN; communications servers may also provide LAN users with connections to other networks.

Communications system A group of electronic devices used with software to provide an electronic pathway on which information is transmitted.

Compiled code Program instructions in the form needed for execution in the computer; most of the software sold commercially is available only in compiled code (machine readable) form.

Computer conferencing Workgroup computing technologies that enable users to compose messages and to have these messages displayed on the screens of other users; this makes it possible for two or more users to have interactive conversations in real time.

Computer network A group of computers that share a communications medium that enables the computers to communicate with one another.

Computer operators Workers in data processing centers who keep the mainframe going, mount tapes and disk units, monitor computer processing, bring the system back up after a crash, and so on.

Concentrator A hardware device usually found in WANs that combines the functions of a controller and a multiplexer; most utilize store-and-forward messaging—storing messages from many remote devices until the quantity of data is sufficient to make it worthwhile to send over a high-speed line.

Conceptual skills Conceptual skills include the ability to see the "big picture" and how all the different parts of the organization fit together, in order to work toward achieving long-term goals.

Concurrent access The ability of multiple network users to simultaneously access the same data files or programs; this is often needed in shared database environments.

Concurrent logons The ability of a single user to logon to the same network from multiple workstations.

Contention A media access control approach found in networks in which workstations must contend for the use of the communications medium. CSMA/CD is one of the most widely used contention-based media access control protocols.

Contract A formal, legally-binding and enforceable business agreement.

Controller (or terminal controller, cluster controller, communications controller) A specialized computer whose purpose is to relieve the host CPU of the burden of communicating with a large number of terminals and peripheral devices.

Corporate license These make it possible for all users at all of an organization's operating locations to have access to certain software (concurrently if necessary); such agreements are structured similarly to site licenses but extend access privileges to all sites.

Crosstalk When the signals in one channel distort or interfere with the signals in another channel; this may occur because guardbands fail to adequately separate the channels.

CSMA/CD Carrier sense multiple access/collision detection; a type of access control protocol used in LANs.

Data communications The sending and receiving of encoded information from one networked device to another.

Data striping An approach used in RAID systems that reads and writes data in parallel; this involves dividing data between disk drives at the block, byte, and bit levels.

Data switch Sub-LAN technology that provides a connection between microcomputers or between microcomputers and peripheral devices.

Database servers Servers that provide network users with shared access to a database.

DCE (Distributed Computing Environment) DCE, from Open Software Foundation, Inc., is a de facto middleware standard for remote procedure calls.

Dedicated line A private communications line, not shared with any other organization, and exclusively available 24 hours a day; these can be leased or purchased from common carriers.

Dedicated server A server that can only perform server functions and cannot be concurrently used as a workstation.

Dial-out capabilities The ability to connect to another network using a public, switched, dial-up telephone line.

Dial-up telephone line Phone lines that are rented from local telephone companies—the type of telephone service that most people have in their homes.

Differential backup Backing up only those files changed since the last backup.

Digital transmission A discrete or discontinuous signal whose various states are discrete intervals apart; the different states represent different data levels, often different bits or bit combinations.

Disaster recovery plan A contingency plan for recreating an organization's entire computing system, both hardware and software, on short notice, in case of destruction of the primary system.

Distributed object management systems (DOMS) These are middleware used to support object-oriented message exchanges in distributed client/server networks.

Distributed processing systems A group of interconnected computing devices that enable data processing at workstations and other nodes located remotely from the central host.

Document coauthoring software This is a type of groupware that enables two or more group members to concurrently work on the same document.

Document review/sharing software A specialized type of workgroup computing software that keeps track of the individuals who must review a document as well as the order in which they must review it; these systems typically include an "electronic routing slip."

Downsizing The movement of applications that have been traditionally run on mainframe computers to smaller computing platforms, especially LANs.

Dumb terminals Terminal devices with very limited processing capabilities; generally used for data input and online retrieval.

Duplexing The process of constructing a parallel configuration so that if the primary component fails, the secondary one takes over automatically; examples include both server duplexing and disk duplexing.

Dynamic routing (or adaptive routing) When intermediate nodes are given the responsibility for identifying and selecting from among alternative message routes.

EBCDIC Extended Binary Coded Decimal Interchange code. A standard 8-bit data encoding and interchange scheme widely used in IBM minicomputer and mainframe environments.

Electronic data interchange (EDI) Allows standard business documents, such as purchase orders and invoices, to be electronically exchanged between companies.

Electronic funds transfer (EFT) A specially formatted data transmission sent to a banking clearinghouse where payment is made to the proper account and a corresponding amount is deducted from the payer's account.

Electronic mail (e-mail) A networking application that enables users to send and receive messages electronically.

Electronic mailbox Space within an electronic mail system is which a particular user's messages are stored.

Electronic meeting systems (EMSs) (or group support systems—GSSs) Typically combine decision support and electronic communications capabilities.

Encryption The modification of a bit stream in a data transmission so that the actual transmission has little resemblance to the original data; analogous to "scrambling" television signals.

Enhanced e-mail services The inclusion of additional features beyond those found in traditional e-mail systems—for example, the ability to develop personal mailing lists, mail forwarding capabilities, delivery confirmation, mass mailing capabilities, and other customization features, such as integration with voice mail.

Enveloping The process of putting a message in one network format inside the format "envelope" needed for transmission over another network.

Error detection The process of checking transmitted data to ensure that it is received in the same form that it was transmitted by the sender.

Event log file A file, often a server file, that enables network managers to record events that exceed preset performance thresholds.

Fault tolerant A network is said to be fault tolerant when it can resist crashing and when it can recover from a crash quickly.

Feasibility study A feasibility study may be done to determine whether or not a proposed network (or network change) can be implemented, operated, and maintained in a cost-effective manner.

Feedback A mechanism included in most systems to provide communications in reverse direction; the feedback might indicate that the message was successfully delivered or that it has been forwarded.

Fiber Distributed Data Interface (FDDI) ANSI standard for token-passing fiber optic LANs and MANs, spanning up to 200 kilometers, with data transmission speeds of 100 Mbps or more.

Fiber optic cable An increasingly popular networking medium that uses light waves to carry data over very thin strands of glass (optical fibers); the most significant benefits of fiber optic cable are its huge bandwidth and its capacity for rapidly transmitting large volumes of data.

Frame (or block) A term used to describe data packets, especially in bit-oriented synchronous data link protocols such as those used in X.25 networks and frame relay systems.

Frame relay Frame relay (also called fast-packet) systems are built around a data link protocol (LAP-D, for Link Access Procedure-D Channel) that defines how frames of data are assembled and routed through a packet distribution network (PDN); these are designed to provide faster data transfer rates (up to 1.544 Mbps) than is possible with traditional X.25 implementations.

Frequency The term used to describe differences in the wavelength of an analog signal.

Front end processor (FEP) A smaller computer that is used to handle communications tasks for a host; this frees the host to handle processing tasks requested by users.

Functional testing Includes testing the operation of individual network components as well as the operations of the entire network (with all components participating).

Gateway Internetworking technology that provides an interface between two networks at or above the network layer of the OSI model; these are capable of converting one network's data frames into the format needed by the other network.

General-purpose application software Software useful to workers throughout the organization, not just to those in particular departments or functional areas; examples include word processing and spreadsheet packages.

Grace logons These allow the user to briefly logon to a network using an old (and technically expired) password, but require the password to be changed before the user is allowed to access any of the network resources.

Group scheduling software A type of workgroup computing software (groupware) that is sometimes called calendaring software; this is used to maintain both individual and group meeting/appointment calendars.

Group support system (GSS) Group-oriented decision support systems used in network environments; many support group voting, ranking, and rating systems, and some support brainstorming, the nominal group technique (NGT), and Delphi decision making approaches.

Groupware Software packages designed to support the collaborative efforts of group members; Lotus Notes is a well-known example.

Half-duplex Communications circuits that allow data transmission to travel in both directions but only in one direction at a time.

Hardware The tangible components of a computing system; examples include servers, workstations, hard disks, printers, and other equipment.

Hertz (Hz) A measure of frequency; one hertz is equal to one cycle per second.

Host A computer that provides processing support and services to terminals or other computers. In WANs, hosts are typically mainframes or minicomputers; in LANs, servers may be hosts.

Host-based network A group of terminals and/or networked computers in which a host plays the central role in communications and processing.

Hot fix Being able to perform maintenance on (or replace) a malfunctioning network component, such as a disk or disk drive, without having to interrupt normal network operations; the later versions of Novell NetWare include this fault-tolerance feature.

Hub A networking device that provides a connection between networked devices. Some examples are multiple access units (MAUs) in token-ring LANs and the active and passive hubs found in Arcnet LANs. Superhubs (collapsed backbones) provide interconnections between different LANs or WANs and often include bridges, routers, brouters, LAN switches, and gateways.

Hybrid networks Networks that combine two or more of the fundamental networking technologies (ring, bus, mesh, hierarchical and star topologies); also, this refers to combinations of public and private data communication networks.

IEEE Project 802 IEEE (Institute of Electrical and Electronics Engineers) Project 802 was started in 1980 to develop standards for LAN implementations and applications. Standards for all of the major types of media access control protocols found in LANs have been developed by the IEEE; these are called "802" standards.

Information float The time it takes to get information from the source into the hands of decision makers; this is often reduced by new telecommunications technology.

Information technology Computing devices with telecommunications capabilities.

Installation contract A legally-binding contract for the installation of a LAN or network; this may include penalty clauses if installation deadlines are missed or if installation activities are omitted.

Integrated services digital system (ISDN) ISDN is a digital network available from common carriers that enables all types of data—fax, video, voice, and data—to be transmitted.

Intelligent terminal A terminal device with limited processing capabilities.

Interpersonal and communication skills The ability to interact and communicate with others; important to the career progress of network managers and personnel.

Intruder detection systems Mechanisms that guard against network and network application intrusions by unauthorized users.

LAN modems These provide remote users with access to a LAN; they are often implemented as a dial-in modem that is connected to the LAN via a network interface card (NIC).

Leadership skills For network managers, these include the ability to create (and sell) a vision for networks within the organization, motivate network personnel toward long-term goals, and exert the influence needed to turn the vision into reality. Network managers who are effective leaders are likely to progress faster and further than network managers who are not.

Leased line Communications circuits rented from common carriers; such circuits are used by organizations when network traffic volumes make leasing cheaper than the long-distance toll charges from using dial-up circuits.

License The right to use particular software or hardware for a specified time period. License agreements specify the extent to which the computing resource can be used by the organization.

Local area network (LAN) A group of interconnected computers, in a geographically restricted area, that share computing resources.

Local middleware Software that enables software packages designed to run on one operating system to run on workstations using another operating system.

Logical user Logical user configurations enable multiple users to share the individual workstations that are attached to the network—such users may logon to the network using the same user identification. In Novell 4.xx LANs, logical users have authorization to access resources in one LAN although their workstation is physically attached to another LAN.

Low bid policy A purchasing/contracting policy that obligates an organization to do business with the vendor who has met all the conditions of the RFQ or RFB at the lowest price; such policies are often found in government and municipal organizations.

Mail-enabled applications Applications that use e-mail systems as the primary means of data exchange between clients and servers; these usually include an API and software similar to middleware that can reformat messages and data as they are passed from one e-mail system to another.

Mainframe A large computing system with extensive input, output, storage, and processing capabilities; such computers have processing capabilities measured in the millions of instructions per second and have traditionally served as hosts in centralized, host-based systems.

Management information base (MIB) A database storing network management data and information; this is often found in large, enterprise-wide network management systems, especially those utilizing SMNP or SMNP.2.

Management skills These consist of skills in the areas of planning, organizing, directing, and controlling and can be factors in a network administrator's effectiveness and career progress.

Medium The component of the communications system that actually carries the data signals. In LANs, twisted pair, coaxial cable, and fiber optic cable are common media; wireless media (including radio waves, microwave transmission, and infrared light) may also be used.

Meetingware Software that allows two or more members in a workgroup to simultaneously communicate electronically via their personal computers; these packages are sometimes also known as interactive work applications and are considered a type of groupware.

Mesh topology A network topology characterized by physical connections between every node in the network.

Message In telecommunications, a message is the actual data that is transmitted from one location to another. The message may be audio, data, image, or video signals (or some combination of these).

Message log A special file in which messages transmitted across a network are stored. Logging makes it possible to store a message so that it can be sent at a later time; it also adds a level of fault tolerance to the network and the ability to recover from network failures.

Message-oriented middleware (MOM) Middleware that allows the client and server segments of an application to communicate with one another via messages; one of the three major types of middleware in popular use.

Metering software Software that monitors and regulates the number of concurrent users of an application.

Metropolitan area network (MAN) A distributed network which spans a metropolitan area (a city and its major suburbs); in terms of geographic scope, a MAN is considered to be between a LAN and a WAN. A MAN is often defined by ANSI's FDDI standard and IEEE standard 802.6.

Micro-to-mainframe link A connection between a microcomputer (often remote) and a host computer. This is often done over dial-up circuits using modems and communications software that supports terminal emulation and microcomputer-oriented communications protocols such as KERMIT, XMODEM, YMODEM; most links include file transfer utilities to facilitate the uploading and downloading of files and data between microcomputers and larger computers.

Micro-to-micro software Software packages that enable the transfer of data and files between two microcomputers.

Microcomputer A fully functioning computer on a microprocessor chip with input, output, processing, and storage capabilities; examples include personal computers (PCs) and LAN workstations.

Microprocessor A chip or chip set designed to function as the central processing unit of a microcomputer or computing device; examples include Intel's Pentium chip and the PowerPC chip.

Microwave transmission Very high frequency signals, typically above 1000 MHz, directed over a line-of-sight path between two specially designed towers; microwave transmission is also used in satellite-based telecommunications.

Middleware Software used in diverse client/server networks that helps establish connections between distributed clients and servers.

Minicomputer A midrange computing system with capabilities that are generally between those of microcomputers and mainframes.

Modem A modulator-demodulator that converts the digital signals of computers or terminal devices to the analog signals needed for data transmission over many telephone circuits, and vice-versa. Modems are used in pairs, one at each end of the communications circuit.

Multimedia applications Applications that combine data, text, graphics, video, and sound.

Multiplexer A communications device that enables several terminal devices to share the same high-speed line. These are often found in WANs; examples include frequency division, time division, and statistical, optical, and T-1 multiplexers.

Multipoint A telecommunications circuit with multiple terminal connections.

Multiuser OS An operating system that enables multiple users at different workstations to concurrently run tasks on the same microcomputer.

NetWare Directory Services An important component of NetWare 4.xx systems that controls user access privileges, trustee rights, and so on, over the entire network.

Network Two or more computers that are capable of communicating with one another through a communications medium.

Network adapter card driver Software that allows the computer to communicate properly with the network interface card (NIC); it must be compatible with both the network adapter card hardware and the network operating system.

Network availability Network availability means that all needed network components are operational when a user needs them.

Network control center The organization (location and networking staff) that is responsible for monitoring the network and for correcting problems after they have been detected.

Network effectiveness The extent to which the network is both available and reliable.

Network interface card (NIC) An adapter (typically an add-in card) that establishes the link between the network medium and the workstation's bus.

Network interface module A network operating system module within the workstation that is responsible for placing data onto the network and for receiving data from the network; this software module interfaces directly with the NIC.

Network maintenance The process of keeping a network operational; this may also include modifying the network over time. Elements of network maintenance include adaptive, corrective, perfective, and preventive maintenance.

Network management The collective set of processes and activities needed to manage a computer network over the long term.

Network management system (NMS) A combination of hardware and software that is used to monitor and administer the network.

Network redirector module A NOS module within workstations that determines which user requests are to be carried out by the workstation's OS and which must be sent through the workstation's NIC and over the network to the appropriate server or device.

Network reliability The extent to which the network remains operational and functions consistently over time.

Network simulation Simulation models enable the network administrator to analyze how the network will perform under a set of prespecified conditions; this can be used to predict response times, processor utilization, line contention, and potential communications bottlenecks.

Network simulation software Software used for capacity planning that enables network managers to change network configurations in hardware or software, modify resource utilization rates and message traffic, and to run tests to determine how network performance would be impacted by these and other changes. This enables network managers to estimate server use, disk use, traffic over the communications medium, message and print queue lengths, message bottlenecks, and response times.

Network tuning This involves analyzing and utilizing network statistics to make appropriate network configuration changes in order to ensure that the network maintains satisfactory levels of performance.

Niche expertise Expertise in a specific, well-defined area—for example, groupware applications, desktop-to-desktop videoconferencing, Novell NetWare, electronic data interchange (EDI) applications, superhubs, middleware, FDDI LANs, and so on.

Node The junction at which two points of the network meet. This term can mean a switching center in the network, or, in LANs, the term may be used to refer to a workstation or microcomputer.

Nondedicated server A computer in a network that concurrently operates both as a server and a workstation.

Null modem cable Enables two microcomputers that are located within a few feet of each other to communicate without modems.

Object request brokers (ORBs) A type of middleware, essentially an offshoot of object-oriented programming systems (OOPS), that enables the exchange of objects in distributed client/server networks.

ODI drivers The Open Data Link (ODI) interface is a network adapter card driver standard jointly developed by Novell and Apple in 1989; it is a commonly used NIC driver in NetWare LANs.

Online processing (or immediate processing or real time processing) Processing a transaction as it occurs in real time.

Open systems Networks of computing devices and software that have been designed to facilitate communications among devices.

Open system drivers Network adapter card drivers that support multiple NOSs; these are particularly well-suited for diverse client/server networks.

Operational network management The day-to-day management of network operations; this includes linking users with one another and with computing resources and troubleshooting problems.

Optical character recognition (OCR) The process of converting scanned documents to a format that can be edited by text processors.

Optical disk servers Provide shared access to data stored on optical disks.

Optical read/write disks Removable optical read/write disks that can hold more than 100 megabytes of information.

OSI reference model The Open Systems Interconnection reference model is a seven-layer framework developed by the International Standards Organization (ISO) to facilitate the interconnection of networks; it provides the framework around which standard protocols can be defined. Basically, this model specifies how software should handle the transmission of a message from one end-user workstation (or application program) to another.

Packet distribution network (PDN) (or packet switching network) Networks, usually WANs such as X.25 and public data networks, in which data is transmitted in packets.

Parallel transmission This mode refers to the transfer of data where all the bits of a character are transferred simultaneously from sender to receiver over distinct circuits. Parallel transmission is used internally within a computer and often between a microcomputer and an attached printer.

Passive hub A wiring connector between networked devices in a LAN environment (especially in Arcnet LANs); these hubs lack signal regeneration capabilities and basically serve as switches.

Password A string of characters that, when input to the computer system, allows a user access to specific hardware, software, or data.

Peer-to-peer A type of network in which no hierarchical relationship exists. In LANs, this means that any workstation may be used as both a client and a server; in peer-to-peer client/server computing, clients and servers play equal roles in processing applications and may switch roles across applications.

Penalty clause A clause in a contract between vendor (or network installer) and customer that specifies what actions may be taken if the vendor fails to satisfy the contract; penalties may include reduction of the total purchase price or monetary damages.

Performance testing Testing all at once of all the applications in a newly installed (or reconfigured) network to ensure that they interact appropriately; this can be a form of stress testing. Such testing is used to confirm that the network will function appropriately when fully loaded with users who are making maximum demands on the network.

Personal digital assistants (PDAs) Small, hand-held computing devices usually used for personal organizational functions such as addresses and electronic note keeping; these may include wireless communications capabilities such as paging and fax.

Personal information management (PIM) software Software packages such as Sidekick and Lotus Organizer that manage a user's personal contact information such as appointments, meeting summaries, addresses, and so on.

Physical security Protecting network and computer resources with such devices as locked doors, workstation locks, badge readers, biometric verification systems (such as retina scanners, fingerprint readers, hand geometry scanners, and voice verification systems), fire detectors, fire extinguishers, sprinklers, burglar alarms, closed-circuit television, and security guards.

Physical user A user whose workstation is physically attached to the network.

Point-to-point A direct connection between a terminal device and a host computer; a dedicated telecommunications circuit between two computing devices.

Political skills Awareness of and expertise in organizational politics; the ability of managers to understand what peers and superiors are really interested in and to use this knowledge to influence them and gain their support.

Polling A line control protocol in which a central computer (called the supervisor) asks (polls), in a round-robin fashion, each of the workstations attached to it if it has data to send.

Post office A shared directory in an e-mail system or groupware package that manages the messages exchanged between members of the work group.

Print queue The set of print jobs waiting to be printed by one or more shared printers; this is usually managed by a print server.

Print server Handles the automated portions of printing operations, including spooling and the management of the print queue, for one or more shared printers; may be a dedicated or nondedicated computer.

Private branch exchange (PBX) An automated, software-controlled switchboard that may integrate both voice and data communications; these also allow users to share the trunks to the central telephone office.

Problem tracking Monitoring the status of problem resolutions and assembling statistics about network problems (including network availability, network reliability, and network effectiveness).

Professional association Groups such as the Association for Computing Machinery (ACM), Data Processing Management Association (DPMA), and Society for Information Management (SIM) that enable people with common interests or goals to share knowledge and experiences.

Project management software Software that enables project team managers to utilize automated forms of the program evaluation and review technique (PERT), the critical path method (CPM), and Gantt charts.

Propagation delay The amount of time required for a signal to travel from an earth station to the satellite and back to another earth station; because the signal must travel the uplink (the path from the earth station to the satellite) and the downlink (the path from the satellite back to the earth), this journey introduces almost a second in total transmission delay when satellite communications are used.

Protocol A communications protocol is a set of specifications that must be met before senders and receivers can physically exchange messages and files; protocols establish rules for delineating data, error detection, message lengths, media access, and so on.

Protocol analyzer A diagnostic tool for capturing and analyzing individual data packets sent over a network. These also provide network managers with a variety of valuable statistics and can be used to test individual workstations and communications channels.

Purchase contract A contract that states the responsibilities and expectations of both buyers and sellers; this can help protect buyers from frustration with what vendors deliver.

Redundant arrays of independent disks (RAID) Storage systems increasing in popularity in LANs because of their high storage capacities and because they provide disk striping, mirroring, and duplexing capabilities that add redundancy and fault tolerance to LANs.

Reader/scanner devices Devices such as those used in the validation of credit cards and at point-of-sale terminals in retail operations and other consumer locations. These types of input technologies have become very common in bar code systems and LAN-based point-of-sale systems.

Remote access Accessing a network from a distant location or accessing a network through a public dial-up telephone network.

Remote procedure calls (RPCs) The oldest and most mature type of middleware for distributed applications. These essentially allow an application running on a client workstation to access needed resources from distributed (and diverse) platforms in much the same manner as a (local) procedure call is used in traditional programs.

Repeater Internetworking technology that recreates signals on a communications circuit in order to correct for attenuation; this may be used to connect network segments or to extend the usual limits of a LAN.

Request for bid (RFB) A document sent to vendors that specifies exactly what the buyer wants to purchase; this is more specific than a request for proposal (RFP) and is sometimes called a request for quotation (RFQ).

Request for quotation (RFQ) See Request for bid (RFB).

Reverse compilation A process by which source code for a compiled program is derived from the object code; this is usually prohibited by software licenses.

Rightsizing A term used to refer to making sure that a particular application is matched to an appropriate computing platform.

Ring topology A closed loop of workstations or nodes with messages passing to each station in succession; this is usually found in FDDI LANs.

RJ-45 jack A plastic plug and twisted pair wire network connection similar to an ordinary telephone jack.

Router Internetworking technology that connects two or more networks at the network level of the OSI model; routers also determine the optimal path for a message through the network by using routing tables and current network traffic statistics.

Security equivalence Users with equivalent rights to other users are said to have security equivalence in NetWare LANs. Security equivalence, when granted by the network manager, can automate the granting of access rights to users.

Sender The person or computing device that initiates a message on the network. A sender transmits a message over the medium to a receiver.

Serial transmission Transmission of data one bit after another in sequence.

Server Computers in a network that provide resources (such as shared applications, disk storage, modems, printers, and so on) to other computers (or terminals) connected to the network.

Server error Log A log file containing error messages from the server.

Server license A type of software licensing agreement that makes it possible for an application to be installed on a single server; once installed, all workstations that are directly attached to that server may use the application.

Service bureau A service bureau is an organization which provides information processing services to subscriber organizations.

Shared directories Directories created on a workgroup member's disk drive that are accessible by other members of the workgroup.

Simple network management protocol (SNMP) Originally developed to control and monitor the operations of network components on networks using the TCP/IP protocol, now it can be used with other network level protocols. It has been endorsed by a variety of vendors including Hewlett-Packard, IBM, and Sun Microsystems.

Simplex A transmission mode in which data can be sent along a circuit in one direction only.

Site license A software agreement that gives an organization the right to install the software on all workstations and servers at a given site.

Smart terminal A terminal device with a limited amount of processing capability.

Software license agreement An agreement that specifies the extent to which buyers or leaseholders may use the software.

Source code The English-like statements that the programmer actually keys into his or her computer using a programming language; these are converted to compiled (object) code by translation software such as compilers.

Special-purpose application software Software packages that are useful to a limited range of workers, such as those in a single functional area of an organization.

Spooling The process used to place print jobs in print queues and to direct them to printers as they become available.

SQL servers These translate Structured Query Language (SQL) commands into database processing functions that can be carried out by the servers; these are likely to be found in distributed database environments.

SQL tools for distributed databases A set of middleware products for SQL application developers in distributed database environments that enable the SQL programmers to make an SQL call from clients to remote servers without having to know the underlying protocols.

Stand-alone system A computer system (or network) that is not connected to or does not communicate with another computer system (or network).

Star topology A type of network topology in which all devices are attached to a central point, or hub.

Static routing A routing algorithm in which messages always use the same path between two nodes.

Strategic network management The aspect of network management that considers the role of networks and networking over the long term. It ensures that the organization has the necessary network infrastructure to help it achieve its long-term goals and objectives, which involves the creation of network plans that specify the networks that will be implemented or interconnected.

Store-and-forward capabilities These enable a node's received messages to be stored and sent at a later time; this is an important part of message logging and introduces a level of fault tolerance to the network.

Sub-LAN A type of LAN without the range of functionality found in full-fledged LANs; the two major services available in such systems are file transfer and peripheral device sharing capabilities.

Superhub (or collapsed backbone) These are usually found in backbone networks for a single building. Superhubs essentially consist of a connection hub with slots for cabling to all LANs as well as slots for internetworking technologies such as bridges, routers, and gateways.

Support/maintenance agreement An agreement between an organization and a vendor or third-party firm to provide technical support and assistance for a specified time period. While some of these services were free in the past, fees for such services are common today.

Synchronous transmission A data transmission mode in which data characters are sent in blocks or frames; synchronization between sender and receiver is accomplished through one or more synchronization characters in each message frame that precede the data characters.

System software Background programs, such as operating systems, that enable application software to run on computer hardware; these programs provide an interface between application software and the computer hardware and carry out many of the "housekeeping" tasks such as writing a file to disk or transmitting a file to a printer for output.

T-connector A network connector with a T-shape that permits the connection with the computer to be severed without interfering with the network circuit. The circuit continues to run through the top of the T even if no network interface card is connected to the lower portion. T-connectors are often used in networks with bus topologies.

Tactical network management The aspect of network management that involves translating strategic network plans into more detailed, feasible actions. Essentially, tactical network management is responsible for making strategic network plans become realities.

TCP/IP The Transmission Control Protocol/Internet Protocol (TCP/IP) is one of the oldest network standards. It was developed for the U.S. Department of Defense as part of its Advanced Research Project Agency Network (ARPANET). ARPANET has since evolved into the Internet. TCP/IP is commonly found in distributed UNIX-based networks.

Technical skills Knowledge, skills, and abilities about particular types of technologies, tools, or products; these skills can be important selection factors for entry-level and specialist positions in organizations. Niche expertise also commonly requires strong technical skills and knowledge.

Telecommunications access programs Programs that reside in host computers or front end processors that provide users with access to applications and databases. These programs handle the routing, scheduling, and movement of messages between terminals and the host.

Teleconferencing See Videoconferencing.

Teleprocessing monitor A set of software programs, usually located in the host that manages incoming and outgoing messages; these programs relieve the operating system of most communications and network management tasks.

Terminal An electronic device used for input/output that may be connected to a local or remote host computer.

Terminal emulation Software that causes a host computer to recognize a microcomputer as a terminal; this is often an important component of communications software used to establish micro-to-mainframe connections.

Terminal server A means to connect LAN terminals to a host over a single communications line (and port).

Throughput For networks, this is the total amount of data (excluding overhead characters/bits) transmitted over the network in a given period of time; this is often measured in kilobytes per second.

Token passing Media access control protocols cover by IEEE standards 802.4 and 802.5. With token passing, a small packet called a *token* is passed from one device to the next in a predefined order. When a workstation has a message to send, it "captures" the token, inserts

its message and the address of the receiving workstation, and then passes the token along to the next device in the LAN.

Transaction tracking system (TTS) A system designed to assist in network and database recovery operations by reconciling modifications to the database.

Transceiver Technology that establishes the physical connection to communications media and implements the transmit and receive functions of the access control method used.

Transponder The component of the satellite that receives the uplink transmission, converts it to a different frequency, and sends the message back to earth over the downlink.

Troubleshooting Finding, diagnosing, and correcting problems experienced by network users.

Turnkey system A complete network (hardware and software) that once installed is ready to be used; vendors of turnkey systems will install the LAN, load all software, train users, and essentially take care of all implementation and conversion chores for the buyer.

Twisted pair Two (or four) wires in an insulator, used for standard telephone wiring or as cabling for LANs. The wires are twisted around one another to minimize interference.

Uninterruptable power supply (UPS) A device that provides backup electrical power if normal power sources fail or fall to unacceptable levels; this adds fault tolerance to a network.

Upsizing 1) The creation of LANs by networking stand-alone workstations; 2) enhancing the power of a network by upgrading servers and workstations; or 3) the migration from stand-alone LANs to interconnected client/server networks.

Value-added networks (VANs) Businesses that provide specialized telecommunications services to the public over common carrier facilities; these networks usually lease channels from common carriers and then re-lease the channels to their customers.

Vendor rating system An assessment system that may be used to evaluate vendor proposals in terms of how well their proposed networks rate against a specific set of criteria.

Videoconferencing (or teleconferencing) Real-time electronic meetings that involve people at different locations who can both see and hear each other.

Viruses One of the major threats to computer system and network security. These are programs that contain code to make copies of and to transmit themselves to other networked devices. Viruses typically cause at least some loss of productivity for network users.

Voice-over-data Capabilities often found in PBXs and Centrex systems that allow users to simultaneously be engaged in voice and data communications over the same medium.

Weighted routing In weighted routing, the path that a message will take from among the available alternatives is randomly selected by using weights determined by past utilization of the different links.

Wide area network (WAN) Networks with significant geographic coverage; a WAN may span a significant region of a state, an entire state, several states, countries, and even the world.

Wireless LAN A LAN that transmits and receives messages over media (such as radio waves, microwave, and infrared light).

Workflow automation software Software that allows geographically dispersed coworkers to work together on teams; this groupware often automatically routes work-in-progress to appropriate group members.

Workgroup A small number of users (usually two to eight) who need to access common information databases, communicate online and offline, schedule meetings, and work collaboratively.

Workload generator Software used by network planners and managers to estimate software and network performance of a proposed network configuration. These programs duplicate the pattern of work expected on the proposed network and assist in network tuning and balancing.

Workstation A terminal, computer or other data communication device attached to a network at which users work.

Zero-slot LAN Zero-slot LANs link PCs and printers with cables connected to the PCs' standard serial or parallel ports. These LANs get their name from the fact that they do not require an additional slot on the workstation's motherboard for a NIC.

Index